PID AND PREDICTIVE CONTROL OF ELECTRICAL DRIVES AND POWER CONVERTERS USING MATLAB®/SIMULINK®

PID AND PREDICTIVE CONTROL OF ELECTRICAL DRIVES AND POWER CONVERTERS USING MATLAB®/SIMULINK®

Liuping Wang, Shan Chai, Dae Yoo, Lu Gan and Ki Ng

IEEE
WILEY

This edition first published 2015

Registered office
John Wiley & Sons Singapore Pte. Ltd., 1 Fusionopolis Walk, #07-01 Solaris South Tower, Singapore 138628.

For details of our global editorial offices, for customer services and for information about how to apply for permission to reuse the copyright material in this book please see our website at www.wiley.com.

Library of Congress Cataloging-in-Publication Data applied for.

Hardback ISBN: 9781118339442

Typeset in 9/11pt TimesLTStd by Laserwords Private Limited, Chennai, India
Printed and bound in Singapore by Markono Print Media Pte Ltd

1 2015

Contents

About the Authors

Liuping Wang received her PhD in 1989 from the University of Sheffield, UK; subsequently, she was an Adjunct Associate Professor in the Department of Chemical Engineering at the University of Toronto, Canada. From 1998 to 2002, she was a Senior Lecturer and Research Coordinator at the Center for Integrated Dynamics and Control, University of Newcastle, Australia before joining RMIT University where she has been Professor of Control Engineering since 2006. She is the author of three books, joint editor of two books, and has published over 180 papers. Liuping Wang has successfully applied PID control and predictive control technologies to many industrial processes. She is a Fellow of Institution of Engineers, Australia.

Shan Chai was born in Shandong, China, in 1981. He received his B.Eng. degree in Electronics Engineering from Shandong University, China in 2004 and his M.Eng. in Electrical Engineering from RMIT University, Australia in 2007. Since 2009 he has been working toward the PhD degree at RMIT that he was awarded in 2013. Dr Shan Chai works in the electrical drives industry.

Dae Yoo received his B.Eng, MSc, and PhD in Electrical Engineering from the RMIT University, Melbourne, Australia in 2003, 2006 and 2013 respectively. From 2006 to 2007, he worked as an embedded software engineer at Orbital Engine Company, WA, Australia, where he was involved in various research oriented automotive projects for developing internal combustion engine control software. During 2008–2009 and since 2012, he has been at the Toyota Technical Center, Melbourne, Australia, where he is a control/software engineer developing advanced control algorithms for motor control units in Toyota's hybrid system.

Lu Gan, was born in Anhui Province of China, in 1987. He received his B.Eng. degree in Electrical Engineering from RMIT University, Australia, in 2009. Since then he has been working at RMIT towards the PhD degree that he received in 2014. Dr Lu Gan aspires to work in the electrical drives industry.

Ki Ng was born in Hong Kong, in 1986. He received his B.Eng. degree in Electrical Engineering from RMIT University, Australia, in 2009. He is currently working toward a PhD degree at RMIT University, Melbourne, Australia.

Preface

About This Book

Electrical drives play a critical role in electromechanical energy conversions. They are seen everywhere in our daily life from the cooling fans, washing machines to computers. They are the fundamental building blocks in manufacturing, transportation, mineral processing, wind energy and many other industries. For the last several decades, the advances of electronically switched semiconductors in the form of power electronics have made *AC* motor drives gain more prominence over the *DC* machines in industries since they allow a direct connection to power grids via grid connected power converters and have a more reliable physical structure. The grid connected three phase power converter has wide applications in renewable energy generation.

This book gives an introduction to the automatic control of electrical drives and grid connected three phase power converters, and to recent developments in design and implementation. When they are combined together as one unit, it will provide a direct connection for the electrical drives to the power grid for electromechanical energy conversions and renewable wind energy applications. In the context of control system design, electrical drives and grid connected three phase power converters share similar characteristics in their dynamic models and use the same type of semiconductors as actuators in the implementation of control systems. Therefore, in this book, electrical drives and power converters will be studied as individual components of the larger system and examined in the same framework.

As electrical drives and power converters have restricted operations imposed by electronically switched semiconductors, their operational constraints are paramount in the design and implementation of the control systems. In this regard, model predictive control has an established reputation in successfully handling the operational constraints in an optimal manner. Two chapters of this book will be devoted to seeking new predictive control technologies that address the specific needs of controlling electrical drives and power converters, and an additional two chapters will apply the existing predictive control technologies to these systems. Since PID control systems are used in the majority of industrial electrical drives and power converters, understanding these control systems and having the capability to design and implement them are important to a control engineer. There are three chapters in the book that will systematically cover PID control system design, PID control system implementation with anti-windup mechanisms and tuning of PID control systems. All control systems presented in this book have been experimentally validated using self-built test-beds with industrial sized motors. To assist the reader, tutorials about the real-time control system implementation and the physical model based simulators are presented in this book.

This book is intended for readers who have completed or are about to complete four years engineering studies with some basic knowledge in electrical and control systems. The targeted readers are students, practitioners, instructors and researchers who wish to learn electrical motor control and power converter control. The book is self-contained with MATLAB/Simulink tutorials and supported with simulation and experimental results. It is worth mentioning that the material contained in the first five chapters is aimed at readers who are working or are going to work in the relevant engineering field.

Outline of This Book

The structure of the book is illustrated by the block diagram as shown in Figure 0.1. There are ten chapters in this book, covering the topics of mathematical modeling, control of semiconductor switches, PID control system design, implementation and tuning, Finite Control Set (FCS)-predictive control in both $d-q$ and $\alpha-\beta$ reference frames, traditional predictive control in both continuous-time and discrete-time. PID controllers (see Chapters 3–5) are implemented using Pulse-Width-Modulation (PWM) technologies introduced in Chapter 2. The traditional model predictive controllers (see Chapters 8–9) use this technology too. However, FCS-predictive controllers (see Chapters 6–7) are implemented without PWM mechanisms by directly optimizing the switching patterns of semiconductors. Hence, this has significantly simplified the implementation procedure of control systems.

This book begins by discussing the physical models of electrical drives and grid connected three phase power converter since mathematical modeling is the first step toward the design and implementation of control systems. In Chapter 1, the mathematical models of machine drives and power converter are derived in a unified way that firstly uses space vector description of physical variables such as voltage, current and flux, and secondly converts the space vector based model to various reference frames. By adopting this unified framework, it is hoped that through the derivations in a similar process, the dynamic models of drives and power converters can be easily understood by a reader who does not have extensive background in *AC* machines and power converters. It must be emphasized, due to the efforts of generations of electrical engineers (see for example Park (1929), Duesterhoeft *et al.* (1951), Vas (1992), Leonhard (2001), Drury (2009), Hughes and Drury (2013), Quang and Dittrich (2008)), that the dynamics models are highly structured and have incredibly high fidelity, which forms the solid basis for control system designs introduced in the book.

From a control engineer's perspective, the next natural question following from mathematical modeling is how to realize manipulated control variables in applications. It has been well established that control of semiconductor switches is the most efficient and convenient means to achieve control of *AC*

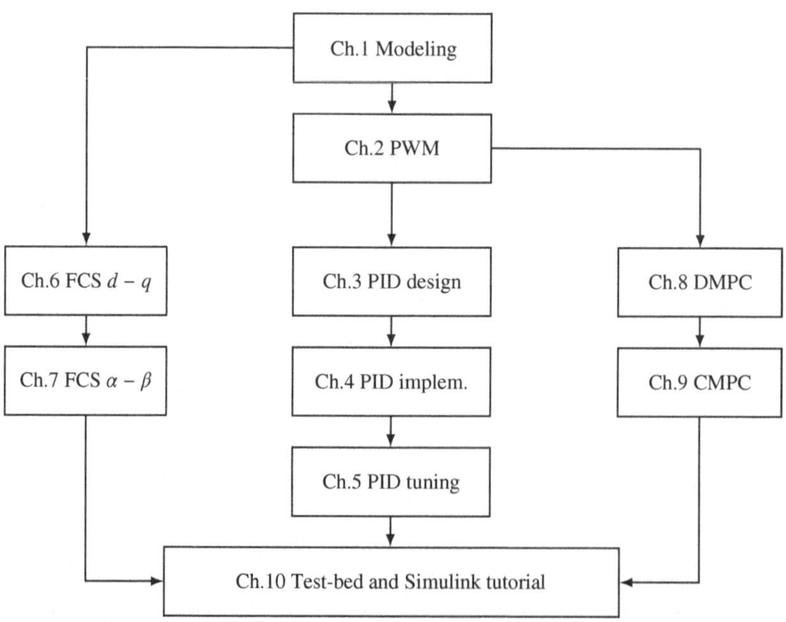

Figure 0.1 Book structure diagram

machine drives and power converters. It is shown in Chapter 2 that they act as actuators in the implementation of control systems where the manipulated control inputs in the form of three phase voltage signals are realized by turning on and off semiconductor switches. Also, the PWM implementation of control systems dictates the operational limits termed linear modulation range, which, in later chapters, will be translated into constraints imposed in the PID and predictive controllers using PWM mechanisms for implementation.

The next three chapters of this book will see the developments of PID control systems for electrical drives and power converter (see Chapters 3–5). In Chapter 3, for *AC* motor control systems, electromechanical torque control is achieved using PI control of currents in the $d-q$ reference frame, followed by achieving further requirements of controlling angular velocity and position via a cascade control system architecture. Identical control strategies are deployed to control the currents of a three phase power converter and its *DC* voltage in a grid connected environment. In all PID controller designs presented in the book, the pole-assignment control method is used. The reasons for this choice of design method is that it is perhaps among the simplest control system design methods and yet offers an effective means of selecting desired closed-loop performance in terms of response to reference signals and to disturbance rejection. In Chapter 4, PI controller implementation is discussed for both current controllers as inner-loop controllers, and velocity and *DC* voltage controllers as outer-loop controllers. In particular, continuous-time controllers are discretized for digital implementation, and operational constraints imposed by PWM operations are taken into consideration in the implementation of PI controllers. In order to avoid integrator wind-up in the presence of control signal reaching saturation limits, anti-windup mechanisms are proposed together with digital implementation, which leads to the so-called velocity form that has naturally embedded anti-windup mechanisms and is convenient for implementation. A MATLAB tutorial is introduced in this chapter to show how an embedded function can be created for the PI controller with its anti-windup mechanism, which has been directly used in the experimental validation. In Chapter 5, sensitivity functions in feedback control systems are introduced to measure the closed-loop control system performance against set-point following, disturbance rejection and noise attenuation in the frequency domain. Current control systems are analyzed for the effects of current sensor errors and harmonics caused by the voltage source inverter used in implementation of the control system. When velocity control, position control or *DC* voltage control is required in a cascade control structure, performance robustness in the outer-loop control system is considered where a weighting function is introduced to quantify the difference between the desired closed-loop performance and the actual closed-loop performance. Parameter variations are also studied using Nyquist plots. A large number of experiments are conducted in this chapter to demonstrate tuning procedures of the PI cascade control systems.

There are two approaches used in this book to generate the gate signal for the semiconductor switches. The first approach uses Pulse Width Modulation (PWM) based on which PID controllers (see Chapters 3–5) and traditional model predictive controllers are implemented (see Chapters 8–9). In control applications, the control signals calculated are the three phase voltage signals that are obtained from one of the controller designs using the model either in the $d-q$ reference frame or $\alpha-\beta$ reference frame. The role of the voltage source inverter with power electronics devices is to realize three phase voltage control signals as closely as possible. Namely, the sinusoidal phase voltage signals created by turning on-off each power switch with PWM technologies are aimed to be closely matched with three phase voltage control signals. The second methodology features a much simpler approach in the implementation of control systems that generates such a gate signal by direct optimization of an error function between the desired control signals and those that can be achieved by semiconductor switches (see Chapters 6–7). In the second approach, there is no need to use the PWM technology; therefore it significantly reduces the complexity of controlling semiconductor switches.

In Chapter 6, in the $d-q$ reference frame, finite control set (FCS) predictive controllers are used to directly optimize inverter states; as a result, PWMs are not required in the implementation of control systems, which simplifies the implementation procedure. The original FCS predictive control systems did not include integrators in their design and implementation. Consequently, there are steady-state errors within control systems. The existence of steady-state errors affects closed-loop performance, particularly when there are parameter uncertainties in the system, which is the main reason why the majority of practical control systems have integrator in the controller structure. By analyzing the original FCS predictive control system without constraints, the discrete-time feedback controller gain and locations of closed-loop eigenvalues are revealed. To embed integrators in the FCS predictive controller, a cascade control system structure is proposed where the inner-loop system is controlled with the original FCS predictive controller and the outer-loop is by an integrated feedback control. There are perhaps many ways to include integrators in the FCS predictive control system; however, the proposed approach has kept the spirit of the original FCS predictive control system and maintained its simplicity both conceptually and computationally. Because the FCS predictive control systems are designed for current control, this chapter will also show how to design velocity and position control for *AC* drives when current controllers are FCS predictive controllers.

In Chapter 7, under investigation is the finite control set (FCS) predictive current control in the $\alpha - \beta$ reference frame (or stationary frame). In the $\alpha - \beta$ reference frame, the currents $i_\alpha(t)$ and $i_\beta(t)$ are linear combinations of three phase currents $i_a(t)$, $i_b(t)$ and $i_c(t)$. Thus, they are sinusoidal functions. So are the voltage variables $v_\alpha(t)$ and $v_\beta(t)$. The current reference signals to FCS predictive control systems are sinusoidal signals, which differentiates current control systems in the $\alpha - \beta$ reference frame from those in the $d - q$ reference frame. It will be shown in this chapter that the original FCS predictive controllers are single-input and single-output controllers in exceptionally simple forms. However, in order to track sinusoidal current reference signals without steady-state errors, a controller with resonant characteristic is required in the $\alpha - \beta$ reference frame. Extensive simulation and experimental results have been presented in these two chapters to show the outstanding closed-loop control performance of FCS predictive control systems.

The next two chapters of this book (see Chapters 8–9) apply the traditional model predictive control algorithms to *AC* machine drives and power converters. These predictive control algorithms were derived for general applications without those restrictions imposed on system dynamics. The MATLAB programs used in applications were given in Wang (2009). Although the traditional predictive control algorithms could be applied to current control, their advantages are perhaps lost to the simpler and more effective FCS predictive control approaches, also to simpler PI controllers. Therefore, in Chapters 8 and 9, velocity control in *AC* drives and *DC* voltage control in power converters are considered, and for these cases, traditional model predictive controllers offer the advantages of designing the control systems using multi-input and multi-output approaches in the presence of constraints.

The final chapter of this book will discuss the test beds used in the experimental evaluations of control systems. For those who wish to know how to perform real-time simulations using the physical models of drives and power converter, Simulink tutorials are given to show the model building process in a step-by-step manner.

References

Drury B 2009 *The Control Techniques Drives and Controls Handbook* 2nd edn. IET.

Duesterhoeft W, Schulz MW and Clarke E 1951 Determination of instantaneous currents and voltages by means of alpha, beta, and zero components. *Transactions of the American Institute of Electrical Engineers* **70**(2), 1248–1255.

Hughes A and Drury B 2013 *Electric Motors and Drives: Fundamentals, Types and Applications* 4th edn. Elsevier.

Leonhard W 2001 *Control of Electrical Drives* 3rd edn. Springer.

Park RH 1929 Two-reaction theory of synchronous machines – part I. *AIEE Transations* **48**(2), 716–739.

Quang NP and Dittrich JA 2008 *Vector Control of Three-Phase AC Machines* 1st edn. Springer.

Vas P 1992 *Electrical Machines and Drives – A Space-Vector Theory Approach*. Oxford University Press, New York, USA.

Wang L 2009 *Model Predictive Control System Design and Implementation Using MATLAB* 1st edn. Springer, London.

Acknowledgment

We wish to thank the Australian Manufacturing Cooperative Research Center (AMCRC) for financial support over the projects of Control Systems and Regenerative Machine Stop. We gratefully acknowledge the help and support of Mr Pat Borland, the owner of ANCA Pty, who gave us the first opportunity to work in the field of *AC* drive control and shared with us many of his valuable experiences in the industry.

We wish to thank Professor Mike Johnson from the University of Strathclyde, UK for carefully reading the manuscript and giving us feedback.

On suggestions and feedback, our thanks go to Professor Stephen Boyd and Dr Nicholas Moehle at Stanford University, USA, Professor Jose Rodriguez and Ms Margarita Norambuena at Santa Maria, Chile, Dr Craig Buhr at MathWorks, USA, and Mr Gerardo Medrano who works in the Australian wind energy industry.

Asking for Feedback

We would like to ask our readers to contact us about any errors or suggestions for future improvement of our book.

Liuping Wang, Shan Chai, Dae Yoo, Lu Gan, Ki Ng
Melbourne, Australia

List of Symbols and Acronyms

Symbols

arg min	Minimizing argument
A	State matrix of state-space model
B	Input-to-state matrix of state-space model
C	State-to-output matrix of state-space model
D	Direct feed-through matrix of state-space model
(A, B, C, D)	State-space realization
ΔU	Parameter vector for the control sequence in discrete time MPC
$\Delta u(k)$	Incremental control at sample k
F_x, Φ	Pair of matrices used in the prediction equation $X = F_x x(k_i) + \Phi\Delta U$
B_v	Viscous friction coefficient in PMSM
f_d	Viscous friction coefficient in induction motor model
$G(s)$	Transfer function model
γ	Tuning parameter for PI controllers
i_α, i_β	Currents of PMSM and power converter in $\alpha - \beta$ reference frame
i_d, i_q	Currents of PMSM and power converter in $d - q$ reference frame
i_{sd}, i_{sq}	Stator currents of induction motor in $d - q$ reference frame
$i_{s\alpha}, i_{s\beta}$	Stator currents of induction motor in $\alpha - \beta$ reference frame
$I_{q\times q}$	Identity matrix with appropriate dimensions
J	Performance index for optimization
J_m	Moment of inertia $(\mathrm{kg}\cdot\mathrm{m}^2)$
K_c	Proportional control gain
K_{lqr}	Feedback control gain using LQR
K_{mpc}	Feedback control gain using MPC
K_{fcs}	Feedback control gain using FCS predictive control in $d - q$ reference frame
$k^\alpha_{fcs}, k^\beta_{fcs}$	Feedback control gain using FCS predictive control in $\alpha - \beta$ reference frame
K_{ob}	Observer gain vector
$l_i(t)$	The ith continuous-time Laguerre function
$L(t)$	Continuous-time Laguerre functions in vector form
L_s	Inductance of power converter and PMSM
L_r, L_s	Inductance of stator / rotor winding of induction motor
L_h	Machine mutual inductance of induction motor
λ	Lagrange multiplier
$\lambda_i(A)$	The ith eigenvalue of matrix A

$\lambda^l, \lambda^m, \lambda^h$	Scheduling parameters
μ	Disturbance vector
N	Number of terms used in Laguerre function expansion in continuous time
N_c	Control horizon
N_p	Prediction horizon
Ω_{mpc}, Ψ_{mpc}	Pair of matrices in the cost of predictive control in either the continuous-time or discrete-time design, $J = \eta^T\Omega_{mpc}\eta + 2\eta^T\Psi_{mpc}x(t) + cons$
η	Parameter vector in the Laguerre expansion
p	Scaling factor for continuous-time Laguerre functions
Q, R	Pair of weight matrices in the cost function of predictive control
R_s	Resistance of stator in PMSM and induction motor, also grid resistance in power converter
R_r	Resistance of rotor winding.
$r(\cdot)$	Set-point signal
q^{-i}	Backward shift operator, $q^{-i}[f(k)] = f(k-i)$
$S(s)$	Sensitivity function
$S^i(s)$	Input sensitivity function
S_i	Switching state of inverter
S_d, S_q	Normalized voltage variables of converter's d-axis voltage v_d and q-axis voltage v_q
$T(s)$	Complementary sensitivity function
T_e	Electromagnetic torque $(\mathrm{N \cdot m})$
T_L	Load torque $(\mathrm{N \cdot m})$
T_p	Prediction horizon in continuous-time
τ_D	Derivative control time constant
τ_f	Derivative control filter time constant
τ_I	Integral control time constant
θ_r	Mechanical position of motor shaft (radian)
θ_e	Electrical position of motor shaft (radian)
θ_s	Position of synchronous flux (radian)
ψ_s	Stator flux of induction motor (Wb)
ψ_{rd}, ψ_{rq}	Rotor flux of induction motor (Wb)
$u(\cdot)$	Control signal
$u_{s\alpha}, u_{s\beta}$	Stator voltages of induction motor (V) in $\alpha - \beta$ reference frame
u_{sd}, u_{sq}	Stator voltages of induction motor (V) in $d - q$ reference frame
v_α, v_β	Voltages of PMSM and power converter (V) in $\alpha - \beta$ reference frame
v_d, v_q	Voltages of PMSM and power converter (V) in $d - q$ reference frame
ω_e	Electrical motor speed (rad/s) (or RPM)
ω_m (or ω_r)	Mechanical motor speed (rad/s) (or RPM)
ω_s	Speed of synchronous flux (rad/s) (or RPM)
ω_g	Grid frequency (rad/s).
ω_{slip}	Slip in induction motor
u^{min}, u^{max}	Minimum and maximum limits for u
w_n	Bandwidth or natural frequency in PI controller design (rad/s)
$x(\cdot)$	State vector
$x(t_i + \tau \| t_i)$	Predicted state vector at time τ given current state vector $x(t_i)$
$\hat{x}(t)$	Estimated state vector in continuous-time
ξ	Damping coefficient in PI controller design
Y	Predicted output data vector
Z_p	Number of pole pairs

Acronyms

CMPC	Continuous-time model predictive control
DMPC	Discrete-time model predictive control
MMF	Magnetic motive force
PMSM	Permanent magnetic synchronous machine
PLL	Phase-locked loop
PID	Proportional, integral and derivative
FCS	Finite control set
FCS-MPC	Finite control set predictive control
I-FCS-MPC	Integral finite control set predictive control

1

Modeling of AC Drives and Power Converter

Building mathematical models of *AC* drives and power converters is the first step towards the design and implementation of control systems. This chapter presents the mathematical models of machine drives and power converters in a uniform way that firstly uses space vector description of the physical variables such as voltage, current and flux, and secondly converts the space vector based model to various reference frames. From Sections 1.1 to 1.4, the Permanent Magnetic Synchronous Machine (PMSM) will be used as an example to illustrate in detail how its dynamic model is established. In Section 1.5, the dynamic model for an induction machine is obtained by following the same thought process used for the PMSM. Section 1.6 derives the dynamic model for a 2-level grid connected voltage source converter, also using the same approaches as electrical drives. In the Summary section 1.7, characteristics of dynamic models are highlighted for future applications.

1.1 Space Phasor Representation

The analysis of a three phase system could be significantly simplified by adopting vector based approaches. Here, the concept of space vector will be introduced first before deriving a model of a PMSM. To simplify the analysis, a 2-pole machine with balanced three phase windings is assumed.

1.1.1 Space Vector for Magnetic Motive Force

Figure 1.1 shows the cross section of stator windings for a 2-pole machine. By Ampere's law, a magnetic motive force (MMF) will be generated when current is flowing in the windings. The peak of magnetic motive force produced by each phase will align with their own magnetic field and is separated by 120° from each other. Here, it is assumed that phase current has a frequency ω, initial angle ϕ_0 and amplitude I_s. When each of the three phase windings is provided with balanced three phase currents, where

$$i_a(t) = I_s \cos(\omega t + \phi_0) \tag{1.1}$$

$$i_b(t) = I_s \cos(\omega t + \phi_0 - 2\pi/3) \tag{1.2}$$

$$i_c(t) = I_s \cos(\omega t + \phi_0 - 4\pi/3). \tag{1.3}$$

PID and Predictive Control of Electrical Drives and Power Converters using MATLAB®/Simulink®, First Edition.
Liuping Wang, Shan Chai, Dae Yoo, Lu Gan and Ki Ng.

Companion Website: www.wiley.com/go/wang/pid

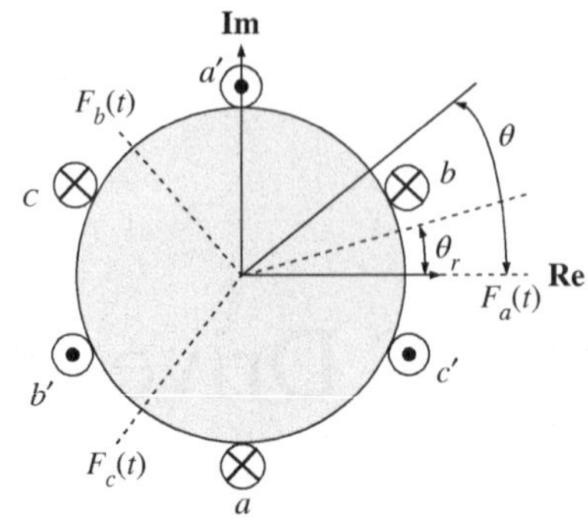

Figure 1.1 Cross section of stator winding. ⊗ and ⊙ denote the cross sections of the wires, $F_a(t)$, $F_b(t)$ and $F_c(t)$ are the peaks of magnetic motive forces for the three phase currents, θ_r is the position of the rotor and θ is an arbitrary position.

each phase current will produce a sinusoidal distributed MMF whose peak aligns with their respective magnetic axis for each phase, which are

$$F_a(t) = N_s i_a(t) = F_m \cos(\omega t + \phi_0) \tag{1.4}$$

$$F_b(t) = N_s i_b(t) = F_m \cos(\omega t + \phi_0 - 2\pi/3) \tag{1.5}$$

$$F_c(t) = N_s i_c(t) = F_m \cos(\omega t + \phi_0 - 4\pi/3), \tag{1.6}$$

where $F_m = N_s I_s$ is the magnitude of the peak MMF, N_s is a constant related to the number of coil turns and winding factor and I_s is the amplitude of the phase current.

At a certain position θ, referred to the magnetic axis of phase **a-a'** in Figure 1.1, the magnetic motive forces contributed from each phase winding are

$$F_a(t)^\theta = F_a(t)\cos(0 - \theta) \tag{1.7}$$

$$F_b(t)^\theta = F_b(t)\cos(2\pi/3 - \theta) \tag{1.8}$$

$$F_c(t)^\theta = F_c(t)\cos(4\pi/3 - \theta), \tag{1.9}$$

which are functions of θ. Therefore, the resultant total MMF at the position θ is the summation of Equations (1.7)–(1.9), which gives

$$F(t)^\theta = F_a(t)\cos(-\theta) + F_b(t)\cos(2\pi/3 - \theta) + F_c(t)\cos(4\pi/3 - \theta). \tag{1.10}$$

Note that

$$\cos(-\theta) = \mathbf{Re}\{e^{-j\theta}\}; \quad e^{-j\theta} = \cos(\theta) - j\sin(\theta).$$

Equation (1.10) can also be represented by

$$\begin{aligned} F(t)^\theta &= \mathbf{Re}\{F_a(t)e^{-j\theta} + F_b(t)e^{-j(\theta-2\pi/3)} + F_c(t)e^{-j(\theta-4\pi/3)}\} \\ &= \frac{3}{2}\mathbf{Re}\left\{\frac{2}{3}\left(F_a(t) + F_b(t)e^{j\frac{2\pi}{3}} + F_c(t)e^{j\frac{4\pi}{3}}\right)e^{-j\theta}\right\}. \end{aligned} \tag{1.11}$$

Based on the calculation of the total MMF at the position θ, the space vector of the three-phase peak MMF is defined by

$$\vec{F}(t) = \frac{2}{3}\left(F_a(t) + F_b(t)e^{j\frac{2\pi}{3}} + F_c(t)e^{j\frac{4\pi}{3}}\right). \tag{1.12}$$

For notational simplicity, the rest of this chapter will use the notation $\overrightarrow{\{.\}}$ to denote a space vector. With this definition, the total MMF at the position θ is expressed as

$$F(t)^{\theta} = \frac{3}{2}\mathbf{Re}\left\{\overrightarrow{F}(t)e^{-j\theta}\right\}. \tag{1.13}$$

Furthermore, it can be verified that by substituting (1.4)–(1.6) into (1.12), the space vector of the three-phase peak MMF $\overrightarrow{F}(t)$ has the following compact expression:

$$\begin{aligned}\overrightarrow{F}(t) &= \frac{2}{3}\left(F_a(t) + F_b(t)e^{j\frac{2\pi}{3}} + F_c(t)e^{j\frac{4\pi}{3}}\right)\\ &= \frac{2}{3}F_m\left(\cos(\omega t+\phi_0) + \cos(\omega t+\phi_0-2\pi/3)e^{j\frac{2\pi}{3}} + \cos(\omega t+\phi_0-4\pi/3)e^{j\frac{4\pi}{3}}\right)\\ &= F_m e^{j(\omega t+\phi_0)}.\end{aligned} \tag{1.14}$$

In the derivation of (1.14), the following equalities are used:

$$\cos(\alpha) = \frac{1}{2}(e^{j\alpha} + e^{-j\alpha})$$

$$1 + e^{j\frac{4\pi}{3}} + e^{j\frac{8\pi}{3}} = 0.$$

Following the compact expression of $\overrightarrow{F}(t)$ given by (1.14), the total MMF at the position θ is also simply expressed as

$$\begin{aligned}F(t)^{\theta} &= \frac{3}{2}\mathbf{Re}(\overrightarrow{F}(t)e^{-j\theta})\\ &= \frac{3}{2}F_m\cos(\omega t+\phi_0-\theta).\end{aligned} \tag{1.15}$$

As shown by (1.14), the space vector $\overrightarrow{F}(t)$ is a rotating vector in the complex plane. As a result, the instantaneous value of $F(t)^{\theta}$ in (1.11) and its equivalent (1.15) can be interpreted as the magnitude of the projection of $\overrightarrow{F}(t)$ on the position θ.

Figure 1.2 gives an example of the vector representation of MMF at $t = 0$ while assuming $\phi_0 = 0$. In this figure, at $t = 0$, the vectors of peak MMF for each phase current are

$$\overrightarrow{F_a} = F_m\cos(0)e^{j0} = F_m$$

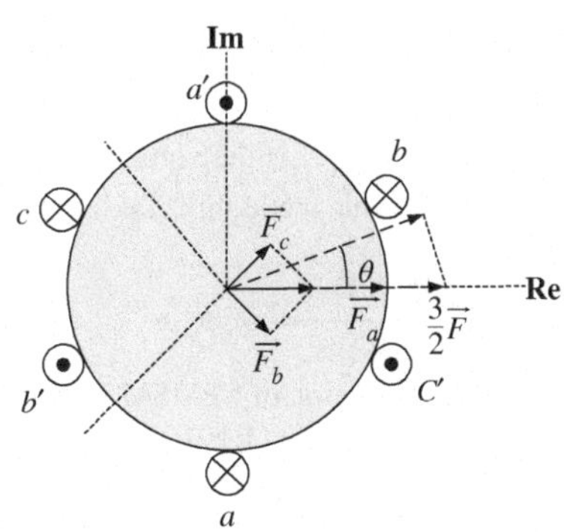

Figure 1.2 Space vector of MMF ($t = 0$ and $\phi_0 = 0$).

$$\vec{F_b} = F_m \cos\left(-\frac{2\pi}{3}\right) e^{j\frac{2\pi}{3}} = \frac{-1}{2} F_m e^{j\frac{2\pi}{3}}$$

$$\vec{F_c} = F_m \cos\left(-\frac{4\pi}{3}\right) e^{j\frac{4\pi}{3}} = \frac{-1}{2} F_m e^{j\frac{4\pi}{3}}$$

and the resulting total peak MMF is,

$$\vec{F} = \frac{2}{3}(\vec{F_a} + \vec{F_b} + \vec{F_c}) = F_m.$$

thus, the total MMF at angle θ is the magnitude of projection $\vec{F}$ onto angle θ,

$$F(0)^{\theta} = F_m \cos(\theta).$$

1.1.2 Space Vector Representation of Voltage Equation

The use of space vector facilitates the derivation of voltage equation for the PMSM with a compact expression. This derivation is the key to form the dynamic model for current control.

With a similar principle, the space vector for three-phase stator current can be written as

$$\vec{i_s} = \frac{2}{3}\left(i_a(t) + i_b(t)e^{j\frac{2\pi}{3}} + i_c(t)e^{j\frac{4\pi}{3}}\right) \tag{1.16}$$

$$\vec{i_s} = I_s e^{j(\omega t + \phi_0)} \tag{1.17}$$

and the space vector of three-phase stator voltage is defined as

$$\vec{v_s} = \frac{2}{3}\left(v_a(t) + v_b(t)e^{j\frac{2\pi}{3}} + v_c(t)e^{j\frac{4\pi}{3}}\right), \tag{1.18}$$

where $v_a(t)$, $v_b(t)$ and $v_c(t)$ are terminal line-to-neutral voltage for each phase, respectively.

When a surface-mounted PMSM is considered, the space vector of stator flux consists of two parts. One is produced by the stator current while the other is produced by the permanent magnets of the rotor:

$$\vec{\varphi_s} = L_s \vec{i_s} + \phi_{mg} e^{j\theta_e}, \tag{1.19}$$

where ϕ_{mg} is the amplitude of the flux induced by the permanent magnets of the rotor in the stator phases and this parameter is assumed constant in the design, θ_e is the electrical angle of the rotor, and L_s is the sum of leakage inductance and mutual inductance.

With the space vector representation of voltage, current and flux, the stator voltage equation in space vector form can be written according to voltage law,

$$\vec{v_s} = R_s \vec{i_s} + \frac{d\vec{\varphi_s}}{dt}, \tag{1.20}$$

where $\vec{v_s}$ is the space vector of stator voltage, $R_s \vec{i_s}$ is the voltage drop on the resistors of the stator and $\frac{d\vec{\varphi_s}}{dt}$ is the induced voltage due to changing of magnetic flux. Taking the derivative of the second term of the flux based on (1.19) gives

$$\frac{d(\phi_{mg} e^{j\theta_e})}{dt} = j\omega_e \phi_{mg} e^{j\theta_e},$$

where $\theta_e(t) = \omega_e t$. Thus, the vector voltage equation of a PMSM is obtained by replacing the flux derivative in (1.20) with the calculated derivative based on (1.19), which leads to

$$\vec{v_s} = R_s \vec{i_s} + L_s \frac{d\vec{i_s}}{dt} + j\omega_e \phi_{mg} e^{j\theta_e}. \tag{1.21}$$

This is the fundamental equation that governs the relationship between the current and voltage of PMSM in a space vector form, based on which the dynamic models will be obtained.

1.2 Model of Surface Mounted PMSM

As seen from the previous sections, the space vectors of three-phase voltages, currents and flux are three rotating vectors in the complex plane. The speed of their rotation depends on the frequency (ω) of the three-phase voltages and currents. In the complex plane, as shown in Figure 1.3, each vector (see the three phase current $\vec{i_s}$) could be decomposed into a component on real axis (see i_α) and a quadrature component on imaginary axis (see i_β). Such a decomposition could be carried out with respect to different reference frames (see the $d-q$ reference frame). Importantly, control strategies will devised according to the models in the relevant reference frames.

This section presents the two most widely adopted reference frames and their relationships, which are the stationary reference frame (also called $\alpha-\beta$ reference frame) and the synchronous reference frame (also called $d-q$ reference frame). Both reference frames are illustrated in Figure 1.3.

1.2.1 Representation in Stationary Reference ($\alpha-\beta$) Frame

One choice of the reference frame is a stationary reference frame with the real (α) axis aligned with the peak MMF ($F_a(t)$, see Figure 1.1) and the imaginary (β) axis in quadrature (see Figure 1.3).

By projecting the space vectors of voltage and current onto real (α) and imaginary (β) axes, these vectors can be represented by the complex notations,

$$\vec{v_s} = v_\alpha + jv_\beta \tag{1.22}$$

$$\vec{i_s} = i_\alpha + ji_\beta. \tag{1.23}$$

Substituting the complex representations (1.22) and (1.23) into the space vector voltage equation (1.21), and equating their real and imaginary parts in both sides, respectively, gives the model of PMSM in the $\alpha-\beta$ reference frame,

$$v_\alpha = R_s i_\alpha + L_s \frac{di_\alpha}{dt} - \omega_e(t)\phi_{mg} \sin\theta_e \tag{1.24}$$

$$v_\beta = R_s i_\beta + L_s \frac{di_\beta}{dt} + \omega_e(t)\phi_{mg} \cos\theta_e. \tag{1.25}$$

This model of PMSM will be used in the Chapter 7 for a current controller design in the $\alpha-\beta$ reference frame.

One might ask what is the relationship between the current and voltage variables in the $\alpha-\beta$ reference frame and the original three phase variables. This unique relationship is given by the Clarke

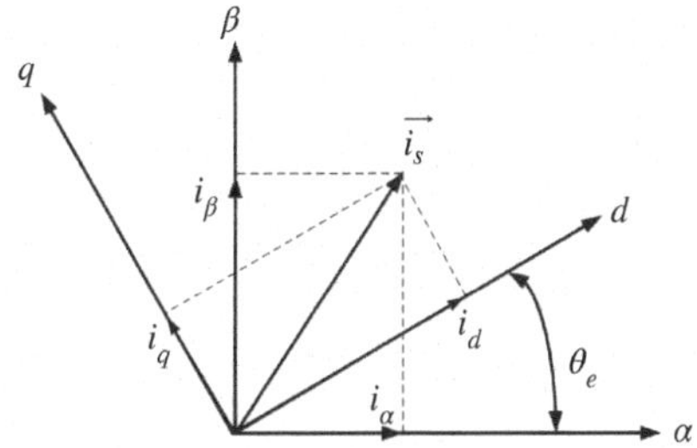

Figure 1.3 Illustration of stationary reference frame and synchronous reference frame. θ_e is the electrical angle of the rotor. Current space vector is projected to two reference frames.

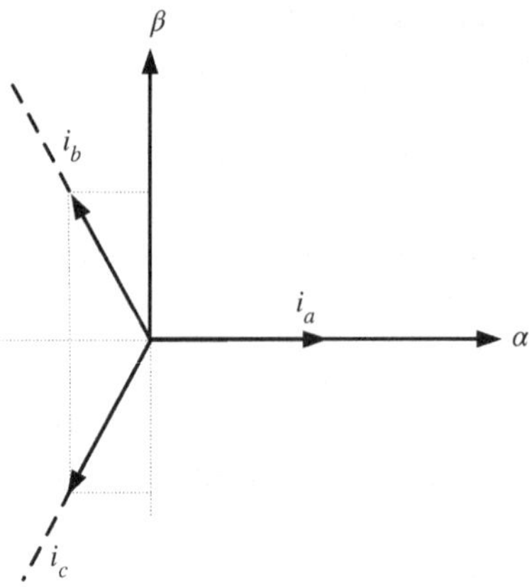

Figure 1.4 Illustration of Clarke transformation of current.

transformation. Taking the three phase currents as an example, the transformation of three phase variables to their components in the $\alpha - \beta$ reference frame is achieved by Clarke transformation, where

$$\begin{bmatrix} i_\alpha \\ i_\beta \\ i_0 \end{bmatrix} = \frac{2}{3} \begin{bmatrix} 1 & -\frac{1}{2} & -\frac{1}{2} \\ 0 & \frac{\sqrt{3}}{2} & -\frac{\sqrt{3}}{2} \\ \frac{1}{2} & \frac{1}{2} & \frac{1}{2} \end{bmatrix} \begin{bmatrix} i_a \\ i_b \\ i_c \end{bmatrix}. \tag{1.26}$$

As shown in Figure 1.4, the transformation matrix is obtained by projecting i_a, i_b and i_c on α and β axes, respectively. Here, it is seen that the α axis is aligned with the direction of i_a current which is also the direction of the peak MMF $F_a(t)$ (see Figure 1.1). The coefficient $\frac{2}{3}$ here is to guarantee the energy conservation. In addition, i_0 represents the zero sequence component of three phase current and is zero for balanced three phase currents. Conversely, the inverse Clarke transformation is defined as

$$\begin{bmatrix} i_a \\ i_b \\ i_c \end{bmatrix} = \begin{bmatrix} 1 & 0 & 1 \\ -\frac{1}{2} & \frac{\sqrt{3}}{2} & 1 \\ -\frac{1}{2} & -\frac{\sqrt{3}}{2} & 1 \end{bmatrix} \begin{bmatrix} i_\alpha \\ i_\beta \\ i_0 \end{bmatrix}. \tag{1.27}$$

It is easy to show that

$$\frac{2}{3} \begin{bmatrix} 1 & -\frac{1}{2} & -\frac{1}{2} \\ 0 & \frac{\sqrt{3}}{2} & -\frac{\sqrt{3}}{2} \\ \frac{1}{2} & \frac{1}{2} & \frac{1}{2} \end{bmatrix} \begin{bmatrix} 1 & 0 & 1 \\ -\frac{1}{2} & \frac{\sqrt{3}}{2} & 1 \\ -\frac{1}{2} & -\frac{\sqrt{3}}{2} & 1 \end{bmatrix} = \begin{bmatrix} 1 & 0 & 0 \\ 0 & 1 & 0 \\ 0 & 0 & 1 \end{bmatrix}.$$

The current and voltage variables in the $\alpha - \beta$ reference frame are all sinusoidal in nature because they are directly related to their original three phase current and voltage variables (see the Clarke transformation (1.26)).

1.2.2 Representation in Synchronous Reference $(d-q)$ Frame

Another reference frame is the $d-q$ reference frame, where the direct axis (d) is always aligned with rotating flux produced by the permanent magnets of the rotor, and the q axis is in quadrature. Because the rotor runs the same speed as the supplying frequency at steady-state, it is also called the synchronous frame for PMSM.

To change the reference frame to the $d-q$ reference frame, as shown in Figure 1.3, it is equivalent to rotating the space vector in $\alpha-\beta$ reference frame clockwise by θ_e. Mathematically, this rotation of space vectors is translated into multiplication by the factor $e^{-j\theta_e}$, which leads to a set of new space vectors, for example, voltage space vector $\vec{v}_s{}'$, current space vector $\vec{i}_s{}'$. By projecting these transformed space vectors into the real and imaginary axes, the current and voltage variables in the $d-q$ reference frame are formed. That is,

$$\vec{v}_s{}' = \vec{v}_s e^{-j\theta_e} = v_d + jv_q \tag{1.28}$$

$$\vec{i}_s{}' = \vec{i}_s e^{-j\theta_e} = i_d + ji_q \tag{1.29}$$

where $\vec{v}_s{}'$ and $\vec{i}_s{}'$ denote the space vectors referred to synchronous $d-q$ reference frame.

Multiplying the the original space vector voltage equation (1.21) by $e^{-j\theta_e}$ leads to the following equation:

$$\vec{v}_s e^{-j\theta_e} = R_s \vec{i}_s e^{-j\theta_e} + L_s \frac{d\vec{i}_s}{dt} e^{-j\theta_e} + j\omega_e \phi_{mg}. \tag{1.30}$$

Noting that

$$\vec{i}_s{}' = \vec{i}_s e^{-j\theta_e}$$

and by taking derivative on both sides of this equality, it can be shown that

$$\frac{d\vec{i}_s}{dt} e^{-j\theta_e} = \frac{d\vec{i}_s{}'}{dt} + j\omega_e \vec{i}_s{}'. \tag{1.31}$$

Therefore, from (1.30), together with (1.31), the voltage equation in terms of the space vectors $\vec{v}_s{}'$ and $\vec{i}_s{}'$ has the following form:

$$\vec{v}_s{}' = R_s \vec{i}_s{}' + L_s \frac{d\vec{i}_s{}'}{dt} + j\omega_e L_s \vec{i}_s{}' + j\omega_e \phi_{mg}. \tag{1.32}$$

This is the fundamental equation that governs the relationship between the voltage and current variables in space vector form that leads to the dynamic model in the $d-q$ reference frame.

Now, substituting (1.28) and (1.29) into (1.32), and because the real and imaginary components from the left-hand side are equal to the corresponding components in their right-hand side, it can be shown that the $d-q$ model of PMSM is,

$$v_d = R_s i_d + L_s \frac{di_d}{dt} - \omega_e L_s i_q \tag{1.33}$$

$$v_q = R_s i_q + L_s \frac{di_q}{dt} + \omega_e L_s i_d + \omega_e \phi_{mg}, \tag{1.34}$$

where (1.33) is obtained with the components from the real part whilst (1.34) is from the imaginary part.

There is a unique relationship between the variables in the $\alpha-\beta$ reference frame and those in the $d-q$ reference frame. The transformation of real and imaginary components in $\alpha-\beta$ frame to its counterparts

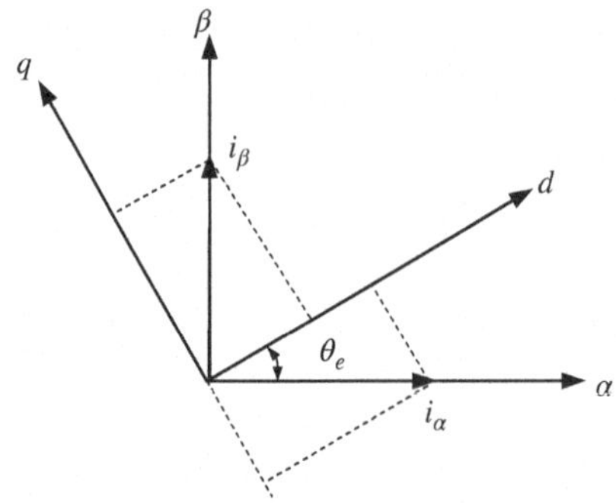

Figure 1.5 Illustration of Park transformation of current from $d-q$ reference frame to $\alpha-\beta$ reference frame.

in the rotating $d-q$ reference frame is achieved by the so called Park transformation, as shown in Figure 1.5, which leads to the current and voltage relations in the two reference frames:

$$\begin{bmatrix} v_d \\ v_q \end{bmatrix} = \begin{bmatrix} \cos\theta_e & \sin\theta_e \\ -\sin\theta_e & \cos\theta_e \end{bmatrix} \begin{bmatrix} v_\alpha \\ v_\beta \end{bmatrix} \tag{1.35}$$

$$\begin{bmatrix} i_d \\ i_q \end{bmatrix} = \begin{bmatrix} \cos\theta_e & \sin\theta_e \\ -\sin\theta_e & \cos\theta_e \end{bmatrix} \begin{bmatrix} i_\alpha \\ i_\beta \end{bmatrix}, \tag{1.36}$$

where θ_e is the angle between the two reference frames and also the electrical angle of the rotor.

Conversely, the inverse Park transformation is defined as

$$\begin{bmatrix} v_\alpha \\ v_\beta \end{bmatrix} = \begin{bmatrix} \cos\theta_e & -\sin\theta_e \\ \sin\theta_e & \cos\theta_e \end{bmatrix} \begin{bmatrix} v_d \\ v_q \end{bmatrix} \tag{1.37}$$

$$\begin{bmatrix} i_\alpha \\ i_\beta \end{bmatrix} = \begin{bmatrix} \cos\theta_e & -\sin\theta_e \\ \sin\theta_e & \cos\theta_e \end{bmatrix} \begin{bmatrix} i_d \\ i_q \end{bmatrix}. \tag{1.38}$$

Combining Clarke transformation (1.26) with Park transformation (1.36) gives the Park-Clarke transformation from three-phase values to their representations in the $d-q$ reference frame:

$$\begin{bmatrix} i_d \\ i_q \end{bmatrix} = \frac{2}{3} \begin{bmatrix} \cos\theta_e & \cos\left(\theta_e - \frac{2\pi}{3}\right) & \cos\left(\theta_e - \frac{4\pi}{3}\right) \\ -\sin\theta_e & -\sin\left(\theta_e - \frac{2\pi}{3}\right) & -\sin\left(\theta_e - \frac{4\pi}{3}\right) \end{bmatrix} \begin{bmatrix} i_a \\ i_b \\ i_c \end{bmatrix}. \tag{1.39}$$

The amazing fact about the mathematical model (see (1.33) and (1.34)) in the $d-q$ reference frame is that the current and voltage variables are no longer sinusoidal signals, instead, they are *DC* signals. In other words, because of this, in the design of a control system, the reference signals to the closed-loop system could be constants or step signals, which explains why PI controllers are widely used for this class of systems.

1.2.3 Electromagnetic Torque

For the surface mounted PMSM, the $d-q$ axis inductance is equal to each other due to the uniform air-gap, that is

$$L_d = L_q = L_s$$

and thus the stator flux in $\alpha - \beta$ reference frame is as in Equation (1.19). Similarly, the stator flux can also be represented in the $d - q$ frame by rotating the flux vector clockwise by θ_e, leading to

$$\vec{\varphi}_s' = \vec{\varphi}_s e^{-j\theta_e} = L_s \vec{i}_s' + \phi_{mg} \tag{1.40}$$

and its real and imaginary parts are

$$\varphi_d = L_s i_d + \phi_{mg} \tag{1.41}$$

$$\varphi_q = L_s i_q \tag{1.42}$$

respectively, where the flux (ϕ_{mg}) induced by the permanent magnets of the rotor is aligned with rotor and its q-axis component is zero.

The electromagnetic torque is computed as the cross product[1] of the space vector of the stator flux with stator current in the $\alpha - \beta$ reference frame as

$$T_e = \frac{3}{2} Z_p \vec{\varphi}_s \otimes \vec{i}_s \tag{1.43}$$

or equivalently, in the $d - q$ reference frame as,

$$T_e = \frac{3}{2} Z_p \vec{\varphi}_s' \otimes \vec{i}_s', \tag{1.44}$$

where Z_p is the number of pole pairs. The cross product is calculated using two three dimensional vectors $[\varphi_d, \varphi_q, 0]$ and $[i_d, i_q, 0]$ and the result is the vector $[0, 0, \varphi_d i_q - \varphi_q i_d]$. Hence,

$$T_e = \frac{3}{2} Z_p (\varphi_d i_q - \varphi_q i_d). \tag{1.45}$$

By substituting Equations (1.41) and (1.42) into (1.45), we obtain

$$T_e = \frac{3}{2} Z_p \phi_{mg} i_q. \tag{1.46}$$

With the flux of permanent magnet assumed to be a constant, the electromagnetic torque can be controlled through varying the q-axis component of stator currents. Therefore, with the electrical model of PMSM in $d - q$ reference frame, the control of PMSM is analogous to the principle of controlling *DC* motors.

However, if one were to use the electrical model in the $\alpha - \beta$ reference frame to compute the electromagnetic torque T_e, the matter would be more complicated. In this case, the space vector of the flux has the relationship with the currents in the $\alpha - \beta$ reference frame:

$$\vec{\varphi}_s = L_s \vec{i}_s + \phi_{mg} e^{j\theta_e} \tag{1.47}$$

$$= L_s i_\alpha + \phi_{mg} \cos\theta_e + j(L_s i_\beta + \phi_{mg} \sin\theta_e). \tag{1.48}$$

This leads to the expression of the electromagnetic torque T_e via the cross product of $\vec{\varphi}_s \otimes \vec{i}_s$ that has the following form:

$$T_e = \frac{3}{2} Z_p [(L_s i_\alpha + \phi_{mg} \cos\theta_e) i_\beta - (L_s i_\beta + \phi_{mg} \sin\theta_e) i_\alpha]. \tag{1.49}$$

One could easily compute the torque T_e from (1.49), when the i_α and i_β currents are given. However, in reverse, it would be difficult to determine the trajectories of i_α and i_β if a desired electromagnetic torque

[1] The cross product of two vectors $\vec{a}$ and $\vec{b}$ is the vector $\vec{c} = \vec{a} \otimes \vec{b}$. Letting $\vec{a} = [a_1, a_2, a_3]$ and $\vec{b} = [b_1, b_2, b_3]$, the vector $\vec{c} = [c_1, c_2, c_3]$ has the components: $c_1 = a_2 b_3 - a_3 b_2$, $c_2 = a_3 b_1 - a_1 b_2$ and $c_3 = a_1 b_2 - a_2 b_1$ (see Kreyszig (2010)).

T_e were given, which is what is required in the control of the electromagnetic torque. This explains why in the control strategies chosen for the later chapters, the mathematical model in the $d-q$ reference frame is predominately used for the reason that there is a simple relation between the electromagnetic torque T_e and the i_q current (see (1.46)).

It is worthwhile to emphasize that in the context of controlling a PMSM drive, if the model used for the control system design is based on the $\alpha-\beta$ reference frame, then the manipulated variables are the voltage variables v_α and v_β. Similarly, if the model in the $d-q$ reference frame is used in the design, then the manipulated variables are the voltage variables v_d and v_q. However, in the implementation of the control law, the control signals v_d and v_q will be converted to v_α and v_β signals using the inverse Park Transform, then to three phase voltage signals v_a, v_b and v_c that will be realized using a voltage source inverter typically consisting of a *DC* power supply and several semiconductor switches (see Chapter 2). The same control law implementation procedure applies to other *AC* machine drives and power converters.

1.3 Model of Interior Magnets PMSM

The main difference between an interior magnets PMSM and a surface mounted motor is that the salience due to the rotor magnets results in a non-uniform air-gap flux. The derivation of its $d-q$ model is very similar to the surface mounted case and is briefly introduced here. The vector voltage equations are the same as those in (1.20) and (1.32) and presented here again for convenience,

$$\vec{v}_s = R_s\vec{i}_s + \frac{d\vec{\varphi}_s}{dt} \tag{1.50}$$

$$\vec{v}_s' = R_s\vec{i}_s' + \frac{d\vec{\varphi}_s'}{dt} + j\omega_e\vec{\varphi}_s'. \tag{1.51}$$

In contrast to a surface mounted PMSM, the stator flux due to the salience of the interior magnets needs to be modeled in the $d-q$ reference frame with different inductances in the $d-q$ axes,

$$\vec{\varphi}_s' = \varphi_d + j\varphi_q, \tag{1.52}$$

where

$$\varphi_d = L_d i_d + \phi_{mg} \tag{1.53}$$

$$\varphi_q = L_q i_q. \tag{1.54}$$

Here the quadrature axis stator inductance L_q is usually smaller than the direct axis inductance L_d for an interior magnets PMSM. Therefore, substituting (1.52), (1.53) and (1.54) into (1.51) yields the $d-q$ model for interior magnets PMSM,

$$v_d = R_s i_d + L_d\frac{di_d}{dt} - \omega_e L_q i_q \tag{1.55}$$

$$v_q = R_s i_q + L_q\frac{di_q}{dt} + \omega_e L_d i_d + \omega_e\phi_{mg}. \tag{1.56}$$

It is apparent that the model of interior PMSM is equivalent to the surface mounted case if $L_d = L_q = L_s$. For the interior magnets PMSM, the nonlinearity of torque is mainly due to the salience of the rotor, which causes the non-uniformity of air-gap.

Its electromagnetic torque is obtained by substituting (1.53), (1.54) into (1.45):

$$T_e = \frac{3}{2}Z_p(\phi_{mg}i_q + (L_d - L_q)i_d i_q). \tag{1.57}$$

In comparison with torque equation (1.46) of a surface mounted machine, the extra component $\frac{3}{2}Z_p(L_d - L_q)i_d i_q$ is the reluctance torque due to the saliency.

1.3.1 Complete Model of PMSM

For the PMSM with multiple pair of poles, the electrical speed relates to the mechanical speed by

$$\omega_e = Z_p \omega_m \tag{1.58}$$

where Z_p denotes the pair of poles of the PMSM. The rotation of motor could be described by the following dynamic equation:

$$J_m \frac{d\omega_m}{dt} = T_e - B_v \omega_m - T_L \tag{1.59}$$

with J_m denoting the total inertia, B_v viscous friction coefficient and T_L load torque. Replacing the mechanical speed (ω_m) with electrical speed ω_e in (1.59) gives

$$\frac{d\omega_e}{dt} = \frac{Z_p}{J_m}\left(T_e - \frac{B_v}{Z_p}\omega_e - T_L\right). \tag{1.60}$$

For the surface mounted PMSM or $i_d = 0$ control, there is no additional torque component. Thus, substituting the torque (1.46) into (1.60) yields

$$\frac{d\omega_e}{dt} = \frac{Z_p}{J_m}\left(\frac{3}{2}Z_p \phi_{mg} i_q - \frac{B_v}{Z_p}\omega_e - T_L\right). \tag{1.61}$$

Together with the electrical model derived in the last section, the complete model of a PMSM is represented by

$$\frac{di_d}{dt} = \frac{1}{L_d}(v_d - R_s i_d + \omega_e L_q i_q) \tag{1.62}$$

$$\frac{di_q}{dt} = \frac{1}{L_q}(v_q - R_s i_q - \omega_e L_d i_d - \omega_e \phi_{mg}) \tag{1.63}$$

$$\frac{d\omega_e}{dt} = \frac{Z_p}{J_m}\left(\frac{3}{2}Z_p \phi_{mg} i_q - \frac{B_v}{Z_p}\omega_e - T_L\right). \tag{1.64}$$

1.4 Per Unit Model and PMSM Parameters

Using the explicit machine model (1.62) to (1.64) with SI unit to design the controller could lead to numerical problems due to different units of machine parameters and variables. For example, in the third equation (1.64), a small inertia value (J_m) in $\text{kg} \cdot \text{m}^2$ may lead to a very large coefficient for i_q in Amp. As a result, the controller gain becomes numerically very small for speed control purpose. Such wide variation of numerical ranges makes the implementation on micro-controllers or Digital Signal Processors (DSP) rather complex. Hence it is more convenient to use the per unit model of the PMSM for the design of controllers.

1.4.1 Per Unit Model and Physical Parameters

In the per unit model, as an example, the base values of parameters and variables of a PMSM are listed in Table 1.1. Here, only three independent base values need to be given. With the three independent

Table 1.1 Base values for the per unit model

Symbol	Description	Base value	SI unit
P_b	Rated power	0.35	kW
U_b	Rated voltage	$150/\sqrt{3}$	V
T_b	Rated torque	1.1	Nm
I_b	Current	8.083	A
R_b	Resistance	10.71	Ω
ω_{eb}	Velocity	630.63	rad/s
L_b	Inductance	0.017	H
Φ_b	Flux	0.1373	Wb
J_b	Inertia	0.0018	$kg \cdot m^2$
B_b	Viscous coefficient	0.0018	$Nm \cdot s$

values P_b, U_b and T_b chosen for this case, the other base values can be obtained by their inherent relationships,

$$I_b = Z_p P_b / U_b,\ R_b = U_b / I_b,\ \omega_{eb} = U_b I_b / T_b$$

$$\phi_b = U_b / \omega_{eb},\ L_b = R_b / \omega_{eb}$$

$$J_b = Z_p P_b / \omega_{eb}^2,\ B_b = T_b / \omega_{eb}.$$

Scaling the parameters and variables in (1.62) to (1.64) with their own base values, the per unit version of a machine model is

$$\frac{di_d}{dt} = \frac{\omega_{eb}}{L_d}(v_d - R_s i_d + \omega_e L_q i_q) \tag{1.65}$$

$$\frac{di_q}{dt} = \frac{\omega_{eb}}{L_q}(v_q - R_s i_q - \omega_e L_d i_d - \omega_e \phi_{mg}) \tag{1.66}$$

$$\frac{d\omega_e}{dt} = \frac{Z_p}{J_m}\left(\frac{3}{2} Z_p \phi_{mg} i_q - \frac{B_v}{Z_p}\omega_e - T_L\right), \tag{1.67}$$

where the notation refers to the per unit value of the machine variables and parameters with the exception that ω_{eb} is in SI units. The numerical values and their per unit counterparts of machine parameters used in obtaining the experimental and simulation results in this book are given in Table 1.2.

1.4.2 Experimental Validation of PMSM Model

The physical model developed is to be validated against the experimental data collected from the test-bed that is described in detail in Chapter 10. The PMSM test-bed has its physical parameters defined by Table 1.2. Using the parameters defined in the table, the differential equations (1.62)–(1.64) are solved to yield the responses of i_d, i_q currents and velocity ω_e. In both simulations and experiments, identical step signals are applied to the v_d and v_q voltages as input to the dynamic system. The load torque $T_L = 0$ Nm in both simulation and experimental validation. Figure 1.6 shows the experimental validation results when the direct axis voltage is a step signal with amplitude 5 and v_q is a step signal with amplitude of 20 (V). With these input voltage signals, the steady-state of the currents i_d (see Figure 1.6(a)) reaches

Table 1.2 Parameters of PMSM

Symbol	Description	SI value	SI unit	Per unit
J_m	Total inertia	0.47×10^{-4}	$kg \cdot m^2$	0.0267
B_v	Viscous coeff.	1.1×10^{-4}	$Nm \cdot s$	0.0625
L_d	d-axis inductance	7.0×10^{-3}	H	0.4120
L_q	q-axis inductance	7.0×10^{-3}	H	0.4120
T_L	Load torque	unknown	Nm	unknown
R_s	Resistance	2.98	Ω	0.2781
ϕ_{mg}	Flux linkage due to permanent magnet	0.125	Wb	0.9102
i_{rated}	Nominal current	2.9	A	0.36
Z_p	No. of pole pairs	2		

1.75 A and i_q (see Figure 1.6(b)) reaches zero from the simulation results and the velocity reaches 73 rad/s (see Figure 1.6(c)). The steady-state of the i_q current is zero because the motor is not loaded and the load torque is zero. In comparison between the simulation and experimental results, the physical model developed here has a high fidelity in both transient responses and steady-state responses. The discrepancies between the responses obtained using the model and those from the experimental test-bed could be explained as the result of the net effects of the PWM harmonics (see Chapter 2), measurement noise, pulsing torque, iron saturation and many other factors, which are ignored in the model.

1.5 Modeling of Induction Motor

Induction motor, or in other words, asynchronous motor generally contains two main components in its structure: stator winding and rotor winding. Stator winding is supplied by an *AC* power source, which will generate a rotating magnetic field, the so-called stator flux. Then, the constantly changing flux will cause a current induced in the rotor winding based on Lenz's Law, subsequently this induced current will generate a magnetic field, the so-called rotor flux. Since both fluxes are opposite to each other, a rotational force is generated to accelerate the rotor until the magnetizing torque is balanced to the load torque. Because the actual rotor's position is always lagging the flux position, in order to ensure the flux cutting through the rotor winding, there is a difference between the angular speed ω_s of the magnetic field and the rotor electrical speed ω_e, which is called *slip* with $\omega_{slip} = \omega_s - \omega_e$. ω_s is also called synchronous flux angular speed.

The traditional three-phase *AC* induction motor has two types, wound and squirrel-cage, which describes the form of the rotor winding. The wound rotor form has a brushed external connection; typical applications include the wind-farm generators. The squirrel-cage rotor is more widely applied and it has the rotor winding connection as a short-circuit without any external connections. The induction motor discussed in this book is of the squirrel-cage type.

1.5.1 Space Vector Representation of Voltage Equation of Induction Motor

The use of the space vector simplifies the derivation of voltage equation for the induction machine and makes the model derivation easier to understand. This derivation is the key to forming the dynamic models for current control.

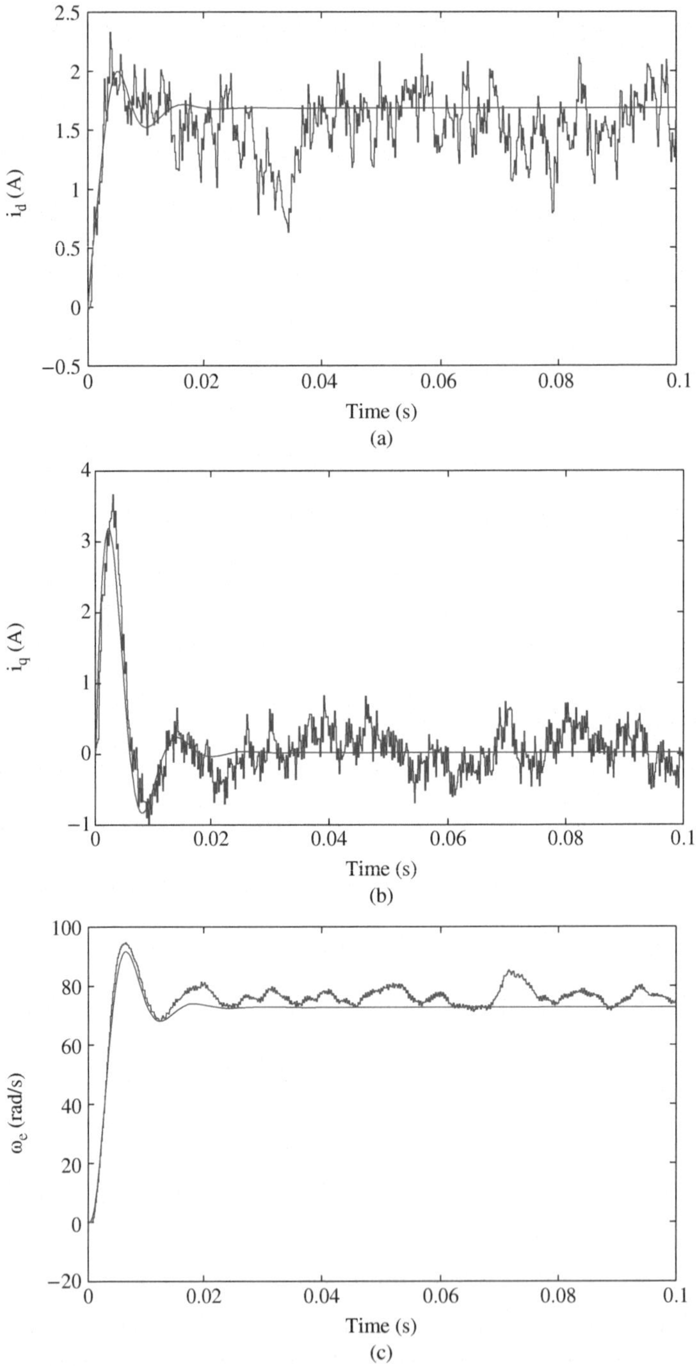

Figure 1.6 Experimental validation (input voltage $v_d = 5$ V and $v_q = 20$ V). Noise free line: model outputs; noisy line: outputs from test-bed. (a) i_d, (b) i_q, and (c) ω_e.

In modeling of the induction motor, the space vector for its three-phase stator current is written as

$$\vec{i}_s = \frac{2}{3}\left(i_a(t) + i_b(t)e^{j\frac{2\pi}{3}} + i_c(t)e^{j\frac{4\pi}{3}^s}\right)$$

$$\vec{i}_s = I_s e^{j(\omega_s t+\phi)},$$

where ω_s is the angular speed of the stator side space vectors, and the space vector of three-phase stator voltage is defined as

$$\vec{u}_s = \frac{2}{3}\left(u_a(t) + u_b(t)e^{j\frac{2\pi}{3}} + u_c(t)e^{j\frac{4\pi}{3}}\right),$$

where $u_a(t)$, $u_b(t)$ and $u_c(t)$ are terminal line-to-neutral voltage for each phase, respectively. Similarly, space vectors of three-phase rotor voltage and current are denoted by $\vec{u}_r$ and $\vec{i}_r$. The rotor dynamics play an important role in the mathematical model of an induction motor.

For a squirrel-cage induction motor, the space vector voltage equation of stators with respect to their own winding systems is

$$\vec{u}_s = R_s\vec{i}_s + \frac{d\vec{\psi}_s}{dt}, \tag{1.68}$$

where R_s is the stator resistance and $\vec{\psi}_s$ is the space vector of stator flux. With respect to its own winding system, the space vector voltage equation of the rotor is

$$\vec{u}_r^{\,*} = R_r\vec{i}_r^{\,*} + \frac{d\vec{\psi}_r^{\,*}}{dt} = 0, \tag{1.69}$$

where R_r is the rotor resistance, $\vec{i}_r^{\,*}$, $\vec{u}_r^{\,*}$ and $\vec{\psi}_r^{\,*}$ are rotor's current, voltage and flux in space vector forms with respect to the rotor reference frame. Due to the short-circuit of the rotor winding, the rotor voltage vector in (1.69) is always equal to zero.

In order to synchronize the reference frame of the rotor windings with the reference frame of the stator windings, a set of space vectors are defined to change the reference frame of the space vectors of the rotor,

$$\vec{u}_r = \vec{u}_r^{\,*}e^{j\theta_e}$$

$$\vec{i}_r = \vec{i}_r^{\,*}e^{j\theta_e}$$

$$\vec{\psi}_r = \vec{\psi}_r^{\,*}e^{j\theta_e},$$

where $\theta_e = \omega_e t$, θ_e is the electrical angle of the rotor, and ω_e is the electrical angular speed of the rotor. This transformation is based on the electrical field of the rotor windings lagging behind that of the stator by θ_e radians; thus, in order to synchronize these two reference frames, the space vectors in the rotor reference frame are advanced with the angle θ_e.

Now, multiplying both sides of (1.69) with the factor $e^{j\theta_e}$ and substituting the transformations into the rotor voltage equation leads to

$$\vec{u}_r = R_r\vec{i}_r + \frac{d\vec{\psi}_r}{dt} - j\omega_e\vec{\psi}_r = 0, \tag{1.70}$$

where the following equality is used,

$$\frac{d\vec{\psi}_r}{dt}e^{j\theta_e} = \frac{d\vec{\psi}_r}{dt} - j\omega_e\vec{\psi}_r.$$

With the space vectors of both currents in stator and rotor, the instantaneous fluxes of both windings are given based on their relationships to currents:

$$\vec{\psi}_s = L_s\vec{i}_s + L_h\vec{i}_r \tag{1.71}$$

$$\vec{\psi}_r = L_h\vec{i}_s + L_r\vec{i}_r \tag{1.72}$$

where L_h is the mutual machine inductance, L_s and L_r are the stator and rotor inductance, respectively.

Note that there are coupling terms in the stator flux (see (1.71)) and rotor flux (see (1.72)). It seems that the derivation of the induction motor model is more complicated than the process used in the PMSM model, however, they follow the same thought process. Here it is to eliminate the rotor current $\vec{i}_r$ from the equations and find the relationship between the stator voltage and current.

Taking derivative of stator flux based on (1.71), and substituting the stator flux with stator and rotor currents, the stator voltage equation (1.68) becomes:

$$\vec{u}_s = R_s\vec{i}_s + L_s\frac{d\vec{i}_s}{dt} + L_h\frac{d\vec{i}_r}{dt}. \tag{1.73}$$

To eliminate the rotor current $\vec{i}_r$, the rotor flux equation (1.72) is used to find

$$\vec{i}_r = \frac{1}{L_r}\vec{\psi}_r - \frac{L_h}{L_r}\vec{i}_s.$$

Substituting this into (1.73) yields

$$\vec{u}_s = R_s\vec{i}_s + L_s\left(1 - \frac{L_h^2}{L_sL_r}\right)\frac{d\vec{i}_s}{dt} + \frac{L_h}{L_r}\frac{d\vec{\psi}_r}{dt}.$$

To eliminate the derivative of the rotor flux from the above equation, the voltage balance equation from the rotor (see (1.70)) is used, which leads to

$$\frac{d\vec{\psi}_r}{dt} = -\frac{R_r}{L_r}\vec{\psi}_r + \frac{R_rL_h}{L_r}\vec{i}_s + j\omega_e\vec{\psi}_r,$$

where the rotor current is replaced with stator flux and current. Finally, it can be verified that the stator voltage equation is expressed in terms of the rotor flux and stator current:

$$\vec{u}_s = \left(R_s + R_r\frac{L_h^2}{L_r^2}\right)\vec{i}_s + L_s\left(1 - \frac{L_h^2}{L_sL_r}\right)\frac{d\vec{i}_s}{dt} + \left(-\frac{L_hR_r}{L_r^2} + j\omega_e\frac{L_h}{L_r}\right)\vec{\psi}_r.$$

Although all the physical parameters are defined in the above model, they could have more compact expressions. More specifically, define the following parameters used in the model, leakage factor:

$$\sigma = 1 - \frac{L_h^2}{L_sL_r}, \tag{1.74}$$

stator time constant:

$$\tau_s = \frac{L_s}{R_s}, \tag{1.75}$$

rotor time constant:

$$\tau_r = \frac{L_r}{R_r}, \tag{1.76}$$

coefficients:

$$k_r = \frac{L_h}{L_r} \tag{1.77}$$

$$r_\sigma = R_s + R_rk_r^2 \tag{1.78}$$

$$\tau'_\sigma = \frac{\sigma L_s}{r_\sigma}. \tag{1.79}$$

With these definitions of parameters, the voltage equation in space vector form is simply expressed as

$$\vec{i}_s + \tau'_\sigma \frac{d\vec{i}_s}{dt} = \frac{k_r}{r_\sigma}\left(\frac{1}{\tau_r} - j\omega_e\right)\vec{\psi}_r + \frac{1}{r_\sigma}\vec{u}_s \tag{1.80}$$

where the rotor flux satisfies the differential equation:

$$\vec{\psi}_r + \tau_r \frac{d\vec{\psi}_r}{dt} = j\omega_e\tau_r\vec{\psi}_r + L_h\vec{i}_s. \tag{1.81}$$

1.5.2 Representation in Stationary $\alpha - \beta$ Reference Frame

Upon obtaining the electrical model in the space vector form, the next step is to convert it to the model in the $\alpha - \beta$ reference frame. The $\alpha - \beta$ reference frame is a stationary reference frame in the stator side with the real (α) axis aligned with the peak MMF ($F_a(t)$, see Figures 1.1 and 1.3) and the imaginary (β) axis in quadrature (see Figure 1.3).

By decomposing the space vector voltage, current and flux onto the real and imaginary axes, they can be represented by the complex notations,

$$\vec{u}_s = u_{s\alpha} + ju_{s\beta} \tag{1.82}$$

$$\vec{i}_s = i_{s\alpha} + ji_{s\beta} \tag{1.83}$$

$$\vec{\psi}_r = \psi_{r\alpha} + j\psi_{r\beta}. \tag{1.84}$$

To obtain the dynamic model in the $\alpha - \beta$ reference frame, the above variables are substituted into the space vector model (1.80) and (1.81).

It can be readily verified that the electrical model of the induction motor in the $\alpha - \beta$ reference frame is described by the following four differential equations:

$$\frac{di_{s\alpha}}{dt} = -\frac{1}{\tau'_\sigma}i_{s\alpha} + \frac{k_r}{r_\sigma\tau'_\sigma\tau_r}\psi_{r\alpha} + \frac{k_r}{r_\sigma\tau'_\sigma}\omega_e\psi_{r\beta} + \frac{1}{r_\sigma\tau'_\sigma}u_{s\alpha} \tag{1.85}$$

$$\frac{di_{s\beta}}{dt} = -\frac{1}{\tau'_\sigma}i_{s\beta} - \frac{k_r}{r_\sigma\tau'_\sigma}\omega_e\psi_{r\alpha} + \frac{k_r}{r_\sigma\tau'_\sigma\tau_r}\psi_{r\beta} + \frac{1}{r_\sigma\tau'_\sigma}u_{s\beta} \tag{1.86}$$

$$\frac{d\psi_{r\alpha}}{dt} = \frac{L_h}{\tau_r}i_{s\alpha} - \frac{1}{\tau_r}\psi_{r\alpha} - \omega_e\psi_{r\beta} \tag{1.87}$$

$$\frac{d\psi_{r\beta}}{dt} = \frac{L_h}{\tau_r}i_{s\beta} + \omega_e\psi_{r\alpha} - \frac{1}{\tau_r}\psi_{r\beta}. \tag{1.88}$$

1.5.3 Representation in $d - q$ Reference Frame

To change the reference frame to the $d - q$ frame is equivalent to rotating the space vector in $\alpha - \beta$ frame clockwise by θ_s, that is

$$\vec{u}_s{}' = \vec{u}_s e^{-j\theta_s} = u_{sd} + ju_{sq} \tag{1.89}$$

$$\vec{i}_s{}' = \vec{i}_s e^{-j\theta_s} = i_{sd} + ji_{sq} \tag{1.90}$$

$$\vec{\psi}_r{}' = \vec{\psi}_r e^{-j\theta_s} = \psi_{rd} + j\psi_{rq}, \tag{1.91}$$

where $\vec{u}_s'$, $\vec{i}_s'$ and $\vec{\psi}_r'$ denote the space vectors referred to rotating $d-q$ frame. $\theta_s = \omega_s t$ where ω_s is the synchronous flux angular speed in the stator. In this rotating $d-q$ reference frame, the rotor flux vector is fixed to the real axis of the coordinate system. Therefore, the quadrature component of $\vec{\psi}_r'$ is zero.

Multiplying (1.80) with the factor $e^{-j\theta_s}$ and substituting in the space vectors in $d-q$ frame gives

$$\vec{i}_s' + \tau_\sigma' \left(\frac{d\vec{i}_s'}{dt} + j\omega_s \vec{i}_s' \right) = \frac{k_r}{r_\sigma} \left(\frac{1}{\tau_r} - j\omega_e \right) \vec{\psi}_r' + \frac{1}{r_\sigma} \vec{u}_s', \tag{1.92}$$

where the following equality is used:

$$\frac{d\vec{i}_s}{dt} e^{-j\theta_s} = \frac{d\vec{i}_s'}{dt} + j\omega_s \vec{i}_s'.$$

Based on the real and imaginary components of (1.92), the dynamic electrical model in the $d-q$ reference frame is obtained:

$$\frac{di_{sd}}{dt} = -\frac{1}{\tau_\sigma'} i_{sd} + \omega_s i_{sq} + \frac{k_r}{r_\sigma \tau_\sigma' \tau_r} \psi_{rd} + \frac{1}{r_\sigma \tau_\sigma'} u_{sd} \tag{1.93}$$

$$\frac{di_{sq}}{dt} = -\omega_s i_{sd} - \frac{1}{\tau_\sigma'} i_{sq} - \frac{k_r}{r_\sigma \tau_\sigma'} \omega_e \psi_{rd} + \frac{1}{r_\sigma \tau_\sigma'} u_{sq}. \tag{1.94}$$

Similarly, it can be shown that the rotor flux in the $d-q$ reference frame satisfies:

$$\frac{d\psi_{rd}}{dt} = \frac{L_h}{\tau_r} i_{sd} - \frac{1}{\tau_r} \psi_{rd} \tag{1.95}$$

$$0 = \frac{L_h}{\tau_r} i_{sq} - (\omega_s - \omega_e)\psi_{rd}, \tag{1.96}$$

where the q component of rotor flux $\psi_{rq} = 0$. Since Equation (1.96) is an algebraic equation, it is not included for control design. However, it yields the relationship used for estimation of ω_s:

$$\omega_s = \omega_e + \frac{L_h}{\tau_r} \frac{i_{sq}}{\psi_{rd}}, \tag{1.97}$$

which is also called slip estimation and ω_e is the electrical angular velocity of the rotor. Since the rotor flux ψ_{rd} is not directly measured, the slip estimation is alternatively performed using the current i_{sd} to replace ψ_{rd} as

$$\omega_s = \omega_e + \frac{1}{\tau_r} \frac{i_{sq}}{i_{sd}}, \tag{1.98}$$

where $\frac{d\psi_{rd}}{dt} = 0$ is assumed in (1.95) to obtain the steady-state solution of i_{sd} in relation to ψ_{rd}.

Relationships exist between the current, voltage and flux variables in the $\alpha-\beta$ reference frame and the $d-q$ reference frame, governed by the Clarke transformation (see Figure 1.4 for illustration), and they are given below:

$$\begin{bmatrix} u_{sd} \\ u_{sq} \end{bmatrix} = \begin{bmatrix} \cos\theta_s & \sin\theta_s \\ -\sin\theta_s & \cos\theta_s \end{bmatrix} \begin{bmatrix} u_{s\alpha} \\ u_{s\beta} \end{bmatrix} \tag{1.99}$$

$$\begin{bmatrix} i_{sd} \\ i_{sq} \end{bmatrix} = \begin{bmatrix} \cos\theta_s & \sin\theta_s \\ -\sin\theta_s & \cos\theta_s \end{bmatrix} \begin{bmatrix} i_{s\alpha} \\ i_{s\beta} \end{bmatrix} \tag{1.100}$$

$$\begin{bmatrix} \psi_{rd} \\ \psi_{rq} \end{bmatrix} = \begin{bmatrix} \cos\theta_s & \sin\theta_s \\ -\sin\theta_s & \cos\theta_s \end{bmatrix} \begin{bmatrix} \psi_{r\alpha} \\ \psi_{r\beta} \end{bmatrix}. \tag{1.101}$$

1.5.4 Electromagnetic Torque of Induction Motor

The electromagnetic torque of induction motor is calculated using the cross product of the space vectors of rotor flux and and stator current in the $d-q$ reference frame, which is

$$T_e = \frac{3}{2} Z_p \frac{L_h}{L_r} (\vec{\psi_r}' \otimes \vec{i_s}'), \tag{1.102}$$

where Z_p is the number of pole pairs. The cross product, defined in Section 1.2.3, is calculated using two three dimensional vectors $[\psi_{rd} \quad 0 \quad 0]$ and $[i_{sd} \quad i_{sq} \quad 0]$ since ψ_{rq} is zero. The result of the cross product is the vector $[0 \quad 0 \quad \psi_{rd} i_{sq}]$. Thus, in the $d-q$ reference frame, the electromagnetic torque is proportional to $\psi_{rd} i_{sq}$, which is

$$T_e = \frac{3}{2} Z_p \frac{L_h}{L_r} \psi_{rd} i_{sq}. \tag{1.103}$$

If the electromagnetic torque is calculated using the space vectors of rotor and stator current in the $\alpha - \beta$ reference frame, then it is proportional to the cross product of the space vectors of rotor flux and stator current in the stationary frame,

$$\begin{aligned} T_e &= \frac{3}{2} Z_p \frac{L_h}{L_r} (\vec{\psi_r} \otimes \vec{i_s}) \\ &= \frac{3}{2} Z_p \frac{L_h}{L_r} (\psi_{r\alpha} i_{s\beta} - \psi_{r\beta} i_{s\alpha}). \end{aligned} \tag{1.104}$$

Note that the expression of electromagnetic torque T_e in the $d-q$ reference frame is only related to ψ_{rd} and i_{sq}. These two variables are *DC* variables, thus the torque control can be achieved by controlling ψ_{rd} and i_{sq} to their specified constant or piece-wise constant reference signals. On the other hand, in the $\alpha - \beta$ reference frame, it is much more difficult to achieve torque control because the expression in (1.104) is associated with the fluxes in $\alpha - \beta$ reference frame that are sinusoidal signals.

The mechanical model of induction motor is derived from the general motion equation of motor rotation, which is given as follows,

$$J_m \frac{d\omega_m}{dt} + f_d \omega_m = T_e - T_L, \tag{1.105}$$

where $\omega_m(t)$ is the rotor's mechanical velocity ($\omega_m = \frac{\omega_e}{Z_p}$), J_m represents the inertia of the motor, f_d is the friction coefficient, T_e and T_L denote the electromagnetic torque and load torque, respectively. The dynamic model of the mechanical equation is obtained by substituting Equation (1.104) into the motion equation (1.105), giving

$$\frac{d\omega_m}{dt} = -\frac{f_d}{J_m}\omega_m + \frac{3}{2}\frac{Z_p L_h}{L_r J_m}\psi_{rd} i_{sq} - \frac{T_L}{J_m}. \tag{1.106}$$

In terms of rotor's electrical velocity,

$$\frac{d\omega_e}{dt} = -\frac{f_d}{J_m}\omega_e + \frac{3}{2}\frac{Z_p^2 L_h}{L_r J_m}\psi_{rd} i_{sq} - \frac{Z_p T_L}{J_m}. \tag{1.107}$$

1.5.5 Model Parameters of Induction Motor and Model Validation

The induction motor has the characteristics given by the nameplate data (see Table 1.3), and its parameters used in the physical model are given in Table 1.4. These physical parameters are used in this book for simulations and experiments.

Table 1.3 Nameplate of the induction motor

Manufacturer	SEW-EURODRIVE
Type	3-phase Cage
Rated power	0.75 kW
Supply frequency	50 Hz
Number of pole pairs	2
Rated Speed	1435 RPM
Rated stator current	1.75 A
Rated RMS phase voltage	415 V
Connection	Y (star connection)

Table 1.4 Motor parameters

$R_s(\Omega)$	$R_r(\Omega)$	L_s(H)	L_r(H)	L_h(H)	$J_m(\text{kg}\cdot\text{m}^2)$	$f_d(\text{Nm}\cdot\text{s})$
11.2	8.3	0.6155	0.6380	0.570	0.00214	0.0041

The induction motor model is validated against the test-bed used in this book (see Chapter 10 for a detailed description of the test-bed). In the $d-q$ reference frame, the responses of stator currents i_{sd}, i_{sq}, rotor flux ψ_{rd} and motor mechanical velocity ω_m are calculated by building an induction motor simulator using the MATLAB/Simulink SimPower Toolbox, followed by entering the physical parameters listed in Table 1.4 into this simulator. Choosing the voltage input signals, u_{sd} and u_{sq}, as step signals with amplitudes of 10 V and 100 V respectively, the simulation results of currents, flux and velocity are shown in Figure 1.7. With the identical conditions as the simulation conditions, experiments are conducted using the test-bed. Figure 1.8 shows the experimental results of the currents, flux and velocity. By comparing these two figures, it is seen that the steady-state responses are very similar; however, the transient responses from the actual motor are faster than those produced by the

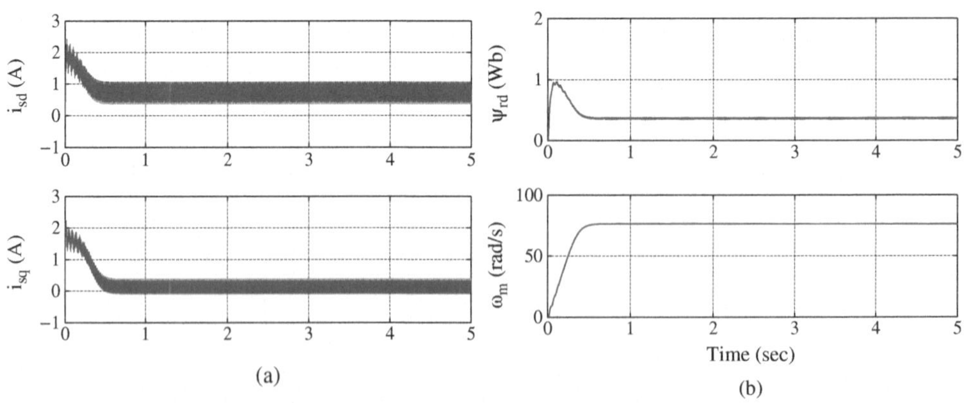

Figure 1.7 Simulation result using SimPower model. (a) i_{sd} and i_{sq} currents and (b) Flux and velocity.

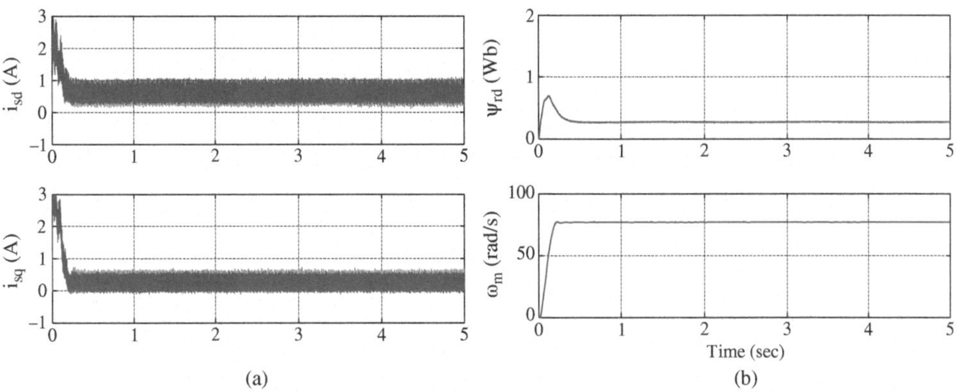

Figure 1.8 Experimental results. (a) i_{sd} and i_{sq} currents and (b) Flux and velocity.

Simulink simulator. Also, due to the PWM mechanisms used in both the simulation and experiment, the stator currents i_{sd} and i_{sq} contain substantial amount of harmonics (see Figures 1.7–1.8), particularly in the steady-state operations. Discussions about harmonics in the *AC* drives can be found in Chapter 2.

For those who are interested in how to build a Simulink simulator using SimPower Toolbox, a step-by-step tutorial for such a practice is given in Chapter 10.

1.6 Modeling of Power Converter

Figure 1.9 shows the block diagram of a grid-connected three phase voltage source converter. The symbols $[S_a, \overline{S_a}]$, $[S_b, \overline{S_b}]$, $[S_c, \overline{S_c}]$ in Figure 1.9 denote six bipolar switching inputs with either value of 0 or 1 for an upper and lower leg of each phase respectively. These switching inputs are conducted in a complementary manner, for example, at any given time $S_a + \overline{S_a} = 1$, $S_b + \overline{S_b} = 1$, $S_c + \overline{S_c} = 1$. This means that only one of the switches is allowed to conduct at any one time.

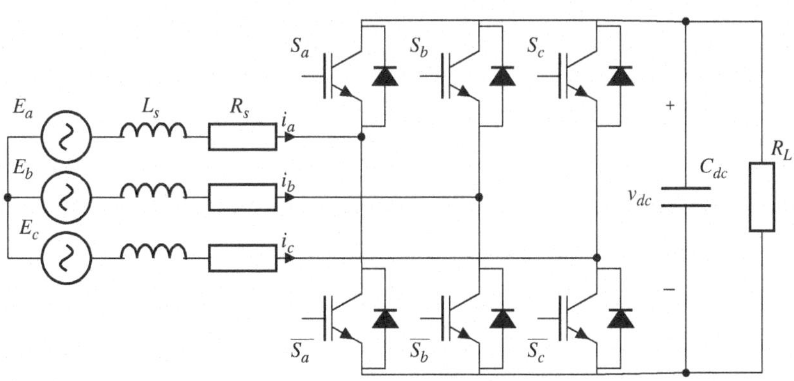

Figure 1.9 Block diagram of two level voltage source converter.

Before proceeding to the model derivation, several assumptions are made about the operation of the converter. Firstly, it is assumed that all switches are ideal and operate in a continuous conduction mode (CCM), and the grid voltage is symmetric and balanced as follows.

$$E_a = E_m \cos(\omega_g t) \tag{1.108}$$

$$E_b = E_m \cos\left(\omega_g t - \frac{2\pi}{3}\right) \tag{1.109}$$

$$E_c = E_m \cos\left(\omega_g t - \frac{4\pi}{3}\right), \tag{1.110}$$

where $\omega_g = 2\pi f, f$ is the grid frequency (50 Hz used in this book). Secondly, the system is assumed to be a three wire system, thus the sum of three phase currents and voltage is equal to zero.

$$i_a + i_b + i_c = 0 \tag{1.111}$$

$$E_{an} + E_{bn} + E_{cn} = 0. \tag{1.112}$$

Thirdly, it is assumed that the inductance and resistance parameters, L_s and R_s, are ideal, therefore, they have the same values for the three phase system. Even though these assumptions are ideal and do not hold entirely in applications, their acceptance ensures the derivation of the dynamic model in a simpler manner.

1.6.1 Space Vector Representation of Voltage Equation for Power Converter

The use of space vector simplifies the derivation of voltage equation for the two level voltage source power converter. The space vector for three-phase grid current can be written as

$$\vec{i_s} = \frac{2}{3}\left(i_a(t) + i_b(t)e^{j\frac{2\pi}{3}} + i_c(t)e^{j\frac{4\pi}{3}}\right), \tag{1.113}$$

while the space vector of three-phase voltage at the three nodes, on the grid side linked with the converter, is defined as

$$\vec{v_s} = \frac{2}{3}\left(v_a(t) + v_b(t)e^{j\frac{2\pi}{3}} + v_c(t)e^{j\frac{4\pi}{3}}\right) \tag{1.114}$$

and the space vector of the three phase grid voltage is described by

$$\vec{E_s} = \frac{2}{3}\left(E_a(t) + E_b(t)e^{j\frac{2\pi}{3}} + E_c(t)e^{j\frac{4\pi}{3}}\right). \tag{1.115}$$

Based on these definitions, the space vector voltage at the nodes of the converter satisfies the voltage equation:

$$\vec{v_s} = -L_s\frac{d\vec{i_s}}{dt} - R_s\overrightarrow{i_s(t)} + \vec{E_s}. \tag{1.116}$$

It is emphasized here that the above space vector voltage equation is obtained at the grid side.

1.6.2 Representation in $\alpha - \beta$ Reference Frame

The next step in the derivation of the model is to convert the model in the space vector form to the model in the $\alpha - \beta$ reference frame. By representing the space vectors, voltage and current, in terms of their real and imaginary components, they can be written in the complex notation,

$$\vec{v_s} = v_\alpha + jv_\beta$$

$$\vec{i_s} = i_\alpha + ji_\beta$$
$$\vec{E_s} = E_\alpha + jE_\beta.$$

To obtain the dynamic model in the $\alpha - \beta$ reference frame, the above variables are substituted into the space vector model (1.116) where the rotating frame in the $\alpha - \beta$ has the same velocity as the rotating speed of the space vector in the grid. Thus, by simply taking the components in the real and imaginary axes, in the $\alpha - \beta$ reference frame, the electrical model of the voltage source converter is described by the following two differential equations:

$$v_\alpha(t) = -L_s \frac{di_\alpha(t)}{dt} - R_s i_\alpha(t) + E_\alpha(t) \tag{1.117}$$

$$v_\beta(t) = -L_s \frac{di_\beta(t)}{dt} - R_s i_\beta(t) + E_\beta(t). \tag{1.118}$$

1.6.3 Representation in $d - q$ Reference Frame

To change the reference frame to the $d - q$ frame is equivalent to rotating the space vectors in (1.116) clockwise by θ_g so that the synchronous reference frame is aligned with the grid voltage at ω_g frequency, where $\theta_g = \omega_g t$ and $\omega_g = 2\pi f$ rad/s, f is the grid frequency in Hz. Thus, the following rotated space vectors are defined:

$$\vec{v_s}' = \vec{v_s} e^{-j\theta_g} = v_d + jv_q \tag{1.119}$$

$$\vec{i_s}' = \vec{i_s} e^{-j\theta_g} = i_d + ji_q \tag{1.120}$$

$$\vec{E_s}' = \vec{E_s} e^{-j\theta_g} = E_d + jE_q, \tag{1.121}$$

where $\vec{v_s}'$, $\vec{i_s}'$ and $\vec{E_s}'$ denote the space vectors referred to rotating $d - q$ frame. The grid voltage in the q-axis, E_q, is zero.

Multiplying (1.116) with the factor $e^{-j\theta_g}$ and replacing the original space vectors by the space vectors ((1.119)–(1.121)) in the rotating frame gives

$$\vec{v_s}' = -L_s \frac{d\vec{i_s}'}{dt} - j\omega_g L_s \vec{i_s}' - R_s \vec{i_s}' + \vec{E_s}', \tag{1.122}$$

where the following equality is used:

$$\frac{d\vec{i_s}}{dt} e^{-j\theta_g} = \frac{d\vec{i_s}'}{dt} + j\omega_g \vec{i_s}'.$$

Based on the real and imaginary components of (1.122), it can be verified that the dynamic model in the $d - q$ reference frame is:

$$v_d = -L_s \frac{di_d(t)}{dt} + \omega_g L_s i_q - R_s i_d + E_d \tag{1.123}$$

$$v_q = -L_s \frac{di_q(t)}{dt} - \omega_g L_s i_d - R_s i_q, \tag{1.124}$$

where $E_q = 0$.

1.6.4 Energy Balance Equation

The output of the voltage source converter is its DC voltage v_{dc} on the capacitor (see Figure 1.9). To derive the dynamic model for the v_{dc}, the energy conservation is investigated, where

$$P_c = P_g - P_L.$$

Here, P_c is the power consumed by the capacitor, P_g is power from the grid and P_L is the power consumed by the load. Assuming that the DC current for the capacitor is $i_c(t)$, then from the energy conservation, the following equality is obtained:

$$v_{dc} i_c = \frac{3}{2} \vec{v}_s \cdot \vec{i}_s - i_L v_{dc}, \tag{1.125}$$

where $\vec{v}_s \cdot \vec{i}_s$ is the inner product[2] of the space vectors of voltage and current, i_L is the load current. In this equation, the term $v_{dc} i_c$ is the energy conserved in the capacitor, the term $i_L v_{dc}$ is the energy consumption of the load, and the term $\frac{3}{2} \vec{v}_s \cdot \vec{i}_s$ is the energy drawn from the grid, consisting of active power and reactive power.

Since in the $\alpha - \beta$ reference frame, the following relationships hold:

$$\vec{v}_s = v_\alpha + j v_\beta$$

$$\vec{i}_s = i_\alpha + j i_\beta.$$

By substituting these equations into (1.125) and calculating the inner product, the capacitor current satisfies:

$$i_c = \frac{3}{2 v_{dc}} (v_\alpha i_\alpha + v_\beta i_\beta) - i_L.$$

Since $i_c = C_{dc} \frac{dv_{dc}}{dt}$, the capacitor voltage v_{dc} is described by the dynamic equation in the $\alpha - \beta$ reference frame:

$$C_{dc} \frac{dv_{dc}}{dt} = \frac{3}{2 v_{dc}} (v_\alpha i_\alpha + v_\beta i_\beta) - i_L, \tag{1.126}$$

where C_{dc} is the capacitance.

Similarly, in the $d - q$ reference frame, (1.125) is expressed in terms of the space vectors of current and voltage $\vec{i}_s'$ and $\vec{v}_s'$, which has the form:

$$\begin{aligned} v_{dc} i_c &= \frac{3}{2} \vec{v}_s' \cdot \vec{i}_s' - i_L v_{dc} \\ &= \frac{3}{2} (v_d i_d + v_q i_q) - i_L v_{dc}, \end{aligned} \tag{1.127}$$

leading to the capacitor voltage v_{dc} described in the $d - q$ reference frame as

$$C_{dc} \frac{dv_{dc}}{dt} = \frac{3}{2 v_{dc}} (v_d i_d + v_q i_q) - i_L. \tag{1.128}$$

Supposing that at steady state operating conditions, the converter maintains a target DC bus voltage with unity power factor, the reference signal to the i_q current is chosen to be zero in order to achieve this control objective.

---

[2] The inner product of two vectors $\vec{a}$ and $\vec{b}$ is the scalar $c = \vec{a} \cdot \vec{b}$. Letting $\vec{a} = [a_1, a_2]$ and $\vec{b} = [b_1, b_2]$, the scalar $c = a_1 b_1 + a_2 b_2$ (see Kreyszig (2010)).

1.7 Summary

This chapter has derived the physical models of PMSM, induction machine and voltage source power converter in both stationary reference frame (also called $\alpha - \beta$ reference frame) and synchronous reference frame (also called $d - q$) reference frame.

There are two steps deployed in the derivation of the mathematical models. The first step is to use space vector description of the physical variables for voltage, current and flux, and the second step is to convert the space vector based model to various reference frames. The mathematical models are presented in terms of differential equations that will be used for control system design in the future chapters.

1.8 Further Reading

General application characteristics of electric motors are discussed in Pillay and Krishnan (1991). Books for electrical drives include Vas (1992), Vas (1993) Hughes and Drury (2013), Leonhard (2001), El-Hawary (2011), Drury (2009), Quang and Dittrich (2008), Linder *et al.* (2010). Modeling and simulation of *AC* motor drive was discussed in Pillay and Krishnan (1988), Pillay and Krishnan (1989), Holtz (1994), Filho and de Souza (1997), Lorenz *et al.* (1994) and Ong (1998). Mathematical modeling and analysis of converters were presented in Wu *et al.* (1991), in Lindgren (1998), in Abdel-Rahim and Quaicoe (1994), in Blasko and Kaura (1997). The Park transformation was described in Park (1929) and the Clarke transformation was described in Duesterhoeft *et al.* (1951). A system theory approach to unify electrical machine analysis was discussed in Willems (1972).

References

Abdel-Rahim N and Quaicoe JE 1994 Modeling and analysis of a feedback control strategy for three- phase voltage-source utility interface systems. *Proceedings of the 29th IAS Annual Meeting* pp. 895–902.

Blasko V and Kaura V 1997 A new mathematical model and control of a three-phase AC-DC voltage source converter. *IEEE Transations on Power Electronics*.

Drury B 2009 *The Control Techniques drives and controls handbook* 2nd edn. IET.

Duesterhoeft W, Schulz MW and Clarke E 1951 Determination of instantaneous currents and voltages by means of alpha, beta, and zero components. *Transactions of the American Institute of Electrical Engineers* **70**(2), 1248–1255.

El-Hawary M 2011 *Principles of Electric Machines with Power Electronic Applications* 2nd edn. IEEE Press.

Filho E and de Souza R 1997 Three-phase induction motor dynamic mathematical model *IEEE International Electric Machines and Drives Conference Record*, pp. MB1/2.1 – MB1/2.3.

Holtz J 1994 The induction motor-a dynamic system *20th International Conference on Industrial Electronics, Control and Instrumentation*, vol. 1, p. P1.

Hughes A and Drury B 2013 *Electric Motors and Drives: Fundamentals, Types and Applications* 4th edn. Elsevier.

Kreyszig E 2010 *Advanced Engineering Mathematics* 10th edn. John Wiley and Sons, New York.

Leonhard W 2001 *Control of Electrical Drives* 3rd edn. Springer.

Linder A, Kanchan R, Kennel R and Stolze P 2010 *Model-Based Predictive Control of Electric Drives*. Cuvillier.

Lindgren M 1998 *Modeling and control of voltage source converters connected to the grid* PhD thesis Chalmers University of Technology.

Lorenz RD, Lipo TA and Novotny DW 1994 Motion control with induction motors. *Proceedings of the IEEE* **82**(8), 1215–1240.

Ong CM 1998 *Dynamic Simulation of Electric Machinery Using MATLAB*. Prentice Hall.

Park RH 1929 Two-reaction theory of synchrobous machines- part I. *AIEE Transations* **48**(2), 716–739.

Pillay P and Krishnan R 1988 Modeling of permanent magnet motor drives. *IEEE Transactions on Industrial Electronics* **35**(4), 537–541.

Pillay P and Krishnan R 1989 Modeling, simulation, and analysis of permanent-magnet motor drives. I. the permanent-magnet synchronous motor drive. *IEEE Transactions on Industry Applications* **25**(2), 265–273.

Pillay P and Krishnan R 1991 Application characteristics of permanent magnet synchronous and brushless DC motors for servo drives. *IEEE Transactions on Industry Applications* **27**(5), 986–996.

Quang NP and Dittrich JA 2008 *Vector Control of Three-Phase AC Machines* 1st edn. Springer.

Vas P 1992 *Electrical Machines and Drives- A Space-Vector Theory Approach*. Oxford University Press, New York, USA.

Vas P 1993 *Parameter Estimation, Condition Monitoring, and Diagnosis of Electrical Machines*. Oxford University Press, New York, USA.

Willems JL 1972 A system theory approach to unified electrical machine analysis. *International Journal of Control* **15**, 401–418.

Wu R, Dewan S and Slemon G 1991 Analysis of an AC-to-DC voltage source converter using PWM with phase and amplitude control. *IEEE Transactions on Industry Applications* **27**(2), 355–364.

2

Control of Semiconductor Switches via PWM Technologies

Control of the semiconductor switches is the most efficient and convenient means to achieve the control of power converters and machine drives. They act as the actuators in the implementation of the control systems where the manipulated control inputs in the form of three phase voltage signals are realized by turning on and off the semiconductor switches.

Depending on the application, a large variety of power electronic devices, different in types of semiconductor switches, construction topologies and concepts, have been developed. The common functionality of these devices is to conduct power flow by varying the on–off duration of each switch. Among them, the 2-Level Voltage Source Inverter (2L-VSI), as shown in Figure 2.1, is the most widely adopted mechanism to control three-phase *AC* machines. A similar topology is illustrated in the previous chapter for the power converter (see Figure 1.9). Based on a *DC* power supply for controlling *AC* motors, the primary concern of using the semiconductor switches for the applications presented in this book is to create sinusoidal phase voltage signals via the fundamental components of rectangular signals that are produced by varying magnitudes and frequencies through turning on–off each power switch for a duration of time. A similar operational principle applies to the voltage source power converter (see Figure 1.9). The command signal for turning on–off each power switch is called a gate signal.

There are two approaches used in this book to generate the gate signal for the semiconductor switches. The first approach uses Pulse Width Modulation (PWM) based on which the PID controllers (see Chapters 3 to 5) and the traditional model predictive controllers (see Chapters 8 to 9) are implemented. In the control applications, the control signals calculated are the three phase voltage signals that are obtained from one of the controller designs using the model either in the $d-q$ reference frame or in $\alpha-\beta$ reference frame. The role of the voltage source inverter with the power electronics devices is to realize the three phase voltage control signals as closely as possible. Namely, the sinusoidal phase voltage signals contained as fundamental components of rectangular wave signals created by turning on–off each power switch with the PWM technologies are aimed to be closely matched with the three phase voltage control signals. The second methodology features a much simpler approach in the implementation of the control systems that generates such a gate signal by direct optimization of an error function between the desired control signals and those achieved by the semiconductor switches (see Chapters 6 to 7). In this second approach, there is no need to use the PWM technology, therefore it significantly reduces the complexity of controlling the semiconductor switches.

PWM technology, originally developed in the telecommunication engineering community, has gained wide popularity and has been the subject of intensive research investigations in the control of power

PID and Predictive Control of Electrical Drives and Power Converters using MATLAB®/Simulink®, First Edition.
Liuping Wang, Shan Chai, Dae Yoo, Lu Gan and Ki Ng.

Companion Website: www.wiley.com/go/wang/pid

electronics over the past several decades (for example, Holtz, 1992, Holmes and Lipo, 2003). With PWM technology and a voltage source inverter, the efficient control of the *AC* motor position, speed and torque by Variable Speed Drive (VSD) becomes possible. Power converters are also controlled by the PWM technology. A wide variety of PWM generation techniques has been developed, which could be categorized into two broad classes: continuous PWM (CPWM) and discontinuous PWM (DPWM) (Hava *et al.*, 1999, Zhou and Wang, 2002). Compared with DPWM, CPWM attains superior performance in the low modulation index range which is within the operating conditions of servo drives and the power converters for most of the time. Therefore, it has gained more popularity in servo drive and converter applications. These methods could be implemented by two methodologies: carrier based PWM and direct digital implementation. For a long time, the carrier based PWM implementation, such as Sine-Triangle intersection techniques, has dominated the industrial applications. With the development of fast Digital Signal Processors (DSPs), direct digital implementation has also gained popularity in more recent years, of which the Space Vector Modulator (SVM) has been well recognized.

In Section 2.1, the topology of a two-level, voltage source inverter is introduced, where the relationship between the semiconductor power switches and the three phase sinusoidal control signal is established. The remainder of this chapter discusses how to manage the semiconductor power switches so that the three phase sinusoidal control signal can be reconstructed. In Section 2.2, the six-step mode is introduced. In Section 2.3, several carrier based PWM techniques are discussed, among which zero-sequence techniques are used to improve the modulation index. Section 2.4 discusses the space vector modulation, which has a direct digital implementation. In Section 2.5, a simulation study of the effect of PWM is conducted to reveal the current ripples.

2.1 Topology of IGBT Inverter

One of the most widely adopted semiconductor power switches for the medium-range power converter/inverter is the Insulated Gate Bipolar Transistor (IGBT), which offers the benefits of both MOSFET and bipolar switches. However, IGBT can only allow the current to flow in one direction, and hence a freewheeling diode in parallel is required to conduct the current flow in the opposite direction. In the three-phase 2L-VSI for controlling *AC* machine, as shown in Figure 2.1, each leg of the inverter has two pairs of such a combination consisting of an IGBT switch and a freewheeling diode, where their middle point is linked to the loads, either a passive load or an *AC* motor. Here, the front-end rectifier is replaced by two *DC* sources connected in series and each supply offers half of the total *DC* bus voltage ($V_{dc}/2$). For the simplification of analysis, it is sufficient to assume that the middle point (denoted by O in Figure 2.1) between two *DC* sources is referred to the ground. Thus, all the voltages can be represented with respect to the ground. For example, the neutral point voltage of *AC* motor is referred to as v_n with respect to the ground.

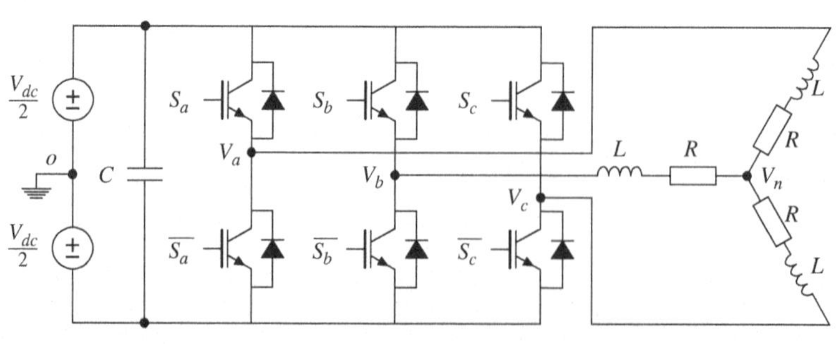

Figure 2.1 Topology of three-phase-leg IGBT.

As seen in Figure 2.1, there are two IGBT switches for each of the three legs. Within each leg of an inverter, only one switch is allowed to turn on (denoted by "1") while the other is off (denoted by "0") at any given time to prevent short circuit. Thus, the switching states of the inverter can be identified by only considering the states of the three upper switches. With the states of three upper switches denoted as S_i $(i = a, b, c)$, the states of their corresponding lower switches can be represented by their negation $\overline{S_i}$ $(i = a, b, c)$. As a result, there are only eight possible switching states by turning on and off all the switches in the inverter. Since the states of upper and lower switches within the same leg are complementary to each other, all eight switching states can be independently identified by the states of the three upper switches, as listed in Table 2.1. Among these, two switching states ($\vec{V}_0$ and $\vec{V}_7$), which represent the cases where either all the upper or all the lower switches are turned on, leading to an open circuit, are called zero vector. In contrast, the other six states that form a closed circuit, are called active vector.

When the *ith* upper switch is on, that is $S_i = 1$ and $\overline{S_i} = 0$, the output of the corresponding phase leg is connected to the top rail of the supplies and thus $v_i = \frac{V_{dc}}{2}$. Conversely, when the lower switch is on, the output is connected to the bottom rail of the supplies and hence $v_i = -\frac{V_{dc}}{2}$. Corresponding to the switching states in Table 2.1, the resulting output voltages v_a, v_b, v_c, are summarized in Table 2.2.

Equivalently, the output voltages could be represented in terms of their switching states,

$$v_i = V_{dc}S_i - \frac{V_{dc}}{2} \quad (i = a, b, c). \tag{2.1}$$

Note that the output voltages are expressed with respect to the ground defined before. It follows that the three-phase voltages with respect to the neutral point of the load are obtained by

$$\begin{aligned} v_{an} &= v_a - v_n \\ v_{bn} &= v_b - v_n \\ v_{cn} &= v_c - v_n. \end{aligned} \tag{2.2}$$

From (2.1), it is seen that the three voltage signals v_a, v_b and v_c generated from the voltage source inverter via the semiconductor switches are in rectangular-wave forms with amplitude changes between

Table 2.1 Switching states of inverter

	$\vec{V}_0$	$\vec{V}_1$	$\vec{V}_2$	$\vec{V}_3$	$\vec{V}_4$	$\vec{V}_5$	$\vec{V}_6$	$\vec{V}_7$
S_a	0	1	1	0	0	0	1	1
S_b	0	0	1	1	1	0	0	1
S_c	0	0	0	0	1	1	1	1

Table 2.2 Output voltage of inverter

	$\vec{V}_0$	$\vec{V}_1$	$\vec{V}_2$	$\vec{V}_3$	$\vec{V}_4$	$\vec{V}_5$	$\vec{V}_6$	$\vec{V}_7$
v_a	$-\frac{V_{dc}}{2}$	$\frac{V_{dc}}{2}$	$\frac{V_{dc}}{2}$	$-\frac{V_{dc}}{2}$	$-\frac{V_{dc}}{2}$	$-\frac{V_{dc}}{2}$	$\frac{V_{dc}}{2}$	$\frac{V_{dc}}{2}$
v_b	$-\frac{V_{dc}}{2}$	$-\frac{V_{dc}}{2}$	$\frac{V_{dc}}{2}$	$\frac{V_{dc}}{2}$	$\frac{V_{dc}}{2}$	$-\frac{V_{dc}}{2}$	$-\frac{V_{dc}}{2}$	$\frac{V_{dc}}{2}$
v_c	$-\frac{V_{dc}}{2}$	$-\frac{V_{dc}}{2}$	$-\frac{V_{dc}}{2}$	$-\frac{V_{dc}}{2}$	$\frac{V_{dc}}{2}$	$\frac{V_{dc}}{2}$	$\frac{V_{dc}}{2}$	$\frac{V_{dc}}{2}$

$\pm\frac{V_{dc}}{2}$ and, therefore, their fundamental components are of primary interest in the realization of the three phase control signals.

2.2 Six-step Operating Mode

One of early technologies to control the power electronic switches is called six-step mode (see Holtz (1992)). In this approach, the maximum magnitude of the fundamental with a 2L-VSI is achieved by sequentially switching the six active vectors, that is starting from $\vec{V_1}$, to $\vec{V_2}$ and the on to $\vec{V_6}$ within one electrical cycle. For one cycle of the six-step operation, each output voltage v_i $(i = a, b, c)$ will have half of cycle connect to top rail of inverter and the other half bottom rail. As an example, Figure 2.2 illustrates the rectangular wave of v_c over one cycle and its associated fundamental where the period of the wave is $\frac{2\pi}{\omega_e}$. Since this rectangular signal v_c is an odd function, it can be approximated using a Fourier sine series as (see Kreyszig (2010)) where

$$v_c(t) = b_1 \sin \omega_e t + b_3 \sin 3\omega_e t + \cdots + b_n \sin n\omega_e t + \cdots,$$

where $n = 1, 3, 5, \ldots$, is an odd number and the coefficient for the *nth* index is

$$\begin{aligned} b_n &= \frac{2\omega_e}{\pi}\int_0^{\frac{\pi}{\omega_e}} \frac{V_{dc}}{2}\sin n\omega_e t dt \\ &= \frac{2}{n\pi}V_{dc}. \end{aligned} \tag{2.3}$$

The fundamental component of the Fourier sine series $b_1 \sin \omega_e t$ is the approximated sinusoidal output of the voltage source inverter for the third phase of the voltage and is the function to be closely matched with the control signal at this phase. The rest of the components in the Fourier sine series are the harmonics. The Fourier coefficient $b_1 = \frac{2}{\pi}V_{dc}$ represents the amplitude of the fundamental component of the rectangular wave $v_c(t)$ generated by one cycle of the six step operation. This amplitude of the fundamental voltage of six-step mode limits the maximum amplitude of the achievable modulation signal. In other words, the maximum amplitude of the sinusoidal voltage signals that is achievable when using the six-step mode is $\frac{2}{\pi}V_{dc}$. From a control system design point of view, this means that if the calculated control signal had an amplitude that exceeded this value, then the output of the inverter would not be able to fully realize it.

One of the drawbacks for the six-step mode technique is that the amplitudes of the harmonics decay in a linear way proportional to the index $\frac{1}{n}$. In particular, the third harmonic still has a relatively large amplitude (see (2.3)) that only becomes one third of the fundamental component. From a control system perspective, the harmonics become part of the noise and disturbances of the system. Therefore, their amplitudes should be minimized. This means that the six-step mode technique will create large noise effects and disturbances in the control system.

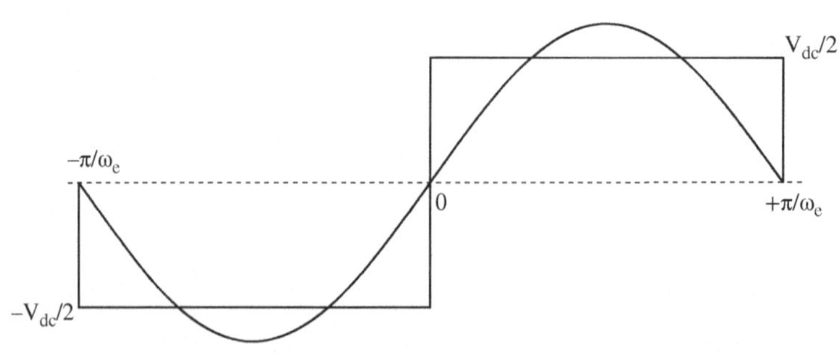

Figure 2.2 Output v_c of six-step mode.

2.3 Carrier Based PWM

Use of the six-step mode to control the power electronic switches is relatively simple in implementation; however, this approach has large harmonic components relative to the amplitude of the fundamental component of the output voltage signal, which is $\frac{2}{\pi}V_{dc}$. To reduce the effects of the harmonics, carrier based Pulse Width Modulation (PWM) techniques with zero sequence injection are introduced.

In general, the carrier based PWM technique is to compare the amplitude of an input voltage signal with that of a carrier signal, usually a periodic triangular signal with frequency f_c. If the input voltage signal (say phase A voltage, v_a^*) is larger than the carrier in amplitude, the switch function S_a outputs high level (logic "1") and otherwise low level (logic "0"), leading to the actual voltage output calculated as

$$v_a = V_{dc}S_a - \frac{V_{dc}}{2}.$$

The carrier frequency (denoted by f_c) is usually chosen to be much higher than the fundamental frequency (denoted by f_1) of the voltage signal and their ratio is the multiple of 3 for better reduction of Total Harmonic Distortion (THD) (see Holmes and Lipo (2003)). For example, a ratio

$$\frac{f_s}{f_1} = 21$$

could be used for this purpose.

2.3.1 *Sinusoidal PWM*

As one of the earliest PWM generation techniques (see Bowes, 1975), the Sinusoidal PWM (SPWM) generates the digital pulses to control the IGBT switches by directly comparing the three-phase voltages with the carrier, commonly a periodic triangular waveform. It was popular due to its simplicity and feasibility to be implemented by analog circuits. As an example to demonstrate the working principle of sinusoidal PWM technique, a carrier signal and three phase sinusoidal signals are shown in Figure 2.3 where the fundamental frequency of the sinusoidal signals is f_1 and the carrier frequency is f_c, and $\frac{f_c}{f_1} = 21$. The carrier signal has amplitude of $\pm\frac{V_{dc}}{2}$, where V_{dc} is the DC voltage of the power supply for the control application of AC drives or V_{dc} as the DC voltage output for the control application of power converter. In this illustration, there are 21 cycles of the carrier signal and one cycle of the three phase sinusoidal signals (see Figure 2.3(a)). When comparing the three phase sinusoidal signals with the carrier signal, the values of S_a, S_b and S_c are taken either 1 or 0, depending on whether the corresponding sinusoidal function is greater or smaller than the carrier signal. The actual output voltages are in rectangular waveforms (see Figure 2.3(b)) with their magnitudes being $\pm\frac{V_{dc}}{2}$ and their switches dependent on the values of S_a, S_b and S_c:

$$v_i = V_{dc}S_i - \frac{V_{dc}}{2} \qquad (i = a, b, c).$$

The fundamental components of the three rectangular waves in Figure 2.3(b) are found through Fourier series analysis and are shown in the corresponding plots. It is seen that they have similar characteristics as the input sinusoidal signals shown in Figure 2.3(a).

However, for the modulation to work within the linear range where its fundamental resembles the desired signal, it requires that the maximum amplitude of the sinusoidal signal be less than $\frac{V_{dc}}{2}$. Otherwise, the exceeding parts will cause the switch states stay at either "1" or "0" and thus the fundamentals lose their linear relationship to the desired reference voltage signals.

As a result, the controller output voltage should limit its value within $\pm\frac{V_{dc}}{2}$ to avoid the nonlinear modulation region if the sinusoidal PWM technique is used. The maximum modulation range could be improved by zero-sequence injection, which has led to several different modulation schemes as introduced below.

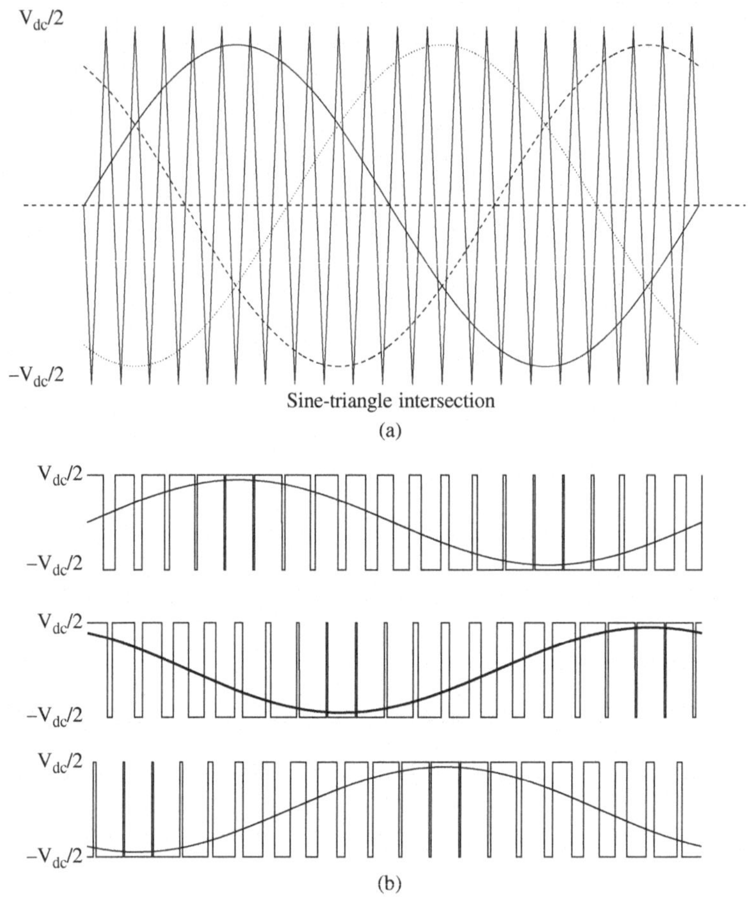

Figure 2.3 Illustration of sinusoidal PWM. Three phase voltage signals in (a) are reconstructed with Pulse-Width Modulation (b). (a) Sine-triangle intersection. (b) Pulses and fundamental sinusoidal signals from Fourier analysis. Top figure corresponds to solid-line wave in (a); middle figure corresponds to dashed-line wave in (a); bottom figure corresponds to dotted-line wave in (a).

2.3.2 Carrier Based PWM with Zero-sequence Injection

To understand the carrier based PWM with zero-sequence injection, the existence of a zero sequence is examined. For a three-phase system, let

$$\begin{aligned} v_a^* &= v_m \sin(\omega_e t) \\ v_b^* &= v_m \sin\left(\omega_e t - \frac{2\pi}{3}\right) \\ v_c^* &= v_m \sin\left(\omega_e t + \frac{2\pi}{3}\right) \end{aligned} \tag{2.4}$$

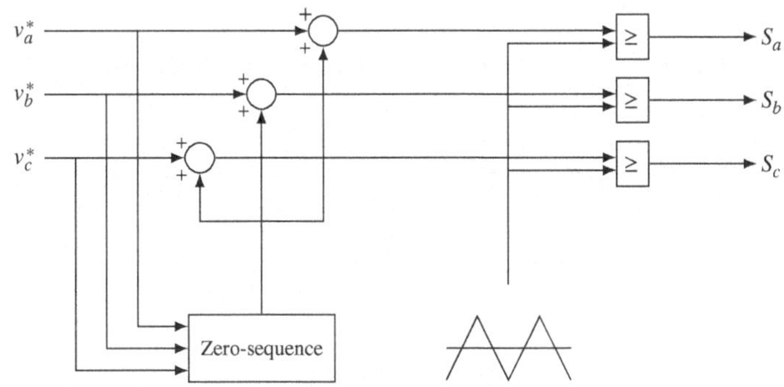

Figure 2.4 Carrier based PWM with zero sequence injection (Hava *et al.*, 1999).

denote the desired reference voltages that will be closely approximated by the outputs of the voltage source inverter. The neutral point voltage (v_n) with respect to ground, shown in Figure 2.1, can be represented by

$$v_n = \frac{v_a^* + v_b^* + v_c^*}{3}. \tag{2.5}$$

If the desired three-phase voltage, v_a^*, v_b^* and v_c^*, are balanced, it means that the neutral point voltage v_n equals 0. Because most balanced three-phase motors are three-wired systems with the isolated neutral point, there is the freedom of adding a nonzero value voltage to the neutral point voltage (v_n), leading to the modified three desired voltage signals,

$$\begin{aligned} v_a^{**} &= v_a^* + v_n \\ v_b^{**} &= v_b^* + v_n \\ v_c^{**} &= v_c^* + v_n. \end{aligned} \tag{2.6}$$

Then, based on the modified three desired voltage signals, the sinusoidal PWM technique explained in the previous section can be applied. Figure 2.4 illustrates the operation of the carrier based PWM technique with zero-sequence injection, which shows that the same "zero sequence" is added to all three reference voltage signals and these signals are compared with the carrier signal to produce the switching signals S_a, S_b and S_c.

There are many approaches in choosing the zero-sequence signal, leading to a variety of carrier based PWM schemes in the literatures (see Hava *et al.*, 1999). Amongst them, a commonly encountered selection of the zero-sequence signal for injection is the third harmonic injection PWM technique.

Third Harmonics Injection PWM (THIPWM) exploits the third harmonic component of the desired reference signal as the injection signal. There are two types of injections with different amplitudes of the third harmonic. With the desired reference voltage defined in (2.4), the THIPWM $1/6$ with one sixth of the reference amplitude is given by

$$v_n = \frac{1}{6} v_m \sin(3\omega_e t).$$

Similarly, the THIPWM $1/4$ is given by

$$v_n = \frac{1}{4} v_m \sin(3\omega_e t).$$

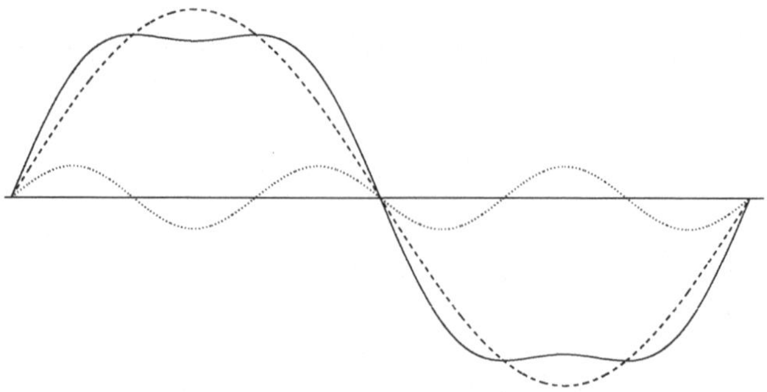

Figure 2.5 Waveform of THIPWM 1/6. Key: dashed-line original sinusoidal wave; dotted-line $\frac{1}{6}v_m \sin(3\omega_e t)$; solid-line sinusoidal wave with third harmonic injection.

With the third harmonics injection, the peak value of modified desired sinusoidal is reduced by the injection, as the waveform in Figure 2.5 demonstrates.

It can shown that the maximum amplitude of the three phase voltage signals is reduced to

$$\frac{2}{\sqrt{3}}\frac{V_{dc}}{2} = \frac{1}{\sqrt{3}}V_{dc}$$

if the $THIPWM1/6$ is used and to

$$\frac{3\sqrt{3}}{\sqrt{7}}\frac{V_{dc}}{2}$$

if the $THIPWM1/4$ technique is used.

The ratio between the maximum amplitude of the three phase voltage signals and the value $\frac{V_{dc}}{2}$ is called the modulation index, which is

$$m^*_{max-THIPWM1/6} = \frac{2/\sqrt{3}V_{dc}/2}{V_{dc}/2} = 1.1547 \tag{2.7}$$

for the THIPWM $1/6$ and

$$m^*_{max-THIPWM1/4} = \frac{3\sqrt{3}/\sqrt{7}V_{dc}/2}{V_{dc}/2} = 1.1223 \tag{2.8}$$

for the THIPWM $1/4$. Both modulation indices indicate that the linear modulation range is larger than the one generated by the original sinusoidal PWM.

Note that THIPWM1/4 is derived for the purpose of the minimization of the Total Harmonic Distortion (THD), whereas THIPWM1/6 is designed based on maximizing the linear modulation range (see Bowes and Lai (1997)).

Now, with the Third Harmonic Injection technique, for the modulation to work within the linear range where its fundamental resembles the desired signal, it requires that the maximum amplitude of the sinusoidal signal be less than $\frac{1}{\sqrt{3}}V_{dc}$ or $\frac{3\sqrt{3}}{\sqrt{7}}\frac{V_{dc}}{2}$ depending on whether THIPWM$1/6$ or THIPWM$1/4$ is being used. Basically, the modulation index will quantify the linear modulation range of the three phase voltage signals with respect to the results from the original sinusoidal PWM. The linear modulation range is translated into operational constraints from the perspective of controller design in later chapters.

2.4 Space Vector PWM

Space Vector PWM (SVPWM), as its name conveys, utilizes the concept of space vector and its geometrical features to derive the on-off time duration for each switch. Similar to the definition of MMF space vector in (1.12), the space vector of three-phase reference voltage is defined as

$$\vec{V_s^*} = \frac{2}{3}\left(v_{an}^* + v_{bn}^* e^{j\frac{2\pi}{3}} + v_{cn}^* e^{j\frac{4\pi}{3}}\right). \tag{2.9}$$

If a balanced three-phase voltage is employed, then $\vec{V_s^*}$ is a rotating vector with electrical speed ω_e, which is the frequency of sinusoidal signal.

The modulation of the desired space vector $\vec{V_s^*}$ is obtained by the time average of its two nearest active vectors and a zero vector, either $\vec{V_0}$ or $\vec{V_7}$. Taking the first sector for example, as illustrated in Figure 2.6, $\vec{V_s^*}$ could be modulated with the time average of the active vector $\vec{V_1}$ and $\vec{V_2}$ within one sampling period T_s by

$$T_s\vec{V_s^*} = T_1\vec{V_1} + T_2\vec{V_2} \tag{2.10}$$

where T_1 and T_2 are the duration of on-time for the active vectors $\vec{V_1}$ and $\vec{V_2}$ respectively. The relationship between the modulated vector $\vec{V_s^*}$ and two nearest active vectors is obtained by applying the geometric properties of the triangle,

$$\frac{T_s \cdot |\vec{V_s^*}|}{\sin\left(\frac{2\pi}{3}\right)} = \frac{T_1 \cdot |\vec{V_1}|}{\sin\left(\frac{\pi}{3} - \theta\right)} = \frac{T_2 \cdot |\vec{V_2}|}{\sin(\theta)} \tag{2.11}$$

that implies the duty cycle ratio of each active vector is

$$\frac{T_1}{T_s} = \frac{|\vec{V_s^*}|}{\frac{2}{3}V_{dc}} \cdot \frac{\sin\left(\frac{\pi}{3} - \theta\right)}{\sin\left(\frac{2\pi}{3}\right)} = \frac{|\vec{V_s^*}|}{V_{dc}/\sqrt{3}} \cdot \sin\left(\frac{\pi}{3} - \theta\right) \tag{2.12}$$

$$\frac{T_2}{T_s} = \frac{|\vec{V_s^*}|}{\frac{2}{3}V_{dc}} \cdot \frac{\sin(\theta)}{\sin\left(\frac{2\pi}{3}\right)} = \frac{|\vec{V_s^*}|}{V_{dc}/\sqrt{3}} \cdot \sin(\theta), \tag{2.13}$$

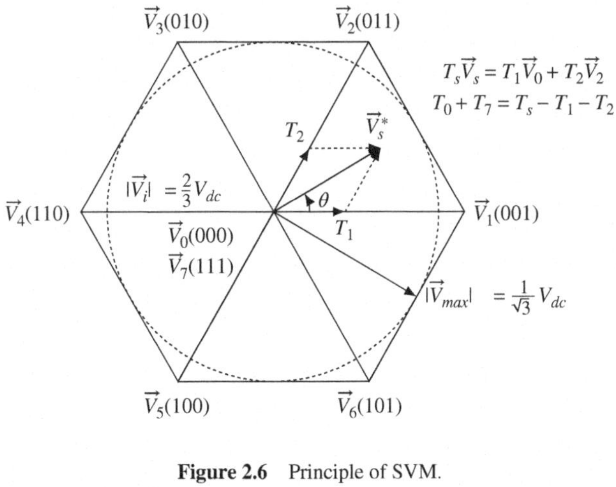

Figure 2.6 Principle of SVM.

where the length of each active vector $|\vec{V}_i| = \frac{2}{3}V_{dc}$ is used. It follows that the duration of the zero vector applied is the remaining time of the sampling period,

$$T_0 + T_7 = T_s - T_1 - T_2. \tag{2.14}$$

Since the hexagon has sixfold symmetry, the geometrical method discussed above can be used for the other five sectors as well by rotating the modulation vector by $(m-1)\frac{\pi}{3}$ *rads* with $m = 1\cdots 6$ denoting the sector in which it locates. The conventional SVPWM symmetrically distributes the four switching vectors, two active vectors and two zero vectors, within one sampling time, as shown in Figure 2.7. Such an arrangement offers the benefits of fixed switching frequency and better harmonics reduction performance.

Figure 2.7 shows a digital implementation of SVPWM within one sampling period for the example illustrated in Figure 2.6. There are four switching vectors denoted by $\vec{V}_0$, $\vec{V}_1$, $\vec{V}_2$, $\vec{V}_7$ corresponding to the four on-time periods, T_0, T_1, T_2 and T_7 calculated using (2.12)–(2.13). The arrangement of the four switching vectors is shown in the top part of Figure 2.7, together with the three switching states S_a, S_b and S_c. It is seen that the arrangement begins with the zero vector $\vec{V}_0$ and ends with the zero vector $\vec{V}_7$ for the first half, and symmetrical with the second half of the graph. This forms a symmetrical pattern from the center of the graph. The individual on-time period is also illustrated in the top part of Figure 2.7.

The bottom part of Figure 2.7 illustrates how SPVPWM is implemented. In the direct digital implementation, a parameter *MAX* is selected and a internal counter is set to count up and down within one sampling period T_s to form two straight lines as illustrated, which can be described by the linear equations:

$$PWM_{\text{time}} = \frac{2MAX}{T_s}t \qquad 0 \le t < \frac{T_s}{2} \tag{2.15}$$

$$PWM_{\text{time}} = 2MAX - \frac{2MAX}{T_s}t \qquad \frac{T_s}{2} \le t < T_s. \tag{2.16}$$

In the bottom part of Figure 2.7, on the vertical axis marked are the parameter *MAX*, and the three IGBT switching counts PWM_{S_a}, PWM_{S_b}, PWM_{S_c}. With the on-time periods T_0, T_1, T_2 and T_7 calculated,

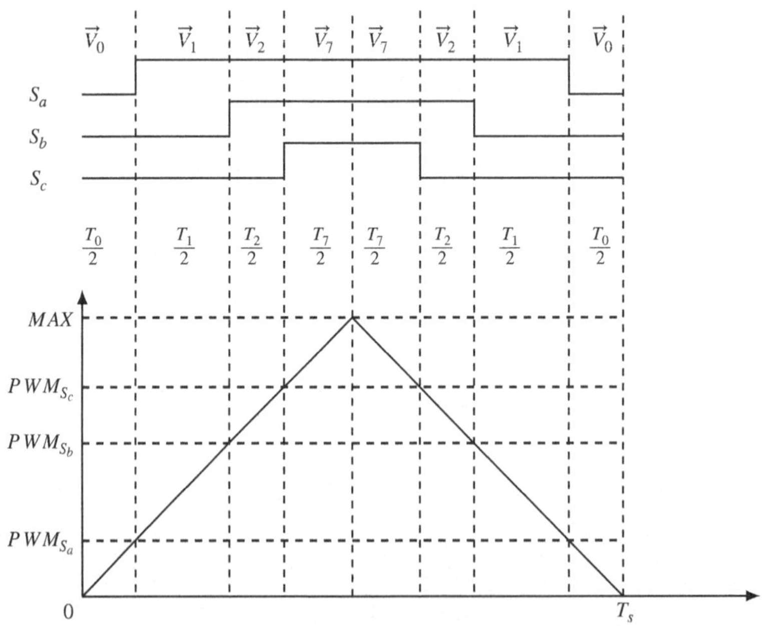

Figure 2.7 Digital implementation of SVPWM.

the corresponding count at which the IGBT switch turn-on period is calculated using the following ratios:

$$\begin{aligned}\frac{PWM_{S_a}}{MAX} &= \frac{T_0}{T_s}\\ \frac{PWM_{S_b}}{MAX} &= \frac{T_0+T_1}{T_s}\\ \frac{PWM_{S_c}}{MAX} &= \frac{T_0+T_1+T_2}{T_s}.\end{aligned} \tag{2.17}$$

In this example, at the beginning of the sample period, $S_a = S_b = S_c = 0$. When $PWM_{\text{time}} > PWM_{S_a}$ and PWM_{time} is increasing, the switching state $S_a = 1$. Similarly, when $PWM_{\text{time}} > PWM_{S_b}$ and PWM_{time} is increasing, the switching state $S_b = 1$. The same conditions apply to the switching state S_c. After the time t reaches $\frac{T_s}{2}$, the operations continue in reverse. Namely, when $PWM_{\text{time}} < PWM_{S_c}$ and PWM_{time} is decreasing, the switching state $S_c = 0$. The same conditions apply to the switching states S_b and S_a to complete one cycle of space vector implementation of PWM. The maximum amplitude of the three phase voltage signals to be realized by the space vector modulation technique is seen from Figure 2.6, which is $\frac{1}{\sqrt{3}}V_{dc}$. Thus, the modulation index is calculated as

$$m^*_{SVPWM} = \frac{2}{\sqrt{3}} = 1.155. \tag{2.18}$$

To ensure that the modulation is within the linear modulation range, it requires that voltages in the $\alpha - \beta$ reference frame satisfy

$$\sqrt{v_\alpha(t)^2 + v_\beta(t)^2} \leq \frac{1}{\sqrt{3}}V_{dc}. \tag{2.19}$$

This is based on the definition of the voltage space vector in relation to the voltage variables in $\alpha - \beta$ reference frame (see (1.22)). Similarly, the voltages in the $d - q$ reference frame are also required to satisfy

$$\sqrt{v_d(t)^2 + v_q(t)^2} \leq \frac{1}{\sqrt{3}}V_{dc}, \tag{2.20}$$

based on the definition of the voltage space vector in relation to those in the $d - q$ reference frame (see (1.28)). The inequalities (2.19) and (2.20) will be used as constraints in the control system design and implementation (see Chapter 4).

Note that the maximum amplitude of the three phase voltage signals when using the PWM with the third harmonic injection technique ($\frac{1}{6}$) is identical to the case when using the space vector modulation technique, which is $\frac{1}{\sqrt{3}}V_{dc}$. Thus, in the applications, the limit on the amplitude of the control signal is often taken as $\frac{1}{\sqrt{3}}V_{dc}$.

2.5 Simulation Study of the Effect of PWM

Since the time varying input voltages for *AC* drives and power converter are to be realized by the PWM generation and inverter, taking the PMSM as an example, the complete plant model, as shown in Figure 2.8, consists of machine model of PMSM, PWM and IGBT inverters. Note that Figure 2.8 is an equivalent representation in $d - q$ frame which does not illustrate the real hardware implementation in three-phase representation. When the modulation signal (v_i^*) works within the linear modulation region of chosen PWM, the fundamental of output voltage (v_i) from the IGBT inverter approximates the desired modulation signal, that is $v_i^* \approx v_i$. With the combination of Park-Clarke and its inverse transformation of which their multiplication is an identity matrix, it is assumed that $v_d^* \approx v_d$ and $v_q^* \approx v_q$ when using IGBT inverters as actuators for controller implementation.

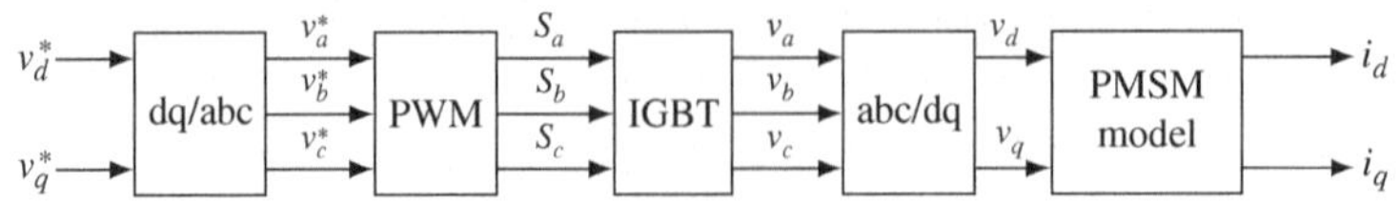

Figure 2.8 Model of PMSM combined with PWM generator.

To study the impact of carrier frequency in PWM, simulation studies are performed using the physical parameters of the PMSM given in Table 1.2. Two sets of simulation results are obtained by using the PMSM MATLAB/Simulink model illustrated in Figure 2.8. The first set of simulation results uses a smaller PWM carrier frequency where $f_c = 1.05$ kHz in contrast with the second set of simulation results where a larger carrier frequency $f_c = 10.5$ kHz is utilized. The voltage input signals to the PMSM model are set as $v_d^* = 3$ V and $v_q^* = 12.1244$ V, and the DC bus voltage is fixed at 100 V so that the modulation signal is within its linear modulation range. Figure 2.9(a) shows the three phase currents

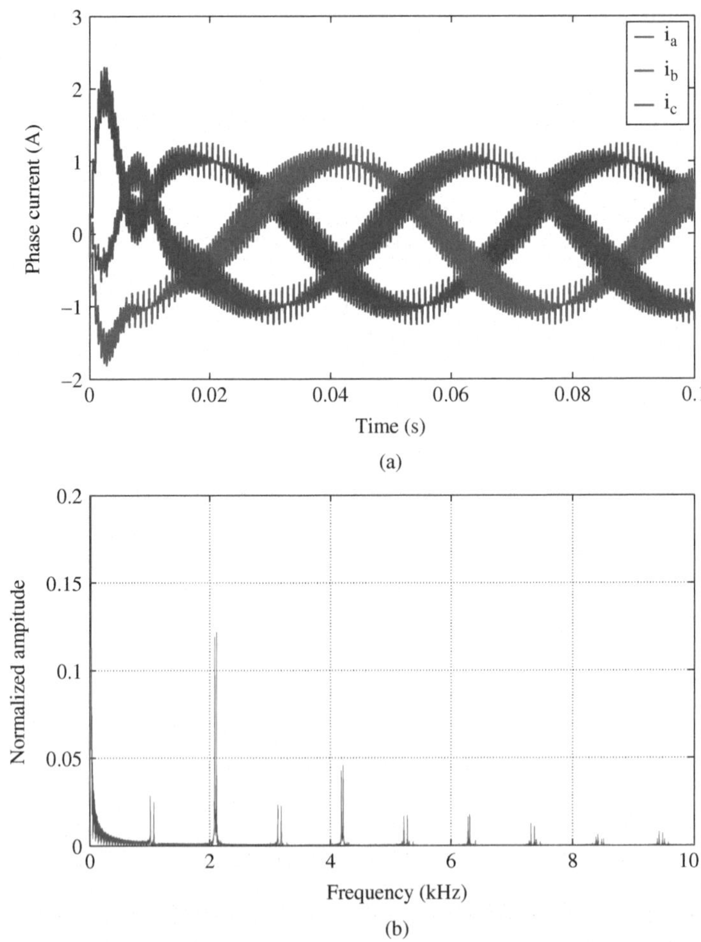

Figure 2.9 Spectrum of phase currents ($f_c = 1.05$ kHz, $v_d = 3$ V and $v_q = 12.1244$ V).

and Figure 2.9(b)shows the corresponding spectrum of A phase current for the lower carrier frequency case. It is seen that with this lower carrier frequency, the i_a, i_b and i_c currents contain a large amount of high frequency harmonic noise (see Figure 2.9(a)) and the harmonics mainly occur around the carrier frequency and its multiples (see Figure 2.9(b)), that is 1.05n kHz with n denoting an integer in this case. When the carrier frequency is increased to f_c = 10.5 kHz, the high frequency harmonic noise in the three phase currents is reduced as seen in Figure 2.10(a) and the current ripple still occurs at the carrier frequency (see Figure 2.10(b)), but the multiples move to the high frequency region. At a certain high frequency region, the current ripple is then attenuated by the limited bandwidth of the PMSM, which is a first order type of system depending on the values of resistance (R_s) and inductance (L_d, L_q).

In summary, the high frequency harmonic ripple with a low carrier frequency is much more severe than the one with a high carrier frequency, and the harmonics occur around the carrier frequency and its multiples. Thus, a higher carrier frequency in PWM offers improved performance in terms harmonics attenuation. However, in practice, the resulting high switching loss, inherent characteristics of switching devices and limited computational power prevent the use of a very high carrier frequency.

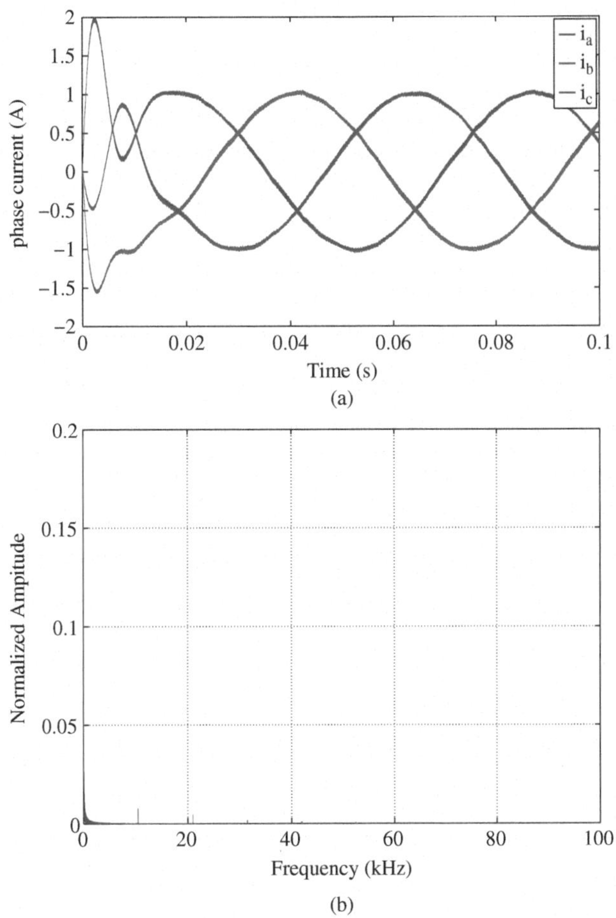

Figure 2.10 Spectrum of phase currents (f_c = 10.5 kHz, v_d = 3 V and v_q = 12.1244 V). (a) three-phase current and (b) Spectrum of i_a at steady-state.

2.6 Summary

Two popular categories of PWM techniques, carrier based PWM and space vector based PWM, have also been briefly revisited. It has been shown in Zhou and Wang (2002) that both implementation approaches can lead to the same type of PWM, such as SVPWM, THIPWM and several variations of Discontinuous PWM (DPWM) schemes (see Hava *et al.* (1999), Zhou and Wang (2002)). The implementation of these PWM schemes is achieved by appropriate zero-sequence injection in the carrier based PWM approach, while its corresponding equivalence is the varieties of the placement of zero-vector in the space vector based PWM approach.

The maximum modulation index of the linear modulation ranges for different PWM is discussed. Simulation results show that the PWM techniques have a reduced effect on the system when the carrier frequency is increased.

2.7 Further Reading

A book was written on PWM by Holmes and Lipo (2003). An early work on sinusoidal PWM generator was presented in Bowes (1975). Survey papers published on PWM technologies included Holtz (1992) and Holtz (1994). PWM schemes based on voltage space vectors were realized and analyzed in van der Broeck *et al.* (1988). Relationships between space-vector modulation and three-phase carrier-based PWM were analyzed in Zhou and Wang (2002), Blasko (1997), Bowes and Lai (1997) and Chai and Wang (2013). Carrier-based PWM-VSI over-modulation strategies were designed, analyzed and compared in Hava *et al.* (1998). Graphic methods were discussed for carrier-based PWM-VSI drives in Hava *et al.* (1999). Space vector modulation implementation with the ADMCF32X was discussed in Analog Devices Inc (2000).

Geyer (2011) made comparative studies of control and modulation schemes for medium-voltage drives between predictive control concepts and PWM-Based Schemes.

References

Analog Devices Inc 2000 *Implementing Space Vector Modulation with the ADMCF32X.*

Blasko V 1997 Analysis of a hybrid PWM based on modified space-vector and triangle-comparison methods. *IEEE Transactions on Industry Applications* **33**(3), 756–764.

Bowes S 1975 New sinusoidal pulsewidth-modulated invertor. *Electrical Engineers, Proceedings of the Institution of* **122**(11), 1279–1285.

Bowes S and Lai YS 1997 The relationship between space-vector modulation and regular-sampled PWM. *IEEE Transactions on Industrial Electronics* **44**(5), 670–679.

Chai S and Wang L 2013 A unified pulse generation approach for 2L-VSI from SVPWM to direct switching *Industrial Electronics Society, IECON 2013–39th Annual Conference of the IEEE*, pp. 346–351.

Geyer T 2011 A comparison of control and modulation schemes for medium-voltage drives: Emerging predictive control concepts versus PWM-based schemes. *IEEE Transactions on Industry Applications* **47**(3), 1380–1389.

Hava A, Kerkman R and Lipo T 1998 Carrier-based PWM-VSI overmodulation strategies: analysis, comparison, and design. *IEEE Transactions on Power Electronics* **13**(4), 674–689.

Hava A, Kerkman R and Lipo T 1999 Simple analytical and graphical methods for carrier-based PWM-VSI drives. *IEEE Transactions on Power Electronics* **14**(1), 49–61.

Holmes D and Lipo T 2003 *Pulse Width Modulation for Power Converters: Principles and Practice* vol. 18. Wiley.

Holtz J 1992 Pulsewidth modulation-a survey. *IEEE Transactions on Industrial Electronics* **39**(5), 410–420.

Holtz J 1994 Pulsewidth modulation for electronic power conversion. *Proceedings of the IEEE* **82**(8), 1194–1214.

Kreyszig E 2010 *Advanced Engineering Mathematics* 10th edn. John Wiley & Sons, Inc., New York.

van der Broeck H, Skudelny HC and Stanke G 1988 Analysis and realization of a pulsewidth modulator based on voltage space vectors. *IEEE Transactions on Industry Applications* **24**(1), 142–150.

Zhou K and Wang D 2002 Relationship between space-vector modulation and three-phase carrier-based PWM: a comprehensive analysis. *IEEE Transactions on Industrial Electronics* **49**(1), 186–196.

3

PID Control System Design for Electrical Drives and Power Converters

Almost all electric drives and power converters in current industrial applications are controlled by PID controllers. The control structures are already defined in commercial drives; however, the controller parameters may be adjustable for performance improvement. In a typical drive, there are a number of PID controllers used, depending on the actual applications.

To our knowledge from working with a number of industries, when *AC* drives are used as servo drives to track velocity and position references, the PID control systems for the drives are in cascade feedback structure in the q-axis, in which inner-loop q-axis current is controlled using a PI and the outer-loop system is either controlled using PI for velocity control or PID for position control. This structure at least can be traced back for the past two decades (see Quang and Dittrich (2008), Lorenz *et al.* (1994), del Blanco *et al.* (1999)). It is necessary to use PID controllers for the outer-loop system because of the closed-loop performance requirement for the disturbance rejection of load torque disturbances. However, in the inner-loop current controller design, a simpler choice for the q-axis current is a proportional controller.

The first section of this chapter presents introductory materials for PID controller design and cascade feedback control systems. Also highlighted in this section is an extremely simple approach to overcome overshoot response to the reference signal of a PID control system. Using a PMSM drive as an example, the next four sections of this chapter are devoted to discussing the design of PI controllers for current control, cascade velocity control and cascade position control (see Sections 3.2–3.5). The similarities between PMSM control and induction motor control are where they both use the same control system architectures for the same control objectives whilst there may be a slight difference in the choice of reference signals. In Section 3.6, the general control system architecture of an induction motor is explained. In Section 3.7, current control and velocity control of induction motor are introduced, where in addition, the estimation of synchronous frequency or slip estimation is introduced. Although the control of power converters also follows the similar control architectures as *AC* motor drives, the cascade control system is used to control the d-axis current in order to achieve voltage control. In Section 3.8, the control system architecture for a two level voltage source power converter is explained. In Section 3.9, PI current and voltage control of a power converter is discussed, where in addition, a phase-locked-loop PI controller is used for estimating the grid frequency.

PID and Predictive Control of Electrical Drives and Power Converters using MATLAB®/Simulink®, First Edition.
Liuping Wang, Shan Chai, Dae Yoo, Lu Gan and Ki Ng.

Companion Website: www.wiley.com/go/wang/pid

3.1 Overview of PID Control Systems Using Pole-assignment Design Techniques

When using pole-assignment controller design technique for PID controllers, the starting point is to assume that the transfer function of a plant model $G(s)$ is available. In order to obtain unique solutions of the controller parameters, its structure is restricted by the order of the transfer function $G(s)$. For instance, if the transfer function model is a first order model, then a PI controller is used, and if it is a second order model, then a PID controller is used. With the model and the controller structure selected, the actual closed-loop polynomial is parameterized using the unknown controller parameters, which is then made to be equal to a desired closed-loop polynomial of the same order, leading to the unique solutions of the PID controller parameters.

3.1.1 PI Controller Design

Assuming that a transfer function is given by

$$G(s) = \frac{b}{s+a} \tag{3.1}$$

and the PI controller is represented by

$$C(s) = K_c\left(1 + \frac{1}{\tau_I s}\right), \tag{3.2}$$

where K_c is the proportional gain and τ_I is the integral time constant. The feedback PI control system is illustrated in Figure 3.1, where $R(s)$, $E(s)$, $U(s)$ and $Y(s)$ denote the Laplace transforms of the reference signal, error signal, control signal and output signal respectively.

To make the design simpler, the PI controller is written in the transfer function form,

$$C(s) = \frac{c_1 s + c_0}{s}, \tag{3.3}$$

where $K_c = c_1$ and $\tau_I = \frac{c_1}{c_0}$. We will first find the coefficients c_1 and c_0 based on the model (3.1), then convert these coefficients into the standard PI controller parameters K_c and τ_I.

The closed-loop transfer function from the reference signal to the output signal is expressed as

$$\begin{aligned}\frac{Y(s)}{R(s)} &= \frac{G(s)C(s)}{1+G(s)C(s)} \\ &= \frac{\frac{b}{s+a}\frac{c_1 s+c_0}{s}}{1+\frac{b}{s+a}\frac{c_1 s+c_0}{s}} \\ &= \frac{b(c_1 s + c_0)}{s(s+a)+b(c_1 s+c_0)}.\end{aligned} \tag{3.4}$$

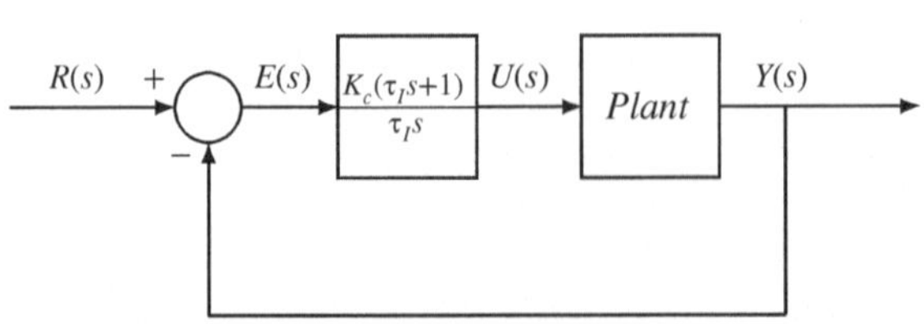

Figure 3.1 Block diagram of PI control system.

The closed-loop poles of the feedback system are the solutions of the polynomial equation with respect to s, that is

$$s(s+a)+b(c_1s+c_0)=0. \tag{3.5}$$

Equation (3.5) is called a closed-loop characteristic equation. It is well understood that the locations of the closed-loop poles determine the closed-loop stability as well as its response speed to reference signal and disturbance rejection.

Since the model parameters a and b are given, the free parameters in (3.5) are the controller parameters c_1 and c_0. To find the controller parameters c_1 and c_0, the following polynomial equation is set,

$$s(s+a)+b(c_1s+c_0)=s^2+2\xi w_n s+w_n^2, \tag{3.6}$$

where the left-hand side of Equation (3.6) is the characteristic polynomial that determines the actual closed-loop poles with the controller and the right-hand side is the desired closed-loop characteristic polynomial that determines the desired closed-loop poles. The discussion on the desired closed-loop characteristic polynomial with respect to the choice of natural frequency w_n (or bandwidth) and damping coefficient ξ is discussed in Section 3.1.2. By equating these two polynomials, the actual closed-loop poles are assigned to the desired closed-loop poles. This controller design technique is called pole-assignment controller design.

Now, we compare the coefficients of the polynomial equation (3.6) on both sides. With the deliberately chosen model structure and the controller structure, the coefficients for s^2 on both sides of (3.6) are equal to 1. Equating the coefficient of s on the left-hand side to the one on the right-hand side gives

$$a+bc_1=2\xi w_n. \tag{3.7}$$

The same procedure is applied to the constant term, leading to

$$bc_0=w_n^2. \tag{3.8}$$

Solving (3.7) gives

$$c_1=\frac{2\xi w_n-a}{b} \tag{3.9}$$

and solving (3.8) gives

$$c_0=\frac{w_n^2}{b}. \tag{3.10}$$

With the relationships between c_1, c_0 and K_c, τ_I (see Equation (3.3)), the PI controller parameters are found as

$$K_c=c_1=\frac{2\xi w_n-a}{b} \tag{3.11}$$

$$\tau_I=\frac{c_1}{c_0}=\frac{2\xi w_n-a}{w_n^2}. \tag{3.12}$$

3.1.2 Selecting the Desired Closed-loop Performance

The natural frequency (or bandwidth) w_n and damping coefficient ξ are the free parameters to be selected by the designer as desired closed-loop performance specification.

The parameter ξ is often chosen as 1 or 0.707 in practice. When $\xi=1$, the poles of the desired closed-loop system are the solutions of the polynomial equation (see (3.6)),

$$s^2+2w_ns+w_n^2=0, \tag{3.13}$$

which are $s_1 = s_2 = -w_n$. Namely, we have two identical poles when $\xi = 1$. With the second choice of $\xi = \frac{1}{\sqrt{2}} \approx 0.707$, the poles are a pair of complex-conjugate numbers determined by

$$s_{1,2} = \frac{-2\xi w_n \pm \sqrt{4\xi^2 w_n^2 - 4w_n^2}}{2} = -0.707w_n \pm j0.707w_n, \tag{3.14}$$

which have identical real and imaginary components.

When the damping coefficient $\xi = 0.707$, the natural frequency w_n is equal to the bandwidth of the desired closed-loop control system that has the transfer function,

$$T(s) = \frac{w_n^2}{s^2 + 2\xi w_n s + w_n^2}, \tag{3.15}$$

where the bandwidth is defined according to the cut-off frequency ω_c with the parameter ω_c being chosen such that $|T(j\omega)| = \frac{T(0)}{\sqrt{2}}$. It can readily verified that when $s = jw_n$, the magnitude of $|T(j\omega)|$ in (3.15) is $\frac{1}{\sqrt{2}}$. Therefore, the larger w_n is, the wider the desired closed-loop bandwidth is.

With the parameter ξ chosen (either 1 or 0.707), the natural frequency w_n becomes a closed-loop performance parameter that the user specifies according to the desired closed-loop response requirement. From the simulation of a step response using (3.15) (see Figure 3.2), the settling time[1] T_s with respect to the parameter w_n is estimated as

$$T_s \approx \frac{5\xi}{w_n}. \tag{3.16}$$

With this estimate, we can convert the desired closed-loop settling time to the actual parameter w_n that will be used in the design of PI controller. For instance, if we want the desired closed-loop settling time to be 4 to 5 seconds, then $w_n = 1$. Figure 3.2 is used to illustrate the relationship between the settling time and the parameter w_n. Figure 3.3 shows the magnitudes of the frequency responses of the closed-loop transfer function (3.15) with $w_n = 1$ and 10. With $\xi = 0.707$, we identify the bandwidth is $\omega_c = 1$ and $\omega_c = 10$ respectively. However, with $\xi = 1$, when $s = jw_n$, simple calculation shows that $|T(j\omega)| = 0.5$, which is less than 0.707, implying that with this choice of damping coefficient, the bandwidth is narrower. In fact, the bandwidth is calculated as $\omega_c \approx 0.645w_n$ as illustrated in Figure 3.3.

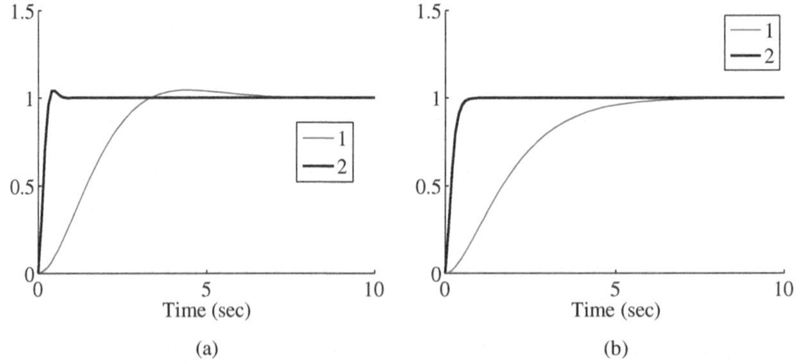

Figure 3.2 Step response of the desired closed-loop transfer function. Key: line (1) $w_n = 1$, line (2) $w_n = 10$. (a) $\xi = 0.707$ and (b) $\xi = 1$.

[1] Settling time is defined as the time required to complete 95 percent of the response.

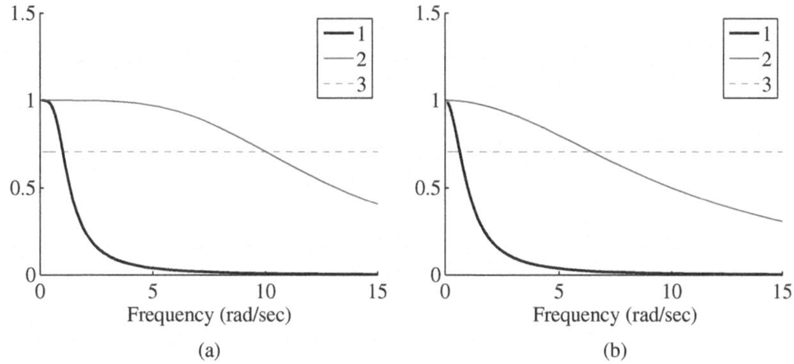

Figure 3.3 Magnitude of the frequency response of transfer function (3.15). Key: line (1) $w_n = 1$; line (2) $w_n = 10$; line (3) indicator $\left(\frac{1}{\sqrt{2}}\right)$ for the bandwidth. (a) $\xi = 0.707$ and (b) $\xi = 1$.

3.1.3 Overshoot in Reference Response

To examine the issue of overshoot in response to reference signal, the following example is used. Assume that the plant is described by the transfer function

$$G(s) = \frac{1}{s(s+1)^3}. \tag{3.17}$$

This model structure is not a first order system, therefore the design methods discussed in the previous sections are not directly applicable. However, it can be designed using the frequency response data generated from the transfer function (3.17) that will capture the low to medium frequency information of the system (see Wang and Cluett (2000)). With the frequency response approach, the PI controller is found to have the parameters: $K_c = 0.56$, $\tau_I = 8.$, leading to a stable closed-loop system. In the following, we will demonstrate that although it is the same PI controller, there are two approaches that can be used for its implementation and the results are very different in terms of reference signal response.

3.1.3.1 The Traditional Approach

In the traditional implementation, the control signal $u(t)$ is calculated using the following expression:

$$u(t) = K_c(r(t) - y(t)) + \frac{K_c}{\tau_I}\int_0^t (r(\tau) - y(\tau))d\tau, \tag{3.18}$$

where $r(t)$ and $y(t)$ are the reference and output signals, respectively. The closed-loop transfer function between the Laplace transform of the reference signal $R(s)$ and the output $Y(s)$ is

$$\frac{Y(s))}{R(s)} = \frac{K_c(\tau_I s + 1)}{\tau_I s^2(s+1)^3 + K_c(\tau_I s + 1)}. \tag{3.19}$$

3.1.3.2 The Alternative Approach

In the alternative implementation, the control signal is calculated using

$$u(t) = -K_c y(t) + \frac{K_c}{\tau_I}\int_0^t (r(\tau) - y(\tau))\, d\tau, \tag{3.20}$$

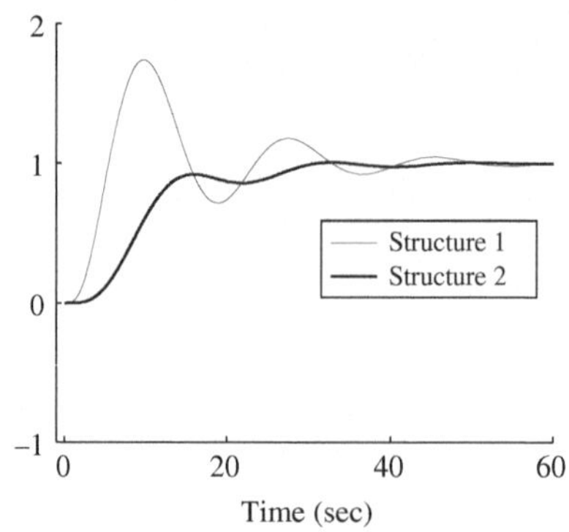

Figure 3.4 Closed-loop step response of PI control system. Key: line 1: the response from the original structure; line 2: response from the alternative structure.

where the proportional controller K_c is implemented on the output signal only. The closed-loop transfer function for the alternative approach is

$$\frac{Y(s)}{R(s)} = \frac{K_c}{\tau_I s^2(s+1)^3 + K_c(\tau_I s + 1)}. \tag{3.21}$$

The closed-loop step responses for both controller implementations are simulated and compared in Figure 3.4, which shows that the original PI closed-loop control system has a large overshoot, in contrast, the alternative PI closed-loop control system has avoided this overshoot.

By comparing the closed-loop transfer function (3.19) from the original PI controller structure with the one (3.21) from the alternative structure, we notice that both transfer functions have the same denominator; however, the one from the original structure has a zero at $-\frac{1}{\tau_I}$. Because of this zero, the original closed-loop step response had an overshoot. In other words, the alternative structure is equivalent to a filter $\frac{1}{\tau_I s+1}$ being added to the reference signal, which is, in fact, a two-degrees of freedom control system illustrated in Chapter 5.

It is important to point out that the implementation of PI controller using the alternative structure (see (3.18)) is useful for the outer-loop control systems where an overshoot response to a reference signal is not desirable. However, if the PI controller is used in an inner-loop control system, then the original implementation structure (see (3.20)) is recommended because it provides a faster dynamic response and overshoot to reference signal is not an issue in this situation.

3.1.4 PID Controller Design

PID control is generally required when controlling the angular position of an *AC* drive because it has a second order transfer function in the following form:

$$\frac{Y(s)}{U(s)} = \frac{b}{s(s+a)}. \tag{3.22}$$

An ideal PID controller has the transfer function

$$C(s) = K_c\left(1 + \frac{1}{\tau_I s} + \tau_D s\right), \tag{3.23}$$

where K_c is the proportional gain, τ_I is the integral time constant and τ_D is the derivative gain.

We rewrite the PID controller given in (3.23) into the transfer function form

$$C(s) = \frac{c_2 s^2 + c_1 s + c_0}{s}. \tag{3.24}$$

By comparing (3.23) with (3.24), we have the relationships,

$$K_c = c_1; \quad \tau_I = \frac{c_1}{c_0}; \quad \tau_D = \frac{c_2}{c_1}. \tag{3.25}$$

In the design, the first step is to find the parameters in (3.24), then convert them into the PID controller parameters required in the implementation stage.

The pole-assignment controller design technique is used here to find the PID controller parameters c_2, c_1 and c_0. The closed-loop transfer function with the PID controller (3.24) and the dynamic transfer function model (3.22) is

$$\begin{aligned} \frac{Y(s)}{R(s)} &= \frac{\frac{b(c_2 s^2 + c_1 s + c_0)}{s^2(s+a)}}{1 + \frac{b(c_2 s^2 + c_1 s + c_0)}{s^2(s+a)}} \\ &= \frac{b(c_2 s^2 + c_1 s + c_0)}{s^2(s+a) + b(c_2 s^2 + c_1 s + c_0)}. \end{aligned} \tag{3.26}$$

Note that the closed-loop polynomial, which is the denominator of (3.26), is of a third order. In the design of a pole-assignment controller, we will choose three desired closed-loop poles for the closed-loop performance specification. The pair of dominant desired closed-loop poles are selected as $s_{1,2} = -\xi w_n \pm j w_n \sqrt{1-\xi^2}$, where $\xi = 0.707$, and the fast pole is then chosen to be $s_3 = -n \times w_n$ ($n \gg 1$, for example, 10). With this specification, the parameter w_n is approximately the bandwidth of the desired closed-loop system. This leads to the desired closed-loop polynomial as

$$A_{cl}^d = (s^2 + 2\xi w_n s + w_n^2)(s + n \times w_n) = s^3 + t_2 s^2 + t_1 s + t_0,$$

where $t_2 = (2\xi + n)w_n$, $t_1 = (2\xi n + 1)w_n^2$, $t_0 = n w_n^3$.

Letting the desired closed-loop polynomial equal to the actual closed-loop polynomial, and as the result, the following polynomial equality is obtained:

$$s^2(s+a) + b(c_2 s^2 + c_1 s + c_0) = s^3 + t_2 s^2 + t_1 s + t_0. \tag{3.27}$$

By comparing the coefficients from both sides of (3.27) and setting them to be equal, the controller parameters are found as

$$c_2 = \frac{t_2 - a}{b} \tag{3.28}$$

$$c_1 = \frac{t_1}{b} \tag{3.29}$$

$$c_0 = \frac{t_0}{b}. \tag{3.30}$$

Accordingly, the PID controller parameters are found for the position control of the outer-loop system:

$$K_c = c_1 = \frac{(2\xi n + 1)w_n^2}{b} \tag{3.31}$$

$$\tau_I = \frac{c_1}{c_0} = \frac{(2\xi n + 1)w_n^2}{n w_n^3} = \frac{(2\xi n + 1)}{w_n} \tag{3.32}$$

$$\tau_D = \frac{c_2}{c_1} = \frac{(2\xi + n)w_n - a}{(2\xi n + 1)w_n^2}. \tag{3.33}$$

In the implementation of the PID controller for position control, it is almost without exception that the implementation of the derivative term should be directly on the output signal to avoid the situation of derivative "kick" due to step reference signal changes on the position. This is because the derivative of a step signal is infinity at the time of change and the derivative action from the reference signal will lead to a large amplitude in the control signal that could cause problems. Therefore, the control signal $u_D(t)$ from the derivative term is

$$u_D(t) = -K_c\tau_D \frac{dy(t)}{dt}. \tag{3.34}$$

In the *AC* drive control, the derivative of the output (angular position), is the angular velocity that is directly available for implementation. Additionally, the proportional control K_c can also be implemented on the output only to avoid an overshoot in the reference response. Thus, the PID control signal $u(t)$ is calculated using the following equation:

$$u(t) = -K_c y(t) + \frac{K_c}{\tau_I}\int_0^t (r(\tau) - y(\tau))d\tau - K_c\tau_D \frac{dy(t)}{dt}. \tag{3.35}$$

3.1.5 Cascade PID Control Systems

Figure 3.5 shows a cascade feedback control structure, which represents the control systems used in electrical drives and power converters. The inner-loop controller C_1 (also called secondary controller) is to control *Plant*1 (secondary plant), where its reference signal is the desired current signal $U_1^*(s)$ that is also the control signal generated from the outer-loop controller. For the cascade control system, the primary objective is to control the *Plant*2 (primary plant) with either a PI or a PID controller, depending on the complexity of *Plant*2.

3.1.5.1 Why a Cascade Control System Is Used

There is always a primary control objective combined with a secondary control objective when deploying a cascade control system structure. For the electrical drives and power converters, the primary objective could be to control an *AC* motor's angular velocity or position or a power converter's voltage, however, the secondary objective is to regulate the current flow. To our best knowledge, all industrial *AC* drives for velocity and position control use a PI cascade control structure. The key to the success of the cascade PI control systems for these applications lies in the large difference between the time constants of the secondary plant and the primary plant. For instance, the response time of the current is much faster than that of a motor's velocity. There are a few distinguishing features when using a cascade control system as discussed below.

1. Simplification of control system design. When using a cascade control system, a complex plant is decomposed into two subsystems. For the electrical drives and power converters, these subsystems are first order systems or second order systems. Because of the large difference between the time

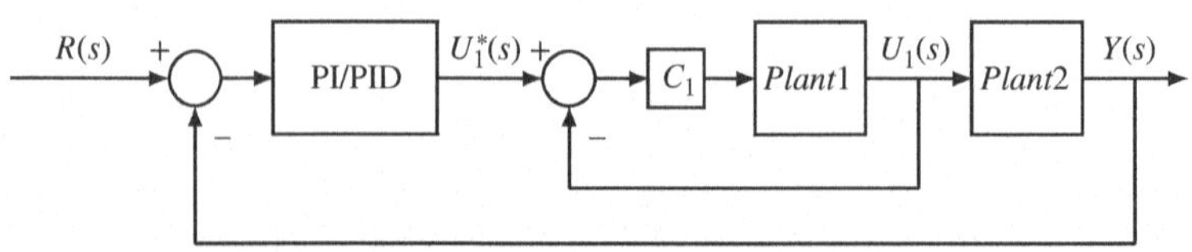

Figure 3.5 Cascade feedback control structure.

constants of the subsystems mentioned above, PI and PID controllers can be designed separately and effectively using the models of the subsystems.

2. Effectively handling nonlinearity of current systems. As shown in Chapter 1, there are nonlinearities in the current systems. With the inner-loop current control, a high gain feedback control is used in practice to overcome the effects of nonlinearities.
3. Effectively handling parameter variations in current systems. The resistance and inductance in the current systems are likely varying with respect to temperature and operating conditions. With high gain feedback control, the impact of their variations on the current regulation is reduced.
4. Implementation of current constraints. Over-current protection is necessary in the control of electrical drives and power converters. Because there is a primary control objective, the calculated current signal may exceed its limits due to a large load. With the cascade control structure, the limits of the current signal can be easily implemented with an anti-windup mechanism as the constraints on the manipulated variable of the primary controller shown in Chapter 4.
5. Reducing computational cost. In the cascade control system design, the inner-loop current control system will have a much larger bandwidth (w_n) than the one used in the outer-loop control system. Dual sampling rate is used in the implementation of the cascade control system in which the outer-loop control system is sampled in a much slower rate. As a result, the computational cost for the entire control system is reduced.

3.1.5.2 Design Principle Used for the Cascade Control System

The design of a cascade control system begins with the secondary controller for the inner-loop system. In principle, the inner-loop current controller can be either a proportional controller or a PI controller. When designing a proportional controller for the inner-loop system, special attention is paid to the steady-state gain of the closed-loop system that will be used in the model of the outer-loop controller design. The bandwidth of the inner-loop control system is selected to be sufficiently large so that its dynamics can be neglected when designing the primary controller.

Because the steady-state error is not a key issue for the inner-loop control system, the deployment of proportional controller could provide the benefit of simplicity. However, the steady-state error needs to be considered in the design of the primary controller and in the implementation of the current constraints (see Chapter 4). Bearing in mind that the steady-state error changes with respect to the parameters of the system when using a proportional controller, it is fair to say that the benefit of simplicity may be lost when taking into consideration performance robustness, which may be one of the key reasons why the majority of the industrial current controllers uses a PI controller.

3.2 Overview of PID Control of PMSM

The starting point is to assume a dynamic model of a PMSM in the $d-q$ reference frame. For those who are interested in the derivation of the dynamic model, Chapter 1 gives a detailed description of PMSM drives.

Assume that a PMSM is described by the differential equations in the $d-q$ rotating reference frame as

$$\frac{di_d(t)}{dt} = \frac{1}{L_d}(v_d(t) - R_s i_d(t) + \omega_e(t) L_q i_q(t)) \tag{3.36}$$

$$\frac{di_q(t)}{dt} = \frac{1}{L_q}(v_q(t) - R_s i_q(t) - \omega_e(t) L_d i_d(t) - \omega_e(t)\phi_{mg}) \tag{3.37}$$

$$\frac{d\omega_e(t)}{dt} = \frac{Z_p}{J_m}\left(T_e - \frac{B_v}{Z_p}\omega_e(t) - T_L\right) \tag{3.38}$$

$$T_e = \frac{3}{2}Z_p[\phi_{mg}i_q + (L_d - L_q)i_d(t)i_q(t)], \tag{3.39}$$

where ω_e is the electrical speed and is related to the rotor speed by $\omega_e = Z_p\omega_r$ with Z_p denoting the number of pole pairs, v_d and v_q represent the stator voltages in the $d-q$ frame, i_d and i_q represent the stator currents in this frame, and T_L is load torque that is assumed to be zero if no load is attached to the motor. The electromagnetic torque T_e consists of two parts: that produced by the flux of the permanent magnet ϕ_{mg} and that by i_d and i_q, respectively.

The physical parameters in the dynamic model of a PMSM are the quadrature axis stator inductance L_q and the direct axis inductance L_d, ϕ_{mg} that is the flux linkage due to permanent magnet, the stator resistance R_s motor inertia J_m, viscous friction coefficient B_v. The physical parameters are measured either through experiments or are given in the nameplate of a PMSM, as discussed in Chapter 1.

In the control system design, a cascade feedback and feedforward control system is configured for velocity control. Figure 3.6 shows the velocity control system configuration of a typical industrial PMSM drive. In the drive control systems, there are two PI controllers to control the d-axis and q-axis currents and one PI controller in the outer-loop to achieve the primary control objective of regulating the velocity (see Figure 3.6). Here, the velocity is controlled using a cascade control system structure for the reasons discussed in Section 3.1.

3.2.1 Bridging the Sensor Measurements to Feedback Signals (See the lower part of Figure 3.6)

The feedback signals to the controllers are d-axis current i_d and q-axis current i_q. If the primary objective is to regulate the electrical velocity, then ω_e will also be one of the feedback signals. There are several procedures used to bridge the sensor measurements and the feedback signals.

There are three current sensors to measure the three phase currents i_a, i_b and i_c as shown in Figure 3.6. The transformation of the three-phase currents to their components in α-β frame is achieved by the Clarke transformation,

$$\begin{bmatrix} i_\alpha \\ i_\beta \end{bmatrix} = \frac{2}{3}\begin{bmatrix} 1 & -\frac{1}{2} & -\frac{1}{2} \\ 0 & \frac{\sqrt{3}}{2} & -\frac{\sqrt{3}}{2} \end{bmatrix}\begin{bmatrix} i_a \\ i_b \\ i_c \end{bmatrix}. \tag{3.40}$$

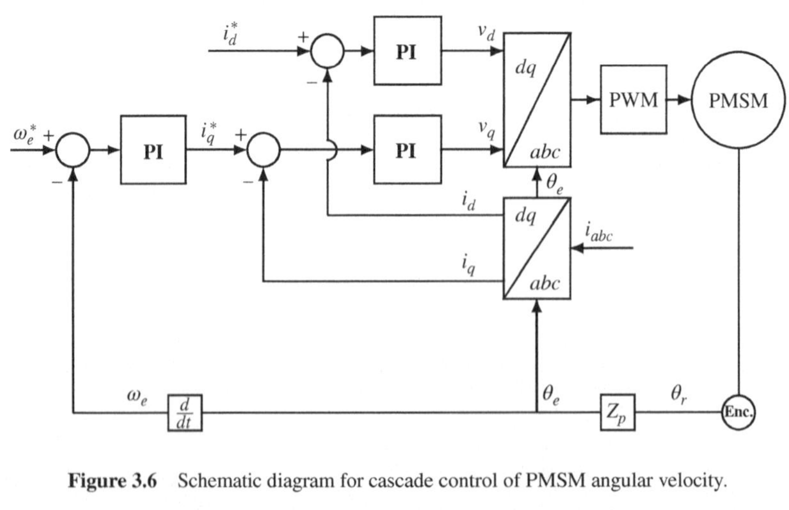

Figure 3.6 Schematic diagram for cascade control of PMSM angular velocity.

This transformation is to reduce the three measurements to two measurements, but the current variables i_α and i_β are sinusoidal signals. To obtain the current feedback signals i_d and i_q in the $d-q$ reference frame or the rotating reference frame, the following Park transformation is used,

$$\begin{bmatrix} i_d \\ i_q \end{bmatrix} = \begin{bmatrix} \cos\theta_e & \sin\theta_e \\ -\sin\theta_e & \cos\theta_e \end{bmatrix} \begin{bmatrix} i_\alpha \\ i_\beta \end{bmatrix}, \tag{3.41}$$

where the sinusoidal signals i_α and i_β are converted to the d-axis current i_d and q-axis current i_q. In a normal steady-state operation of PMSM, the i_d and i_q currents are expected to be constants. Hence, PI controllers are used to regulate the i_d and i_q currents.

To obtain the electrical velocity signal ω_e used in the velocity controller, an encoder is used to measure the mechanical angle of the PMSM, θ_r. This mechanical angle θ_r is converted to electrical angle θ_e by multiplying the number of pair of poles Z_p (see Figure 3.6). By differentiating the electrical angle θ_e, the electrical velocity ω_e is obtained for the PI of the velocity. Note that the parameter θ_e is used in the inverse Clarke transformation in (3.41), thus, the measurement of angular position is important even for the torque control of the PMSM.

3.2.2 Bridging the Control Signals to the Inputs to the PMSM (See the top part of Figure 3.6)

The controllers' outputs are the d-axis voltage v_d and q-axis voltage v_q. There are several mechanisms used to transform the v_d and v_q voltages to the three phase voltages used to drive the PMSM. The steps are reversed operations when dealing with the current measurements. Firstly, the voltages v_α and v_β in the $\alpha-\beta$ reference frame are computed

$$\begin{bmatrix} v_\alpha \\ v_\beta \end{bmatrix} = \begin{bmatrix} \cos\theta_e & -\sin\theta_e \\ \sin\theta_e & \cos\theta_e \end{bmatrix} \begin{bmatrix} v_d \\ v_q \end{bmatrix}. \tag{3.42}$$

Note that the transformation matrix is an inversion of Park transformation. The v_α and v_β are sinusoidal signals with their frequency dependent on the electrical speed of the motor (ω_e). To convert the voltages in the $\alpha-\beta$ reference frame to the three phase voltages required to drive the PMSM, the inverse Clarke transform is used. To do so, let v_0 represent the zero sequence component of three phase voltage which is zero for a balanced three phase voltage. Then, the inverse Clarke transformation leads to

$$\begin{bmatrix} v_a \\ v_b \\ v_c \end{bmatrix} = \begin{bmatrix} 1 & 0 & 1 \\ -\frac{1}{2} & \frac{\sqrt{3}}{2} & 1 \\ -\frac{1}{2} & -\frac{\sqrt{3}}{2} & 1 \end{bmatrix} \begin{bmatrix} v_\alpha \\ v_\beta \\ v_0 \end{bmatrix}. \tag{3.43}$$

At this point, the control system has computed the three phase voltage signals needed to operate the PMSM. The actual implementation of the control signals requires the devices of the semiconductor switches connected with a *DC* power supply (see Chapter 2). The common functions of these devices are to conduct power flow by varying the on–off duration of each switch. In the *AC* motor control case, the primary concern is to achieve the desired sinusoidal phase voltage with varying magnitudes and frequencies through turning on–off each power switch. The technology to generate such a gate control signal is called Pulse Width Modulation (PWM) (see Chapter 2).

For those who are interested in how the gate control signal is generated in the implementation of the PID control systems, Chapter 2 has described the PWM implementation procedure. In addition, Chapters 6–7 will discuss the predictive control technologies that directly generate this gate control signal without deploying PWM.

3.3 PI Controller Design for Torque Control of PMSM

Torque control is among the mainstream applications of PMSM. One of the well known examples is the case when a PMSM is used in an electric car where a mimic to pedal control of a petrol car is designed to make the driver feel more in control. In this application, the set-point signal to the electrical torque could be obtained from the interpretation of the force to the foot pedal, then closed-loop control of the electrical torque is utilized to achieve the control objective. This section will discuss how to design PI control systems to achieve torque control.

Note that from (3.39), the electromagnetic torque $T_e(t)$ comprises two terms where the first term is proportional to the q-axis current and the second term is proportional to the $i_d(t)i_q(t)$ product. Because of this relationship, the closed-loop control of electromagnetic torque is achieved via closed-loop control of the d-axis and q-axis currents.

3.3.1 *Set-point Signals to the Current Control Loops*

In the application of torque control, the set-point signals to the current loops are either constant or piece-wise constant. In normal operation, the reference signal to the d-axis current is selected to be zero for producing a maximum torque given the same amount of currents in the three phase windings. By doing so, the second term in (3.39) is controlled to be zero at steady-state operation. Otherwise, the electromagnetic torque T_e would have a nonlinear relationship to i_q. Thus, at the steady-state operation, the influence on the electromagnetic torque is reflected by the reference signal to the q-axis current.

For a surface mounted PMSM, the inductances L_d and L_q are identical, thus the second term in (3.39) becomes zero from the dynamic structure of the PMSM. Then the relationship between electromagnetic torque and the q-axis current for a surface mounted PMSM is

$$T_e(t) = \frac{3}{2}Z_p\phi_{mg}i_q(t). \tag{3.44}$$

Therefore, in order to control the electromagnetic torque, the control of q-axis current is required. The set-point signal of the current i_q^* is computed using the steady-state relation,

$$i_q^* = \frac{2}{3}\frac{T_e^*}{Z_p\phi_{mg}}, \tag{3.45}$$

where T_e^* is the reference signal to the electromagnetic torque.

If it is an interior mounted PMSM, the second term in (3.39) is not zero, which is a bilinear function of the d-axis and q-axis currents. By approximating it using a first order Taylor series expansion, the bilinear term becomes

$$i_d(t)i_q(t) = i_d^*i_q^* + i_q^*(i_d(t) - i_d^*) + i_d^*(i_q(t) - i_q^*), \tag{3.46}$$

where i_d^* and i_q^* are the steady-state values for the d- axis and q-axis currents, respectively. In the operation of a PMS motor, it is a common practice to choose the steady-state value for the d-axis current to be zero, which is enforced by choosing its set-point signal to be zero, leading to the set-point signal for the q-axis current being calculated with (3.45). However, if the set-point signal of the d-axis current is not zero ($i_d^* \neq 0$), then with its given value T_e^*, the set-point signal to the q-axis current is calculated using

$$i_q^* = \frac{2}{3}\frac{T_e^*}{Z_p\phi_{mg}(1 + i_d^*)}. \tag{3.47}$$

3.3.2 Decoupling of the Current Control Systems

There are nonlinear cross-coupling terms in (3.36) and (3.37) through $\omega_e i_q$, $\omega_e i_d$ and ω_e and these cross-coupling terms can be eliminated using a technique called input-and-output linearization and also decoupling using feedforward manipulation.

The central idea is to use auxiliary variables $\hat{v}_d$ and $\hat{v}_q$ such that

$$\frac{1}{L_d}\hat{v}_d = \frac{1}{L_d}(v_d + \omega_e L_q i_q) \tag{3.48}$$

$$\frac{1}{L_q}\hat{v}_q = \frac{1}{L_q}(v_q - \omega_e L_d i_d - \omega_e \phi_{mg}). \tag{3.49}$$

By substituting these equations into (3.36) and (3.37), we obtain the first order models for the electrical part of the machine dynamics as

$$\frac{di_d}{dt} = -\frac{R_s}{L_d}i_d + \frac{1}{L_d}\hat{v}_d \tag{3.50}$$

$$\frac{di_q}{dt} = -\frac{R_s}{L_q}i_q + \frac{1}{L_q}\hat{v}_q. \tag{3.51}$$

Based on (3.50) and (3.51), two feedback controllers can be designed for the stator current control by manipulating the auxiliary stator voltages in the $d - q$ frame.

Because the set-point signals to the d and q current loops are constant or piece-wise constant as stated previously, in order to ensure zero steady-state errors to a given reference signal, for the first order dynamics exhibited in the current dynamic models (3.50) and (3.51), PI controllers are best suited to the applications.

Using a proportional gain K_c^d and an integral time constant τ_I^d, the PI controller for the d-axis current control has the form,

$$\hat{v}_d(t) = K_c^d(i_d^*(t) - i_d(t)) + \frac{K_c^d}{\tau_I^d}\int_0^t (i_d^*(\tau) - i_d(\tau))d\tau. \tag{3.52}$$

A similar PI controller form is assumed for the q-axis current control:

$$\hat{v}_q(t) = K_c^q(i_q^*(t) - i_q(t)) + \frac{K_c^q}{\tau_I^q}\int_0^t (i_q^*(\tau) - i_q(\tau))d\tau. \tag{3.53}$$

Using the relationship (3.48) between $\hat{v}_d$ and v_d, the actual control signal, which is the d-axis voltage, is calculated:

$$v_d(t) = K_c^d(i_d^*(t) - i_d(t)) + \frac{K_c^d}{\tau_I^d}\int_0^t (i_d^*(\tau) - i_d(\tau))d\tau - \omega_e(t)L_q i_q(t). \tag{3.54}$$

Figure 3.7 shows the d-axis current control using nonlinear feedforward compensation.

A similar procedure is applied for the q-axis current control using nonlinear feedforward compensation (see (3.49)) leading to the computation of the control signal, which is the q-axis voltage, as

$$v_q(t) = K_c^q(i_q^*(t) - i_q(t)) + \frac{K_c^q}{\tau_I^q}\int_0^t (i_q^*(\tau) - i_q(\tau))d\tau$$
$$+\omega_e(t)L_d i_d(t) + \omega_e(t)\phi_{mg}. \tag{3.55}$$

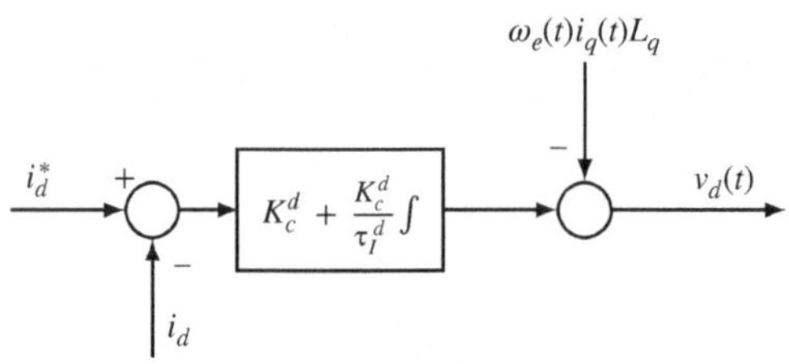

Figure 3.7 PI control of d-axis current using nonlinear feedforwarcontroller structure for controlling thed compensation.

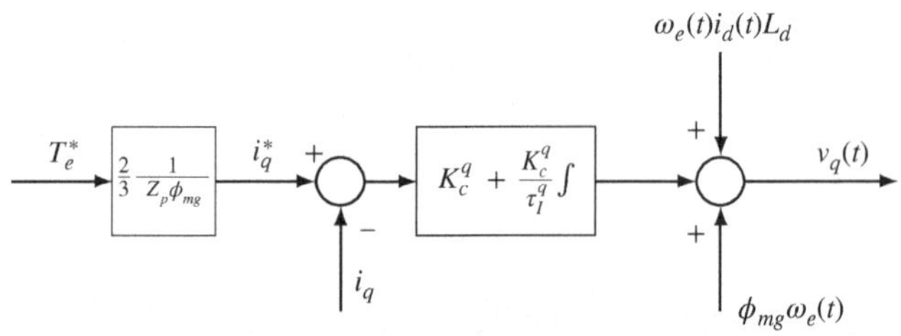

Figure 3.8 PI control of electromagnetic torque using nonlinear feedforward compensation.

Figure 3.8 illustrates the PI controller structure for controlling the electromagnetic torque using nonlinear feedforward compensation.

3.3.3 PI Current Controller Design

In the design of PI controllers for the current control loops, the Laplace transfer functions for the electric system are obtained based on (3.50)–(3.51) as

$$\frac{I_d(s)}{\hat{V}_d(s)} = \frac{\frac{1}{L_d}}{s + \frac{R_s}{L_d}} \tag{3.56}$$

$$\frac{I_q(s)}{\hat{V}_q(s)} = \frac{\frac{1}{L_q}}{s + \frac{R_s}{L_q}}, \tag{3.57}$$

where $I_d(s)$ and $I_q(s)$ are the Laplace transforms of the d-axis and q-axis currents; $\hat{V}_d(s)$ and $\hat{V}_q(s)$ are the Laplace transforms of the auxiliary voltage variables.

The transfer function of the PI controller for the d-axis current loop has the following form:

$$C(s) = K_c^d\left(1 + \frac{1}{\tau_I^d s}\right). \tag{3.58}$$

In the controller design, the proportional gain K_c^d (or K_c^q) and the integral time constant τ_I^d (or τ_I^q) are determined using pole-assignment controller design. Based on the design approach introduced in

Section 3.1, by letting $a = \frac{R_s}{L_d}$ and $b = \frac{1}{L_d}$, the proportional control gain and the integral time constant are calculated as

$$K_c^d = L_d\left(2\xi w_n - \frac{R_s}{L_d}\right) = 2\xi w_n L_d - R_s \tag{3.59}$$

$$\tau_I^d = \frac{2\xi w_n - \frac{R_s}{L_d}}{w_n^2} = \frac{2\xi w_n L_d - R_s}{L_d w_n^2}. \tag{3.60}$$

Similarly, the PI controller parameters for the control of q-axis current are calculated as:

$$K_c^q = 2\xi w_n L_q - R_s \tag{3.61}$$

$$\tau_I^q = \frac{2\xi w_n L_q - R_s}{L_q w_n^2}. \tag{3.62}$$

In the design, the damping coefficient ξ is selected to be 0.707 and the natural frequency w_n is selected to determine the desired closed-loop settling time, which also corresponds to the desired bandwidth of the closed-loop system. The larger w_n is, the shorter the desired closed-loop settling time is. It is also useful to choose this parameter relative to the bandwidth of the open-loop system, which is either $\frac{R_s}{L_d}$ or $\frac{R_s}{L_q}$. Here, a normalized parameter $0 < \gamma < 1$ is proposed to be used where the desired closed-loop bandwidth is calculated as

$$w_n = \frac{1}{1-\gamma}\frac{R_s}{L_d} \tag{3.63}$$

for use in the d-axis current control, or

$$w_n = \frac{1}{1-\gamma}\frac{R_s}{L_q} \tag{3.64}$$

for the q-axis current control. With this type of formulae, the parameter γ is often selected around 0.9 to give a satisfactory performance. The desired closed-loop performance in the d-axis may differ from the one used in the specification of q- axis current control system.

3.4 Velocity Control of PMSM

There are two configurations for the PID current control systems. The d-axis current i_d is always controlled by using a PI controller to ensure there is no steady-state error for the d-axis current. In the majority of industrial electrical drives, the q-axis current control loop is also based on a PI controller. However, from a control engineer's point of view, for the q-axis current, because there is an outer-loop PI control system, as shown in this chapter, it could use a high gain proportional control as the inner-loop controller when using the nonlinear feedforward compensation proposed here.

3.4.1 Inner-loop Proportional Control of q-axis Current

In a cascade control system, the design begins with the inner-loop control systems. Because the inner-loop is a proportional feedback controlled system, there is a steady-state error between the desired set-point signal and the actual output signal. The steady-state error should be taken into consideration in the design of the outer-loop control system. In order to make a suitable choice of the proportional gain, we examine the characteristics of the proportional closed-loop inner-loop system. For a proportional gain K_c^q, the closed-loop transfer function between the set-point signal $I_q^*(s)$ and the actual current $I_q(s)$ is written as

$$\frac{I_q(s)}{I_q^*(s)} = \frac{\frac{K_c^q}{L_q}}{s + \frac{R_s}{L_q} + \frac{K_c^q}{L_q}}. \tag{3.65}$$

The closed-loop pole for the q-axis current control is at $-\frac{R_s}{L_q} - \frac{K_c^q}{L_q}$. The larger K_c^q is, the faster the inner-loop current response will be. In this particular case, the steady-state gain of the current control-loop is calculated by setting $s = 0$ in (3.65) as

$$\frac{I_q(0)}{I_q^*(0)} = \frac{\frac{K_c^q}{L_q}}{\frac{R_s}{L_q} + \frac{K_c^q}{L_q}}.$$

The factors affecting the choice of the proportional gain for the current control loop include the dynamic response speed, the closed-loop steady-state gain, and the noise level in the system. On one hand, we desire a faster closed-loop response speed and a higher closed-loop steady-state gain, and on the other hand, we will try to avoid amplification of the noise in the inner-loop system which will be the consequence of higher gain and faster response speed.

Because the steady-state gain in the inner-loop control system will be used in the design of outer-loop control system, it is convenient to directly specify its desired value, then incorporate it later in the design. For this purpose, for the q-axis current control, we let the parameter $0 < \alpha < 1$ represent the steady-state gain for the current control loop, so that

$$\alpha = \frac{\frac{K_c^q}{L_q}}{\frac{R_s}{L_q} + \frac{K_c^q}{L_q}}. \tag{3.66}$$

By solving the steady-state equation, we obtain the proportional gain for the q-axis current control loop:

$$K_c^q = \frac{\alpha}{1-\alpha} R_s, \tag{3.67}$$

where $0 < \alpha < 1$.

By substituting the proportional controller gain (3.67) into the closed-loop transfer function (3.65), we obtain

$$\frac{I_q(s)}{I_q^*(s)} = \frac{\frac{\alpha}{1-\alpha}\frac{R_s}{L_q}}{s + \frac{1}{1-\alpha}\frac{R_s}{L_q}}, \tag{3.68}$$

Because the open-loop pole is located at $-\frac{R_s}{L_q}$, and from this equation it is seen that the closed-loop pole is at $-\frac{1}{1-\alpha}\frac{R_s}{L_q}$, thus the ratio between the closed and open-loop poles is $\frac{1}{1-\alpha}$. If the desired steady-state gain $\alpha = 0.9$, the controller gain is $K_c^q = 9R_s$, the closed-loop pole is 10 times that of the open-loop pole $-10\frac{R_s}{L_q}$. The implication for this choice is that the steady-state error for the current control loop is 10 percent. Namely, at the steady-state,

$$\lim_{t\to\infty} \frac{i_q^* - i_q(t)}{i_q^*} = 0.1, \tag{3.69}$$

where i_q^* is the reference signal for the q-axis current.

Upon deciding the value of the proportional gain, the control signal for the q-axis current-loop is calculated using the feedback and feedforward configurations:

$$v_q(t) = K_c^q(i_q^*(t) - i_q(t)) + \omega_e(t)L_d i_d(t) + \omega_e(t)\phi_{mg}. \tag{3.70}$$

The closed-loop performance of the q-axis current system is mainly determined by the steady-state gain α. When $\alpha = 0$, the current control gain K_c^q becomes zero; and when $\alpha = 1$, the current control gain K_c^q becomes infinity. Thus, the value of α should be chosen away from these two extreme values. Although the theoretical value of α is between 0 and 1, but in practice, α is found to be in the range between 0.7 and 0.9 in order to achieve high gain control for the inner-loop current system, which is the basis for the cascade closed-loop control systems to work well. More discussion on the choice of α with respect to closed-loop performance will be given in Chapter 5. Figure 3.9 shows the configuration of the inner-loop current control with nonlinear feedforward compensation.

3.4.2 Cascade Feedback Control of Velocity:P Plus PI

Figure 3.10 shows the cascade feedback control structure for velocity control of PMSM, where the inner-loop controller is a proportional controller and outer-loop controller is a PI controller.

The design of outer-loop control system is based on the following equations:

$$\frac{d\omega_e(t)}{dt} = \frac{Z_p}{J_m}\left(T_e - \frac{B_v}{Z_p}\omega_e(t) - T_L\right) \tag{3.71}$$

$$T_e = \frac{3}{2}Z_p[\phi_{mg}i_q + (L_d - L_q)i_d(t)i_q(t)], \tag{3.72}$$

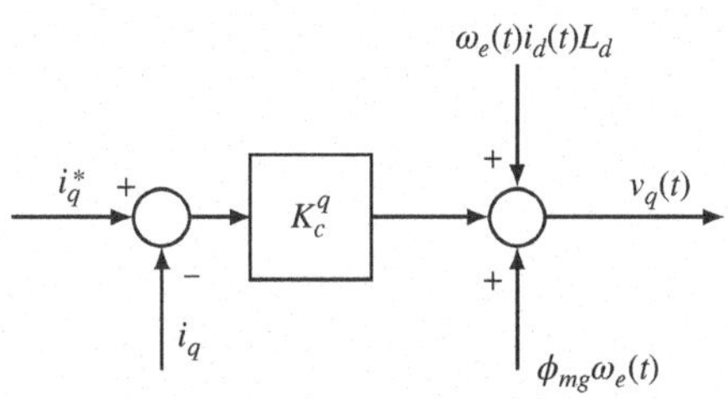

Figure 3.9 Expression of P controller with nonlinear feedforward compensation.

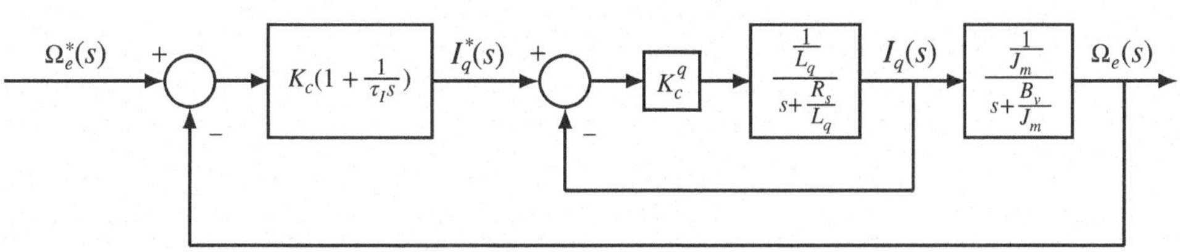

Figure 3.10 Cascade feedback control of velocity using P plus PI structure.

which are part of the dynamic equations from the model of PMSM (see (3.36)–(3.39)). These equations have been used in describing the mechanical part of the system and the link between the mechanical and electrical systems. By substituting (3.72) into (3.71), we obtain

$$\begin{aligned}\frac{d\omega_e(t)}{dt} &= \frac{Z_P}{J_m}T_e(t) - \frac{B_v}{J_m}\omega_e(t) - \frac{Z_P}{J_m}T_L \\ &= \frac{3}{2}\frac{Z_p^2\phi_{mg}}{J_m}i_q(t) + \frac{3}{2}\frac{Z_p^2}{J_m}(L_d - L_q)i_d(t)i_q(t) - \frac{B_v}{J_m}\omega_e(t) - \frac{Z_p}{J_m}T_L.\end{aligned} \tag{3.73}$$

Note that the second term on the right-hand side of (3.73) is bilinear and contains a factor $L_d - Lq$. For the class of surface mounted PMSM, $L_d = L_q$, thus this bilinear term vanishes. However, if $L_d \neq L_q$, the set-point signal for the current control of d-axis is chosen to be zero in the majority of the applications, namely $i_d^* = 0$, then in the steady-state, this term equals zero. Therefore, in the control system design for the outer-loop system, this bilinear term is neglected. The fourth term in (3.73) is proportional to the load torque, which is considered as a disturbance in control system design and should be completely rejected in the steady-state by the outer-loop control system as long as it is a constant or varies in a step signal manner. It is worthwhile to emphasis that because of the existence of load torque, without exception, the outer-loop controller should contain an integrator in order to completely reject the disturbance caused by the load torque.

By neglecting the bilinear term, we rewrite (3.73) in a first order differential equation:

$$\frac{d\omega_e(t)}{dt} = -\frac{B_v}{J_m}\omega_e(t) + \frac{3}{2}\frac{Z_p^2\phi_{mg}}{J_m}i_q(t) - \frac{Z_p}{J_m}T_L. \tag{3.74}$$

From the control system design point of view, the output variable is $\omega_e(t)$ and the input variable is current $i_q(t)$. However, because $i_q(t)$ is the output variable for the inner-loop control system, it is not available for the manipulation needed for the outer-loop. What is available and free is the set-point signal i_q^* to the inner-loop control of the q-axis current. The relationship between i_q and i_q^* is characterized by the inner-loop control of the q-axis current and is, in Laplace transform,

$$I_q(s) = \frac{\frac{\alpha}{1-\alpha}\frac{R_s}{L_q}}{s + \frac{1}{1-\alpha}\frac{R_s}{L_q}}I_q^*(s). \tag{3.75}$$

The Laplace transform of (3.74) in regarding the relationship between $\Omega_e(s)$ and $I_q(s)$ is

$$\left(s + \frac{B_v}{J_m}\right)\Omega_e(s) = \frac{3}{2}\frac{Z_p^2\phi_{mg}}{J_m}I_q(s). \tag{3.76}$$

By substituting (3.75) into (3.76), we obtain the transfer function between $\Omega_e(s)$ and $I_q^*(s)$ as,

$$\Omega_e(s) = \left(\frac{\frac{3}{2}\frac{Z_p^2\phi_{mg}}{J_m}}{s + \frac{B_v}{J_m}}\right)\left(\frac{\frac{K_c^q}{L_q}}{s + \frac{R_s}{L_q} + \frac{K_c^q}{L_q}}\right)I_q^*(s) \tag{3.77}$$

$$= \left(\frac{\frac{3}{2}\frac{Z_p^2\phi_{mg}}{B_v}}{\frac{J_m}{B_v}s + 1}\right)\left(\frac{\alpha}{(1-\alpha)\frac{L_q}{R_s}s + 1}\right)I_q^*(s). \tag{3.78}$$

This is the model for the design of the outer-loop velocity control system. Because it is a second order, a PID controller is appropriate. However, if we closely examine the model, then we find that the closed-loop time constant for the electrical system $\frac{L_q}{R_s}$ is far smaller than the time constant for the mechanical system

$\frac{J_m}{B_v}$. In addition, with the proportional feedback control gain K_c^q being large (see $K_c^q = \frac{\alpha}{1-\alpha}R$ with a large α), the dynamics from the inner-loop control of the q-axis current is ensured to be much faster than the dynamics from the mechanical system $((1-\alpha)\frac{L_q}{R_s} \gg \frac{J_m}{B_v}$. Therefore, the second order model (3.78) is simplified to a first order system by neglecting the dynamics from the inner-loop control of q-axis current by letting $(1-\alpha)\frac{L_q}{R_s} = 0$, so that

$$\frac{\Omega_e(s)}{I_q^*(s)} = \frac{\frac{3}{2}\frac{Z_p^2\phi_{mg}}{B_v}\alpha}{\frac{J_m}{B_v}s+1} = \frac{\frac{3}{2}\frac{Z_p^2\phi_{mg}}{J_m}\alpha}{s+\frac{B_v}{J_m}}. \tag{3.79}$$

With this first order model, the design of a PI controller leads to analytical solutions of the controller parameters using the technique of pole-assignment controller design (see Section 3.1). To simplify the notation, we let

$$a = \frac{B_v}{J_m}; \quad b = \frac{3}{2}\frac{Z_p^2\phi_{mg}}{J_m}\alpha.$$

Here, by choosing a desired closed-loop poles $s_{1,2} = -\xi w_n \pm w_n j\sqrt{1-\xi^2}$, where the damping coefficient $\xi = 0.707$, the proportional gain K_c is calculated as

$$K_c = \frac{2\xi w_n - a}{b} \tag{3.80}$$

and the integral time constant is calculated as

$$\tau_I = \frac{2\xi w_n - a}{w_n^2}. \tag{3.81}$$

Typically, the damping coefficient is chosen to be either 0.707 or 1, and the parameter w_n is used as the parameter for the desired closed-loop performance specification. A larger w_n will correspond to a wider closed-loop bandwidth, hence faster closed-loop response speed, but also more sensitivity to measurement noise. It is worthwhile to note that the closed-loop steady-state gain α from the inner-loop control of q-axis current has been carried over to the model in the outer-loop, which will affect the value of K_c for the outer-loop PI controller. More discussion on the natural frequency w_n with respect to closed-loop performance will be given in Chapter 5.

3.4.3 Simulation Example for P Plus PI Control System

In order to test the PID control system, we will first simulate the closed-loop performance using the nonlinear model built using the process description (3.36–3.39). A Simulink model is constructed based on the nonlinear model and the PI controllers are used with feedforward functions in the control system architecture.

In this simulation example, the parameters for the nonlinear model are given as $\phi_{mg} = 0.125$ Wb, $L_d = 7\times 10^{-3}$ H, $L_q = 7\times 10^{-3}$ H, $R_s = 2.98\,\Omega$, $B_v = 11\times 10^{-5}$ Nm · s, $Z_p = 2$, $J_m = 0.47\times 10^{-4}$ kg · m^2. We choose the steady-state gain for the q-axis inner-loop proportional control system $\alpha = 0.9$, which means that the controller gain K_c^q is $\frac{\alpha}{1-\alpha}R_s = 10R_s$ and the closed-loop pole for the q-axis inner-loop current control system is located at $s = -10\frac{R_s}{L_q} = -4257$.

The design of PI controller for the outer-loop is then based on the first order model

$$\frac{\Omega_e(s)}{I_q^*(s)} = \frac{b}{s+a}, \tag{3.82}$$

where $a = \frac{B_v}{J_m} = 2.34$ and $b = \frac{3}{2}\frac{Z_p^2 \phi_{mg}}{J_m} \times 0.9 = 14362$. By choosing the damping coefficient $\xi = 0.707$ and the natural frequency $w_n = 100$ rad/s, we obtain the PI controller parameters using (3.91) and (3.92) as

$$K_c = \frac{2\xi w_n - a}{b} = 0.0097; \quad \tau_I = \frac{2\xi w_n - a}{w_n^2} = 0.0139.$$

In the design of PI controller for the d-axis current, the closed-loop performance is specified by using the damping coefficient $\xi = 0.707$ and the natural frequency $w_n = 100$ Wb.

With sampling interval $\Delta t = 20 \times 10^{-6}$ sec, the closed-loop response of the cascade PI control of the nonlinear system is simulated, where the proportional controller is implemented on the output only to reduce the overshoot. In the simulation, the load torque T_L is chosen to be zero. Figure 3.11(a) shows the closed-loop response of the angular electrical velocity ω_e in comparison with its set-point signal.

The closed-loop responses of the inner-loop control of the d-axis and q-axis currents are shown in Figure 3.11(b) in comparison with their set-point signals. Finally, the inner-loop control signals (v_d and v_q) are shown in Figure 3.11(c).

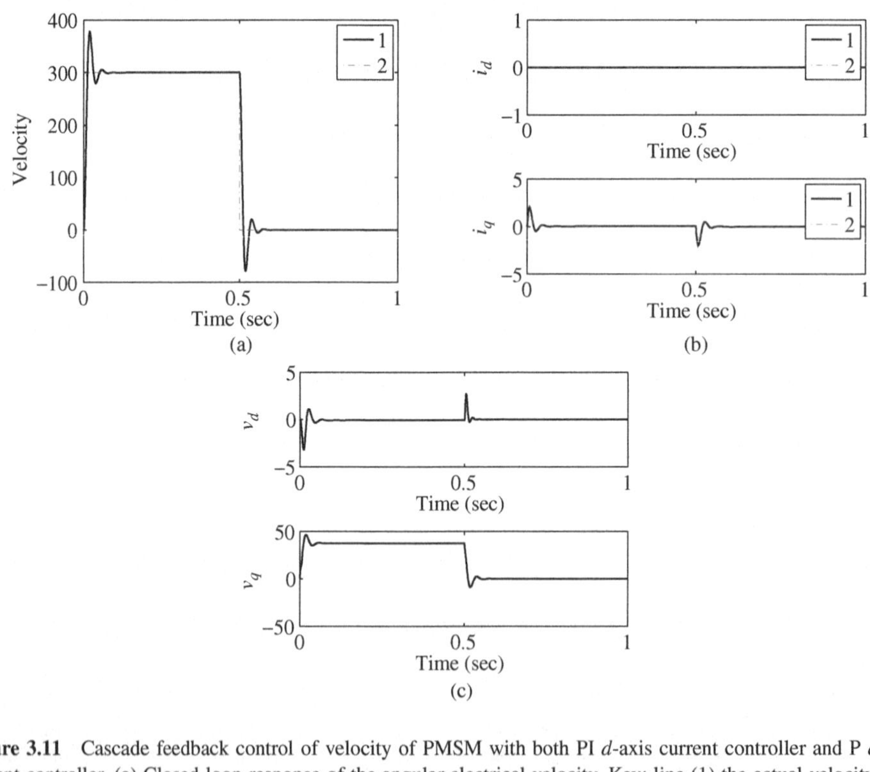

Figure 3.11 Cascade feedback control of velocity of PMSM with both PI d-axis current controller and P q-axis current controller. (a) Closed-loop response of the angular electrical velocity. Key: line (1) the actual velocity; line (2) the reference velocity. (b) Closed-loop responses of the d-axis and q-axis currents. Key: line (1) the actual current; line (2) the reference current. (c) Closed-loop control signal responses (the d-axis and q-axis voltages).

It is interesting to note that despite of the implementation using the alternative PI controller structure with the proportional gain K_c, there is an overshoot, approximately 20%, in the closed-loop velocity response (see Figure 3.11(a)). This overshoot occurred because there is a large overshoot in the i_q current response (see Figure 3.11(b)) due to the load torque T_L being zero. It is confirmed by experiments in Section 5.6 of Chapter 5 that when a motor is free of load, there is an overshoot in the output velocity response because of a large overshoot in the q-axis current.

From these figures, it is seen that the combination of proportional inner-loop control and PI outer-loop produces stable closed-loop response with satisfactory transient performance. In addition, there is no steady-state error between the desired electrical velocity and the actual velocity due to the integral control action in the outer-loop control system. Because of the simple proportional control for the inner-loop systems, very fast dynamic responses have been achieved for the inner-loop systems.

3.4.4 Cascade Feedback Control of Velocity:PI Plus PI

When using PI current controller, the steady-state error of current control system is eliminated, leading to performance robustness in the presence of parameter variations. The cascade feedback control structure when using PI plus PI configuration is similar to the diagram illustrated in Figure 3.10, except that the inner-loop current controller K_c^q is replaced by the PI current controller designed in Section 3.3. We will only describe the design of outer-loop velocity controller here.

Note that the dynamic system for the q-axis current is described by the first order differential equation:

$$\frac{di_q}{dt} = -\frac{R_s}{L_q} i_q + \frac{1}{L_q}\hat{v}_q, \tag{3.83}$$

where $\hat{v}_q$ is related to the original q-axis voltage by the following equation,

$$v_q = \hat{v}_q + \omega_e L_d i_d + \omega_e \phi_{mg}. \tag{3.84}$$

In order to design the outer-loop PI controller, we investigate the closed-loop transfer function between the reference signal $I_q^*(s)$ and the feedback signal $I_q(s)$. By assuming the cancelation of the nonlinear terms using the feedforward term $f^q(t)$, we substitute the PI controller given by (3.55) into the original dynamic model for the q-axis current (see 3.83), leading to the closed-loop differential equation:

$$\frac{di_q}{dt} = -\frac{R_s}{L_q} i_q(t) + \frac{1}{L_q} K_c^q (i_q^*(t) - i_q(t)) + \frac{K_c^q}{\tau_I^q L_q}\int_0^t (i_q^*(\tau) - i_q(\tau))d\tau. \tag{3.85}$$

By taking Laplace transform on both sides, we obtain

$$sI_q(s) = -\frac{R_s}{L_q} I_q(s) + \frac{1}{L_q}(K_c^q(I_q^*(s) - I_q(s)) + \frac{K_c^q}{\tau_I^q L_q s}(I_q^*(s) - I_q(s)). \tag{3.86}$$

Note that from the Section on PI current control (see (3.61) and (3.62)) the following relationships are true:

$$\frac{K_c^q}{\tau_I^q} = L_q w_n^2; \quad \frac{K_c^q}{L_q} = 2\xi w_n - \frac{R_s}{L_q},$$

which gives the closed-loop transfer function,

$$\frac{I_q(s)}{I_q^*(s)} = \frac{(2\xi w_n - \frac{R_s}{L_q})s + w_n^2}{s^2 + 2\xi w_n s + w_n^2}. \tag{3.87}$$

The inner-loop closed-loop poles are located at $s_{1,2} = -\xi w_n \pm w_n j\sqrt{1-\xi^2}$ with $\xi = 0.707$ and w_n being a tuning parameter for the closed-loop performance. The steady-state gain of the closed-loop transfer function is unity setting $s = 0$ in (3.87).

In order to design the outer-loop PI control system, from Section 3.4.1, we recall the transfer function model between the q-axis current $I_q(s)$ and the electrical velocity $\Omega_e(s)$is governed by

$$\left(s + \frac{B_v}{J_m}\right)\Omega_e(s) = \frac{3}{2}\frac{Z_p^2\phi_{mg}}{J_m}I_q(s). \tag{3.88}$$

By substituting (3.87) into (3.88), we obtain the transfer function between the reference for the q-axis current $I_q^*(s)$ and the electrical velocity $\Omega_e(s)$ as,

$$\frac{\Omega_e(s)}{I_q^*(s)} = \left(\frac{\frac{3}{2}\frac{Z_p^2\phi_{mg}}{J_m}}{s + \frac{B_v}{J_m}}\right)\left(\frac{\left(2\xi w_n - \frac{R_s}{L_q}\right)s + w_n^2}{s^2 + 2\xi w_n s + w_n^2}\right). \tag{3.89}$$

We understand that in order to design a PI controller using pole-assignment controller design method, a first order model is needed (see Section 3.1). Because (3.89) is a third order model, it needs to be approximated by a first order model. Notice that the parameter w_n is the natural frequency for the inner-loop q-axis current control, and we select this parameter in the design. To obtain a first order approximation, we choose $w_n \gg \frac{B_v}{J_m}$, for instance, $w_n = 10\frac{B_v}{J_m}$ rad/s. With the selection of large w_n by neglecting the dynamics from the inner-loop current control, we obtain the first order model for the design of PI controller:

$$\frac{\Omega_e(s)}{I_q^*(s)} \approx \frac{\frac{3}{2}\frac{Z_p^2\phi_{mg}}{J_m}}{s + \frac{B_v}{J_m}}. \tag{3.90}$$

In comparison with the proportional inner-loop control, here, the steady-state gain of the closed-loop current control system is unity, thus, the model for the design of outer-loop controller is based on the mechanical part of the PMSM without adding the steady-state component from the inner-loop current control. This means that even if the physical parameters R_s and L_q changed due to temperature and operating condition variations, the model used in the design for velocity control would still be free from their effects.

From the first order model (3.90), the PI controller parameters are calculated using pole-assignment controller design method. Here, we let

$$a = \frac{B_v}{J_m}; \quad b = \frac{3}{2}\frac{Z_p^2\phi_{mg}}{J_m}.$$

Then, by choosing desired closed-loop poles $s_{1,2} = -\xi w_n \pm j w_n\sqrt{1-\xi^2}$, where the damping coefficient $\xi = 0.707$ or 1, the proportional gain K_c is calculated as

$$K_c = \frac{2\xi w_n - a}{b} \tag{3.91}$$

and the integral time constant is calculated as

$$\tau_I = \frac{2\xi w_n - a}{w_n^2}. \tag{3.92}$$

The natural frequency w_n for the outer-loop PI control system is selected to be much smaller than that of the inner-loop PI control system.

3.4.5 Simulation Example for PI Plus PI Control System

With the same system parameters as in Section 3.4.3, we simulate the closed-loop responses when using PI inner-loop current control and PI outer-loop velocity control. In the design of inner-loop PI controllers, the closed-loop performance parameters are selected as $\xi = 1$ and $w_n = 100$ rad/s for both d-axis current and q-axis current controllers. In the design of outer-loop PI controller, the damping coefficient $\xi = 0.707$ and the bandwidth $w_n = 20$ rad/s. In this application, $\frac{B_v}{J_m} = 2.34$, so that $w_n = 100 \gg 2.34$ rad/s ensures a first order approximation for the outer-loop PI controller design.

In the closed-loop simulation, the load torque T_L is chosen to be zero. The proportional control is implemented on the output signal only. Figure 3.12(a) shows the closed-loop response of the electrical velocity. The closed-loop responses of the inner-loop control of the d-axis and q-axis currents are shown in Figure 3.12(b) in comparison with their set-point signals. Finally, the inner-loop control signals (v_d and v_q) are shown in Figure 3.12(c). It is seen that with the inner-loop PI control of q-axis current, the overshoot is about 10 %. This overshoot is due to the large overshoot of i_q current.

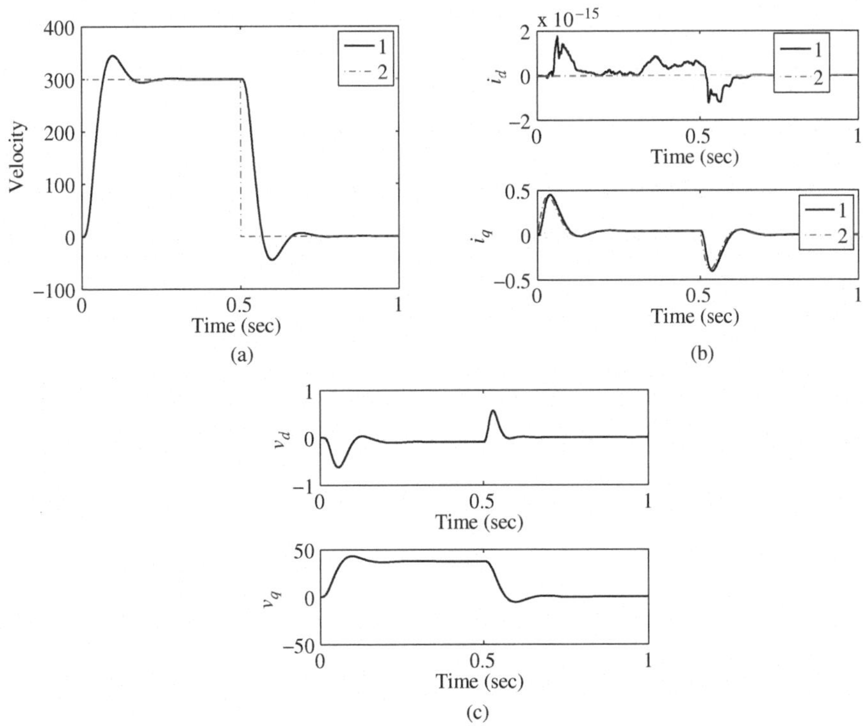

Figure 3.12 Cascade feedback control of velocity of PMSM with both PI current controllers. (a) Closed-loop response of the angular electrical velocity. Key: line (1) the actual velocity; line (2) the reference velocity. (b) Closed-loop responses of the d-axis and q-axis currents. Key: line (1) the actual current; line (2) the reference current. (c) Closed-loop control signal responses (the d-axis and q-axis voltages).

3.5 PID Controller Design for Position Control of PMSM

Angular position of a PMSM is required to be controlled in some applications. If the position control is used to track a desired reference position signal and if the load torque should change during the operation, the controller must have an integrator as part of its structure in order to reject the torque disturbance.

Similar to velocity control, the position control of PMSM uses the cascade control structure. The design of inner-loop current control system for both d-axis and q-axis currents is identical to the previous velocity control case (see Sections 3.3 and 3.4.4), thus it will not be repeated here. Instead, we will focus on the design of outer-loop position control.

Note that the angular position $\theta_e(t)$ is the integral of the velocity $\omega_e(t)$ with

$$\theta_e(t) = \int_0^t \omega_e(\tau)d\tau. \tag{3.93}$$

Therefore, the Laplace transfer function between the velocity $\Omega_e(s)$ and the angular position $\Theta_e(s)$ is

$$\frac{\Theta_e(s)}{\Omega_e(s)} = \frac{1}{s}. \tag{3.94}$$

Recall from the original dynamic model that governs the relationship between the q-axis current $i_q(t)$ and the angular velocity $\omega_e(t)$ (see 3.74), the relationship between $\Omega_e(s)$ and $I_q(s)$ is

$$\left(s + \frac{B_v}{J_m}\right)\Omega_e(s) = \frac{3}{2}\frac{Z_p^2\phi_{mg}}{J_m}I_q(s), \tag{3.95}$$

which leads to

$$\frac{\Theta_e(s)}{I_q(s)} = \frac{3}{2}\frac{Z_p^2\phi_{mg}}{J_m}\frac{1}{s\left(s+\frac{B_v}{J_m}\right)}. \tag{3.96}$$

By substituting $I_q(s)$ with the reference signal $I_q^*(s)$ to the inner-loop current control, we obtain the transfer function model,

$$\frac{\Theta_e(s)}{I_q^*(s)} = \frac{3}{2}\frac{Z_p^2\phi_{mg}}{J_m}\frac{1}{s\left(s+\frac{B_v}{J_m}\right)}\frac{\alpha}{(1-\alpha)\frac{L_q}{R_s}s+1}. \tag{3.97}$$

As before, we take an approximation of (3.97) by neglecting the inner-loop dynamics. As a result, the second order model is obtained for the design of the position control of a motor, where

$$\begin{aligned}\frac{\Theta_e(s)}{I_q^*(s)} &= \frac{3}{2}\frac{Z_p^2\phi_{mg}\alpha}{J_m}\frac{1}{s\left(s+\frac{B_v}{J_m}\right)}\\ &= \frac{b}{s(s+a)},\end{aligned} \tag{3.98}$$

with $a = \frac{B_v}{J_m}$ and $b = \frac{3}{2}\frac{Z_p^2\phi_{mg}\alpha}{J_m}$, $\alpha = 1$ if PI controller is used for controlling of the q-axis current. This is a second order model, which can be used for the design of a PID controller (see Section 3.1). An ideal PID controller has the transfer function

$$C(s) = K_c\left(1 + \frac{1}{\tau_I s} + \tau_D s\right), \tag{3.99}$$

where K_c is the proportional gain, τ_I is the integral time constant and τ_D is the derivative gain. Following the design steps outlined in Section 3.1, the PID controller parameters are found for the position control of the outer-loop system:

$$K_c = c_1 = \frac{(2\xi n + 1)w_n^2}{b} \tag{3.100}$$

$$\tau_I = \frac{c_1}{c_0} = \frac{(2\xi n + 1)w_n^2}{nw_n^3} = \frac{(2\xi n + 1)}{w_n} \tag{3.101}$$

$$\tau_D = \frac{c_2}{c_1} = \frac{(2\xi + n)w_n - a}{(2\xi n + 1)w_n^2}. \tag{3.102}$$

The control signal for the purpose of position control is computed using the combination of proportional, integral and derivative terms:

$$i_q^*(t) = K_c(\theta_e^*(t) - \theta_e(t)) + \frac{K_c}{\tau_I}\int_0^t (\theta_e^*(\tau) - \theta_e(\tau))d\tau - K_c\tau_D\omega_e(t). \tag{3.103}$$

With the implementation of derivative action on the output only, the configuration of the position control algorithm using cascade feedback control is illustrated in Figure 3.13.

3.6 Overview of PID Control of Induction Motor

When parasitic effects such as eddy currents and magnetic field saturation are neglected, the dynamic model of an induction motor is governed by the following differential equations (see Chapter 1):

$$i_{sd}(t) + \tau_\sigma \frac{di_{sd}(t)}{dt} = \omega_s(t)\tau_\sigma i_{sq}(t) + \frac{k_r}{r_\sigma \tau_r}\psi_{rd}(t) + \frac{1}{r_\sigma}u_{sd}(t) \tag{3.104}$$

$$i_{sq}(t) + \tau_\sigma \frac{di_{sq}(t)}{dt} = -\omega_s(t)\tau_\sigma i_{sd}(t) - \frac{k_r}{r_\sigma}\omega_e(t)\psi_{rd}(t) + \frac{1}{r_\sigma}u_{sq}(t) \tag{3.105}$$

$$\psi_{rd}(t) + \tau_r \frac{d\psi_{rd}(t)}{dt} = L_h i_{sd}(t) \tag{3.106}$$

$$\frac{d\omega_m}{dt} = -\frac{f_d}{J_m}\omega_m + \frac{3}{2}\frac{Z_p L_h}{L_r J_m}\psi_{rd} i_{sq} - \frac{T_L}{J_m} \tag{3.107}$$

$$\omega_s(t) = \omega_e(t) + \frac{L_h i_{sq}(t)}{\tau_r \psi_{rd}(t)}, \tag{3.108}$$

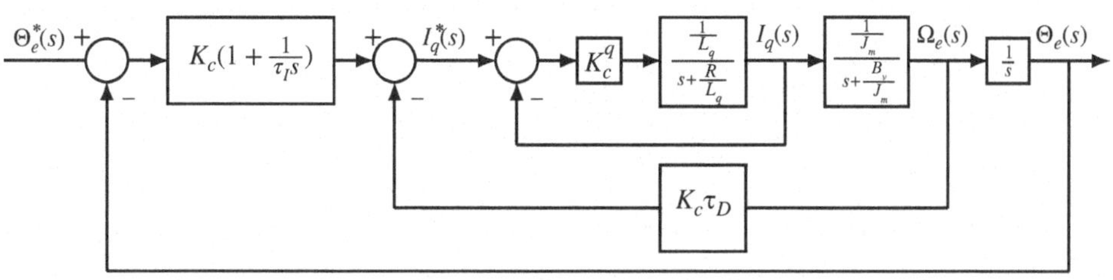

Figure 3.13 Cascade feedback control of angular position of PMSM.

where $\omega_e(t) = Z_p\omega_m(t)$. In the induction motor model, $i_{sd}(t)$ and $i_{sq}(t)$ are the stator currents in the $d-q$ reference frame, $\psi_{rd}(t)$ is the rotor flux in d-axis. The input variables $u_{sd}(t)$ and $u_{sq}(t)$ represent the stator voltages in the $d-q$ reference frame. $\omega_s(t)$ and $\omega_m(t)$ are the synchronous and rotor velocity respectively. $T_L(t)$ is the load torque that might change with respect to time. This model was derived and discussed in Chapter 1.

It is seen from the dynamic model of the induction motor (3.104–3.108) that the control problem is similar to that of PMSM, however, to a certain degree, it is more complex. From the control system design point of view, the manipulated variables in the induction motor control problem are the stator voltages, $u_{sd}(t)$ and $u_{sq}(t)$ (see (3.104) and (3.105)). Similar to PMSM control, there will be current controllers to regulate the i_{sd} and i_{sq} stator currents by manipulating the stator voltages u_{sd} and u_{sq} in the $d-q$ reference frame. The design task of the PI current controllers is not new to us because it simply follows the same current controller design procedures outlined in Section 3.3. However, the difference might be the reference signal to the i_{sd} that is not zero, but a constant depending on the operating condition of $\psi_{rd}(t)$ that is the rotor flux in d-axis. If the control objective is to regulate the velocity of the induction motor, it is seen from (3.107) that this objective will be achieved by manipulating the stator current i_{sq} while maintaining the rotor flux $\psi_{rd}(t)$ as a constant specified as operating condition of the induction motor.

The additional complexity of the induction motor control problem originates from the discrepancy between the synchronous velocity $\omega_s(t)$ and the rotor's electrical velocity $\omega_e(t)$ (see (3.108)). In addition, $\omega_s(t)$ is not measured directly, however, it is used in the Park transformations to convert the current and voltage variables in the $\alpha-\beta$ reference frame to the $d-q$ reference frame. Thus, the estimation of $\omega_s(t)$ is required as part of the control systems, and this is called slip estimation (see Section 3.7.3).

In the control system design, similar to PMSM control, a cascade feedback and feedforward control system is configured for velocity control. Figure 3.14 shows the velocity control system configuration of a typical industrial induction machine drive. In the drive control systems, there are two PI controllers to control the d-axis and q-axis stator currents and one PI controller in the outer-loop to achieve the velocity control of the motor (see Figure 3.14). Nonlinear compensation terms are used in the current control loops and slip estimation is used to generate θ_s that is used in the Park transformation.

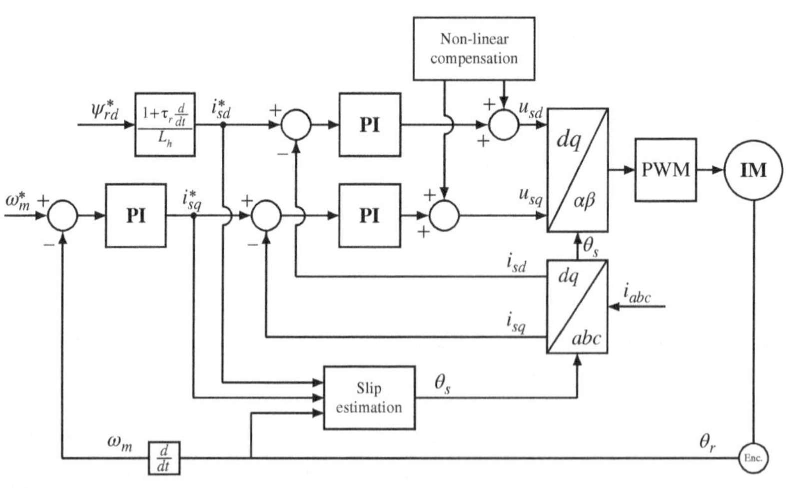

Figure 3.14 Schematic diagram for nonlinear cascade control of induction motor.

3.6.1 Bridging the Sensor Measurements to Feedback Signals

The feedback signals to the controllers are d-axis stator current i_{sd} and q-axis stator current i_{sq}, and the electrical velocity $\omega_e(t)$.

There are three current sensors to measure the three phase stator currents i_a, i_b and i_c as shown in Figure 3.14. The transformation of the three-phase currents to their components in α-β frame is achieved by the Clarke transformation,

$$\begin{bmatrix} i_{s\alpha} \\ i_{s\beta} \end{bmatrix} = \frac{2}{3} \begin{bmatrix} 1 & -\frac{1}{2} & -\frac{1}{2} \\ 0 & \frac{\sqrt{3}}{2} & -\frac{\sqrt{3}}{2} \end{bmatrix} \begin{bmatrix} i_a \\ i_b \\ i_c \end{bmatrix}. \tag{3.109}$$

To obtain the current feedback signals i_{sd} and i_{sq} in the $d-q$ stator reference frame, the following Park transformation is used,

$$\begin{bmatrix} i_{sd} \\ i_{sq} \end{bmatrix} = \begin{bmatrix} \cos\theta_s & \sin\theta_s \\ -\sin\theta_s & \cos\theta_s \end{bmatrix} \begin{bmatrix} i_{s\alpha} \\ i_{s\beta} \end{bmatrix}, \tag{3.110}$$

where the sinusoidal signals $i_{s\alpha}$ and $i_{s\beta}$ are converted to the d-axis stator current i_{sd} and q-axis stator current i_{sq}. Similar to the operation of PMSM, in a normal steady-state operation of an induction motor, the i_{sd} and i_{sq} currents are regulated to their operating points.

Similar to PMSM, an encoder is used to measure the angle of the rotor θ_r, its differentiation leads to the angular velocity of the induction motor, ω_m, obtained as the feedback signal for the PI of the velocity. Using the information θ_r to calculate θ_e $(= Z_p\theta_r)$, and the reference signals i^*_{sd} and i^*_{sq}, based on the integrated variables in (3.108), the estimation of synchronous angle θ_s is performed using

$$\theta_s(t) = \theta_e(t) + \frac{1}{\tau_r}\int_0^t \frac{i^*_{sq}(\tau)}{i^*_{sd}(\tau)} d\tau. \tag{3.111}$$

As explained in the later section (see Section 3.7.3), the use of integrated variables and the reference signals in the calculation will reduce the effect of noise.

3.6.2 Bridging the Control Signals to the Inputs to the Induction Motor

The controllers' outputs are the d-axis stator voltage u_{sd} and q-axis stator voltage u_{sq}. Identical mechanisms are used to transform the u_{sd} and u_{sq} voltages to the three phase voltages as inputs to the induction motor. Basically, the $u_{s\alpha}$ and $u_{s\beta}$ are obtained via the calculation,

$$\begin{bmatrix} u_{s\alpha} \\ u_{s\beta} \end{bmatrix} = \begin{bmatrix} \cos\theta_s & -\sin\theta_s \\ \sin\theta_s & \cos\theta_s \end{bmatrix} \begin{bmatrix} u_{sd} \\ u_{sq} \end{bmatrix}. \tag{3.112}$$

Also, let u_{s0} represent the zero sequence component of three phase voltage which is zero for a balanced three phase voltage. Then, the inverse Clarke transformation leads to

$$\begin{bmatrix} v_a \\ v_b \\ v_c \end{bmatrix} = \begin{bmatrix} 1 & 0 & 1 \\ -\frac{1}{2} & \frac{\sqrt{3}}{2} & 1 \\ -\frac{1}{2} & -\frac{\sqrt{3}}{2} & 1 \end{bmatrix} \begin{bmatrix} u_{s\alpha} \\ u_{s\beta} \\ u_{s0} \end{bmatrix}. \tag{3.113}$$

At this point, the control system has computed the three phase voltage signals needed to operate the induction motor. The actual implementation of the control signals requires the devices of the semiconductor switches connected with a *DC* power supply (see Chapter 2).

3.7 PID Controller Design for Induction Motor

From the dynamic relationships (see (3.104)–(3.108)), it is noticed that there are nonlinear coupled terms in the dynamic model. Therefore, feedforward compensation will be used in the current control system design.

3.7.1 PI Control of Electromagnetic Torque of Induction Motor

The strategy to control the electromagnetic torque of an induction motor is based on PI control of the stator currents i_{sd} and i_{sq} in order to eliminate the steady-state errors for the current control systems. The first step in the control system design is to determine the reference signals to the current control loops.

3.7.1.1 Reference Signal to *q* Axis Stator Current

Given a desired electromagnetic torque signal, T_e^*, and a desired rotor flux signal ψ_{rd}^*, the desired stator current in the q axis is calculated as

$$i_{sq}^* = \frac{2}{3}\frac{L_r}{L_h Z_p}\frac{T_e^*}{\psi_{rd}^*}, \tag{3.114}$$

on the basis of the relationship given in (1.103). The desired stator current i_{sq}^* is then used as the reference signal to PI control of q-axis current.

3.7.1.2 Reference Signal to *d* Axis Stator Current

From the model equation (3.106), the dynamic response of ψ_{rd} from i_{sd} is a first-order system with time constant τ_r. The differential equation (3.106) can be expressed in terms of the set-point values ψ_{rd}^* and i_{sd}^*:

$$i_{sd}^*(t) = \frac{1}{L_h}\psi_{rd}^*(t) + \frac{\tau_r}{L_h}\frac{d\psi_{rd}^*(t)}{dt}. \tag{3.115}$$

If the operation of field-weakening is not involved, the set-point for rotor flux ψ_{rd}^* will be constant. Therefore, the derivative of the of reference rotor flux signal, $\frac{d\psi_{rd}^*(t)}{dt}$, is taken to be zero. The reference signal to the d axis stator current is determined via the steady-state relationship:

$$i_{sd}^*(t) = \frac{1}{L_h}\psi_{rd}^*, \tag{3.116}$$

where L_h is the mutual machine inductance. However, if field-weakening is involved, the reference signal to rotor flux ψ_{rd} may change. If the trajectory of the ψ_{rd}^* is chosen to be a combination of constant signals, interspaced with ramp signals between the transient periods, then the reference signal i_{sd}^* is calculated using (3.115).

3.7.1.3 PI Current Controller Design

The PI controller design for the $d-q$ axes current controllers follows the same procedure as outlined in Section 3.1. For illustration purpose, only the design of d-axis current is presented here. To design the PI controller for the i_{sd} current based on (3.104), let the auxiliary control variable $\approxeq_{sd}(t)$ be defined by

$$\frac{1}{\tau'_\sigma r_\sigma}\hat{u}_{sd}(t) = \frac{1}{\tau'_\sigma r_\sigma}u_{sd}(t) + \frac{k_r}{\tau'_\sigma r_\sigma \tau_r}\psi_{rd}(t) + \omega_s i_{sq}(t). \tag{3.117}$$

By substituting this auxiliary variable into model equation (3.104), a linear time-invariant (LTI) model is obtained with respect to the auxiliary control variable $\hat{u}_{sd}(t)$:

$$\frac{di_{sd}(t)}{dt} = -\frac{1}{\tau'_\sigma}i_{sd}(t) + \frac{1}{\tau'_\sigma r_\sigma}\hat{u}_{sd}(t). \tag{3.118}$$

The Laplace transfer function of (3.118) is

$$\frac{I_{sd}(s)}{\hat{U}_{sd}(s)} = \frac{\frac{1}{\tau'_\sigma r_\sigma}}{s + \frac{1}{\tau'_\sigma}}.$$

Let $a = \frac{1}{\tau'_\sigma}$ and $b = \frac{1}{\tau'_\sigma r_\sigma}$. By choosing the closed-loop bandwidth parameter w_n and a damping coefficient ξ, the PI controller parameters are calculated as (see Section 3.1)

$$K_c^d = \frac{2\xi w_n - a}{b}; \quad \tau_I^d = \frac{2\xi w_n - a}{w_n^2}.$$

The manipulated variable u_{sd} is computed using the position form of the PI controller where

$$u_{sd} = K_c^d(i_{sd}^*(t) - i_{sd}(t)) + \frac{K_c^d}{\tau_I^d}\int_0^t (i_{sd}^*(\tau) - i_{sd}(\tau))d\tau$$

$$-\frac{k_r}{\tau_r}\psi_{rd}(t) - \tau'_\sigma r_\sigma \omega_s(t) i_{sq}(t).$$

Figure 3.15 shows the configuration of the d-axis current controller (PI), where a steady-state pre-compensator $\frac{1}{L_h}$ is used to convert the set-point of ψ_{rd}^* to the set-point of i_{sd}^*.

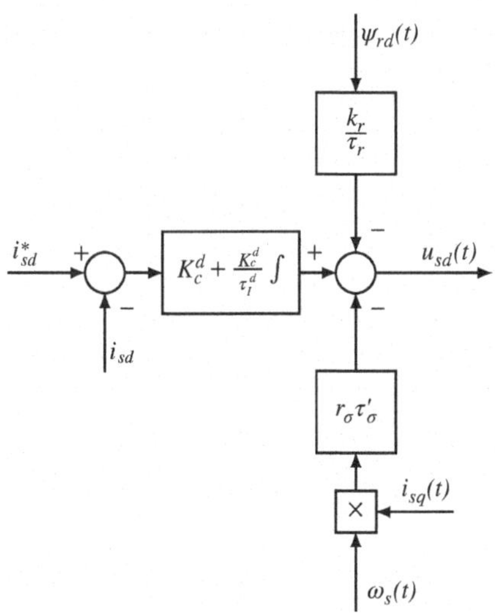

Figure 3.15 Block diagram of the PI controller with nonlinear feedforward compensation for d-axis stator current control.

3.7.2 Cascade Control of Velocity and Position

Similar to the design of PI control systems for a PMSM, the inner-loop control of i_{sq} uses either PI or P controller with feedforward compensation because of the nonlinearity and the coupling terms, and the outer-loop velocity control by manipulating the set-point of the current i_{sq} achieves the control objective. However, the electrical-mechanical equation (3.107) reveals the dependence of velocity ω_m on both i_{sq} and ψ_{rd}. In order for the outer-loop velocity control to produce desirable closed-loop performance, the variation of ψ_{rd} needs to be minimized so that it can be closely approximated by a linear time-invariant system. In other words, its value needs to converge to an expected steady-state value in a closed-loop feedback control. The set-point signal for ψ_{rd} is not zero. In an ideal situation, there would be a cascade feedback and feedforward control of ψ_{rd}, where an inner-loop proportional controller with feedforward linearization is used to control i_{sd} by manipulating u_{sd} and an outer-loop PI controller is used to control ψ_{rd} by manipulating the set-point signal i^*_{sd}. However, this cascade control strategy for the rotor flux ψ_{rd} requires the signal ψ_{rd} as the feedback signal, and this signal is not measured, although in Chapter 9, it is shown that ψ_{rd} could be estimated using an observer.

The cascade control of velocity involves the dynamics of the current i_{sq} and the dynamics of the velocity ω_m, which are described by models (3.105) and (3.107). The i_{sq} current model will be used for the inner-loop design of proportional feedback linearization control system and the motion model for velocity will be used for the design of a PI controller for the outer-loop system.

In the first stage of the design of a cascade control system, the inner-loop proportional controller or PI controller needs to be considered. How to design a PI current controller is illustrated in Section 3.1, here, it will be shown how a proportional controller will be designed.

From (3.105), an auxiliary variable $\hat{u}_{sq}(t)$ is defined by the relationship:

$$\frac{1}{r_\sigma \tau'_\sigma}\hat{u}_{sq}(t) = \frac{1}{r_\sigma \tau'_\sigma}u_{sq}(t) - \omega_s(t)i_{sd}(t) - \frac{k_r}{r_\sigma \tau'_\sigma}\omega_e(t)\psi_{rd}(t). \tag{3.119}$$

By substituting (3.119) into (3.105), we obtain the first order differential equation:

$$\frac{di_{sq}(t)}{dt} = -\frac{1}{\tau'_\sigma}i_{sq}(t) + \frac{1}{r_\sigma \tau'_\sigma}\hat{u}_{sq}(t), \tag{3.120}$$

which will be used as the basis for the design of the proportional controller for the inner-loop system. The Laplace transfer function of (3.120) is

$$\frac{I_{sq}(s)}{\hat{U}_{sq}(s)} = \frac{\frac{1}{\tau'_\sigma r_\sigma}}{s + \frac{1}{\tau'_\sigma}}. \tag{3.121}$$

Since the proportional control will lead to a steady-state error in the closed-loop system, its closed-loop steady-state gain is not unity, and this steady-state gain should be taken into consideration in the design of the outer-loop control system. Consequently, we specify the desired inner-loop control performance via the steady-state gain of the the proportionally controlled system. Let K_c^q denote the proportional feedback controller gain. The closed-loop transfer function between the $I_{sq}(s)$ and set-point signal $I^*_{sq}(s)$ is

$$\frac{I_{sq}(s)}{I^*_{sq}(s)} = \frac{\frac{K_c^q}{\tau'_\sigma r_\sigma}}{s + \frac{1}{\tau'_\sigma} + \frac{K_c^q}{\tau'_\sigma r_\sigma}}. \tag{3.122}$$

The closed-loop pole is found by setting the denominator $s + \frac{1}{\tau'_\sigma} + \frac{K_c^q}{\tau'_\sigma r_\sigma} = 0$ leading to

$$s = -\frac{1}{\tau'_\sigma} - \frac{K_c^q}{\tau'_\sigma r_\sigma}.$$

For $K_c^q > 0$, the closed-loop pole is located at the left-hand side of the open-loop pole $-\frac{1}{\tau'_\sigma}$ in the complex plane, which means that the closed-loop dynamic response is always faster than the open-loop response. The steady-state gain of the closed-loop transfer function (3.122) is calculated by letting $s = 0$, leading to the definition of variable α, as

$$\alpha = \frac{\frac{K_c^q}{\tau'_\sigma r_\sigma}}{\frac{1}{\tau'_\sigma} + \frac{K_c^q}{\tau'_\sigma r_\sigma}}. \tag{3.123}$$

In the design of the proportional controller, we specify the desired closed-loop performance by choosing a desired value for α, which from (3.123) gives the solution of the proportional controller gain K_c^q,

$$K_c^q = \frac{\alpha}{1-\alpha} r_\sigma. \tag{3.124}$$

If we choose $\alpha = 0.9$, then the proportional controller gain is $K_c^q = 9r_\sigma$, and the closed-loop pole is at $s = -\frac{10}{\tau_\sigma}$.

Assuming the set-point signal to the current control loop is i_{sq}^*, the auxiliary control signal $\hat{u}_{sq}(t)$ is calculated using the feedback error signal $i_{sq}^*(t) - i_{sq}(t)$

$$\hat{u}_{sq}(t) = K_c^q(i_{sq}^*(t) - i_{sq}(t)).$$

Thus, from the relationship between $\hat{u}_{sq}(t)$ and $u_{sq}(t)$ given by (3.119), we obtain the feedback control signal $u_{sq}(t)$ with linearization as

$$u_{sq}(t) = K_c^q(i_{sq}^*(t) - i_{sq}(t)) + r_\sigma \tau'_\sigma \omega_s i_{sd}(t) + k_r \omega_m \psi_{rd}(t). \tag{3.125}$$

Figure 3.16 shows the computation of the control signal $u_{sq}(t)$ with feedforward compensations.

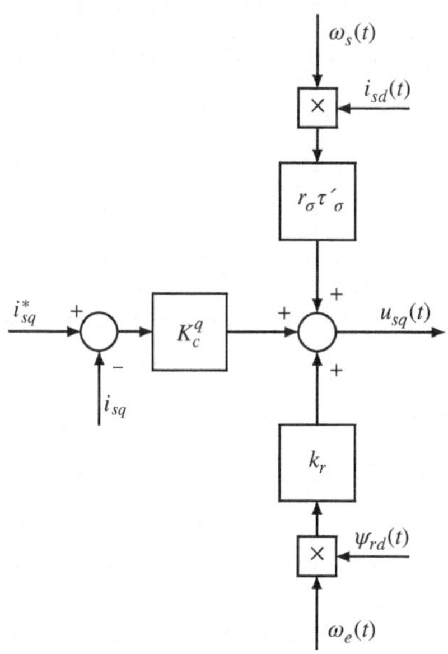

Figure 3.16 Block diagram of P controller with nonlinear feedforward compensation for q-axis stator current control.

In the second stage of the design of a cascade control system, the outer-loop PI control system is considered for velocity control and PID control system is for the position control. The controller structures are selected according to the complexity of the system dynamics (see Section 3.1).

3.7.2.1 Velocity Control

The mathematical model for the outer-loop control of $\omega_m(t)$ is based on the motion model (3.107). In order to linearize the bilinear term $i_{sq}(t)\psi_{rd}(t)$, the product is approximated using a first order Taylor series expansion (see Goodwin *et al.*, 2001, Bay, 1999):

$$i_{sq}(t)\psi_{rd}(t) \approx -i_{sq}^{ss}\psi_{rd}^{ss} + \psi_{rd}^{ss}i_{sq}(t) + i_{sq}^{ss}\psi_{rd}(t).$$

Substituting this approximation into (3.107), we obtain the linearized model for the design of PI control of the velocity,

$$\frac{d\omega_m(t)}{dt} = -\frac{f_d}{J_m}\omega_m(t) + \frac{3Z_pL_h}{2L_rJ_m}(-i_{sq}^{ss}\psi_{rd}^{ss} + \psi_{rd}^{ss}i_{sq}(t) + i_{sq}^{ss}\psi_{rd}(t)) - \frac{T_L}{J_m},$$

which is

$$\frac{d\omega_m(t)}{dt} = -\frac{f_d}{J_m}\omega_m(t) + \frac{3Z_pL_h}{2L_rJ_m}\psi_{rd}^{ss}i_{sq}(t) + \frac{3Z_pL_h}{2L_rJ_m}i_{sq}^{ss}\psi_{rd}(t) - \frac{3Z_pL_h}{2L_rJ_m}i_{sq}^{ss}\psi_{rd}^{ss} - \frac{T_L}{J_m}, \tag{3.126}$$

where $\kappa_t = \frac{3Z_pL_h}{2L_rJ}$. The last three terms on the right-hand side of (3.126) are considered as disturbances in the feedback control system. In particular, the constant term $\kappa_t i_{sq}^{ss}\psi_{rd}^{ss}$ is an input constant disturbance and if the load torque $\frac{T_L}{J_m}$ changes in a step signal manner, then both disturbances will be completely rejected without steady-state error by the PI controller. The term $\kappa_t i_{sq}^{ss}\psi_{rd}(t)$ is a linear function of the rotor flux $\psi_{rd}(t)$. Because $i_{sd}(t)$ is under PI control, in the steady-state, $\psi_{rd}(t)$ will converge to its set-point signal ψ_{rd}^*, where $\psi_{rd}^* = L_h i_{sd}^*$, hence becoming a constant. This disturbance term will also be completely rejected without steady-state error. In short, with PI control of velocity, all the constant terms on the right-hand side of (3.126) are regarded as constant disturbances that will be completely rejected by the action of integrator in the controller.

Based on (3.126), the Laplace transfer function between the velocity $\Omega_r(s)$ and the current $I_{sq}(s)$ is given by

$$\frac{\Omega_r(s)}{I_{sq}(s)} = \frac{\kappa_t\psi_{rd}^{ss}}{s + \frac{f_d}{J_m}}.$$

By substituting the current feedback signal $I_{sq}(s)$ with its set-point signal $I_{sq}^*(s)$ using the closed-loop transfer function (3.122), the relationship between the velocity $\Omega_r(s)$ and the set-point signal for the inner-loop current control is established

$$\frac{\Omega_r(s)}{I_{sq}^*(s)} = \frac{\kappa_t\psi_{rd}^{ss}}{s + \frac{f_d}{J_m}} \frac{\frac{K_c^q}{\tau_\sigma' r_\sigma}}{s + \frac{1}{\tau_\sigma'} + \frac{K_c^q}{\tau_\sigma' r_\sigma}}, \tag{3.127}$$

where $K_c^q = \frac{\alpha}{1-\alpha}r_\sigma$, and α is the desired steady-state gain which is specified in the design of the inner-loop proportional controller. Generally, α is chosen close to unity in the range of $0 < \alpha < 1$. Then, the assumption can be made that the time constant $\frac{J_m}{f_d}$ is far larger than the time constant $(1-\alpha)\tau_\sigma$ from the inner-loop current control. The transfer function (3.127) is approximated as following

$$\frac{\Omega_r(s)}{I_{sq}^*(s)} = \frac{\kappa_t\psi_{rd}^{ss}}{s + \frac{f_d}{J_m}}. \times \alpha \tag{3.128}$$

Let $a = \frac{f_d}{J_m}$ and $b = \kappa_t \psi_{rd}^{ss} \times \alpha$ and the PI controller be represented by

$$C(s) = \frac{c_1 s + c_0}{s} = K_c \left(1 + \frac{1}{\tau_I s}\right),$$

where K_c is the proportional gain and τ_I is the integral time constant. Following the design procedure outlined in Section 3.1, the PI controller parameters are calculated as:

$$K_c = \frac{2\xi w_n - a}{b}; \quad \tau_I = \frac{2\xi w_n - a}{w_n^2}.$$

The control signal $i_{sq}^*(t)$ is computed using the position form from the outer-loop PI controller, where

$$i_{sq}^*(t) = K_c(\omega_m^*(t) - \omega_m(t)) + \frac{K_c}{\tau_I} \int_0^t (\omega_m^*(\tau) - \omega_m(\tau))d\tau.$$

If the inner-loop q-axis current controller is a PI controller, then the steady-state gain of the closed-loop current control system is unity. In this case, the calculations of the K_c and τ_I for the velocity controller are the same except the α value in (3.128) is taken as unity.

3.7.2.2 Position Control

When considering angular position control of an induction motor, a PID controller is used in the outer-loop control system. The design procedure follows the one stated in Section 3.5. The mathematical model for outer-loop PID controller design is described by the following relation:

$$\frac{\Theta(s)}{I_{sq}^*(s)} = \frac{\kappa_t \psi_{rd}^{ss}}{\left(s + \frac{f_d}{J_m}\right) s} \times \alpha. \tag{3.129}$$

With the parameters $a = \frac{f_d}{J_m}$ and $b = \kappa_t \psi_{rd}^{ss} \times \alpha$, the calculations of the PID controller gains are obtained as

$$K_c = \frac{(2\xi n + 1)w_n^2}{b} \tag{3.130}$$

$$\tau_I = \frac{(2\xi n + 1)w_n^2}{n w_n^3} = \frac{(2\xi n + 1)}{w_n} \tag{3.131}$$

$$\tau_D = \frac{(2\xi + n)w_n - a}{(2\xi n + 1)w_n^2}, \tag{3.132}$$

where the desired closed-loop polynomial is chosen to be $A_{cl}^d(s) = (s^2 + 2\xi w_n s + w_n^2)(s + n w_n)$. Here, the pair of dominant desired closed-loop poles are selected as $s_{1,2} = -\xi w_n \pm j w_n \sqrt{1 - \xi^2}$, where $\xi = 0.707$, and the fast pole is then chosen to be $s_3 = -n \times w_n$ ($n \gg 1$). The parameter α is chosen to be unity if the inner-loop q-axis current controller is a PI controller.

3.7.3 *Slip Estimation*

The position of the field is required in the vector control of an induction motor, the angle θ_s is estimated for its use in the Park transformation. Here, the simple open-loop slip estimation is used, where the model equation is defined as

$$\omega_s(t) = Z_p \omega_m(t) + \frac{i_{sq}(t)}{\tau_r i_{sd}(t)}. \tag{3.133}$$

By assuming the motor is started from rest and the initial values ω_s and ω_m are equal to zero, then integrating both sides of the equation (3.133) leads to

$$\theta_s(t) = Z_p\theta_r(t) + \frac{1}{\tau_r}\int_0^t \frac{i_{sq}(\tau)}{i_{sd}(\tau)}d\tau, \tag{3.134}$$

where $\theta_r(t)$ is the rotor position calculated from the encoder measurement. Because the current feedback signals i_{sd} and i_{sq} typically contain a significant amount of noise, direct application of (3.134) leads to inaccurate estimation of $\theta_s(t)$. The traditional approach is to use the reference signals i^*_{sd} and i^*_{sq} to replace the actual i_{sd} and i_{sq} in order to reduce the effect of noise. In addition, if a proportional controller is used for the q-axis current control, because the inner-loop proportional control causes a steady-state error between the reference current i^*_{sq} and the feedback current i_{sq}, the steady-state gain for the inner-loop current control system ($= \alpha$) needs to be considered, which yields the estimation of θ_s,

$$\theta_s(t) \approx Z_p\theta_r(t) + \frac{1}{\tau_r}\int_0^t \frac{i^*_{sq}(\tau) \times \alpha}{i^*_{sd}(\tau)}d\tau, \tag{3.135}$$

where $\alpha = 1$ if the inner-loop q-axis current is controlled using a PI controller.

3.8 Overview of PID Control of Power Converter

For a three phase system, the dynamic model of the grid connected voltage source power converter in the $d-q$ reference frame is expressed as

$$L_s\frac{di_d}{dt} = -R_s i_d + \omega_g L_s i_q + E_d - v_d \tag{3.136}$$

$$L_s\frac{di_q}{dt} = -v_q - R_s i_q - \omega_g L_s i_d \tag{3.137}$$

$$C_{dc}\frac{dv_{dc}}{dt} = \frac{3}{4}(S_d i_d + S_q i_q) - i_L, \tag{3.138}$$

where ω_g is the grid frequency, E_d is a grid source voltage that is a constant or a slowly varying quantity, i_d and i_q are the grid currents in the $d-q$ reference frame. The variable v_{dc} is the output voltage of the power converter. The d-axis voltage v_d and q-axis voltage v_q are the voltage control variables as the inputs to the converter. The variables S_d and S_q are related to the voltages v_d and v_q via the relationship below,

$$v_d = S_d \times (v_{dc}/2) \tag{3.139}$$

$$v_q = S_q \times (v_{dc}/2). \tag{3.140}$$

It is seen from these equations that the S_d and S_q are dimensionless, and with a v_{dc} being regulated to a constant value, they are the normalized voltage variables of d-axis voltage v_d and q-axis voltage v_q. Because they are related to v_{dc} through (3.138), if the control objective is to regulate the output voltage of the converter v_{dc}, the normalized voltage variables S_d and S_q are selected as the manipulated variables in a cascade control structure (see Figure 3.17). The disturbances in this control system are the slow variations of the grid voltage E_d and the load current i_L, and disturbance rejection is one of the main concerns in the design of the closed-loop control system.

There are three inductors acting as filters between the grid currents and the measured three phase currents i_a, i_b and i_c. The parameters for the inductor filters are the inductance L_s and the resistance R_s. The role of the capacitor is to store the energy and to reduce voltage ripples and its capacitance is C_{dc}.

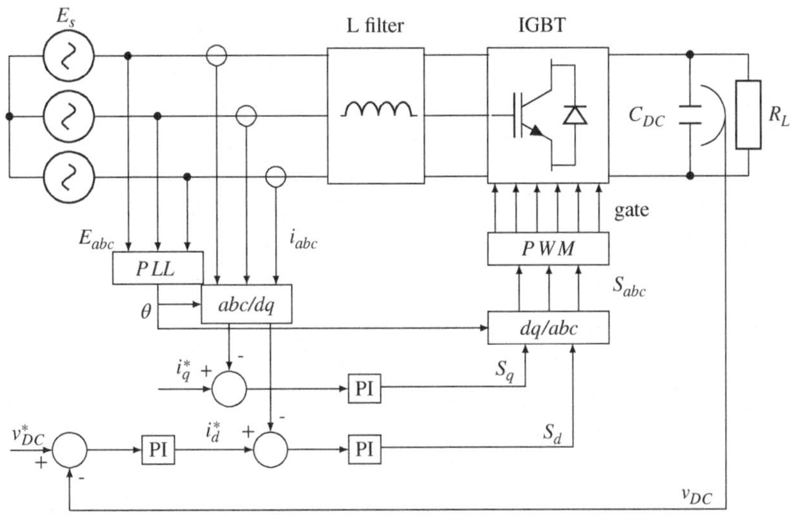

Figure 3.17 Schematic diagram for a grid connected three phase power converter.

There are two control objectives for closed-loop control of the power converter. The first is to maintain a constant output voltage v_{dc} in the presence of load current (i_L) variation, and the second is to achieve a desired power factor of unity, which is translated to the control system specification of maintaining the q-axis current to be at zero value in the presence of disturbance. Namely, the set-point signal $i_q^*(t)$ in the power converter control problem is selected to be unity.

In the control system design, similar to the *AC* motor control problems, a cascade feedback and feed-forward control system is configured for controlling the output *DC* voltage v_{dc}. Figure 3.17 shows the voltage control system configuration of a typical grid connected power converter. In the control system configuration, there are two PI controllers to control the d-axis and q-axis currents and one PI controller in the outer-loop to achieve the output voltage control (see Figure 3.17). A phase locked loop is used to estimate the angle θ_g used in the Park transformation, which is the electrical angle of the grid currents.

3.8.1 Bridging Sensor Measurements to Feedback Signals

The feedback signals to the controllers are d-axis stator current i_d and q-axis stator current i_q, and the output voltage of the converter v_{dc}.

There are three current sensors to measure the three phase filtered grid currents i_a, i_b and i_c as shown in Figure 3.14. The transformation of the three-phase currents to their components in α-β frame is achieved by the Clarke transformation,

$$\begin{bmatrix} i_\alpha \\ i_\beta \end{bmatrix} = \frac{2}{3}\begin{bmatrix} 1 & -\frac{1}{2} & -\frac{1}{2} \\ 0 & \frac{\sqrt{3}}{2} & -\frac{\sqrt{3}}{2} \end{bmatrix}\begin{bmatrix} i_a \\ i_b \\ i_c \end{bmatrix}. \tag{3.141}$$

To obtain the current feedback signals i_d and i_q in the $d-q$ reference frame, the following inverse Park transformation is used,

$$\begin{bmatrix} i_d \\ i_q \end{bmatrix} = \begin{bmatrix} \cos\theta_g & \sin\theta_g \\ -\sin\theta_g & \cos\theta_g \end{bmatrix}\begin{bmatrix} i_\alpha \\ i_\beta \end{bmatrix}, \tag{3.142}$$

where the sinusoidal signals i_α and i_β are converted to the d-axis filtered grid current i_d and q-axis filtered grid current i_q. In a normal steady-state operation of a power converter, the i_d and i_q currents are regulated to their desired operating points. The electrical angle of the grid currents θ_g is estimated using a phase-locked-loop (PLL) to be discussed in Section 3.9.5.

3.8.2 Bridging the Control Signals to the Inputs of the Power Converter

The controllers' outputs are the normalized d-axis and q-axis voltages, S_d and S_q (see (3.140)). To obtain the input signals to the power converter, the S_α and S_β are obtained via the calculation,

$$\begin{bmatrix} S_\alpha \\ S_\beta \end{bmatrix} = \begin{bmatrix} \cos\theta_g & -\sin\theta_g \\ \sin\theta_g & \cos\theta_g \end{bmatrix} \begin{bmatrix} S_d \\ S_q \end{bmatrix}, \tag{3.143}$$

which are sinusoidal signals with frequency ω_g equal to the grid frequency. Also, let S_0 represent the zero sequence component of three phase normalized voltage which is zero for a balanced three phase voltage. Then, the inverse Clarke transformation leads to,

$$\begin{bmatrix} S_a \\ S_b \\ S_c \end{bmatrix} = \begin{bmatrix} 1 & 0 & 1 \\ -\frac{1}{2} & \frac{\sqrt{3}}{2} & 1 \\ -\frac{1}{2} & -\frac{\sqrt{3}}{2} & 1 \end{bmatrix} \begin{bmatrix} S_\alpha \\ S_\beta \\ S_0 \end{bmatrix}. \tag{3.144}$$

At this point, the control system has computed the three phase normalized voltage signals as inputs to the power converter in order to achieve the control objectives. The actual implementation of the control signals requires the devices of the semiconductor switches connected with the electrical grid (see Chapter 2).

3.9 PI Current and Voltage Controller Design for Power Converter

3.9.1 P Control of d-axis Current

The d-axis current controller is designed based on the differential equation:

$$\frac{di_d(t)}{dt} = -\frac{R_s}{L_s} i_d(t) + \omega_g i_q(t) + \frac{E_d}{L_s} - \frac{S_d(t) v_{dc}(t)}{2L_s}. \tag{3.145}$$

Let the auxiliary control variable $\hat{S}_d$ be defined by

$$-\frac{1}{2L_s}\hat{S}_d(t) = -\frac{S_d(t) v_{dc}(t)}{2L_s} + \omega_g i_q(t) + \frac{E_d}{L_s}. \tag{3.146}$$

Substituting (3.146) into (3.145), we obtain the linear model for the design of inner-loop current controller:

$$\frac{di_d(t)}{dt} = -\frac{R_s}{L_s} i_d(t) - \frac{1}{2L_s}\hat{S}_d(t). \tag{3.147}$$

This is a first-order dynamic model. Because the controller operates in the inner loop system, the accuracy of the controlled variable is not as important as the one from the outer loop system. Thus, a proportional controller is used for controlling the d-axis current. Note that the open-loop transfer function based on (3.147) is given by

$$\frac{I_d(s)}{\hat{S}_d(s)} = \frac{-\frac{1}{2L_s}}{s + \frac{R_s}{L_s}}. \tag{3.148}$$

With a proportional feedback controller K_c^d, the closed-loop transfer function between the reference signal $I_d^*(s)$ and the feedback signal $I_d(s)$, is

$$\frac{I_d(s)}{I_d^*(s)} = \frac{-\frac{K_c^d}{2L_s}}{s + \frac{R_s}{L_s} - \frac{K_c^d}{2L_s}}. \tag{3.149}$$

The steady-state gain of (3.149) is calculated by setting $s = 0$, which is

$$\alpha = \frac{-\frac{K_c^d}{2L_s}}{\frac{R_s}{L_s} - \frac{K_c^d}{2L_s}}. \tag{3.150}$$

In the design, we specify the desired closed-loop performance by choosing a value of the steady-state gain $0 < \alpha < 1$ for the inner-loop closed-loop control. With the desired value α, the proportional feedback controller gain K_c^d is found from the solution of (3.150):

$$K_c^d = -\frac{2\alpha}{1-\alpha}R_s. \tag{3.151}$$

Substituting (3.151) into (3.149) forms the closed-loop transfer function for the design. The closed-loop pole is calculated using this closed-loop transfer function at $s = -\frac{1}{1-\alpha}\frac{R_s}{L_s}$. A typical choice for α is 0.9, and with this choice, the closed-loop pole is $s = -10\frac{R_s}{L_s}$, which is 10 times the original open-loop pole at $s = -\frac{R_s}{L_s}$. The steady-state gain for the closed inner-loop system will be considered in the design for the outer-loop system. By substituting the controller gain K_c^d given by (3.151) into the closed-loop transfer function of the d-axis current, we obtain

$$\frac{I_d(s)}{I_d^*(s)} = \frac{\alpha}{(1-\alpha)\frac{L_s}{R_s}s + 1}, \tag{3.152}$$

which confirms that the time constant for the closed-loop system is $\tau_c = (1-\alpha)\frac{L_s}{R_s}$ and steady-state gain is α. This relation will be used in the design of outer-loop control system for the voltage.

The feedback realization for the auxiliary control variable $\hat{S}_d(t)$ is based on feedback error $i_d^*(t) - i_d(t)$, where $i_d^*(t)$ is the reference signal to the inner-loop d-axis current control loop, namely,

$$\hat{S}_d(t) = K_c^d(i_d^*(t) - i_d(t)). \tag{3.153}$$

Based on this, the actual inner-loop control signal $S_d(t)$ is computed by substituting (3.153) into (3.146), which is,

$$\begin{aligned} S_d(t) &= \frac{1}{v_{dc}(t)}\hat{S}_d(t) \\ &= \frac{1}{v_{dc}(t)}(K_c^d(i_d^*(t) - i_d(t)) + 2L_s\omega_g i_q(t) + 2E_d). \end{aligned} \tag{3.154}$$

Figure 3.18 shows the inner-loop d-axis current proportional control using a nonlinear feedforward compensation.

3.9.2 PI Control of q-axis Current

To design the input and output feedback linearization controller for the q-axis current control, we consider the dynamic model for the q-axis current:

$$\frac{di_q(t)}{dt} = -\frac{R_s}{L_s}i_q(t) - \omega_g i_d(t) - \frac{S_q(t)v_{dc}(t)}{2L_s}. \tag{3.155}$$

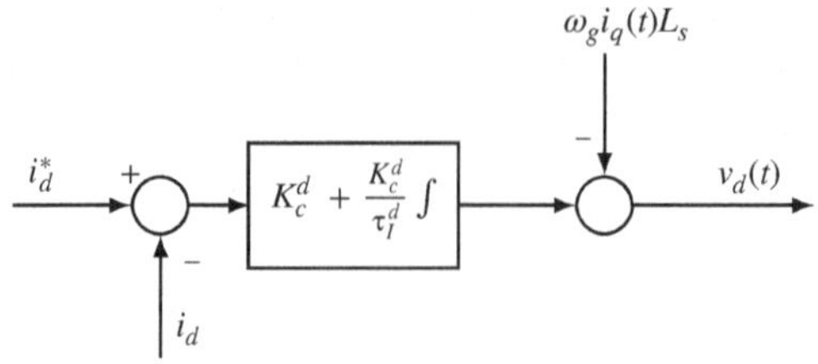

Figure 3.18 PI control of d-axis current using nonlinear feedforward compensation, where $v_d(t) = S_d(t) v_{dc}(t)$.

Similar to the controller design for the d-axis current loop, we choose an auxiliary control variable $\hat{S}_q(t)$ as

$$-\frac{1}{2L_s}\hat{S}_q(t) = -\frac{S_q(t) v_{dc}(t)}{2L_s} - \omega_g i_d(t). \tag{3.156}$$

By substituting this auxiliary control variable into (3.155), we obtain the linearized differential equation that describes the q-axis current dynamics:

$$\frac{di_q(t)}{dt} = -\frac{R_s}{L_s} i_q(t) - \frac{1}{2L_s}\hat{S}_q(t). \tag{3.157}$$

In the q-axis current control, a PI controller is needed to ensure zero steady-state error in the closed-loop system. The auxiliary control signal $\hat{S}_q(t)$ is expressed using the feedback error $i_q^*(t) - i_q(t)$ as,

$$\hat{S}_q(t) = K_c^q (i_q^*(t) - i_q(t)) + \frac{K_c^q}{\tau_I^q} \int_0^t (i_q^*(\tau) - i_q(\tau)) d\tau,$$

where $i_q^*(t)$ is the set-point signal to the q-axis current, which is chosen to be zero in most applications.

Upon obtaining the auxiliary control signal, the actual control signal $S_q(t)$ is found based on (3.156):

$$\begin{aligned} S_q(t) &= \frac{1}{v_{dc}(t)} (\hat{S}_q(t) - 2\omega_g L_s i_d(t)) \\ &= \frac{1}{v_{dc}(t)} \left(K_c^q (i_q^*(t) - i_q(t)) + \frac{K_c^q}{\tau_I^q} \int_0^t (i_q^*(\tau) - i_q(\tau)) d\tau - 2\omega_g L_s i_d(t) \right). \end{aligned}$$

Figure 3.19 shows the inner-loop current control using PI with nonlinear feedforward compensation.

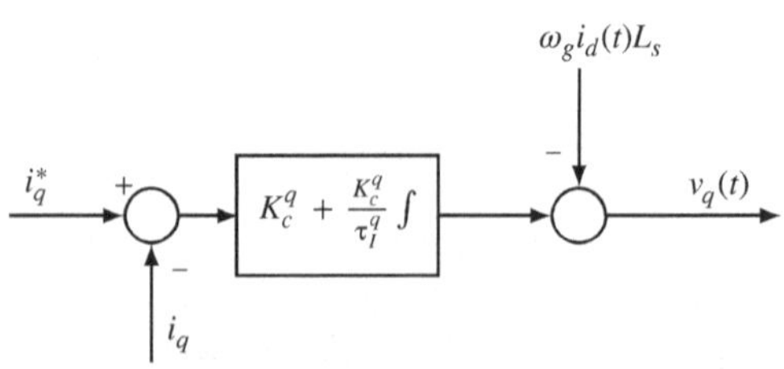

Figure 3.19 PI control of q-axis current using nonlinear feedforward compensation, where $v_q(t) = S_q(t) v_{dc}(t)$.

3.9.3 PI Cascade Control of Output Voltage

The objective of the outer-loop control system is to maintain steady-state operation of the output voltage $v_{dc}(t)$ in the presence of load current variation. The accuracy of the output response at steady-state is an important factor so a PI controller is chosen for this control loop. The outer-loop PI controller is designed based on the dynamic model that describes the relationship between the output variable v_{dc} and the input variable i_d given by (3.138), which is:

$$C_{dc}\frac{dv_{dc}}{dt} = \frac{3}{4}(S_d i_d + S_q i_q) - i_L, \tag{3.158}$$

This is a bilinear model with respect to the control variables S_d and S_q. We first seek its linearization, where a steady-state operation for the power converter is assumed with the defined parameters S_d^{ss}, S_q^{ss}, i_d^{ss} and i_q^{ss}. The bilinear terms are approximated by

$$S_d(t)i_d(t) \approx -S_d^{ss}i_d^{ss} + S_d^{ss}i_d(t) + i_d^{ss}S_d(t) \tag{3.159}$$

$$S_q(t)i_q(t) \approx -S_q^{ss}i_q^{ss} + S_q^{ss}i_q(t) + i_q^{ss}S_q(t). \tag{3.160}$$

In the majority of the applications, the reference signal to the q-axis current loop control system is zero, the three quantities in (3.160) are very small in the steady-state; hence, they will be neglected in the design, and their variations will be treated as part of the disturbance in the outer-loop control system. By substituting (3.159) into (3.158), with the assumption that $S_q(t)i_q(t) \approx 0$, we obtain the linearized model for the design of outer-loop PI controller:

$$C_{dc}\frac{dv_{dc}}{dt} = \frac{3}{4}(-S_d^{ss}i_d^{ss} + S_d^{ss}i_d(t) + i_d^{ss}S_d(t)) - i_L(t). \tag{3.161}$$

On the right-hand side of (3.161), all the terms, except $\frac{3}{4}S_d^{ss}i_d(t)$, are regarded as disturbances to the outer-loop control system. Based on (3.161), the Laplace transfer function between $I_d(s)$ and $V_{dc}(s)$ is given by

$$\frac{V_{dc}(s)}{I_d(s)} = \frac{3}{4}\frac{S_d^{ss}}{C_{dc}}\frac{1}{s}. \tag{3.162}$$

Now, with the relationship between the set-point signal $I_d^*(s)$ and $I_d(s)$ given by (3.152), we obtain the transfer function between $V_{dc}(s)$ and $I_d^*(s)$:

$$\frac{V_{dc}(s)}{I_d^*(s)} = \frac{3}{4}\frac{S_d^{ss}}{C_{dc}}\left(\frac{1}{s}\right)\left(\frac{\alpha}{(1-\alpha)\frac{L_s}{R_s}s+1}\right). \tag{3.163}$$

This is a second order model, however, in comparison with the integrator, the time constant $(1-\alpha)\frac{L_s}{R_s}$ is very small and is taken to be zero as an approximation. Therefore, the model for the design of outer-loop PI controller is approximated by the first order model:

$$\frac{V_{dc}(s)}{I_d^*(s)} \approx \frac{3}{4}\frac{S_d^{ss}}{C_{dc}}\frac{\alpha}{s}. \tag{3.164}$$

For notational simplicity, we let $a = 0$ and $b = \frac{3}{4}\frac{S_d^{ss}}{C_{dc}} \times \alpha$. By choosing a damping coefficient $\xi = 0.707$, the bandwidth of the outer-loop control system w_n,the PI controller parameters are calculated as (see Section 3.1),

$$K_c = \frac{2\xi w_n - a}{b} = \frac{2\xi w_n}{b} \tag{3.165}$$

$$\tau_I = \frac{2\xi w_n - a}{w_n^2} = \frac{2\xi}{w_n}. \tag{3.166}$$

With the values of K_c and τ_I calculated, the reference signal to the inner-loop current control, $i_d^*(t)$, is calculated using the PI controller:

$$i_d^*(t) = K_c(v_{dc}^*(t) - v_{dc}(t)) + \frac{K_c}{\tau_I}\int_0^t (v_{dc}^*(\tau) - v_{dc}(\tau))d\tau, \tag{3.167}$$

where $v_{dc}^*(t)$ is the set-point signal for the voltage.

Here, because of the linearization, the variables $v_{dc}(t)$, $i_d(t)$ and $S_d(t)$ are deviation variables. Special care should be taken at the implementation of the PI controller (see Chapter 4).

3.9.4 Simulation Example

The nonlinear model for a two level power converter given by (3.136–3.138) is used in the control system simulation. The parameters in the simulation are chosen as $L_s = 2\times10^{-3}$ H, $C_{dc} = 10^{-3}$ F, $R_s = 1\ \Omega$, $\Delta t = 100\times10^{-6}$ sec, $E_d = 60$ V, $I_{dss} = 7$ A, $vdc_{ss} = 80$ V. The control objective is to maintain the constant voltage by rejecting a load current disturbance i_L that is a square wave signal with its magnitude varying between 1 A and 0 A as shown in Figure 3.20(a). The inner-loop i_d current is controlled with a proportional controller with the proportional gain $K_c^d = -\frac{2*\alpha}{1-\alpha}R_s$ with $\alpha = 0.9$, which is also used to control the q-axis current for simplicity and outer-loop of voltage control is a PI controller with the closed-loop specification of the damping coefficient $\xi = 0.707$ and $w_n = 5$ rad/s. From Figure 3.20(a), it is seen that when the load current i_L is reduced to zero, the output voltage v_{dc} is increased first; however, with the cascade current and voltage control, the output voltage v_{dc} is regulated to its set-point value of 100 V. When the load current i_L increases, the output voltage v_{dc} is decreased first, then with the cascade control system, this disturbance is rejected and the voltage is regulated back to its set-point. Figure 3.20(b) shows the control actions observed from the S_d and S_q signals.

3.9.5 Phase Locked Loop

In the power converter control system, the seven pairs of candidate voltage variables are generated using the electrical angle $\theta_g(t)$ where $\theta_g(t)$ is estimated using a Phase Locked Loop (PLL). Figure 3.21 shows the

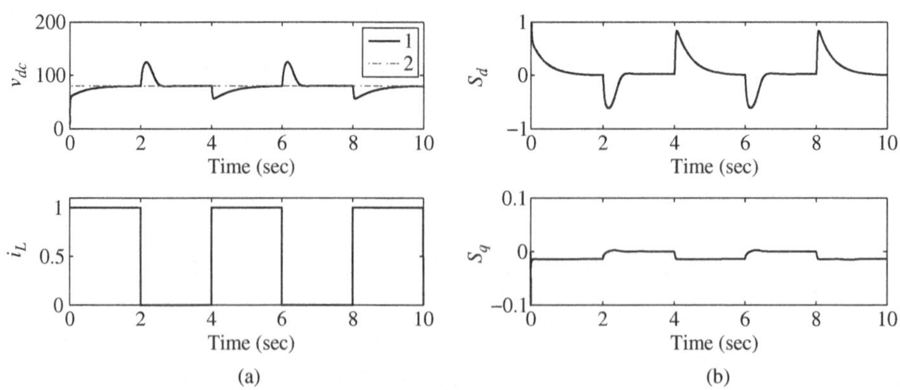

Figure 3.20 Cascade feedback control of with P q-axis grid current controller and P d-axis grid current controller and outer-loop PI voltage controller. (a) Closed-loop response of converter. Top figure: key: line (1) the actual voltage; line (2) the reference voltage. Bottom figure: current load disturbance i_L. (b) Closed-loop control signal responses, the normalized d-axis voltage S_d and q-axis voltage S_q).

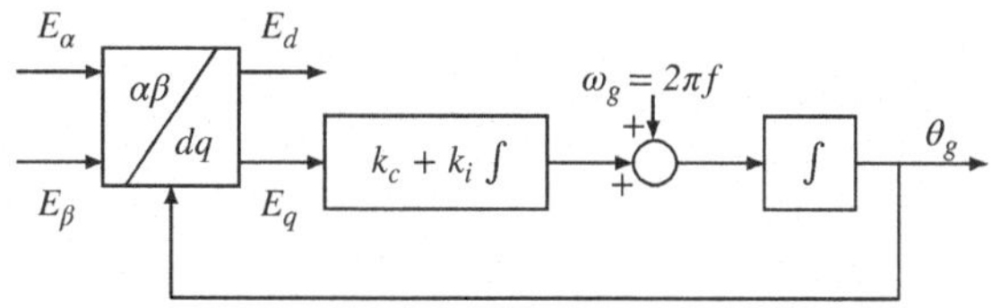

Figure 3.21 Phase Locked Loop for grid connected power converter.

block diagram of a phase locked loop for estimating the grid electrical angle $\theta_g(t)$ using the grid frequency $\omega_g = 2\pi f$ (f is 50 Hz used in the simulations and experiments) together with the Park transformation. The basic idea in a PLL is to use a PI feedback controller to automatically find the grid electrical angle $\theta_g(t)$. When the correct $\theta_g(t)$ is found, in the $\alpha - \beta$ reference frame, $E_\alpha(t) = \cos\theta_g E$ and $E_\beta(t) = \sin\theta_g E$, where E is the magnitude of grid voltage and is assumed constant. Then from the Park transformation, we have the following relationships

$$\begin{aligned} E_d(t) &= \cos\theta_g E_\alpha(t) + \sin\theta_g E_\beta(t) \\ &= (\cos^2\theta_g + \sin^2\theta_g)E = E \end{aligned} \tag{3.168}$$

$$\begin{aligned} E_q(t) &= -\sin\theta_g E_\alpha(t) + \cos\theta_g E_\beta(t) \\ &= (-\sin\theta_g\cos\theta_g + \cos\theta_g\sin\theta_g)E = 0, \end{aligned} \tag{3.169}$$

Therefore, when the correct θ_g is found, the input signal to the PLL is zero as $E_q(t)$ is zero. The input signal $\omega_g = 2\pi f$ provides the bias signal to the PLL.

The design model for the PI controller of the PLL system takes consideration of the integrator dynamics together with a gain:

$$G(s) = \frac{\sqrt{E_d^2 + E_q^2}}{s} = \frac{k_p}{s}. \tag{3.170}$$

By letting $a = 0$, $b = k_p$, following the same PI controller design procedure as discussed in Section 3.1, the proportional and integral control gains for the PI controller are found:

$$k_c = \frac{2\xi w_n}{k_p}; \quad k_i = \frac{w_n^2}{k_p}, \tag{3.171}$$

where w_n and ξ are chosen by the user. The parameters are also expressed in terms of PI controller proportional gain $K_c = k_c$ and integral time constant $\tau_I = \frac{2\xi}{w_n}$. Since the grid frequency is unchanged, the phase locked system could be turned on first to estimate the correct electrical angle θ_g, followed by the switching on the PI control systems.

The key equations in computation of the $\theta_g(t)$ are summarized as follows. A MATLAB embedded function is written in Chapter 10 to generate the grid electrical angle θ_g.

Algorithm 1 *PLL*

1. *At sampling time t, assuming a grid electrical angle $\theta_g(t)$, use $E_\alpha(t)$ and $E_\beta(t)$ to compute the $E_d(t)$ and $E_q(t)$ by the Park transformation:*

$$\begin{bmatrix} E_d(t) \\ E_q(t) \end{bmatrix} = \begin{bmatrix} \cos\theta_g(t) & \sin\theta_g(t) \\ -\sin\theta_g(t) & \cos\theta_g(t) \end{bmatrix} \begin{bmatrix} E_\alpha(t) \\ E_\beta(t) \end{bmatrix}.$$

2. *Calculate the estimated grid frequency $\hat{\omega}_g$ which is the output of the PI controller together with a constant bias $2\pi f = \omega_g$*

$$\hat{\omega}_g(t) = k_c E_q(t) + k_i \int_0^t E_q(\tau)d\tau + 2\pi f.$$

3. *Calculate the estimated grid electrical angle using*

$$\theta_g(t) = \int_0^t \hat{\omega}_g(\tau)d\tau.$$

4. *Go back to Step 1 with the estimated $\theta_g(t)$ and repeat the computational steps.*

3.10 Summary

This chapter has presented PID controller design methodologies for closed-loop feedback control of *AC* electrical drives and two level power converters. Although these electrical systems are different in terms of their dynamics and control objectives, a unified framework is used in the design of PID controllers.

1. All mathematical models are described in the $d-q$ reference frame. In particular, the manipulated variables are d-axis voltage and q-axis voltage, and the controlled variables are the d-axis current and q-axis current.
2. If only current control is required (or torque control in the *AC* drive case), then two PI controllers are used to control the d-axis currents and q-axis currents respectively, where the current reference signals are known constants.
3. If velocity or position control is required in the *AC* drives, then cascade PI control system is needed, which basically manipulates the current reference signal to the q-axis current (i_q^*) using the output of the outer-loop velocity or position controller. PI controller is sufficient for velocity control while PID is used for position control.
4. If voltage control is required for the power converter, then cascade PI control system is needed. The manipulated variable in the outer-loop voltage controller is the reference signal to the d-axis current control (i_d^*).
5. When the cascade PI control system is designed, it is simpler to use a proportional controller for the inner-loop current control. However, there could be some performance discrepancies in terms of disturbance rejections as discussed in Chapter 5.
6. The proportional controller in general is expressed as

$$u(t) = K_c e(t) + f(t),$$

   where $e(t)$ is the feedback error and $f(t)$ is the nonlinear feedforward signal. The proportional controller is limited to the applications of inner-loop current control because it results in steady-state errors. Since the steady-state value of the inner-loop current control system is considered in the outer-loop control system design, it is suggested that the steady-state value of the inner-loop system is selected as α and the feedback controller gain K_c is chosen in relation to α. α is selected to be in the range of 0.85 to 0.95.
7. The PI controller in general is expressed as

$$u(t) = K_c e(t) + \frac{K_c}{\tau_I}\int_0^t e(\tau)d\tau + f(t),$$

   where $e(t)$ is the feedback error and $f(t)$ is the nonlinear feedforward signal. The PI controller can be used either in the controlling of d-axis current and q-axis current, or as a controller used in the outer-loop system. Assuming that the transfer function model for the design is a first order, having the form

$$G(s) = \frac{b}{s+a},$$

the PI controller parameters have the expressions:

$$K_c = \frac{2\xi w_n - a}{b}; \quad \tau_I = \frac{2\xi w_n - a}{w_n^2},$$

where the controller performance is specified by choosing the damping coefficient ξ and the bandwidth w_n. It is suggested that the damping coefficient is selected to be 0.707 and the bandwidth $w_n = \frac{1}{1-\gamma}a$ rad/s with the parameter γ in the range of 0.85 to 0.95.

3.11 Further Reading

Books on control system design and analysis include Goodwin *et al.* (2001), Astrom and Murray (2008), Astrom and Hagglund (1995). General application characteristics of electric motors are discussed in Pillay and Krishnan (1991). Books for *AC* electric drives are Vas (1992), Vas (1993), Hughes and Drury (2013), Leonhard (2001), El-Hawary (2011), Drury (2009), Trzynadlowski (2000), Quang and Dittrich (2008). PID controllers for electric drives were discussed and compared with model predictive control in Thomsen *et al.* (2011). FOC and DTC were studied in Casadei *et al.* (2002) for induction motors. A survey paper was published on current control techniques for three-phase voltage-source PWM converters by Kazmierkowski and Malesani (1998). Direct torque control for PMSM drives was analyzed in Zhong *et al.* (1997). Torque minimization techniques for permanent magnet *AC* motor drives were discussed and reviewed in Jahns and Soong (1996). A comparative study was presented for current control of PMSM drives by Morel *et al.* (2009). Internal model control method was used for *AC* machine drive control by Harnefors and Nee (1998). A survey paper was published on direct torque control of PWM inverter-fed *AC* motors by Buja and Kazmierkowski (2004). PMS machines in the application of electric vehicles were discussed in Liu and Liu (2012), Estima and Marques Cardoso (2012), Dai *et al.* (2007), Nasiri (2007), induction motor was used in electric vehicles (see Dilmi (2005)). Renewable source applications using power converter for the grid integration were discussed in Casadei *et al.* (2006), Alepuz *et–al.* (2006). Induction motor control strategies were proposed in Takahashi and Noguchi (1986), Depenbrock (1988), Lorenz *et al.* (1994). Current control of *AC*-drives using complex vector approaches was analyzed in del Blanco *et al.* (1999). Slip estimation for induction motor was discussed in Rubin *et al.* (1992) and Abbondanti and Brennen (1975). Induction motor control using dynamic feedback linearization was discussed in Chiasson (1998). Adaptive flux observer of induction motor was discussed in Kubota *et al.* (1990), Maes and Melkebeek (2000) and Kubota *et al.* (1993).

Mathematical modeling and analysis of converters were presented in Wu *et al.* (1991), in Lindgren (1998), in Abdel-Rahim and Quaicoe (1994), in Blasko and Kaura (1997). Selected problems in control of power electronics were given in Kazmierkowski *et al.* (2002), and selected control techniques in Liserre *et al.* (2001), in Liu and Liu (2004), in Svensson (2001). Limitations of voltage-oriented PI current control of grid-connected PWM rectifiers with LCL filters were discussed in Dannehl *et al.* (2009). Current control of power converters was presented in Brod and Novotny (1985), in Kazmierkowski *et al.* (1991), in B. H. Kwon and Youn (1998), in Pan and Chang (1994).

The phase locked loop was discussed in Chung (2000).

References

Abbondanti A and Brennen M 1975 Variable speed induction motor drives use electronic slip calculator based on motor voltage and currents. *IEEE Transactions on Industry Applications* **IA-11**(5), 483–8.

Abdel-Rahim N and Quaicoe JE 1994 Modeling and analysis of a feedback control strategy for three-phase voltage-source utility interface systems. *Proceedings of the 29th IAS Annual Meeting* pp. 895–902.

Alepuz S, Busquets-Monge S, Bordonau J, Gago J, Gonzalez D and Balcells J 2006 Interfacing renewable energy sources to the utility grid using a three-level inverter. *IEEE Transactions on Industrial Electronics* **53**(5), 1504–1511.

Astrom KJ and Hagglund T 1995 *PID Controllers:Theory, Design, and Tuning*. Instrument Society of America, Research Triangle Park, NC.

Astrom KJ and Murray RM 2008 *Feedback Systems: An Introduction for Scientists and Engineers*. Princeton University Press.

B. H. Kwon TWK and Youn JH 1998 A novel SVM-based hysteresis current controller. *IEEE Transactions on Power Electronics* **13**, 297–307.

Bay JS 1999 *Fundamentals of Linear State Space Systems*. McGraw-Hill.

Blasko V and Kaura V 1997 A new mathematical model and control of a three-phase AC-DC voltage source converter. *IEEE Transactions on Power Electronics*. **12**: 116–123.

Brod DM and Novotny DW 1985 Current control of VSI-PWM inverters. *IEEE Transactions on Industry Application* **21**, 562–570.

Buja G and Kazmierkowski M 2004 Direct torque control of PWM inverter-fed AC motors – a survey. *IEEE Transactions on Industrial Electronics* **51**(4), 744–757.

Carrasco J, Franquelo L, Bialasiewicz J, Galvan E, Guisado R, Prats M, Leon J and Moreno-Alfonso N 2006 Power-electronic systems for the grid integration of renewable energy sources: A survey. *IEEE Transactions on Industrial Electronics* **53**(4), 1002–1016.

Casadei D, Profumo F, Serra G and Tani A 2002 FOC and DTC: two viable schemes for induction motors torque control. *IEEE Transactions on Power Electronics* **17**(5), 779–787.

Chiasson J 1998 A new approach to dynamic feedback linearization control of an induction motor. *Automatic Control, IEEE Transaction* **43**(3), 391–397.

Chung SK 2000 Phase-locked loop for grid-connected three phase power conversion systems. *IEE Proceedings – Electric Power Applications* **147**, 213–219.

Dai Y, Song L and Cui S 2007 Development of PMSM drives for hybrid electric car applications. *IEEE Transactions on Magnetics* **43**(1), 434–437.

Dannehl J, Wessels C and Fuchs F 2009 Limitations of voltage-oriented PI current control of grid-connected PWM rectifiers with LCL filters. *IEEE Transactions on Industrial Electronics* **56**(2), 380–388.

del Blanco F, Degner M and Lorenz R 1999 Dynamic analysis of current regulators for AC motors using complex vectors. *IEEE Transactions on Industry Applications* **35**(6), 1424–1432.

Depenbrock M 1988 Direct self-control (DSC) of inverter-fed induction machine. *IEEE Transactions on Power Electronics* **3**(4), 420–429.

Dilmi S and Yurkovich S 2005 Nonlinear torque control of the induction motor in hybrid electric vehicle applications *American Control Conference*, vol. 5, pp. 3001–3006.

Drury B 2009 *The Control Techniques Drives and Controls Handbook* 2nd edn. IET.

El-Hawary M 2011 *Principles of Electric Machines with Power Electronic Applications* 2nd edn. IEEE Press.

Estima J and Marques Cardoso A 2012 Efficiency analysis of drive train topologies applied to electric/hybrid vehicles. *IEEE Transactions on Vehicular Technology* **61**(3), 1021–1031.

Goodwin GC, Graebe SF and Salgado ME 2001 *Control System Design*. Prentice Hall, New Jersey.

Harnefors L and Nee HP 1998 Model-based current control of AC machines using the internal model control method. *IEEE Transactions on Industry Applications* **34**(1), 133–141.

Hughes A and Drury B 2013 *Electric Motors and Drives: Fundamentals, Types and Applications* 4th edn. Elsevier.

Jahns T and Soong W 1996 Pulsating torque minimization techniques for permanent magnet AC motor drives-a review. *IEEE Transactions on Industrial Electronics* **43**(2), 321–330.

Kazmierkowski M and Malesani L 1998 Current control techniques for three-phase voltage-source PWM converters: a survey. *IEEE Transactions on Industrial Electronics* **45**(5), 691–703.

Kazmierkowski M, Dzieniakowski M and Sulkowski W 1991 Novel space vector based current controllers for PWM inverters. *IEEE Transactions on Power Electronics* **6**(1), 158–166.

Kazmierkowski M, R. K and Blaajerg F 2002 *Control in Power Electronics: Selected Problems*. Elsevier.

Kubota H, Matsuse K and Nakano T 1990 New adaptive flux observer of induction motor for wide speed range motor drives *Industrial Electronics Society, 1990. IECON '90., 16th Annual Conference of IEEE*, vol. 2, pp. 921–926.

Kubota H, Matsuse K and Nakano T 1993 DSP-based speed adaptive flux observer of induction motor. *IEEE Transactions on Industry Applications* **29**(2), 344–348.

Leonhard W 2001 *Control of Electrical Drives* 3rd edn. Springer.

Lindgren M 1998 *Modeling and control of voltage source converters connected to the grid* PhD thesis Chalmers University of Technology.

Liserre M 2001 *Innovative control techniques of power converters for industrial automation* PhD thesis Politechnico di Bari, Italy.

Liserre M, Blaabjerg F and Dell Aquila A 2004 Step-by-step design procedure for a grid-connected three-phase PWM voltage source converter. *International Journal of Electronics* **91**(8), 445–460.

Liu P and Liu HP 2012 Permanent-magnet synchronous motor drive system for electric vehicles using bidirectional z-source inverter. *Electrical Systems in Transportation, IET* **2**(4), 178–185.

Lorenz RD, Lipo TA and Novotny DW 1994 Motion control with induction motors. *Proceedings of the IEEE* **82**(8), 1215–1240.

Maes J and Melkebeek J 2000 Speed-sensorless direct torque control of induction motors using an adaptive flux observer. *IEEE Transactions on Industry Application* **36**(3), 778–785.

Morel F, Lin-Shi X, Retif JM, Allard B and Buttay C 2009 A comparative study of predictive current control schemes for a permanent-magnet synchronous machine drive. *IEEE Transactions on Industrial Electronics* **56**(7), 2715–2728.

Nasiri A 2007 Full digital current control of permanent magnet synchronous motors for vehicular applications. *IEEE Transactions on Vehicular Technology* **56**(4), 1531–1537.

Pan C and Chang T 1994 An improved hysteresis current controller for reducing switching frequency. *IEEE Transactions on Power Electronics* **9**(1), 97–104.

Pillay P and Krishnan R 1991 Application characteristics of permanent magnet synchronous and brushless DC motors for servo drives. *IEEE Transactions on Industry Applications* **27**(5), 986–996.

Quang NP and Dittrich JA 2008 *Vector Control of Three-Phase AC Machines* 1st edn. Springer.

Rubin N, Harley R and Diana G 1992 Evaluation of various slip estimation techniques for an induction machine operating under field-oriented control conditions. *IEEE Transactions on Industry Applications* **28**(6), 1367–1375.

Svensson J 2001 Synchronization methods for grid-connected voltage source converters. *IEE Proceedings-Generation, Transmission and Distribution* **148**(3), 229–235.

Takahashi I and Noguchi T 1986 A new quick-response and high-efficiency control strategy of an induction motor. *IEEE Transactions on Industry Applications* **IA-22**(5), 820–827.

Thomsen S, Hoffmann N and Fuchs F 2011 PI control, PI-based state space control, and model-based predictive control for drive systems with elastically coupled loads - a comparative study. *IEEE Transactions on Industrial Electronics* **58**(8), 3647–3657.

Trzynadlowski A 2000 *Control of Induction Motors*. Academic Press.

Vas P 1992 *Electrical Machines and Drives–A Space-Vector Theory Approach*. Oxford University Press, New York, USA.

Vas P 1993 *Parameter Estimation, Condition Monitoring, and Diagnosis of Electrical Machines*. Oxford University Press, New York, USA.

Wang L and Cluett WR 2000 *From Plant Data to Process Control: Ideas for Process Identification and PID Design*. Taylor and Francis, London.

Wu R, Dewan S and Slemon G 1991 Analysis of an AC-to-DC voltage source converter using PWM with phase and amplitude control. *IEEE Transactions on Industry Applications* **27**(2), 355–364.

Zhong L, Rahman M, Hu WY and Lim KW 1997 Analysis of direct torque control in permanent magnet synchronous motor drives. *IEEE Transactions on Power Electronics* **12**(3), 528–536.

4

PID Control System Implementation

After the design of PID control systems for drives and power converters, in this chapter we will discuss the implementation of PID cascaded control systems. Because the control systems for electrical drives are configured as cascaded control systems and there are inner-loop and outer-loop PI controllers, there may be differences between the implementation algorithms. Furthermore, there are operational limits that will be imposed in the control system implementation. For instance, the operational limits of voltages and currents are necessary for the protection of electronic equipment in the presence of a large load and in the transient process. With these limits imposed, if a PI controller is used in the control system, anti-windup mechanisms are required to ensure a satisfactory closed-loop performance.

In Section 4.1, we will first address the implementation of P or PI controllers in the inner-loop control system, where operational constraints for the d-axis and q-axis voltages are introduced and two methods are discussed on how to implement these voltage constraints, followed by detailed discussions on the implementation of current controllers for PMSM in Section 4.2, induction motors in 4.3 and power converters in Section 4.4. In Section 4.5, we will address the implementation of PI controllers in the outer-loop control system, where the operational limit is the current constraint that provides protection for the electronic equipment in the case of a large load. In Section 4.6, a MATLAB tutorial is given to demonstrate how to implement the PI controllers using the velocity form with an anti-windup mechanism.

4.1 P and PI Controller Implementation in Current Control Systems

There are two types of controllers we discussed in the previous chapter that can be used in the current-loop control system. One is the P controller and the other is the PI controller. Because of the operational constraints, special care is needed for saturation and anti-windup schemes.

4.1.1 Voltage Operational Limits in Current Control Systems

The current control signals are the d-axis and q-axis voltages, v_d and v_q. Due to inverter operation, the $d-q$ axes voltages for the electrical drives and power converter are limited by the PWM and DC voltage (see Chapter 2). Suppose that the DC-bus voltage is V_{dc} (V), and using the PWM with harmonic injection or the Space vector PWM, the amplitude of the space vector voltage signal is restricted to $\frac{V_{dc}}{\sqrt{3}}$.

PID and Predictive Control of Electrical Drives and Power Converters using MATLAB®/Simulink®, First Edition.
Liuping Wang, Shan Chai, Dae Yoo, Lu Gan and Ki Ng.

Companion Website: www.wiley.com/go/wang/pid

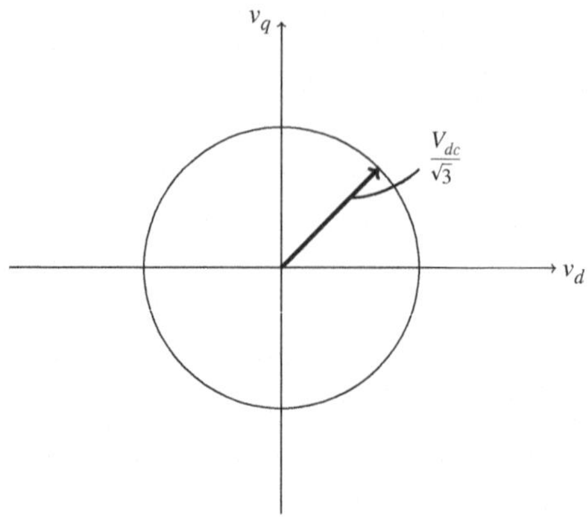

Figure 4.1 Constraints on d-axis and q-axis voltages.

As discussed in Chapter 2, this limit is translated to the constraints on the amplitudes of the d axis and q axis voltages as

$$\sqrt{v_d^2 + v_q^2} \leq \frac{V_{dc}}{\sqrt{3}}. \tag{4.1}$$

This is a nonlinear constraint equation. Figure 4.1 illustrates that the constraint equation is a circle with radius $\frac{V_{dc}}{\sqrt{3}}$, where the inside the circular area is the region where the constraint is satisfied.

Because the constraint equation is a quadratic function of both v_d and v_q, linear approximation is proposed to obtain the limits for the d-axis and q-axis voltages.

4.1.1.1 Rectangular Approximation

The first approach is to approximate the circular area with a rectangular area that is also called boxed constraints, as shown in Figure 4.2. The idea for using this rectangular approximation is to assume a parameter $0 \leq \epsilon \leq 1$, where the maximum value of v_q is set to

$$v_q^{max} = \epsilon \frac{V_{dc}}{\sqrt{3}}. \tag{4.2}$$

Accordingly, the maximum value of v_d is calculated as

$$v_d^{max} = \sqrt{1 - \epsilon^2} \frac{V_{dc}}{\sqrt{3}}. \tag{4.3}$$

For instance, if ϵ is chosen to be 0.8, then $v_q^{max} = 0.8\frac{V_{dc}}{\sqrt{3}}$ and $v_d^{max} = 0.6\frac{V_{dc}}{\sqrt{3}}$. With the operational limits specified, the box constraints in the current loop control system are specified as

$$-v_d^{max} \leq v_d(t) \leq v_d^{max} \tag{4.4}$$

$$-v_q^{max} \leq v_q(t) \leq v_q^{max}. \tag{4.5}$$

The advantage of using this rectangular approximation to the circular area is that the operational constraints are constants once the DC power supply voltage V_{dc} is determined. The constant constraints

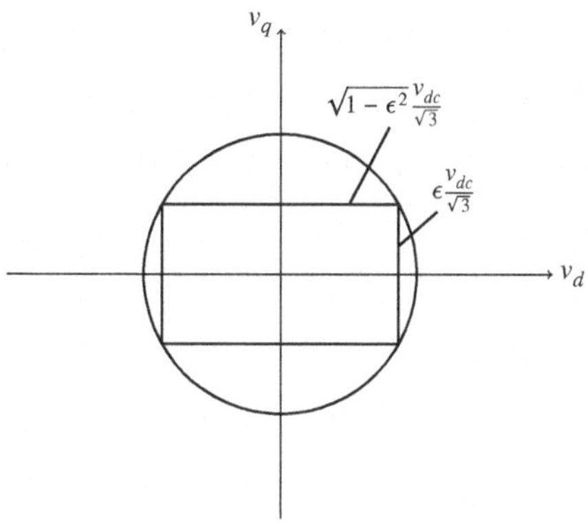

Figure 4.2 Constraints on d-axis and q-axis voltages with a rectangular approximation.

are easier to implement in a real-time control system. However, the shortcoming of this approach is its conservativeness due to the area of rectangular being much smaller than the circular area as shown in Figure 4.2.

4.1.1.2 Maintaining the Same Ratio

Another approach is to determine the maximum and minimum values of v_d and v_q in real-time by scaling the values of actual v_d and v_q proportionally. Assume that at the sampling instant t_i, the two current control-loops have the computed control signals $\bar{v}_d(t_i)$ and $\bar{v}_q(t_i)$ and the voltage constraint equation (4.1) is violated so that

$$\sqrt{\bar{v}_d(t_i)^2 + \bar{v}_q(t_i)^2} > \frac{V_{dc}}{\sqrt{3}}. \tag{4.6}$$

If the violation occurs, the solution is then to find the $v_d(t_i)$ and $v_q(t_i)$ such that

$$\sqrt{v_d(t_i)^2 + v_q(t_i)^2} = \frac{V_{dc}}{\sqrt{3}}. \tag{4.7}$$

This is indeed what the model predictive control tries to do as shown in later chapters of this book. Because there are two degrees of freedom (v_d and v_q) in the single equality constraint, a close approximation is to maintain the same ratio

$$\frac{\bar{v}_d(t_i)}{\bar{v}_q(t_i)} = \frac{v_d(t_i)}{v_q(t_i)}, \tag{4.8}$$

while satisfying the constraint (4.7).

Figure 4.3 illustrates a possible solution that will satisfy the constraint (4.7) while maintaining the same voltage ratio. From the similar triangles shown in Figure 4.3, it is seen that the following equalities are true:

$$\frac{v_d(t_i)}{\frac{V_{dc}}{\sqrt{3}}} = \frac{\bar{v}_d(t_i)}{\sqrt{\bar{v}_d(t_i)^2 + \bar{v}_q(t_i)^2}}$$

$$\frac{v_q(t_i)}{\frac{V_{dc}}{\sqrt{3}}} = \frac{\bar{v}_q(t_i)}{\sqrt{\bar{v}_d(t_i)^2 + \bar{v}_q(t_i)^2}}.$$

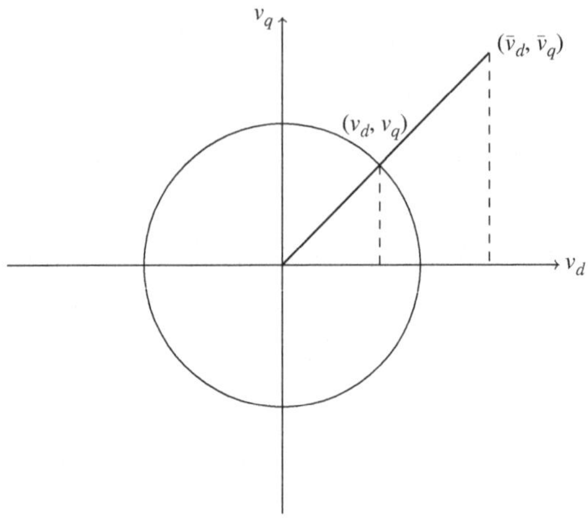

Figure 4.3 Constraints on d-axis and q-axis voltages with an approximation by maintaining the same ratio.

Hence, the values of $v_d(t_i)$ and $v_q(t_i)$ are determined by

$$v_d(t_i) = \frac{\bar{v}_d(t_i)}{\sqrt{\bar{v}_d(t_i)^2 + \bar{v}_q(t_i)^2}} \times \frac{V_{dc}}{\sqrt{3}} \tag{4.9}$$

$$v_q(t_i) = \frac{\bar{v}_q(t_i)}{\sqrt{\bar{v}_d(t_i)^2 + \bar{v}_q(t_i)^2}} \times \frac{V_{dc}}{\sqrt{3}}. \tag{4.10}$$

It is easily verified that from (4.9) and (4.10) that

$$\sqrt{v_d^2 + v_q^2} = \frac{V_{dc}}{\sqrt{3}}.$$

Without confusion, it is emphasized that the calculation of v_d and v_q using (4.9) and (4.10) is based on the assumption that the constraint was violated. Thus in the implementation, the values of $v_d(t_i)$ and $v_q(t_i)$ calculated using (4.9) and (4.10) are the saturated control signals at time t_i. Thus, the limits of the d-axis and q-axis voltages are varying with respect to time, depending on the computed control signals.

Although the implementation of the constraints is slightly more complicated than the constant constraints (see (4.4) and (4.5)), it will be shown in the next section that with a velocity form PI control implementation, it is a straightforward task to convert the time-varying constraints into the operational limits.

4.1.2 *Discretization of Current Controllers*

The first step in the implementation process is to discretize the continuous-time controllers. We will first describe the general discretization process, then list the implementation equations for the current controllers.

For a proportional controller, assuming that the control system is sampled using a sampling interval Δt, at time t_i, the control signal $u(t_i)$ is computed based on the feedback error $e(t_i)$ and the feedforward nonlinear compensation $f(t_i)$:

$$u(t_i) = K_c e(t_i) + f(t_i). \tag{4.11}$$

Thus, the discretization is straightforward because there are no other dynamics involved in the proportional controller except the feedback gain control. In the proportional control implementation, we will take the actual reference signal $r(t)$ and the actual physical feedback signal $y(t_i)$ to form the feedback error $e(t_i) = r(t_i) - y(t_i)$. Also $f(t_i)$ consists of the actual physical measurements.

For a PI controller, because there is the integrator dynamic, we need to pay attention to how this controller is discretized. Here, a PI controller in velocity form is suggested to be used in the implementation of current controllers for the reason that this form takes the actual system input and output signals in the computation and its flexibility in implementing the saturation and anti-windup mechanisms later on.

4.1.2.1 PI Controller in Position Form

In the general case, we assume that the feedback error is $e(t)$ and the feedforward function is $f(t)$, then with K_c the proportional gain and τ_I the integral time constant, the control signal $u(t)$ from a PI controller is defined as

$$u(t) = K_c e(t) + \frac{K_c}{\tau_I} \int_0^t e(\tau) d\tau + f(t). \tag{4.12}$$

When at the sampling instant t_i, the integral term is approximated by the finite sum,

$$\int_0^{t_i} e(\tau) d\tau \approx \sum_{k=0}^{M-1} e(t_k)\Delta t,$$

where $t_i = t_0 + (M - 1)\Delta t$ with M being the number of samples, the control signal is discretized as

$$u(t_i) = K_c e(t_i) + \frac{K_c}{\tau_I} \sum_{k=0}^{M-1} e(t_k)\Delta t + f(t_i). \tag{4.13}$$

This is the computational equation for implementation of a PI controller using the so called position form. The name of position form refers to the computation that is based on the control signal itself. Because the control signal $u(t_i)$ computed using (4.13) is meant for the deviation variable, not the actual control signal used for the input to the physical system, there is a constant term that corresponds to the steady-state of the control signal. This constant term needs to be considered in the implementation. Supposing that this constant term is u_{ss}, which is assumed to be the steady-state value of the control signal, the actual control signal to be used as the input to the physical system is

$$u_{act}(t_i) = u(t_i) + u_{ss}. \tag{4.14}$$

To understand the idea of the steady-state term, we consider the situation where an *AC* machine is stopped in a stand-by operation with output velocity and the reference velocity being zero, however, the currents and voltages are not zero. Here, the nonzero values of the voltages are the steady-state terms required in the implementation of the current controllers. One of the main drawbacks when using the position form is the requirement of these terms. The other drawback is that the position form will lead to an oscillatory closed-loop response when the saturation limits are reached, so called integrator windup. Although anti-windup mechanisms can be incorporated in the implementation, it is not a straightforward approach unlike the velocity form as introduced below.

4.1.2.2 PI Controller in Velocity Form

The velocity form implementation of PI controller is different from the position form implementation. With the velocity form implementation, we consider the derivative of the control signal. Differentiating (4.12), the derivative of the control signal is expressed as

$$\frac{du(t)}{dt} = K_c \frac{de(t)}{dt} + \frac{K_c}{\tau_I} e(t) + \frac{df(t)}{dt}. \tag{4.15}$$

Taking the approximation of the derivatives by the first order difference, at sample time t_i, the derivatives in (4.15) are written as

$$\frac{du(t)}{dt} \approx \frac{u(t_i) - u(t_i - \Delta t)}{\Delta t} = \frac{u(t_i) - u(t_{i-1})}{\Delta t} \tag{4.16}$$

$$\frac{de(t)}{dt} \approx \frac{e(t_i) - e(t_i - \Delta t)}{\Delta t} = \frac{e(t_i) - e(t_{i-1})}{\Delta t} \tag{4.17}$$

$$\frac{df(t)}{dt} \approx \frac{f(t_i) - f(t_i - \Delta t)}{\Delta t} = \frac{f(t_i) - f(t_{i-1})}{\Delta t}, \tag{4.18}$$

where for notational simplicity,

$$t_i - \Delta t = t_{i-1}.$$

Substituting these approximations into (4.15) with the simplified notations leads to the control signal $u(t_i)$ at the sampling time t_i:

$$u(t_i) = u(t_{i-1}) + K_c(e(t_i) - e(t_{i-1})) + \frac{K_c}{\tau_I} e(t_i)\Delta t + f(t_i) - f(t_{i-1}). \tag{4.19}$$

To embed the steady-state value in the implementation, we define the actual control signal

$$u_{act}(t_i) = u(t_i) + u_{ss}$$

$$u_{act}(t_{i-1}) = u(t_{i-1}) + u_{ss}.$$

Thus, by adding the same steady-state value u_{ss} to both sides of (4.19), we obtain the computational equation for the actual control signal used for the input to the system as

$$u_{act}(t_i) = u_{act}(t_{i-1}) + K_c(e(t_i) - e(t_{i-1})) + \frac{K_c}{\tau_I} e(t_i)\Delta t + f(t_i) - f(t_{i-1}). \tag{4.20}$$

The computation of the actual control signal has embedded the steady-state value of the control signal in the implementation. In the implementation, the algorithm requires the initialization of the actual control signal and the other signals from the system. Assuming that the closed-loop control occurs at time $t_i = 0$, the initial conditions for the control signal $u_{act}(-\Delta t)$, $e(-\Delta t)$ and $f(-\Delta t)$ are chosen to be equal to the actual measurements of the control signal and other signals before the closed-loop control, and the steady-state values are correctly identified. Consequently, all the variables in (4.20) use the actual physical measurements.

4.1.3 Anti-windup Mechanisms

The two key points in the implementation of an anti-windup mechanism in controllers are to have a stable controller structure and to make sure that the actual control signal equals the control signal computed (see Goodwin *et al.*, 2001). In the current control systems, if a proportional controller is used in conjunction with the input-output linearization, then the controller structure is stable (namely, with no pole on the imaginary axis or the right-hand side of the complex plane), and the actual control signal equals the control signal computed because of the nonlinear compensations. Thus, for the current-loop control systems, if a proportional controller is used, we will only use a nonlinear saturation to limit the amplitude of control signal, which is straightforward in the implementation. However, if a PI controller is used for the current-loop control system, then an anti-windup mechanism is needed for the implementation, because the PI controller has an integrator (namely, with a pole at the origin of the complex plane).

It is much more straightforward to implement anti-windup mechanisms in velocity form of PID controllers, mainly because the velocity form of PID controller is derived to use the actual control signals (see (4.20)). In addition, to achieve a stable controller structure when the saturation is reached, the derivative of the control signal is taken to be zero in the velocity form implementation, that is $\frac{du(t)}{dt} = 0$, leading to $u_{act}(t_i) = u_{act}(t_{i-1})$. This means that the control signal becomes a constant when the saturation is reached and it stops growing.

The saturation with anti-windup mechanism is summarized as follows. We assume that the actual control variable is limited by U^{min} and U^{max}. Namely, the actual control signal must satisfy the constraints:

$$U^{min} \leq u_{act}(t) \leq U^{max}.$$

The actual control signal $u_{act}(t_i)$ at sample time t_i is calculated using the equation,

$$u_{act}(t_i) = u_{act}(t_{i-1}) + K_c(e(t_i) - e(t_{i-1})) + \frac{K_c}{\tau_I}e(t_i)\Delta t + f(t_i) - f(t_{i-1}). \tag{4.21}$$

In order to stop the integration, when the actual control signal reaches the limit, we impose the limits on the actual control signal with the computation that if $u_{act}(t_i) < U^{min}$, then $u_{act}(t_i) = U^{min}$; if $u_{act}(t_i) > U^{max}$, then $u_{act}(t_i) = U^{max}$. When the sample time t_i moves one step forward to t_{i+1}, the signal $u_{act}(t_i)$ carries the information of saturation at the previous sample time and the control signal computation is automatically informed of the saturation, thus both requirements in an anti-windup mechanism are satisfied.

4.2 Implementation of Current Controllers for PMSM

When implementing current control for PMSM, we have two options for the control system configuration. The first is to use a proportional controller for the q-axis current control system and the second is to use a PI controller for the same loop. PI controller is required for the d-axis current control because it is the integral term that will eliminate steady-state error in the closed-loop system and there is no outer-loop system for the d-axis current control. There are also two methods to calculate the saturation values for the d-axis and q-axis voltages. One is to use box constraints (see (4.4) and (4.5)) and the other is to use the nonlinear constraint calculated using (4.7) and (4.10). To simplify the expressions, we let

$$f_d(t) = -\omega_e(t)L_q i_q(t) \tag{4.22}$$

$$f_q(t) = \omega_e(t)L_d i_d(t) + \omega_e(t)\phi_{mg}. \tag{4.23}$$

The computational steps are summarized as follows.

Algorithm 2 *PI current control!anti-windup mechanism*

1. *Initialize the control system parameters before the closed-loop control. The initialization will lead to the parameters in sampling time t_{i-1} where the sampling index $i = 0$.*
   - *Initialization of the control signals. At the sampling instant $i = 0$, take the measurements of d-axis voltage v_d and q-axis voltage v_q and assign these values to $v_d(t_{i-1})$ and $v_q(t_{i-1})$.*
   - *Initialization of the feedforward term. Take the measurements of ω_e, i_q and i_d, and calculate $f_d(t_{i-1})$ and $f_q(t_{i-1})$ using (4.22) and (4.23) with the measured ω_e and i_q values.*

- *Initialization of the feedback errors. Use the reference signal from outer-loop control system,* i_q^* *to form the feedback error* $e_q(t_{i-1}) = i_q^*(t_{i-1}) - i_q(t_{i-1})$ *and use the set-point signal for the d-axis current control* i_d^* *to form the feedback error* $e_d(t_{i-1}) = i_d^*(t_{i-1}) - i_d(t_{i-1})$.

After the initialization, the clock ticks to the next sample period (say $i = 1$ *at time* $t = t_i$*) and the closed-loop control begins.*

2. *At sampling time* t_i*, take the measurements of* $\omega_e(t_i)$*,* $i_q(t_i)$ *and* $i_d(t_i)$*, and calculate* $f_d(t_i)$ *and* $f_q(t_i)$ *using the following equations:*

$$f_d(t_i) = -\omega_e(t_i)L_q i_q(t_i) \tag{4.24}$$

$$f_q(t_i) = \omega_e(t_i)L_d i_d(t_i) + \omega_e(t_i)\phi_{mg}. \tag{4.25}$$

3. *Use the reference signal from outer-loop control system,* $i_q^*(t_i)$ *to form the feedback error:*

$$e_q(t_i) = i_q^*(t_i) - i_q(t_i)$$

and use the set-point signal for the d-axis current control $i_d^*(t_i)$ *to form the feedback error:*

$$e_d(t_i) = i_d^*(t_i) - i_d(t_i).$$

4. *Calculate the actual control signal using*

$$v_d(t_i) = v_d(t_{i-1}) + K_c^d(e_d(t_i) - e_d(t_{i-1})) + \frac{K_c^q}{\tau_I} e_d(t_i)\Delta t + f_d(t_i) - f_d(t_{i-1})$$

$$v_q(t_i) = v_q(t_{i-1}) + K_c^q(e_q(t_i) - e_q(t_{i-1})) + \frac{K_c^q}{\tau_I} e_q(t_i)\Delta t + f_q(t_i) - f_q(t_{i-1}).$$

5. *Check if the control signals are within the limits*

$$\sqrt{(v_d(t_i))^2 + (v_q(t_i))^2} \leq \frac{V_{dc}}{\sqrt{3}}.$$

6. *If this is satisfied, then* $v_d(t_i)_{act} = v_d(t_i)$ *and* $v_q(t_i)_{act} = v_q(t_i)$ *are the actual voltage input signals at the sample time* t_i*, which will be implemented. However, if this constraint is violated, then scale the control signals to satisfy the constraint by computing:*

$$v_d^0 = \frac{v_d(t_i)}{\sqrt{(v_d(t_i))^2 + (v_q(t_i))^2}} \times \frac{V_{dc}}{\sqrt{3}}$$

$$v_q^0 = \frac{v_q(t_i)}{\sqrt{(v_d(t_i))^2 + (v_q(t_i))^2}} \times \frac{V_{dc}}{\sqrt{3}}$$

and updating $v_d(t_i)_{act} = v_d^0$*;* $v_q(t_i)_{act} = v_q^0$.

7. *Implement the voltage control signals* $v_d(t_i)_{act}$ *and* $v_q(t_i)_{act}$ *to the system.*
8. *Update the control signals with the implemented values,*

$$v_d(t_i) = v_d(t_i)_{act}$$

$$v_q(t_i) = v_q(t_i)_{act}.$$

9. *When the next sample period arrives, repeat the computation from Step 2.*

If we use the box constraints for v_d and v_q, we will first assign parameter $0 \leq \epsilon \leq 1$, then set maximum value of

$$v_q^{max} = \epsilon \frac{V_{dc}}{\sqrt{3}} \tag{4.26}$$

$$v_d^{max} = \sqrt{1-\epsilon^2}\frac{V_{dc}}{\sqrt{3}}. \tag{4.27}$$

At Step 5, we will check if

$$-v_d^{max} \leq v_d(t_i) \leq v_d^{max}$$

$$-v_q^{max} \leq v_q(t_i) \leq v_q^{max}.$$

If all the four constraints are satisfied, then $v_d(t_i)$ and $v_q(t_i)$ are the actual voltage input signals to the system. Otherwise, if $v_d(t_i) > v_d^{max}$, then $v_d(t_i) = v_d^{max}$; if $v_d(t_i) < -v_d^{max}$ then $v_d(t_i) = -v_d^{max}$; if $v_q(t_i) > v_q^{max}$, then $v_q(t_i) = v_q^{max}$; if $v_q(t_i) < -v_d^{max}$ then $v_q(t_i) = -v_d^{max}$.

4.3 Implementation of Current Controllers for Induction Motors

The implementation of PI current controllers for induction motors uses a similar procedure as the one used in the controller implementation for PMSM. The key issues in the implementation of current control for the induction motor are the estimation of some unmeasured variables. For induction motor control, based on the current controller design stated in Section 3.6, the current-loop nonlinear compensation functions are

$$f_d(t) = -\frac{k_r}{\tau_r}\psi_{rd}(t) - \tau'_\sigma r_\sigma \omega_s(t) i_{sq}(t) \tag{4.28}$$

$$f_q(t) = r_\sigma \tau'_\sigma \omega_s(t) i_{sd}(t) + k_r \omega_e(t) \psi_{rd}(t), \tag{4.29}$$

where $f_d(t)$ is the nonlinear feedforward function for the d-axis current control and $f_q(t)$ for the q-axis current control. Note that these compensators require the rotor flux $\psi_{rd}(t)$ and the synchronous speed $\omega_s(t)$.These signals are not measured by sensors, thus, estimation schemes are needed in order to access the relevant information.

4.3.1 *Estimation of ω_s and θ_s*

The estimation of $\omega_s(t)$ is achieved using the slip estimation:

$$\omega_s(t) = Z_p \omega_m(t) + \frac{i_{sq}(t)}{\tau_r i_{sd}(t)}. \tag{4.30}$$

Because of the large amount of noise in the d-axis and q-axis currents $i_{sd}(t)$ and $i_{sq}(t)$, direct application of (4.30) will not lead to an accurate estimate of $\omega_s(t)$. Instead, the reference signals $i_{sd}^*(t)$ and $i_{sq}^*(t)$ are used to replace the actual signals. If a PI controller is used to control the q-axis current, then the estimate $\hat{\omega}_s(t)$ at the sampling time t_i is calculated as

$$\hat{\omega}_s(t_i) = Z_p \omega_m(t_i) + \frac{i_{sq}^*(t_i)}{\tau_r i_{sd}^*(t_i)}. \tag{4.31}$$

Because integrators are used to control the currents, at the steady-state, $i_{sd}(t_i) = i^*_{sd}(t_i)$ and $i_{sq}(t_i) = i^*_{sq}(t_i)$. Therefore, (4.31) provides a good estimate of $\omega_s(t)$ after the transient period.

$\theta_s(t)$ is required in the computation of Park-Clarke transforms. The estimate of $\theta_s(t)$ is performed using

$$\hat{\theta}_s(t) = Z_p\theta_r(t) + \frac{1}{\tau_r}\int_0^t \frac{i^*_{sq}(\tau)}{i^*_{sd}(\tau)}d\tau. \tag{4.32}$$

Discretization of (4.32) gives the numerical solution of $\hat{\theta}_s(t)$ that can be implemented in a control system, where

$$\hat{\theta}_s(t_i) = Z_p\theta_r(t_i) + \frac{1}{\tau_r}\sum_{k=0}^{M-1} \frac{i^*_{sq}(t_k)}{i^*_{sd}(t_k)}\Delta t, \tag{4.33}$$

where $t_i = t_0 + (M-1) * \Delta t$ and M is the number of samples from time t_0 to the time t_i. The potential problem of directly using the integral equation is that as time t_i increases, the finite sum approximation in (4.33) could become excessively, hence causing numerical overflow.

Instead of using the integral expression, the difference approximation of

$$\frac{d\hat{\theta}_s(t)}{dt} = \hat{\omega}_s(t)$$

leads to

$$\hat{\theta}_s(t_i) = \hat{\theta}_s(t_{i-1}) + \hat{\omega}_s(t_i)\Delta t. \tag{4.34}$$

In addition, to avoid numerical overflow occurring, θ_s is converted to a value between 0 and 2π. The advantage of using the difference approximation is that the calculation is in a recursive manner; however, it requires the initial guess of $\hat{\theta}_s$ at the first sample. This could be taken approximately at the first sample, $\hat{\theta}_s = \theta$, but this approximation may have an error. Another approach is to combine the estimation (4.34) with the one using (4.33). The idea is that at the initial period, (4.33) is used to correctly identify the $\hat{\theta}_s$, which does not have the problem of numerical overflow, and after this initial period, the difference approximation (4.34) is used with the correct initialization of θ_s, which then will avoid the numerical problem.

When the proportional controller is used to control the q-axis current, there is a steady-state error between the $i_{sq}(t)$ and the $i^*_{sq}(t)$. In the ideal case when there is no mismatch of the resistance τ_σ between the system and model, the steady-state gain for the q-axis current control is α. Thus, the estimation of $\omega_s(t)$ will take into account the steady-state discrepancy, which leads to :

$$\hat{\omega}_s(t_i) = Z_p\omega_m(t_i) + \frac{i^*_{sq}(t_i) \times \alpha}{\tau_r i^*_{sd}(t_i)}. \tag{4.35}$$

Similarly, the estimation of $\theta_s(t)$ is performed using

$$\theta_s(t) \approx Z_p\theta_r(t) + \frac{1}{\tau_r}\int_0^t \frac{i^*_{sq}(\tau) \times \alpha}{i^*_{sd}(\tau)}d\tau.$$

If a PI controller is used to control the q-axis current, then the parameter α is unity.

4.3.2 *Estimation of ψ_{rd}*

Because the rotor flux $\psi_{rd}(t)$ is not measured in the system, $\psi_{rd}(t)$ will be estimated in the implementation. The rotor flux is related to the $i_{sd}(t)$ current via the differential equation:

$$\psi_{rd}(t) + \tau_r\frac{d\psi_{rd}(t)}{dt} = L_h i_{sd}(t). \tag{4.36}$$

Using the similar approach to the estimation of ω_s, because of the high noise level in the feedback signal $i_{sd}(t)$, the set-point signal to ψ_{rd}, which is $\psi^*_{rd}(t) = L_h i^*_{sd}(t)$, is used in the feedforward function.

4.3.3 The Implementation Steps

By substituting the estimated variables into (4.28) and (4.29), the estimated feedforward compensation functions are

$$\hat{f}_d(t) = -\frac{k_r}{\tau_r}\psi^*_{rd}(t) - \tau'_\sigma r_\sigma \hat{\omega}_s(t) i_{sq}(t) \tag{4.37}$$

$$\hat{f}_q(t) = r_\sigma \tau'_\sigma \hat{\omega}_s(t) i_{sd}(t) + k_r \omega_e(t) \psi^*_{rd}(t). \tag{4.38}$$

The implementation steps for the d-axis and q-axis current control of an induction motor are identical to those outlined in Sections 4.1.1 to 4.2, where the topics of constraints handling, discretization of the PI current controllers and velocity implementation of the current controllers are discussed.

4.4 Current Controller Implementation for Power Converter

The control signals in the power converter are the normalized d-axis voltage $S_d(t)$ and normalized q-axis voltage $S_q(t)$, and they are related to the voltage of the converter $v_{dc}(t)$ via

$$v_d = S_d \times (v_{dc}/2) \tag{4.39}$$

$$v_q = S_q \times (v_{dc}/2). \tag{4.40}$$

With S_d and S_q as the control variables, the relationships become bilinear. Although the implementation procedure of the current controllers is basically the same as the current controllers for the PMSM and the induction motor, there are some differences caused by the bilinearities of the control variables given in (4.39) and (4.40).

4.4.1 Constraints on the Control Variables

The constraint on the voltages due to the physical constraints from the inverter is

$$\sqrt{v_d^2 + v_q^2} \leq \frac{2v_{dc}}{\sqrt{3}}. \tag{4.41}$$

This constraint is the same as the inverter constraint in the AC machines. By substituting (4.39) and (4.40) into (4.41), we obtain the constraint on the control variables S_d and S_q:

$$\sqrt{S_d^2 + S_q^2} \leq \frac{2}{\sqrt{3}}. \tag{4.42}$$

This is a circular constraint with radius $\frac{2}{\sqrt{3}}$, and it is handled in the same manner as discussed in Sections 4.1.1 to 4.2.

Because of the bilinear relationship between the control variables and the output voltage that is to be controlled, the discretization of the PI controllers needs a small modification.

Letting

$$f_d(t) = 2L_s\omega_g i_q(t) + 2E_d(t) \tag{4.43}$$

$$f_q(t) = -2\omega_g L_s i_d(t), \tag{4.44}$$

the voltage control variables are $v_d(t) = S_d v_{dc}(t)$ and $v_q(t) = S_q v_{dc}(t)$. Then from the design of the PI current controllers, the control variables are expressed in terms of the feedback errors and the feedforward variables $f_d(t)$ and $f_q(t)$ where

$$v_d(t) = K_c^d(i_d^*(t) - i_d(t)) + \frac{K_c^d}{\tau_I^d}\int_0^t (i_d^*(\tau) - i_d(\tau))d\tau + f_d(t) \tag{4.45}$$

$$v_q(t) = K_c^q(i_q^*(t) - i_q(t)) + \frac{K_c^q}{\tau_I^q}\int_0^t (i_q^*(\tau) - i_q(\tau))d\tau + f_q(t). \tag{4.46}$$

The velocity forms of the PI controllers will be derived in terms of the auxiliary variables v_d and v_q using (4.45) and (4.46). They have the following forms:

$$v_d(t_i) = v_d(t_{i-1}) + K_c^d(e_d(t_i) - e_d(t_{i-1})) + \frac{K_c^d}{\tau_I^d} e_d(t_i)\Delta t + f_d(t_i) - f_d(t_{i-1})$$

$$v_q(t_i) = v_q(t_{i-1}) + K_c^q(e_q(t_i) - e_q(t_{i-1})) + \frac{K_c^q}{\tau_I^q} e_q(t_i)\Delta t + f_q(t_i) - f_q(t_{i-1}),$$

where $e_d(t_i) = i_d^*(t_i) - i_d(t_i)$ and $e_q(t_i) = i_q^*(t_i) - i_q(t_i)$. Upon obtaining the values of v_d and v_q, the control variables are obtained using

$$S_d(t_i) = \frac{v_d(t_i)}{v_{dc}(t_i)}$$

$$S_q(t_i) = \frac{v_q(t_i)}{v_{dc}(t_i)}.$$

Note that the division operations are used in reconstructing the control variables. In order to ensure that division is valid, the voltage $v_{dc}(t_i)$ must satisfy $v_{dc}(t_i) \neq 0$. Thus, the closed-loop current control systems are switched on after the *DC* bus voltage is charged to a certain value. This means that the steady-state values of the control signals and voltage signal must be considered in the calculation of the actual control signals to be implemented. However, by using the velocity form of the PI controller, the initial conditions of the $S_d(t_{i-1})$, $S_q(t_{i-1})$, $v_{dc}(t_{i-1})$ can be chosen to be equal to their measured values before the closed-loop control begins. Thus, their steady-state values are not required explicitly in the computation of the control signals. This shows one of the advantages in using the velocity form of PI controller implementation because it is much easier to handle steady-state values in the initialization stage of the control system.

4.5 Implementation of Outer-loop PI Control System

4.5.1 Constraints in the Outer-loop

The manipulated variable in the outer-loop system is the reference current signal to the inner-loop system. Thus, the constraints are imposed on the manipulated variable i_q^* for the *AC* machines or i_d^* for the power converters. The constraints provide over current protection in the event of overload or a large transient response in the event of rapid start-up and shut-down of the machines.

4.5.2 Over Current Protection for AC Machines

As we know, it is the actual current i_q on which we wish to impose the constraints for the over current protection. However, because it is the feedback variable already used in the inner-loop system, it is not available for manipulation in the outer-loop system. Instead, the outer-loop manipulated variable i_q^* is utilized to handle the constraints. Therefore, the relationship between the feedback current i_q and the reference i_q^* is important for the implementation of the over current protection.

4.5.2.1 PI Control of *q*-Axis Current

When the current control-loop is controlled by a PI controller, the constraints are simply imposed as

$$-I_q^{max} \le i_q^*(t) \le I_q^{max}, \tag{4.47}$$

where I_q^{max} is the maximum current allowed. This is because when an integrator is used in the inner-loop controller, in the steady-state

$$\lim_{t\to\infty} i_q^*(t) - i_q(t) = 0,$$

the constraints given by (4.47) are the same as the actual constraints on i_q. However, this also means that the simplified relationship only imposes accurate constraints on the steady-state over current protection, and it does not provide the exact constraints in the transient period. If the constraints during the transient period are imposed, then the current feedback error $e_q(t_i)$ should be considered similar to the proportional control case given below.

4.5.2.2 Proportional Control of *q*-Axis Current

If the inner-loop system is controlled by a proportional controller, then the steady-state current feedback error $e_q(t)$ is not equal to zero:

$$\lim_{t\to\infty} e_q(t) = \lim_{t\to\infty} i_q^*(t) - i_q(t) \neq 0. \tag{4.48}$$

Here, the inner-loop feedback error $e_q(t)$ is governed by the difference of the two currents:

$$e_q(t) = i_q^*(t) - i_q(t). \tag{4.49}$$

Thus, the feedback current is related to its reference via

$$i_q(t) = i_q^*(t) - e_q(t). \tag{4.50}$$

Assuming that the maximum current allowed is I_q^{max}, then the constraints using (4.50) are

$$-I_q^{max} \le i_q^*(t) - e_q(t) \le I_q^{max}, \tag{4.51}$$

which leads to the constraints imposed on the reference current $i_q^*(t)$ as

$$-I_q^{max} + e_{iq}(t) \le i_q^*(t) \le I_q^{max} + e_q(t). \tag{4.52}$$

In the case of proportional control of q-axis current, we will utilize the inner-loop feedback error $e_{iq}(t)$ in the implementation of the outer-loop over current protection based on (4.52).

4.5.3 Implementation of Outer-loop PI Control of Velocity

As an example, we examine the implementation of outer-loop PI control of velocity for PMSM. The control signal $i_q^*(t)$ is calculated using the PI controller as

$$i_q^*(t) = K_c(\omega_e^*(t) - \omega_e(t)) + \frac{K_c}{\tau_I}\int_0^t (\omega_e^*(\tau) - \omega_e(\tau))d\tau, \tag{4.53}$$

where $\omega_e^*(t)$ is the set-point signal for the electrical velocity. Equation (4.53) is the position form of the PI controller implementation. For convenience of the implementation, the velocity form is preferred in the applications as it is easier to deal with the steady-state, also integrator windup as well as ensuring a smooth transition when the task of switching controller is performed. Following the same derivation of the velocity form of the PI controller as discussed in Section 4.1.2, the outer-loop PI control signal calculation is obtained as

$$\begin{aligned} i_q^*(t_i) &= i_q^*(t_{i-1}) + K_c(\omega_e^*(t_i) - \omega_e^*(t_{i-1})) - K_c(\omega_e(t_i) - \omega_e(t_{i-1})) \\ &+ \frac{K_c \Delta t}{\tau_I}(\omega_e^*(t_i) - \omega_e(t_i)). \end{aligned} \tag{4.54}$$

We often prefer to implement the proportional control on the feedback signal only, which has an effect of reducing overshoot in the closed-loop set-point response (see Section 3.1). This avoids the effect known as "proportional kick". The integral control must be implemented with the difference between the desired set-point signal and the actual measurement to ensure zero steady-state error when tracking a constant set-point signal. Thus, an alternative implementation of the PI controller for the outer-loop is

$$i_q^*(t_i) = i_q^*(t_{i-1}) - K_c(\omega_e(t_i) - \omega_e(t_{i-1})) + \frac{K_c \Delta t}{\tau_I}(\omega_e^*(t_i) - \omega_e(t_i)). \tag{4.55}$$

4.5.4 Over Current Protection for Power Converters

Similar principles apply to power converters. In the control of power converters, the constraints are imposed on the i_d current. If a proportional controller is used to control the d-axis current, assuming the the maximum current allowed is

$$-I_d^{max} \leq i_d(t) \leq I_d^{max},$$

then because

$$i_d(t) = i_d^*(t) - e_{id}(t),$$

the constraints on the reference current $i_d^*(t)$ are expressed as

$$-I_d^{max} + e_{id}(t) \leq i_d^*(t) \leq I_d^{max} + e_{id}(t). \tag{4.56}$$

4.6 MATLAB Tutorial on Implementation of PI Controller

The embedded PI controller function will be used in the simulation and implementation through the entire book. It is written in a general form and small modification may be needed for individual applications. One may wish to test this embedded function using the simulation examples given in Chapter 3.

Tutorial 1 *This tutorial is to illustrate how to implement the PI control algorithm in real-time. The core of this activity is to produce a MATLAB embedded function that can be used in a Simulink simulation as well as in xPC Target implementation. This embedded function is based on the PI controller velocity form discussed in the previous sections. The entire embedded MATLAB completes one cycle of computation for the control signal. For every sampling period, it will repeat the same computation procedure. The embedded function is written in a set of general variables and it is suitable for all PI control applications such as current, velocity and voltage controls.*

Step by Step

1. *Create a new Simulink file called PIV.mdl*
2. *In Simulink's directory of User-Defined Functions, find the icon of embedded MATLAB function and copy it to the PIV model.*
3. *Click on the icon of the embedded function, and define the input and output variables to the PIV model so that the embedded function has the following form:*

```
function uCur= PIV(yCur,rCur,Kc,tauI,Ts,uLmt)
```

   where uCur is the calculated control signal at the sampling time t_i, the first two elements (yCur and rCur) among the input variables are the measurement of the output and the reference signals at sampling time t_i, Kc and tauI are the proportional gain and integral time constant, Ts is the sampling interval and uLmt is the upper limit imposed on the control signal uCur. Here, we assume symmetric constraints on the control signal for simplicity of the expressions.
4. *At the top of the embedded function, find "Model Explorer" among the "Tools." When opening the Model Explorer, select "discrete" for the "update method" and input "Ts" into the "sample time"; select "Support variable-size arrays"; select "Saturate on integer overflow"; select "Fixed point". Click "Apply" to save the changes.*
5. *We need to edit the input and output data ports in order to let the embedded function know which input ports are the real-time variables and which are the parameters. This editing task is performed using Model Explorer.*
   - *click on "yCur", on Scope, select "input", assign port "1" and size "−1", complexity "Inherited", type "Inherit: Same as Simulink". Repeat the same editing procedure for the reference signal "rCur".*
   - *The rest of 4 inputs to the embedded function are the parameters required in the computation. Click on "Kc", on Scope, select "Parameter" and click "Tunable" and click "Apply" to save the changes. Repeat the same editing procedure for the rest of the parameters, "tauI", "Ts", "uLmt".*
   - *To edit the output port from the embedded function, click on "uCur", on Scope, select "Output", Port "1", Size "−1", Sampling Model "Sample based", Type "Inherit: Same as Simulink", and click on "Apply" to save the changes.*
6. *In the following, the program will declare those variables that are stored in the embedded function during each iteration for their dimensions and initial values. "uPast" is the past control signal ($u(t_{i-1})$), "yPast" is the past output signal ($y(t_i - Deltat)$). Because the PI controller implemented using velocity form takes the actual measured variables, the past input and output variables should be initialized to the actual measured physical variables before the closed-loop control. These initial values can enter the computation as the extra parameters in the embedded function. Here, for simplicity of the programming, they are assigned to zero. Enter the following program into the file:*

```
persistent uPast
if isempty(uPast)
    uPast=0;
end
```

```
persistent yPast
if isempty(yPast)
    yPast=0;
end
persistent rPast
if isempty(rPast)
    rPast=0;
end
```

7. *Calculate the actual control signals. Enter the following program into the file:*

```
uCur=uPast+Kc*((rCur-yCur)-(rPast-yPast))...
+(Kc*Ts/tauI)*(rCur-yCur);
```

Alternatively, if one wishes to reduce overshoot in the set-point response by putting the proportional control on output only, the following computation is used instead,

```
uCur=uPast+Kc*(-yCur+yPast)+(Kc*Ts/tauI)*(rCur-yCur);
```

8. *Impose constraints on the control signal. Enter the following program into the file:*

```
if (uCur>uLmt)
    uCur=uLmt;
end
if (uCur<-uLmt)
    uCur= -uLmt;
end
```

9. *Update the past control and output signals. Updating uPast is part of the anti-windup implementation for the constraints so that when the control signal reaches its limit, the integral action is stopped. Enter the following program into the file:*

```
uPast=uCur;
yPast=yCur;
rPast=rCur;
```

10. *Test this program and compare the results with those presented in Section 3.4.5. The test should be performed using the nonlinear model of PMSM to avoid any other errors caused by the physical system simulation. You can test the two different ways to implement the proportional control as indicated in Step 7.*

4.7 Summary

In this chapter, we have discussed the implementation of PID cascade control systems, which have imposed limits on the amplitudes of the voltage control signals and have anti-windup mechanisms. In the implementation procedure, the continuous-time PID control algorithms are discretized and velocity form expressions for the PID controllers are obtained. The velocity form of the PID controller calculates the control signal using its incremental value that has included the steady-state of the control signals. This characteristic is particularly useful when the control system begins its closed-loop operation with nonzero initial conditions. Furthermore, the anti-windup mechanism is implemented with the velocity form of PI controller based on an exceptionally simple approach.

4.8 Further Reading

PID algorithms with anti-windup strategies were discussed and compared in Goodwin *et al.* (2001), Bohn and Atherton (1995), Peng *et al.* (1996), Shin (1998). DSP solution for high position resolution with sin/cos encoder was discussed in Staebler (2000). Clarke transformation was written in Duesterhoeft *et al.* (1951). Slip estimation for induction motor was discussed in Rubin *et al.* (1992) and Abbondanti and Brennen (1975).

References

Abbondanti A and Brennen M 1975 Variable speed induction motor drives use electronic slip calculator based on motor voltage and currents. *IEEE Transactions on Industry Applications* **IA-11**(5), 483–8.

Bohn C and Atherton D 1995 An analysis package comparing PID anti-windup strategies. *Control Systems Magazine, IEEE* **15**(2), 34–40.

Duesterhoeft W, Schulz MW and Clarke E 1951 Determination of instantaneous currents and voltages by means of alpha, beta, and zero components. *Transactions of the American Institute of Electrical Engineers* **70**(2), 1248–1255.

Goodwin GC, Graebe SF and Salgado ME 2001 *Control System Design*. Prentice Hall, New Jersey.

Peng Y, Vrancic D and Hanus R 1996 Anti-windup, bumpless, and conditioned transfer techniques for PID controllers. *IEEE Control Systems* **16**(4), 48–57.

Rubin N, Harley R and Diana G 1992 Evaluation of various slip estimation techniques for an induction machine operating under field-oriented control conditions. *IEEE Transactions on Industry Applications* **28**(6), 1367–1375.

Shin H 1998 New antiwindup PI controller for variable-speed motor drives. *IEEE Transactions on Industrial Electronics* **45**(3), 445–450.

Staebler M 2000 TMS320F240 dsp-solution for high-resolution position with sin/cos-encoders. Technical report, Texas Instruments.

5

Tuning PID Control Systems with Experimental Validations

After the cascade nonlinear PID control system is designed and implemented, tuning of the controllers is essential for the operation of electric drives and power converters. This task is often performed against actual systems, because there are modeling errors, noise, parameter variations, and other undesired factors that arise from the real system. The control systems for the electric drives and power converters could become unstable when the parameters in the PID control system are not chosen suitably. This chapter shows how to tune the cascade nonlinear controllers by using laboratory test-beds as examples.

This chapter begins with an introduction to sensitivity functions in feedback control systems (see Section 5.1) explaining how the closed-loop control system performances against set-point following, disturbance rejection and noise attenuation are measured by the key sensitivity functions in frequency domain. Taking PMSM as an example, the next four sections analyze current control systems, the design of outer-loop controller and the effect of time-delay in the current control loop. In Sections 5.2–5.3, current controllers are analyzed against current sensor errors and harmonics caused by the voltage source inverter used in the implementation of the control system. The proportional current controller is addressed in an independent section because this structure is to be used in Chapters 6–7 as part of a predictive control system solution. In Section 5.4, performance robustness in an outer-loop control system is considered where a weighting function is introduced to quantify the difference between the desired closed-loop performance and the actual closed-loop performance. Sections 5.6–5.7 analyze PI cascade control systems for induction motors and power converters with experimental validation results.

5.1 Sensitivity Functions in Feedback Control Systems

This section will briefly give the background to the analysis of feedback control systems that will be used as the basis for tuning the closed-loop PI control systems.

5.1.1 Two-degrees of Freedom Control System Structure

Assuming that the system to be controlled has a transfer function denoted by $G(s)$ and the controller has a transfer function denoted by $C(s)$, a feedback control system of a one-degree of freedom structure is

PID and Predictive Control of Electrical Drives and Power Converters using MATLAB®/Simulink®, First Edition.
Liuping Wang, Shan Chai, Dae Yoo, Lu Gan and Ki Ng.

Companion Website: www.wiley.com/go/wang/pid

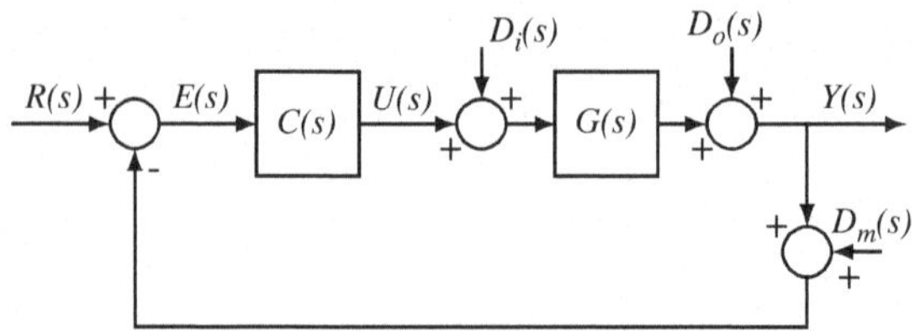

Figure 5.1 One-degree of freedom control system structure.

represented by the block diagram, shown in Figure 5.1, where $R(s)$ is the set-point signal, $Y(s)$ is the output and $U(s)$ is the control signal. There is also an output disturbance in the system, denoted as $D_o(s)$. At this point, for simplicity of the expressions, the measurement error $D_m(s)$ representing both sensor bias and sensor noise and the input disturbance $D_i(s)$ are assumed to be zero.

The Laplace transform of the error signal $E(s)$ is expressed as

$$\begin{aligned} E(s) &= R(s) - Y(s) = R(s) - G(s)U(s) - D_o(s) \\ &= R(s) - G(s)C(s)E(s) - D_o(s). \end{aligned} \tag{5.1}$$

Therefore, the error signal is expressed as

$$E(s) = \frac{1}{1+G(s)C(s)}R(s) - \frac{1}{1+G(s)C(s)}D_o(s). \tag{5.2}$$

Then, the output of the control system is

$$\begin{aligned} Y(s) &= R(s) - E(s) = (1 - \frac{1}{1+G(s)C(s)})R(s) + \frac{1}{1+G(s)C(s)}D_o(s) \\ &= \frac{G(s)C(s)}{1+G(s)C(s)}R(s) + \frac{1}{1+G(s)C(s)}D_o(s) \end{aligned} \tag{5.3}$$

and the control signal is

$$U(s) = C(s)E(s) = \frac{C(s)}{1+G(s)C(s)}R(s) - \frac{C(s)}{1+G(s)C(s)}D_o(s). \tag{5.4}$$

The transfer function between the set-point signal and the plant output is

$$\frac{Y(s)}{R(s)} = \frac{G(s)C(s)}{1+G(s)C(s)} \tag{5.5}$$

and the set-point signal and the control signal is

$$\frac{U(s)}{R(s)} = \frac{C(s)}{1+G(s)C(s)}. \tag{5.6}$$

Similarly, by assuming $R(s) = 0$, we derive the transfer functions between the output disturbance and the output, and the output disturbance and the control signal:

$$\frac{Y(s)}{D_o(s)} = \frac{1}{1+G(s)C(s)} \tag{5.7}$$

$$\frac{U(s)}{D_o(s)} = -\frac{C(s)}{1+G(s)C(s)}. \tag{5.8}$$

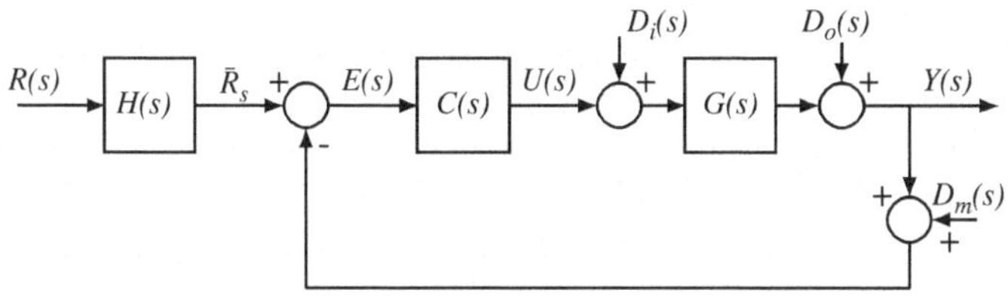

Figure 5.2 Two-degrees of freedom control system structure.

The closed-loop performance is directly related to the properties of these closed-loop transfer functions that determine the behavior of the output signal and the control signal in relation to reference signal $R(s)$ and the output disturbance $D_o(s)$. In this controller structure, once the controller $C(s)$ is selected, all four closed-loop transfer functions are fixed, only one-degree of freedom is available to influence the output response $Y(s)$ to the reference signal $R(s)$ and to the disturbance $D_o(s)$. This is called one-degree of freedom design.

A two-degrees of freedom control system is shown in Figure 5.2. In this structure, an extra component $H(s)$ is placed after the reference signal $R(s)$, which will be used in the design. How does this structure offer a two-degrees of freedom in the design? For this, with the assumption that $D_i(s) = 0$ and $D_m(s) = 0$, we calculate the output response $Y(s)$ in relation to the reference signal $R(s)$ and the output disturbance $D_o(s)$,

$$Y(s) = \frac{G(s)C(s)H(s)}{1 + G(s)C(s)}R(s) + \frac{1}{1 + G(s)C(s)}D_o(s). \tag{5.9}$$

From this, we have the two transfer functions

$$\frac{Y(s)}{R(s)} = \frac{G(s)C(s)H(s)}{1 + G(s)C(s)} \tag{5.10}$$

$$\frac{Y(s)}{D_o(s)} = \frac{1}{1 + G(s)C(s)}. \tag{5.11}$$

Transfer function $H(s)$ provides one more degree of freedom to shape the output response to the reference signal $R(s)$. This extra degree of freedom plus the original one-degree of freedom from the selection of the controller gives the two degrees of freedom in the design. If the control system is configured as a two degrees of freedom, then we can shape, independently, the output response to the reference signal and to the disturbance.

In the industrial applications of electrical drive control, the overshoot response to a step reference signal in velocity control using PI controller or position control using PID controller is not desirable. The same restriction on overshoot response in voltage control of power converter also exists. The overshoot response to a reference signal is often considered an unsafe practice, which could cause equipment damages and accident. The remedy to overcome this overshoot is to use a two-degrees of freedom control structure where an appropriate reference filter $H(s)$ is deployed.

For a PI control system, it is seen that in the one-degree of freedom control system (see Figure 5.1), the closed-loop transfer function that relates the reference signal $R(s)$ to $Y(s)$ is

$$\frac{Y(s)}{R(s)} = \frac{(c_1 s + c_0)B(s)}{A(s)s + (c_1 s + c_0)B(s)}, \tag{5.12}$$

where we have assumed that the system transfer function is $G(s) = \frac{B(s)}{A(s)}$ and the PI controller has the transfer function $C(s) = \frac{c_1 s + c_0}{s}$. From (5.12), it is seen that the zero of the closed-loop transfer function is located at $s = -\frac{c_0}{c_1} = -\frac{1}{\tau_I}$. With the assumption that the PI controller is designed to produce a stable closed-loop system, this stable zero is the main cause of the overshoot response to a step reference signal when using the PI controller. Here, we assume that the system does not have its own stable zeros to cause the overshoot response. As illustrated in Section 3.1, a simple and effective approach is to implement the proportional controller K_c on the output only, that is to calculate the control signal as

$$u(t) = \frac{K_c}{\tau_I} \int_0^t (r(\tau) - y(\tau))d\tau - K_c y(t). \tag{5.13}$$

With this implementation, it can be readily verified that the closed-loop transfer function between the reference signal $R(s)$ and $Y(s)$ is

$$\frac{Y(s)}{R(s)} = \frac{c_0 B(s)}{A(s)s + (c_1 s + c_0)B(s)}. \tag{5.14}$$

By comparing (5.14) with (5.12), we realize that these expressions are identical if the closed-loop transfer function (5.12) is multiplied with the filter $H(s)$ having the following form

$$H(s) = \frac{1}{\frac{c_1}{c_0}s + 1} = \frac{1}{\tau_I s + 1}, \tag{5.15}$$

where $\tau_I = \frac{c_1}{c_0}$. Thus, the second realization given by the closed-loop transfer function (5.14) is effectively a two-degrees of freedom control system with the reference filter $H(s)$ (see Figure 5.2). The requirement for the filter $H(s)$ to be stable is $\tau_I > 0$.

Similarly, for PID controller implementation using the one-degree of freedom control system structure (see Figure 5.1), the closed-loop transfer function between the reference signal and the output signal is

$$\frac{Y(s)}{R(s)} = \frac{(c_2 s^2 + c_1 s + c_0)B(s)}{A(s)s + (c_2 s^2 + c_1 s + c_0)B(s)} \tag{5.16}$$

and the cause of overshoot response to a step reference signal is due to the factor $c_2 s^2 + c_1 s + c_0$ in the numerator of the closed-loop transfer function. In the implementation, we apply the same procedure by calculating the control signal using the following equation,

$$u(t) = \frac{K_c}{\tau_I} \int_0^t (r(\tau) - y(\tau))d\tau - K_c y(t) - K_c \tau_D \frac{dy(t)}{dt}. \tag{5.17}$$

It can be verified that the closed-loop transfer function is

$$\frac{Y(s)}{R(s)} = \frac{c_0 B(s)}{A(s)s + (c_2 s^2 + c_1 s + c_0)B(s)}, \tag{5.18}$$

which effectively is a two-degrees of freedom control system (see Figure 5.2) with the reference filter $H(s)$ given by

$$H(s) = \frac{1}{\frac{c_2}{c_0}s^2 + \frac{c_1}{c_0}s + 1} = \frac{1}{\tau_I \tau_D s^2 + \tau_I s + 1}, \tag{5.19}$$

where the following relationships are used (see Section 3.1):

$$K_c = c_1; \quad \tau_I = \frac{c_1}{c_0}; \quad \tau_D = \frac{c_2}{c_1}.$$

The requirements for the reference filter $H(s)$ to be stable are $\tau_I > 0$ and $\tau_D > 0$.

In summary, these simple modifications on the PID controller implementations lead to, in fact, two-degrees of freedom control systems without the actual implementation of reference signal filters. As a consequence, the overshoot response to a step reference signal is overcome.

5.1.2 Sensitivity Functions

To understand the sensitivity functions and their roles in feedback control, we examine the block diagram of a closed-loop feedback control system illustrated Figure 5.2. Based on the block diagram, we calculate the feedback error of the closed-loop system firstly as

$$\begin{aligned} E(s) &= H(s)R(s) - (Y(s) + D_m(s)) \\ &= H(s)R(s) - [G(s)(U(s) + D_i(s)) + D_o(s) + D_m(s)] \\ &= H(s)R(s) - G(s)C(s)E(s) - G(s)D_i(s) - D_o(s) - D_m(s). \end{aligned} \tag{5.20}$$

By re-arranging (5.20), the closed-loop feedback error is

$$\begin{aligned} E(s) = \frac{H(s)}{1+G(s)C(s)}R(s) - \frac{G(s)}{1+G(s)C(s)}D_i(s) \\ - \frac{1}{1+G(s)C(s)}D_o(s) - \frac{1}{1+G(s)C(s)}D_m(s). \end{aligned} \tag{5.21}$$

Note that the feedback error is in relation to the output via,

$$E(s) = H(s)R(s) - D_m(s) - Y(s). \tag{5.22}$$

By substituting (5.22) into (5.21), we obtain the expression of the closed-loop output $Y(s)$ as

$$\begin{aligned} Y(s) = \frac{G(s)C(s)H(s)}{1+G(s)C(s)}R(s) + \frac{1}{1+G(s)C(s)}D_o(s) + \frac{G(s)}{1+G(s)C(s)}D_i(s) \\ - \frac{G(s)C(s)}{1+G(s)C(s)}D_m(s). \end{aligned} \tag{5.23}$$

Also, from the feedback error (5.21), we calculate the closed-loop control signal as

$$\begin{aligned} U(s) = C(s)E(s) = \frac{C(s)H(s)}{1+G(s)C(s)}R(s) - \frac{G(s)C(s)}{1+G(s)C(s)}D_i(s) \\ - \frac{C(s)}{1+G(s)C(s)}D_o(s) - \frac{C(s)}{1+G(s)C(s)}D_m(s). \end{aligned} \tag{5.24}$$

Based on these relationships, the following sensitivity functions are defined:

The sensitivity function is defined as,

$$S(s) = \frac{1}{1+G(s)C(s)}. \tag{5.25}$$

The complementary sensitivity function is,

$$T(s) = \frac{G(s)C(s)}{1+G(s)C(s)}. \tag{5.26}$$

The input disturbance sensitivity function is defined as,

$$S^i(s) = \frac{G(s)}{1+G(s)C(s)}. \tag{5.27}$$

The control sensitivity function is,

$$S^u(s) = \frac{C(s)}{1+G(s)C(s)}. \tag{5.28}$$

Note that the superscripts of i and u are used for the input disturbance sensitivity ($S^i(s)$) and control sensitivity ($S^u(s)$) functions in order to avoid a potential confusion with the manipulated variables S_d and S_q used in the converter control case. The sensitivity functions are related to each other in the following ways.

The sensitivity plus complementary sensitivity equals to one:

$$S(s) + T(s) = \frac{1}{1 + G(s)C(s)} + \frac{G(s)C(s)}{1 + G(s)C(s)} = 1. \tag{5.29}$$

The input disturbance sensitivity is related to the sensitivity via the following equality:

$$S^i(s) = \frac{G(s)}{1 + G(s)C(s)} = S(s)G(s). \tag{5.30}$$

The control sensitivity is related to the sensitivity as described by the following equality:

$$S^u(s) = \frac{C(s)}{1 + G(s)C(s)} = S(s)C(s). \tag{5.31}$$

With the sensitivity functions, we rewrite the output of the closed-loop system (5.23) as

$$Y(s) = T(s)H(s)R(s) + S(s)D_o(s) + S^i(s)D_i(s) - T(s)D_m(s) \tag{5.32}$$

and the control signal (5.24) as

$$U(s) = S^u(s)H(s)R(s) - S^u(s)D_o(s) - S^u(s)G(s)D_i(s) - D_m(s). \tag{5.33}$$

From these relationships, we can see that

1. The complementary sensitivity function $T(s)$ represents the effects of both reference signal and measurement errors on the output. If we want a fast response speed to a reference signal, then the closed-loop bandwidth will be wider (larger w_n). As a consequence, the closed-loop control system will amplify the measurement noise. However, for the measurement error caused by sensor bias, the closed-loop system will neither amplify it nor attenuate it because with PI control, $|T(j\omega)| \approx 1$ at the lower frequency region.
2. The sensitivity $S(s)$ represents the effect of output disturbance on the output.
3. The input sensitivity $S^i(s)$ represents the effect of input disturbance on the output.

5.1.3 Disturbance Rejection and Noise Attenuation

There are both noise and disturbance existing in a physical system. A good closed-loop performance requires the minimization of the effects of both disturbance and noise. We call these disturbance rejection and noise attenuation respectively.

For minimization of the effects of both input and output disturbances, we will make the magnitude of the output in frequency response

$$|Y_d(j\omega)| = |S(j\omega)(D_o(j\omega) + G(j\omega)D_i(j\omega))| \tag{5.34}$$

as small as possible. For minimization of the measurement error, we will make the magnitude of the output in frequency response

$$|Y_m(j\omega)| = |T(j\omega)D_m(j\omega)| \tag{5.35}$$

as small as possible. We can not alter the disturbances and noise, because they already exist in the system. Thus, what we will do is to make

- the magnitude of sensitivity $S(j\omega)$ ($|S(j\omega)|$) small for disturbance rejection;
- the magnitude of complementary sensitivity $T(j\omega)$ ($|T(j\omega)|$) small for noise attenuation.

These are the desirable design objectives for control systems. However, noting that the relationship between the sensitivity and complementary sensitivity is constrained by

$$S(j\omega) + T(j\omega) = 1, \tag{5.36}$$

which says that we can not make both $|S(j\omega)|$ and $|T(j\omega)|$ small over the same frequency bands. In other words, if the disturbance is minimized in a given frequency region where $|S(j\omega)|$ is small, then inevitably the measurement noise is not attenuated in the same frequency region where $|T(j\omega)|$ is large. The frequency constraint (5.36) in the feedback control system leads to the following trade-off relationships between disturbance rejection and noise attenuation.

Note that the disturbances existing in the system correspond to slow movement of the variables or slow changes, therefore, the frequency contents of the disturbance term $|D_o(j\omega) + G(j\omega)D_i(j\omega)|$ are concentrated in the low frequency region. In contrast, the measurement noise corresponds to fast movement of the variables or fast and frequent changes of the variables, therefore, the frequency contents of the measurement noise $|D_m(j\omega)|$are concentrated in the higher frequency region. This means that we can achieve disturbance rejection by choosing the sensitivity function $S(j\omega) \approx 0$ at the low frequency region, which implies $T(j\omega) \approx 1$ at the low frequency region, because $S(j\omega) + T(j\omega) = 1$. This is not too bad for noise attenuation because $|D_m(j\omega)|$ is small in the low frequency region. At the high frequency region, to avoid the amplification of measurement noise, we choose $|T(j\omega)| \approx 0$, which implies $|S(j\omega)| \approx 1$. This is not too bad for disturbance rejection because $|D_o(j\omega) + G(j\omega)D_i(j\omega)|$ is small in the high frequency region.

Since the measurement error from sensor bias, which is a unknown constant quantity or a slowly time-varying quantity, has predominantly low frequency contents, $|T(j\omega)| \approx 1$ does not reduce the effect of sensor bias. The sensor bias error will be directly passed to the control system output. As a consequence, sensor bias is not addressed in the feedback control system design. However, the sensor bias error can be treated as an input disturbance that will be rejected in the design of outer-loop feedback control (see Section 5.2.5).

5.2 Tuning Current-loop q-axis Proportional Controller (PMSM)

In principle, the bandwidth of current-loop control system is selected to be as wide as possible to ensure a fast closed-loop response speed against the uncertainties due to inaccuracy of the feedforward compensation as well as the fact that the inner-loop dynamics are neglected in the design of outer-loop control system. This is translated into selecting a large α value when using proportional control or a large γ value when using PI control. In addition, it is shown in this section that the errors from the PWM form the input disturbance in the current control system and this kind of disturbance will be compensated by the large bandwidth in the current control. However, there are two key factors that will put an upper limit on the closed-loop bandwidth, which are the unmodeled dynamics in the current-loop systems and the measurement noise from the current sensors. These undesirable factors will affect the choice of bandwidth of the current-loop closed-loop control system.

The sensitivity functions provide a good indication about how the compromise should be reached. More specifically, we will investigate how the output current from the current-loops will respond to the noise and disturbance. Since both the proportional controller and the proportional plus integral controller have been proposed for the q-axis current control, we will investigate these two cases separately and draw interesting conclusions. In both cases, d-axis current control is performed using a PI controller, therefore, its analysis follows identically from the case of q-axis current PI control (see Section 5.3). Since the current control loops for the electrical drives and power converters are identical in design and analysis, for additional simplicity, PMSM model is chosen as an example for illustrative purpose.

5.2.1 Performance Factor and Proportional Gain

The current-loop proportional controller for a PMSM is

$$K_c^q = \frac{\alpha}{1-\alpha} R_s, \tag{5.37}$$

where α is the steady-state gain of the inner-loop control system. In theory, $0 < \alpha < 1$, but in practice, α is found to be in the range between 0.8 and 0.95 in order to achieve high gain current control, which is also fundamentally important for the cascade closed-loop control systems to work well. For simplicity, we choose $\alpha = 0.8$ for a slow response speed, $\alpha = 0.9$ for a median response speed and $\alpha = 0.95$ for a fast response speed in the current control systems. Figure 5.3 shows how the proportional gain in the inner-loop changes as α increases. It is seen that the proportional controller gain increases rapidly when α exceeds 0.9 and it approaches ∞ when α approaches 1.

5.2.2 Complementary Sensitivity Function

The complementary sensitivity function for the inner-loop system when using the proportional controller ($C(s) = \frac{\alpha}{1-\alpha} R_s$) is calculated,

$$\begin{aligned} T(s) &= \frac{G(s)C(s)}{1 + G(s)C(s)} \\ &= \frac{\alpha}{(1-\alpha)\frac{L_q}{R_s}s + 1}, \end{aligned} \tag{5.38}$$

where $G(s) = \frac{\frac{1}{L_q}}{s + \frac{R_s}{L_q}}$. The complementary sensitivity function when using a proportional control is a first order transfer function, having a steady-state gain of α and time constant of $(1 - \alpha)$ multiplying the

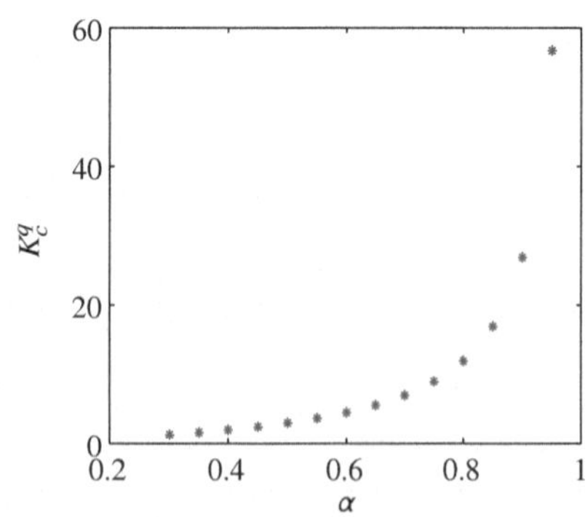

Figure 5.3 Variation of proportional gain for inner-loop current control (PMSM). The resistance $R = 2.98\ \Omega$.

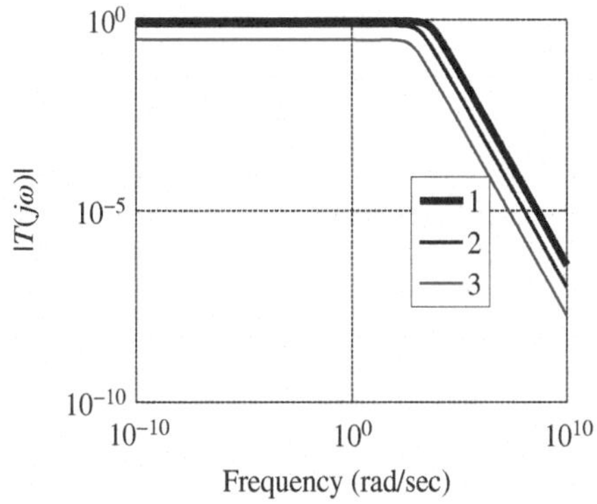

Figure 5.4 Complementary sensitivity function for inner-loop current control (PMSM). Key: line (1) $\alpha = 0.9$, line (2) $\alpha = 0.7$, line (3) $\alpha = 0.3$.

corresponding open-loop time constant $\frac{L_q}{R_s}$. Thus, the ratio between the closed-loop and open-loop time constants is $1 - \alpha$ and the ratio of closed-loop pole in magnitude to the open-loop pole is $\frac{1}{1-\alpha}$. Figure 5.4 shows the variation of the magnitude of the complementary sensitivity function with respect to α, where the resistance $R_s = 2.98\ \Omega$ and inductance $L_q = 0.007$ H. It is seen that with the increase of α, the magnitude of the complementary sensitivity function increases, in particular, the $|T(j\omega)|$ approaches to 1 as α approaching to 1. It is emphasized here that with a proportional controller, $|T(j\omega)|$ will not be equal 1 because when $\alpha = 1$, the proportional controller gain becomes infinity. The behavior of the complementary sensitivity function confirms that in order for the current control system to track a reference signal, a larger value of α is required.

Since the steady-state gain of the closed-loop system is α, the bandwidth of the q-axis current control system is the frequency ω_b such that

$$|T(j\omega)|_{\omega=\omega_b} = \frac{T(0)}{\sqrt{2}} = \frac{\alpha}{\sqrt{2}}, \tag{5.39}$$

where we have used the definition of the bandwidth (see Goodwin *et al.* (2001)). To find the value of ω_b, letting the magnitude of the complementary sensitivity function equal to $\frac{\alpha}{\sqrt{2}}$ and $s = j\omega$ and using (5.38), the bandwidth of the closed-loop system for the q-axis current control is calculated by solving

$$\frac{\alpha}{\sqrt{(1-\alpha)^2\omega^2(L_q/R)^2 + 1}} = \frac{\alpha}{\sqrt{2}}, \tag{5.40}$$

which leads to the bandwidth parameter ω_b:

$$\omega_b = \frac{1}{1-\alpha}\frac{R_s}{L_q}. \tag{5.41}$$

Here the closed-loop bandwidth for proportional current control is expressed as the function of the parameter α and the open-loop bandwidth that is $\frac{R_s}{L_q}$. Say, if we choose $\alpha = 0.9$, then $\frac{1}{1-\alpha} = 10$, and the closed-loop bandwidth is 10 times of the open-loop bandwidth. Similarly, if $\alpha = 0.8$, then $\frac{1}{1-\alpha} = 5$, and the closed-loop bandwidth is 5 times of the open-loop bandwidth.

5.2.3 Sensitivity and Input Sensitivity Functions

The sensitivity function is calculated as

$$S(s) = 1 - T(s) = \frac{(1-\alpha)\left(\frac{L_q}{R_s}s + 1\right)}{(1-\alpha)\frac{L_q}{R_s}s + 1}. \tag{5.42}$$

The input sensitivity function is related to the transfer function of inner-loop electrical system, which is calculated as

$$S^i(s) = S(s)G(s) = \frac{(1-\alpha)\frac{1}{R_s}}{(1-\alpha)\frac{L_q}{R_s}s + 1}. \tag{5.43}$$

Against several choices of α, the magnitude of sensitivity function is shown in Figure 5.5(a) whilst the magnitude of the input sensitivity function is shown in 5.5(b). It is seen that at the lower frequency, the magnitude of the sensitivity is larger when α is smaller, while at the higher frequency, the magnitude of the sensitivity is smaller when α is smaller. In comparison, the magnitude of input sensitivity behaves similarly to that of sensitivity in the lower frequency region, but in the higher frequency region, the magnitude of the input sensitivity becomes very small. The behavior of the sensitivity and input sensitivity functions means that when α is small, the proportional current control system will have a poor disturbance rejection property. This again enforces that in order for the current control system to work well, a larger α value is required for disturbance rejection and reference following.

5.2.4 Effect of PWM Noise on Current Proportional Control System

Let $\epsilon_1(t)$ denote the error for the d-axis voltage and $\epsilon_2(t)$ for the q-axis voltage, then the actual control signals $\bar{v}_d(t)$ and $\bar{v}_q(t)$ to the modulators are expressed as

$$\bar{v}_d(t) = v_d(t) + \epsilon_1(t) \tag{5.44}$$

$$\bar{v}_q(t) = v_q(t) + \epsilon_2(t). \tag{5.45}$$

Here the PWM error signals, $\epsilon_1(t)$ and $\epsilon_2(t)$, are input disturbances to the current-loop control systems, and they are, in general, periodic signals with their frequencies determined by the carrier frequency of the PWM generator. The error signals, $\epsilon_1(t)$ and $\epsilon_2(t)$, are not observed directly from the control signals, however, they are observed from the current measurements $i_q(t)$ and $i_d(t)$. Here, we simulate the PWM

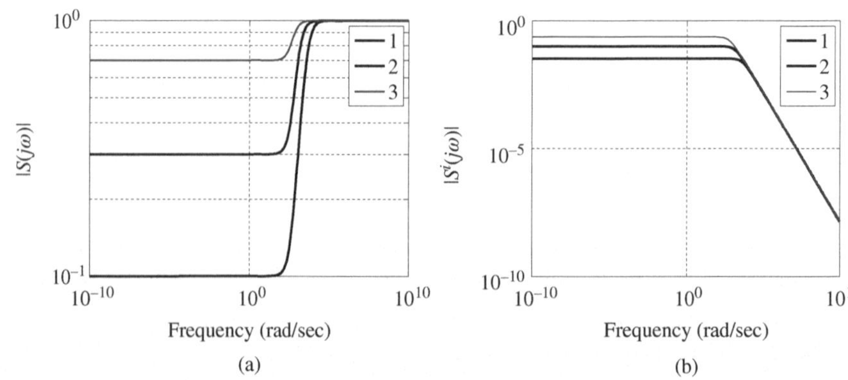

Figure 5.5 Sensitivity function and input sensitivity function for inner-loop current control (PMSM). Key: line (1) $\alpha = 0.9$, line (2) $\alpha = 0.7$, line (3) $\alpha = 0.3$. (a) Sensitivity function, and (b) Input sensitivity function.

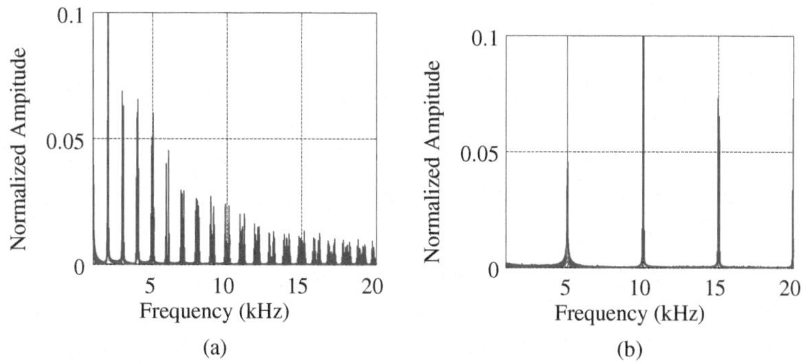

Figure 5.6 Spectrum of PWM error signal. (a) Carrier frequency: 1 kHz, and (b) Carrier frequency: 5 kHz.

error signals by using constant input signals $v_d(t)$ and $v_q(t)$ to the Simulink PMSM simulator without feedback control. The frequency contents of the PWM noise for the d-axis and q-axis are analyzed using Fourier transforms and found to be identical. The fundamental frequency of the PWM error signal is equal to the carrier frequency used. For instance, if the carrier frequency is 5 kHz, then the fundamental frequency of the input disturbance due to PWM is 5 kHz, followed by the remaining harmonic frequencies (see Figure 5.6(a)). If the carrier frequency for the PWM is reduced to 1 kHz, then the fundamental frequency of the input disturbance due to PWM is 1 kHz followed by the multiple harmonic frequencies. It is seen that the carrier frequency is essential to the frequency contents of the input disturbance due to thc PWM. As the carrier frequency increases, the spectrum of the input disturbance shifts to the higher frequency band. The effect of the PWM errors on the current output signal is characterized through the input sensitivity function in the frequency domain:

$$|Y_d(j\omega)| = |S^i(j\omega)||D_i(j\omega)|, \tag{5.46}$$

where $Y_d(j\omega)$ is the frequency response of the current signal, $S^i(j\omega)$ is the frequency response of the input sensitivity function and $D_i(j\omega)$ is the frequency response of a PWM error signal. To investigate how the choice of α affects the current output in the presence of PWM error, $|Y_d(j\omega)|$ is calculated with $\alpha = 0.7$ and $\alpha = 0.9$ with respect to the carrier frequency being 1 and 5 kHz, shown in Figures 5.7–5.8. It is seen from these figures that $|Y_d(j\omega)|$ has values at the fundamental carrier frequency and its

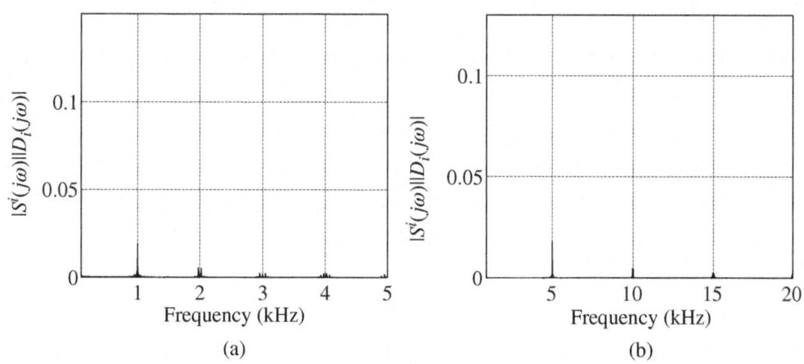

Figure 5.7 $|Y_d(j\omega)| = |S^i(j\omega)||D_i(j\omega)|$, effect of PWM noise on the current output using P current control system with varying carrier frequency. The proportional gain is $K_c = \frac{\alpha}{1-\alpha}R_s$, $\alpha = 0.9$. (a) Carrier frequency: 1 kHz, and (b) Carrier frequency: 5 kHz.

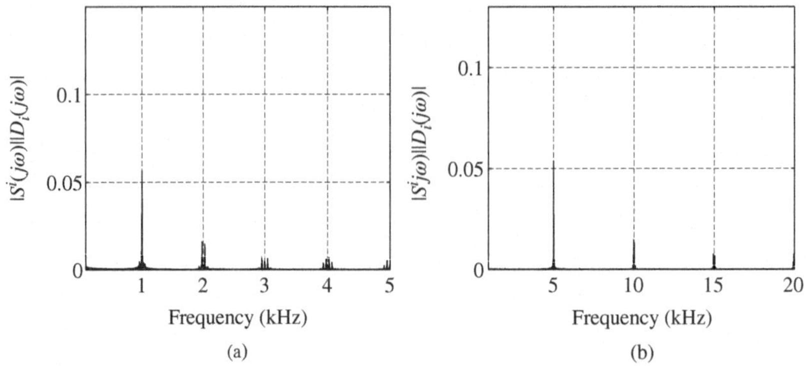

Figure 5.8 $|Y_d(j\omega)| = |S^i(j\omega)||D_i(j\omega)|$, effect of PWM noise on the current output using P current control system with varying carrier frequency. The proportional gain is $K_c = \frac{\alpha}{1-\alpha}R_s$, $\alpha = 0.7$. (a) Carrier frequency: 1 kHz and (b) Carrier frequency: 5 kHz.

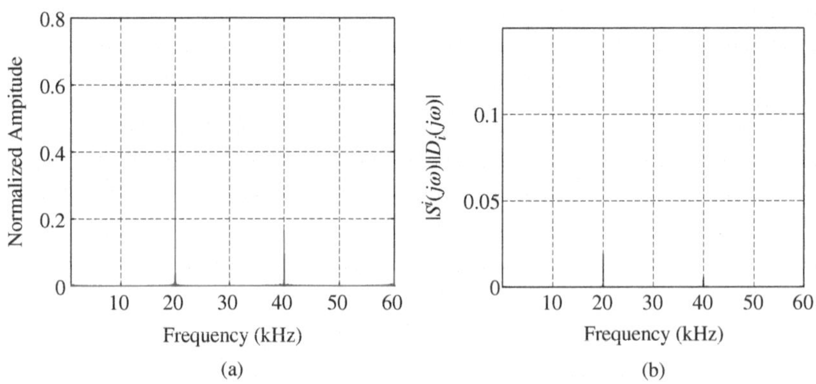

Figure 5.9 Effect of PWM noise on the current output with carrier frequency 20 kHz. (a) Spectrum of PWM error signal and (b) $|Y_d(j\omega)|$ with $\alpha = 0.9$.

harmonic frequencies. Its values between these harmonic frequencies are negligible. In addition, by comparing these two figures, it is seen that when α reduces, $|Y_d(j\omega)|$ increases, implying that it is less effective for compensating the PWM errors when a smaller value of α is used in the design. Namely, a larger closed-loop bandwidth in current control will be beneficial to reduce the errors originating from the PWMs. If the carrier frequency for a PWM is much larger, say approximately 20 kHz, Figure 5.9(a) shows the spectrum of the error signal from PWM as well as $|Y_d(j\omega)|$ (see Figure 5.9(b)), where the closed-loop performance is specified with the parameter $\alpha = 0.9$. It is seen from these figures that with this much larger carrier frequency, the effects of the PWM errors are much reduced.

5.2.5 Effect of Current Sensor Noise and Bias

Current sensors are another source that generates high frequency noises. These noises from the current sensors are measurement noises in the current-loop control. Let $\epsilon_3(t)$ and $\epsilon_4(t)$ denote the measurement noises, then the d-axis and q-axis currents are expressed as

$$i_d(t) = i_{d0}(t) + \epsilon_3(t) \tag{5.47}$$

$$i_q(t) = i_{q0}(t) + \epsilon_4(t), \tag{5.48}$$

where $i_{d0}(t)$ and $i_{q0}(t)$ denote the actual d-axis and q-axis currents. It is relatively easy to understand the existence of noises in the current-loop control systems. However, for the existence of disturbance in the current-loop control systems, we need some explanation. The disturbances are caused by possible current sensor bias errors. Current sensor bias error often occurs when the motor has been in operation for a long time and the electronic components are ageing. It is known that the feedback d-q axis currents i_d and i_q are obtained through the Park-Clarke transformations of measured three phase currents i_a, $i_b$and i_c. In practice, this measurement is often performed by two current sensors, which carry the sinusoidal measurement on a *DC* offset. The *DC* offset voltage is sometimes drifting and thus causes the measured three phase current being carried on a varied *DC* offset value (Gan and Qiu (2004)). This phenomenon can be modeled by

$$i_a = i'_a + I_a;$$

$$i_b = i'_b + I_b;$$

$$i_c = -(i_a + i_b).$$

where i'_a and i'_b are the actual three phase currents, i_a, i_b and i_c are measured current values and I_a and I_b are *DC* offset errors. The abc/dq transformation of the three phase measured currents would produce sinusoidal oscillations on the actual d-q axis currents. The feedback variables i_d and i_q are given by,

$$i_d(t) = i_{d0}(t) + i_{doff}(t) \tag{5.49}$$

$$i_q(t) = i_{q0}(t) + i_{qoff}(t), \tag{5.50}$$

where $i_{doff}(t)$ and $i_{qoff}(t)$ are the sinusoidal disturbances due to the offset errors,

$$i_{doff} = I_d \cos(\omega_e t + \varphi) \tag{5.51}$$

$$i_{qoff} = I_d \sin(\omega_e t + \varphi), \tag{5.52}$$

where

$$I_d = \frac{2}{3}\sqrt{3(I_a^2 + I_a I_b + I_b^2)}, \quad \varphi = \tan^{-1}(-\frac{I_a + 2I_b}{\sqrt{3}}).$$

Combining together the measurement noise and sinusoidal disturbances, the actual feedback variables $i_d(t)$ and $i_q(t)$ are expressed as

$$i_d(t) = i_{d0}(t) + i_{doff}(t) + \epsilon_3(t) \tag{5.53}$$

$$i_q(t) = i_{q0}(t) + i_{qoff}(t) + \epsilon_4(t). \tag{5.54}$$

The current sensor errors including bias and the sensor noise described in (5.53) and (5.54) affect the closed-loop performance of the inner-loop current control systems. Their effects are analyzed using the complementary sensitivity function $T(s)$. The frequency response analysis is based on the relationship,

$$|Y_m(j\omega)| = |T(j\omega)D_m(j\omega)|, \tag{5.55}$$

where $T(j\omega)$ is the complementary sensitivity function and $D_m(j\omega)$ is the frequency response of current sensor errors that include both measurement noise and measurement bias. From Figure 5.4, it is seen that with the increase of α, the magnitude of $T(j\omega)$ increases. Hence, the effect of current sensor noise will increase. Furthermore, the sinusoidal disturbance caused by the bias of the current sensor will also be amplified more by a larger αvalue for the current-loop control system. The frequency of the disturbance caused by current sensor bias error is dependent on the electrical velocity of the motor. If the motor is running at a low speed, then the sinusoidal disturbance will appear in the low frequency region and the current control system will amplify this disturbance because of the large amplitude of the complementary sensitivity function at the lower frequency region (see (5.55) and Figure 5.4). This is counter-intuitive. Although in general the effect of low frequency disturbances in the control system will be reduced by a closed-loop system with wider bandwidth, in the case of current sensor bias error, the low frequency disturbance occurred at the measurement, and it is part of measurement error, the closed-loop current control with a wider bandwidth will amplify this sinusoidal disturbance. Therefore, it is important to emphasize that the sinusoidal disturbances due to measurement errors can not be reduced by increasing the current-loop controller gain. Instead, from sensitivity analysis, a larger controller gain in the current-loop will amplify this sinusoidal disturbance. This sinusoidal disturbance can be rejected in the outer-loop control system as part of the input disturbance. Because the sensor bias errors are transformed into sinusoidal disturbances to the outer-loop system, in order to overcome this type of periodic disturbances, repetitive control strategies were suitable for the outer-loop control system design. A successful demonstration of this control principle with experimental evaluations was presented in Chai *et al.* (2013).

5.2.6 Experimental Case Study of Current Sensor Bias Using P Control

To examine how the current sensor bias affects the current control system, the PMSM test bed is setup with an unknown current sensor bias. In practice, this sensor bias could be caused by the aging of electronics. Because the bias error through the Park-Clarke transformation becomes a sinusoidal disturbance with its frequency centered at the electrical speed of the PMSM, in the experiment, the set-point signal for velocity control (electrical velocity) is chosen to be 420 RPM. This is translated to a measurement disturbance with center frequency at $\frac{420}{60} = 7$ rad/s. Its effect on the current output is quantified in the frequency domain using (5.55) via the complementary sensitivity function $T(j\omega)$. From Figure 5.4, it is seen that as α increases, at the low frequency region (here, we look at the frequency of 7 rad/s), the magnitude of the complementary sensitivity function increases. Hence, if there is a current sensor bias error, this error will be amplified when α increases.

In the experimental study, the closed-loop bandwidth for the outer-loop velocity control is chosen to be 117 rad/s for all the cases. The inner-loop current proportional control is varied with two different α values. Corresponding to $\alpha = 0.6, 0.8$ are the proportional control gains $K_c^q(= K_c^d)$, 7.45 and 14.9 respectively. The PMSM is running without load, thus the current is quite small. In Case A (see Figure 5.10), when α is small, it is seen that there is a small amount of oscillation in both current and velocity. In Case B (see Figure 5.11), when α is increased to 0.8 where the proportional feedback control gain is doubled from the previous one, the oscillations in current and velocity become clearly visible and the magnitudes of the oscillations have increased significantly.

This experimental study indicates that the current oscillation due to current sensor bias is a factor that limits the bandwidth of the current control system. If the current sensors do not produce high quality measurement, it is better to use a smaller value of α or use a narrower desired bandwidth for closed-loop current control. This current oscillation due to sensor bias is more severe when the motor is operating at a low speed, which translates into a low frequency measurement disturbance that has more effect on the current output because of the larger magnitude of the complementary sensitivity function at the low frequency region.

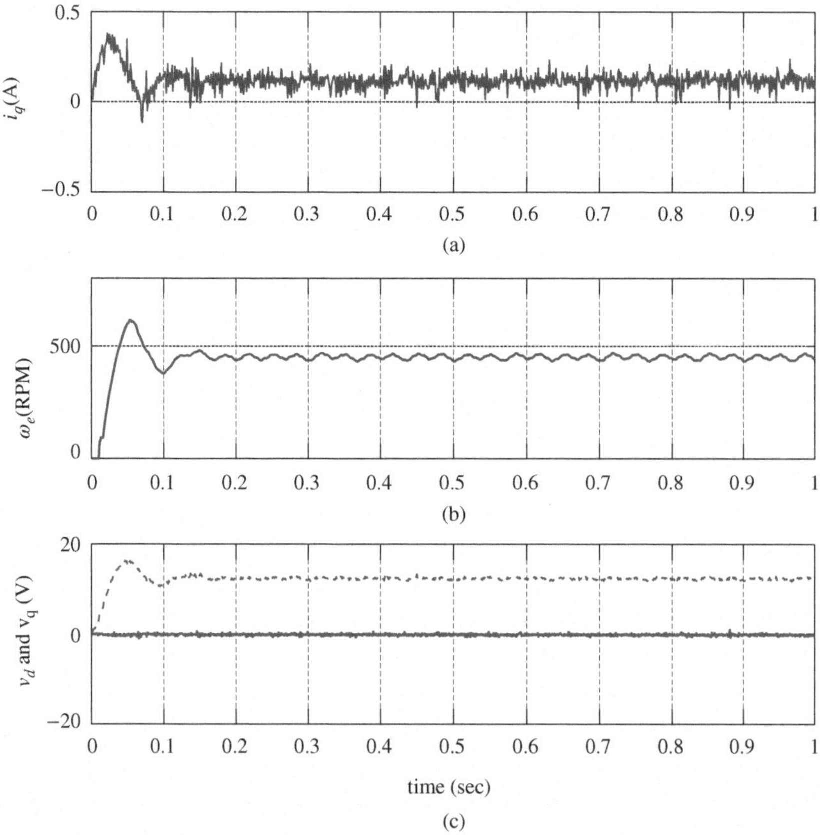

Figure 5.10 Case A. Current control with $\alpha = 0.6$, outer-loop velocity control with closed-loop bandwidth: $w_n = \frac{1}{1-\gamma}a$ rad/s, $\gamma = 0.98$, $a = 2.3404$. Top plot: i_q current; middle plot: output velocity; bottom plot: control variables(solid-line: v_d, dashed-line: v_q).

5.2.7 Experimental Case Study of Current Loop Noise

Another factor that puts a restriction on the bandwidth of the current control system is the various noises existing in the system. These noises could come from the high frequency components of the PWM errors or the current sensor measurement noise. Their effects on the control signals (v_d and v_q) and the current output will increase when the parameter α increases or the closed-loop bandwidth of the current control system increases.

Experimental evaluation of the current control system is performed using a PI controller for the d-axis current and a proportional controller for q-axis current, and PI velocity control in the outer-loop system. The PWM switching frequency is set as 2 kHz, and sampling interval for the current control loops is selected as $\Delta t = 100$ μs and the velocity control loop as $\Delta t = 200$ μs. The test-bed with xPC Target for implementation was used in the experimental study. The remaining control system parameters are shown in Table 5.1. The experimental results for the q-axis proportional controller using $\alpha = 0.8$ are shown in Figure 5.12, with comparison to the experimental results obtained using $\alpha = 0.5$ shown in Figure 5.13. The first thing observed from the experimental results is that with the proportional control of q-axis current there is a steady error between the reference signal i_q^* and the current feedback signal i_q. With the

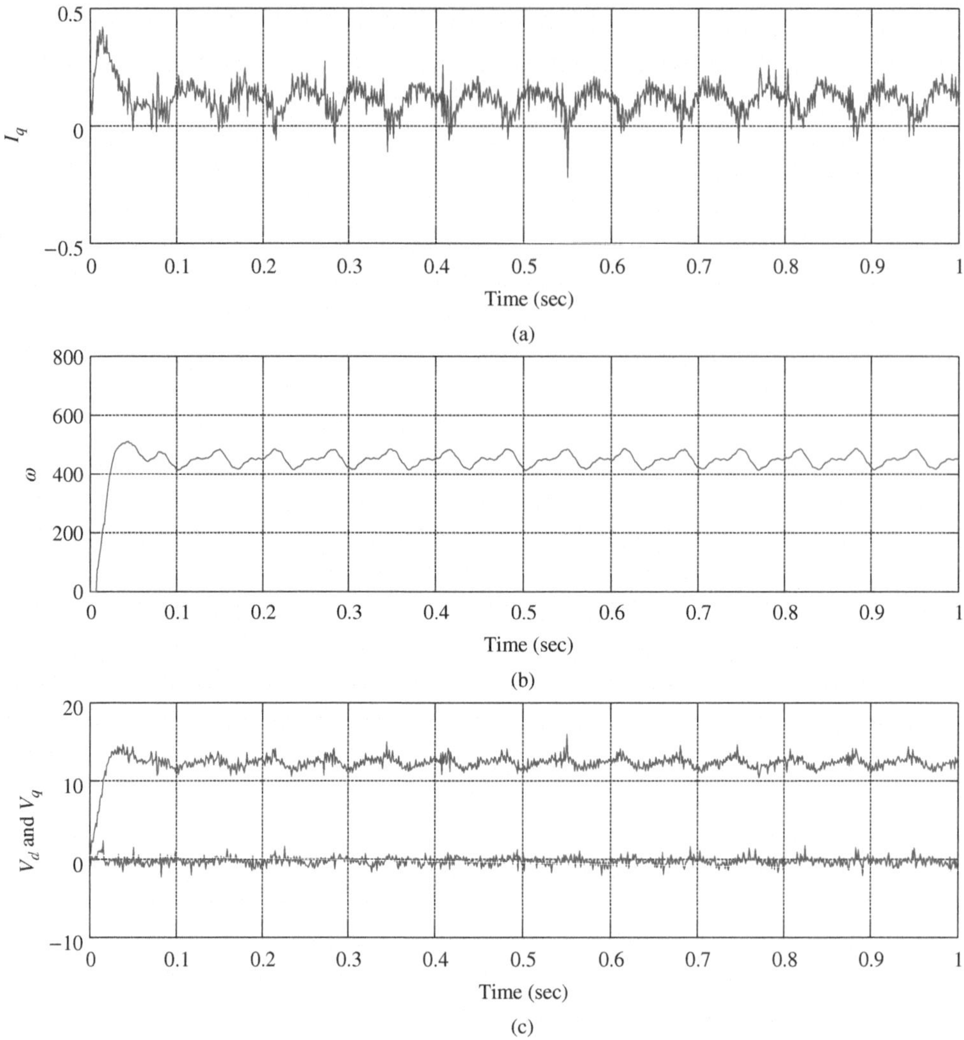

Figure 5.11 Case B. Current control with $\alpha = 0.8$, outer-loop velocity control with closed-loop bandwidth: $w_n = \frac{1}{1-\gamma} a$ rad/s, $\gamma = 0.98$, $a = 2.3404$. Top plot: i_q current; middle plot: output velocity; bottom plot: control variables.

Table 5.1 Performance parameters and operational constraints (proportional control)

	Velocity controller			i_d controller			i_q controller	
	w_n	ξ	$i_{q_{max}}$	w_n	ξ	$v_{d_{max}}$	α	$v_{q_{max}}$
Fig. 5.12	150	0.7	8 A	400	0.7	25.2 V	0.8	52.0 V
Fig. 5.13	150	0.7	8 A	400	0.7	25.2 V	0.5	52.0 V

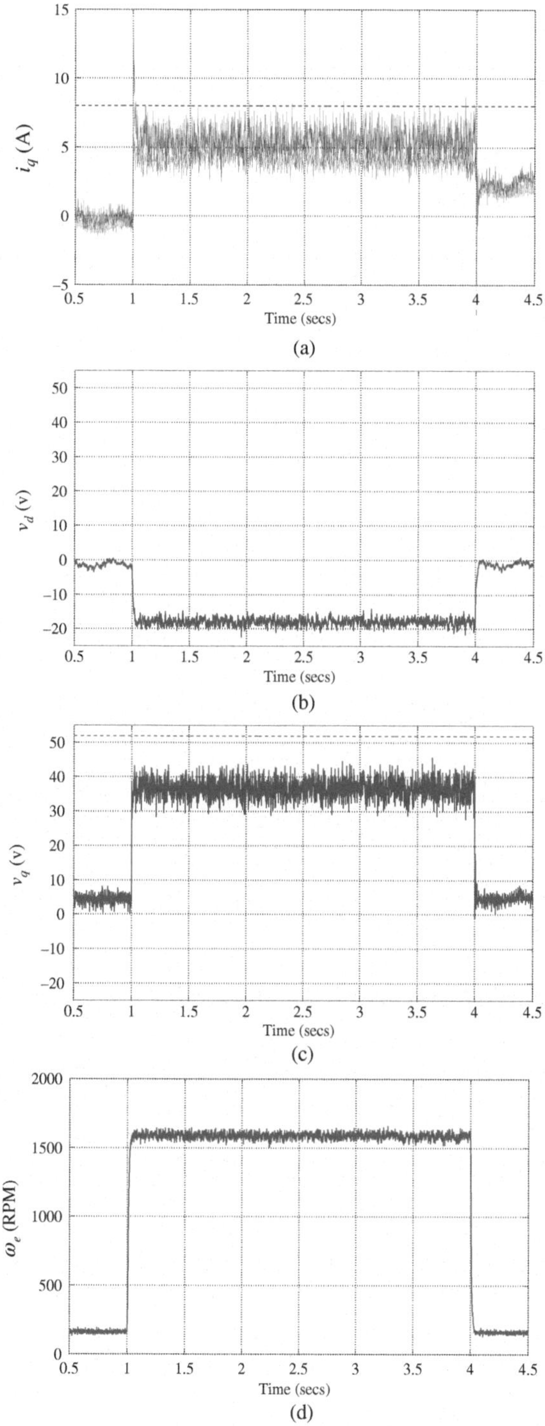

Figure 5.12 Case A. $K_c^q = \frac{\alpha}{1-\alpha} R_s$, $R_s = 2.98\ \Omega$, $\alpha = 0.8$. (a) Current i_q and i_q^*, (b) Voltage v_d, (c) Voltage v_q, and Velocity ω_e.

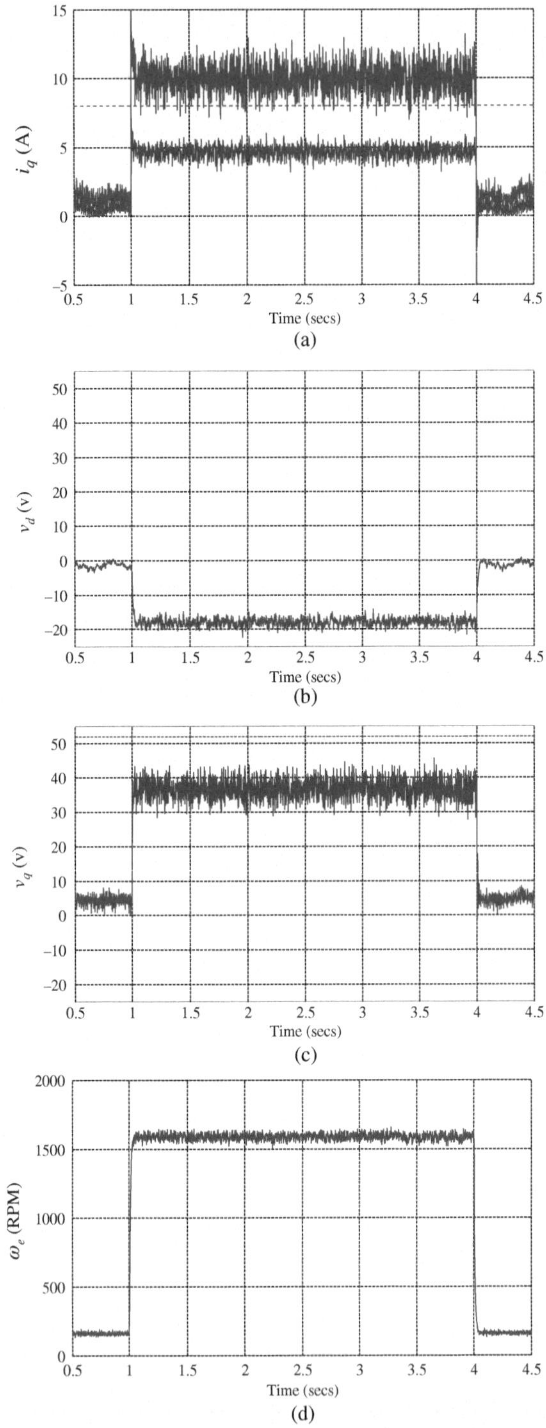

Figure 5.13 Case B. $K_c^q = \frac{\alpha}{1-\alpha}R_s$, $R_s = 2.98\ \Omega$, $\alpha = 0.5$. (a) Current i_q and i_q^*, (b) Voltage v_d, (c) Voltage v_q, and Velocity ω_e.

smaller α value, the steady-state error is larger. Because of the smaller current control gain K_c^q for the q-axis, the current reference signal i_q^* generated from the outer-loop velocity controller is much larger in magnitude in order to achieve the same velocity (see Figure 5.13(b)). It is also observed that the control signal v_q contains much less noise when the smaller α is used. The constraint on v_q became activated because of the larger amplitude of the noise.

5.3 Tuning Current-loop PI Controller (PMSM)

In this section, PI controllers are used to control the currents in $d-q$ reference frame. Since the analysis for the d axis current is identical to that of q-axis current, the presentation is only given for the q-axis current control system. Because torque control belongs to the case where the current control systems use PI controllers (see Section 3.3), the analysis of the PI current control in this section is extended to torque control without any complications.

5.3.1 PI Controller Parameters in Relation to Performance Parameter γ

If a PI controller is used for the q-axis current control, the controller parameters are

$$K_c^q = \frac{2\xi w_n - \frac{R_s}{L_q}}{\frac{1}{L_q}} = 2\xi w_n L_q - R_s \tag{5.56}$$

$$\tau_I^q = \frac{2\xi w_n - \frac{R_s}{L_q}}{w_n^2} = \frac{2\xi w_n L_q - R_s}{L_q w_n^2}, \tag{5.57}$$

where ξ is the damping coefficient, which is chosen to be either 0.707 or 1, and w_n is natural frequency of the closed-loop current control system. The closed-loop control performance is selected based on the parameter w_n. For a larger w_n, the closed-loop system will have a faster response speed.

Since the open-loop pole for the q-axis current dynamic system is at $-\frac{R_s}{L_q}$, we can select the parameter w_n in a relative value to the magnitude of the open-loop pole. Furthermore, if we select the damping coefficient $\xi = 0.707$, the closed-loop bandwidth w_n is relative to the open-loop bandwidth via:

$$w_n = \frac{1}{1-\gamma}\frac{R_s}{L_q}, \tag{5.58}$$

where γ is chosen between zero and one. When γ is zero, the closed-loop bandwidth equals the open-loop bandwidth; when $\gamma \to 1$, the closed-loop bandwidth $w_n \to \infty$. In the majority of applications, the parameter γ is selected in the range between 0.8 and 0.95. When γ is 0.8, the closed-loop bandwidth is five times of the open-loop bandwidth; when γ is 0.9, the closed-loop bandwidth is ten times of the open-loop bandwidth; and when γ is 0.95, the closed-loop bandwidth is twenty times of the open-loop bandwidth.

With this choice of w_n, the proportional control gain of the PI controller is expressed from (5.56) as

$$K_c^q = \left(\frac{2\xi}{1-\gamma} - 1\right) R_s. \tag{5.59}$$

If $\xi = 0.707$,

$$K_c^q = \frac{0.414+\gamma}{1-\gamma} R_s. \tag{5.60}$$

If $\xi = 1$,

$$K_c^q = \frac{1+\gamma}{1-\gamma} R_s. \tag{5.61}$$

It is seen that the proportional gain is dependent on the resistance R_s with this choice of closed-loop bandwidth, and the proportional gain is smaller when the damping coefficient is 0.707.

In a similar manner, the integral time constant τ_I is expressed from (5.57) as

$$\tau_I = \frac{2\xi w_n - \frac{R_s}{L_q}}{w_n^2} = (2\xi - 1 + \gamma)(1 - \gamma)\frac{L_q}{R_s}. \tag{5.62}$$

If $\xi = 0.707$, then

$$\tau_I = (0.414 + \gamma)(1 - \gamma)\frac{L_q}{R_s}. \tag{5.63}$$

If $\xi = 1$, then

$$\tau_I = (1 + \gamma)(1 - \gamma)\frac{L_q}{R_s}. \tag{5.64}$$

It is seen that the integral time constant is also smaller for the case of $\xi = 0.707$. Figure 5.14 shows the variation of proportional gain and integral time constant with respect to the parameter γ. With γ increasing, the proportional gain increases while the integral time constant decreases.

5.3.2 Sensitivity in Relation to Performance Parameter γ

With the expressions of proportional control and integral time constant in terms of the parameter γ, we can derive the sensitivity function as

$$\begin{aligned} S(s) &= \frac{1}{1 + G(s)C(s)} \\ &= \frac{s(s+a)}{s^2 + 2\xi w_n s + w_n^2} \\ &= \frac{(1-\gamma)^2 a^{-1} s(a^{-1}s + 1)}{(1-\gamma)^2 a^{-2} s^2 + 2\xi(1-\gamma)a^{-1}s + 1}, \end{aligned} \tag{5.65}$$

where $a = \frac{R_s}{L_q}$ and $w_n = \frac{a}{1-\gamma} = \frac{1}{1-\gamma}\frac{R_s}{L_q}$ rad/s. It is seen from (5.65) that the magnitude of the sensitivity function is weighted by the factor $(1-\gamma)^2$. The smaller this factor is, the smaller the magnitude of

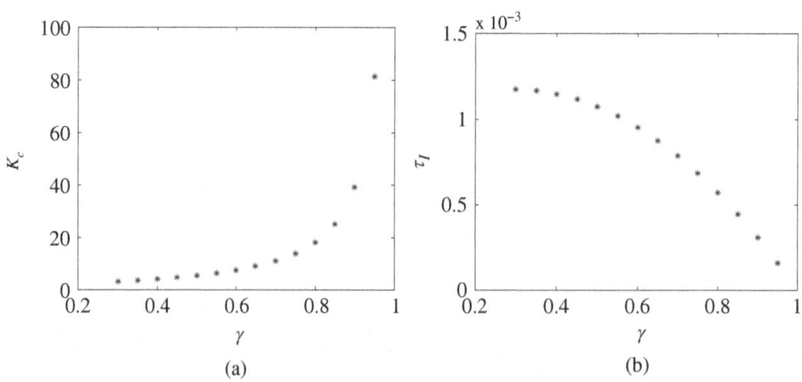

Figure 5.14 Variation of proportional gain and integral time constant for inner-loop current control (PMSM) using PI controller ($\xi = 0.707$, $R_s = 2.98\ \Omega$, and $L_q = 7 \times 10^{-3}$ H). (a) Proportional gain, and (b) Integral time constant.

the sensitivity function will be. In addition, at the zero frequency, with the PI controller the sensitivity function $S(0) = 0$. The smaller magnitude of the sensitivity function will produce a closed-loop control system that has a faster response for disturbance rejection.

Similarly, the complementary sensitivity function for the PI current control system is

$$\begin{aligned} T(s) &= \frac{G(s)C(s)}{1+G(s)C(s)} = 1 - S(s) \\ &= \frac{(1-\gamma)(2\xi - 1 + \gamma)a^{-1}s + 1}{(1-\gamma)^2a^{-2}s^2 + 2\xi(1-\gamma)a^{-1}s + 1}. \end{aligned} \tag{5.66}$$

Figure 5.15(a) and 5.15(b) show the magnitudes of the complementary sensitivity and sensitivity functions respectively. It is seen from these plots that with the increase of parameter γ, the bandwidth of the closed-loop system is increased. From (5.66) and Figure 5.15(a), it is seen that at the zero frequency, with the PI controller the complementary sensitivity function is characterized by $T(0) = 1$. In addition, because $S(s) + T(s) = 1$, for a smaller $(1-\gamma)^2$, the magnitude of complementary sensitivity function is larger because of the smaller magnitude of sensitivity function. A larger magnitude of complementary sensitivity function will produce a closed-loop control system that has a faster reference response, but is more sensitive to the measurement noise and model uncertainty in the current-loop control system. The key characteristic here is that with $T(0) = 1$ the output of the closed-loop system (i_q current) will follow the reference signal (i_q^* current) without steady-state error.

The input sensitivity function is calculated using the sensitivity (5.65) with a multiplication of the transfer function of the current system, leading to

$$\begin{aligned} S^i(s) &= \frac{(1-\gamma)^2a^{-1}s(a^{-1}s+1)}{(1-\gamma)^2a^{-2}s^2 + 2\xi(1-\gamma)a^{-1}s + 1} \times \frac{b}{s+a} \\ &= \frac{(1-\gamma)^2a^{-1}s}{(1-\gamma)^2a^{-2}s^2 + 2\xi(1-\gamma)a^{-1}s + 1} \times \frac{b}{a}, \end{aligned} \tag{5.67}$$

where $a = \frac{R_s}{L_q}$ and $b = \frac{1}{L_q}$. From (5.67), it is seen that at the zero frequency, $|S^i(0) = 0|$, meaning that with a PI controller, a constant input disturbance will be completely rejected. Figure 5.16 shows the magnitude of the input sensitivity function. It is observed from these plots that with the increase of the parameter γ, the magnitude of the input sensitivity function decreases in the low and medium frequency region.

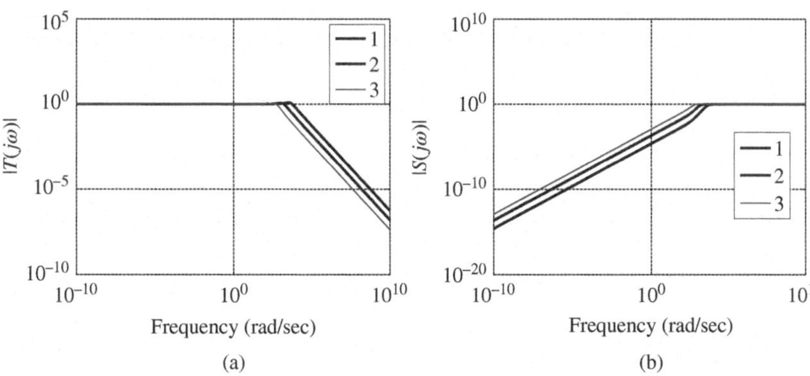

Figure 5.15 Complementary and sensitivity functions for inner-loop current control (PMSM) using PI controller ($\xi = 0.707$, $R_s = 2.98\ \Omega$, and $L_q = 7 \times 10^{-3}$ H) Key: line (1) $\gamma = 0.9$, line (2) $\gamma = 0.7$, line (3) $\gamma = 0.3$. (a) Complementary sensitivity function, and (b) Sensitivity function.

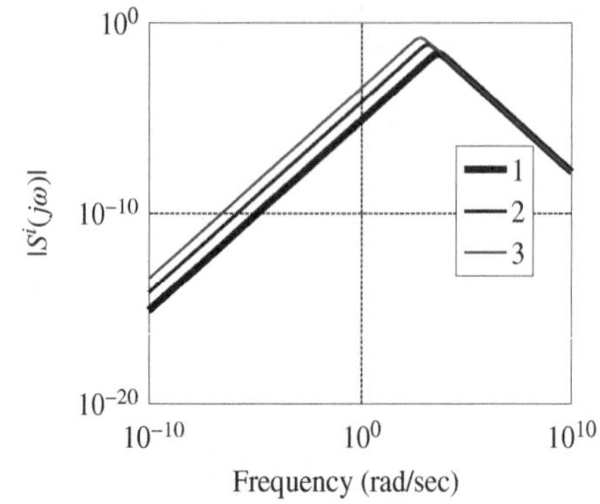

Figure 5.16 Input sensitivity function for inner-loop current control (PMSM) using PI controller ($\xi = 0.707$, $R_s = 2.98\ \Omega$, and $L_q = 7 \times 10^{-3}$ H) Key: line (1) $\gamma = 0.9$, line (2) $\gamma = 0.7$, line (3) $\gamma = 0.3$.

5.3.3 Effect of PWM Error in Relation to γ

Based on the input sensitivity function, the effect of PWM error on the current output is investigated based on the following relation:

$$|Y_d(j\omega)| = |S^i(j\omega)||D_i(j\omega)|, \tag{5.68}$$

where $|S^i(j\omega)|$ is the magnitude of the input sensitivity defined by (5.67) and $D_i(j\omega)$ is the frequency response of a PWM error signal (see Figure 5.6). Figures 5.17–5.18 show the magnitudes of the output frequency response ($|Y_d(j\omega)|$) due to PWM error via the frequency response of the input sensitivity function (see 5.68) with respect to three choices of carrier frequencies. The results show that the PWM error is very small when a PI controller is used for current control. This is because with the integral action, the input sensitivity function has a magnitude near zero in the frequency region that corresponds to the fundamental carrier frequency. By comparing the results presented in Figure 5.17 with those in Figure 5.18, it is seen that the higher the closed-loop bandwidth, the less effect of the PWM error will be.

Furthermore, in comparison with the proportional current control, the PI current controller has significantly reduced the effect of PWM error on the current output. The effect is approximately 5000 times less when the carrier frequency is 1 kHz and about 250 times less when the carrier frequency is 20 kHz (see Figures 5.7, 5.8 and Figure 5.9(b)). These comparative results may illustrate that it is important to use the integral function for current control because the integral action can significantly reduce the errors caused by the PWM implementation.

5.3.4 Experimental Case Study of Current Loop Noise Using PI Control

Experimental evaluation of the current control system is performed using PI controllers for the d-axis and q-axis currents and PI velocity control in the outer-loop system. The PWM switching frequency is set as 2 kHz, and sampling interval for the current control loops is selected as $\Delta t = 100\ \mu$s and the velocity control loop as $\Delta t = 200\ \mu$s. The test-bed with xPC Target for implementation was used in the experimental study. The remaining PI control system parameters are shown in Table 5.2. There are two cases investigated here. Both cases have identical settings of i_d current PI controller and outer-loop velocity controller (see Table 5.2), as well as identical constraints on both q-axis over-current protection and voltage constraints. The difference is the closed-loop performance specification of the q-axis current. In Case A,

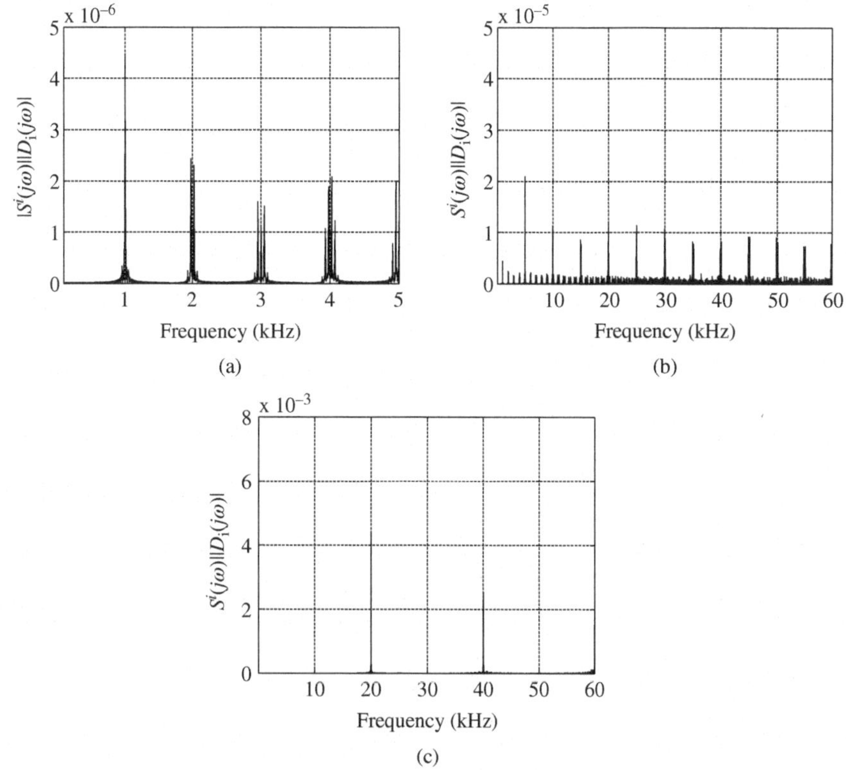

Figure 5.17 $|Y_d(j\omega)| = |S^i(j\omega)||D_i(j\omega)|$, effect of PWM noise on the current output using PI current control system with varying carrier frequency. The bandwidth $w_n = \frac{1}{1-\gamma}\frac{R_s}{L_q}$ rad/s, $R_s = 2.98\ \Omega$, and $L_q = 7 \times 10^{-3}$ H, $\gamma = 0.9$. (a) Carrier frequency: 1 kHz, (b) Carrier frequency: 5 kHz, and (c) Carrier frequency: 20 kHz.

Table 5.2 Performance parameter and operational constraints (PI control)

	Velocity controller			i_d controller			i_q controller		
	w_n	ξ	$i_{q_{max}}$	w_n	ξ	$v_{d_{max}}$	γ	ξ	$v_{q_{max}}$
Fig. 5.19	150	0.7	8 A	400	0.7	25.2 V	0.8	0.7	52.0 (V)
Fig. 5.20	150	0.7	8 A	400	0.7	25.2 V	0.5	0.7	52.0 (V)

a larger $\gamma = 0.8$ is used, which translates into the parameter w_n being $5\frac{R_s}{L_q} = 2128.6$ rad/s. By comparison, in Case B, a smaller $\gamma = 0.5$ is selected for the closed-loop performance specification that gives the parameter $w_n = 2\frac{R_s}{L_q} = 851.4$ rad/s. Thus, the closed-loop bandwidth in Case A is 2.5 times wider than the one used in Case B. The experimental results for Case A are shown in Figure 5.19, and for Case B are shown in Figure 5.20. By comparing the experimental results, the obvious conclusion is that with a wider bandwidth for the q-axis current control, the noise level in the control signal v_q has significantly increased. This means that the noises due to the PWM switching errors and current sensor errors are amplified when a higher current controller gain is used. However, the output velocity is not affected

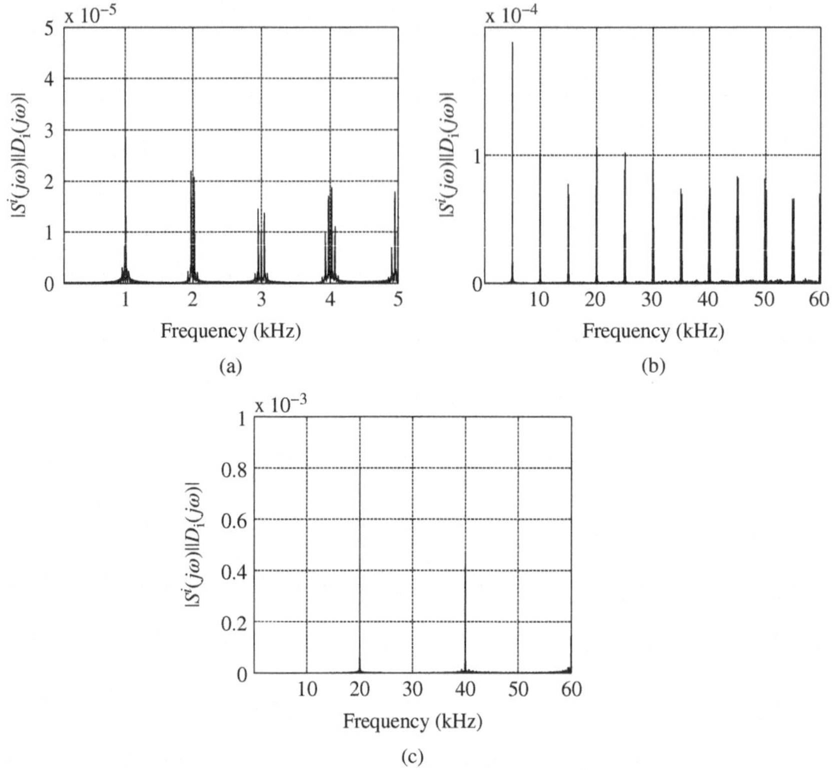

Figure 5.18 $|Y_d(j\omega)| = |S^i(j\omega)||D_i(j\omega)|$, effect of PWM noise on the current output using PI current control system with varying carrier frequency. The bandwidth $w_n = \frac{1}{1-\gamma}\frac{R_s}{L_q}$ rad/s, $R_s = 2.98\ \Omega$, and $L_q = 7e-3$ H, $\gamma = 0.7$. (a) Carrier frequency: 1 kHz, (b) Carrier frequency: 5 kHz, and (c) Carrier frequency: 20 kHz.

by the high current noise level as observed by comparing Figure 5.19(e) with Figure 5.20(e). It is also observed that the constraint on the v_q voltage became activated because of the current noise. When using PI control of q-axis current, it is seen that the actual current output i_q is closely tracking its reference signal i_q^* (see Figures 5.19(b)–5.20(b)), both in dynamic response and steady-state response. In contrast, the steady-state errors exist when using proportional control of q-axis current (see Figures 5.12(b)–5.13(b)).

5.4 Performance Robustness in Outer-loop Controllers

The tuning process for the outer-loop controllers has to reach a compromise between the closed-loop bandwidth and the effect of measurement noise from position sensors, load disturbance, PWM noise and possible sinusoidal disturbance from inner-loop current control system due to current sensor bias, and the neglected dynamics in the design. A faster closed-loop response in the outer-loop control system will require a larger desired closed-loop bandwidth, which is necessary for high performance to set-point response, load disturbance rejection and the current sensor bias compensation. On the other hand, the measurement noise, PWM noise and the neglected dynamics in the design restrict the size of the closed-loop bandwidth. One main factor that affects the bandwidth of the outer-loop control system is

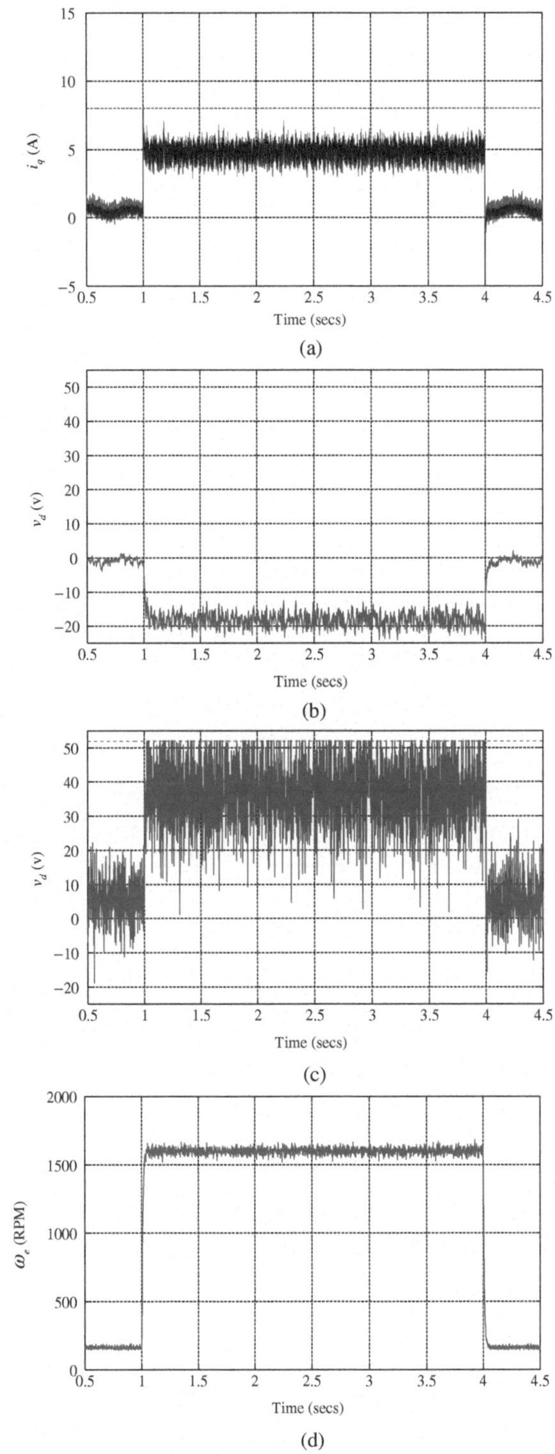

Figure 5.19 Case A. $w_n = \frac{1}{1-\gamma}\frac{R_s}{L_q}$ rad/s, $R_s = 2.98\ \Omega$, and $L_q = 7 \times 10^{-3}$ H, $\gamma = 0.8$. (a) Current i_q and i_q^*, (b) Voltage v_d, (c) Voltage v_q, and Velocity ω_e.

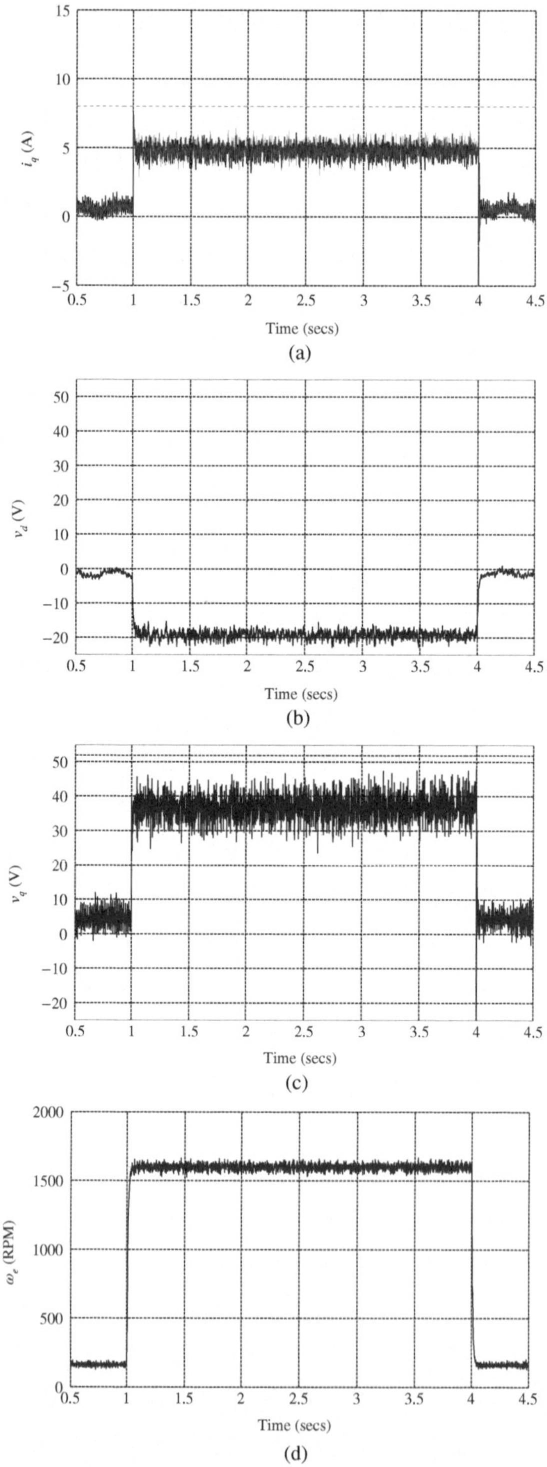

Figure 5.20 Case B. $w_n = \frac{1}{1-\gamma}\frac{R_s}{L_q}$ rad/s, $R_s = 2.98\ \Omega$, and $L_q = 7 \times 10^{-3}$ H, $\gamma = 0.5$. (a) Current i_q and i_q^*, (b) Voltage v_d, (c) Voltage v_q, and Velocity ω_e.

the neglected dynamics from the inner-loop control system. This raises the issue of performance robustness. This section shows that there is a difference between the desired and actual sensitivity functions for the outer-loop control system and this difference is quantified through a weighting function in the frequency domain.

5.4.1 Sensitivity Functions for Outer-loop Control System

Similar to the inner-loop current control systems, there are sensitivity functions used to describe the properties of outer-loop control systems. With these sensitivity functions, we will determine the closed-loop response to set-point change, disturbance rejection, measurement noise attenuation, and closed-loop stability of the system.

There are two types of sensitivity functions for the outer-loop systems. One involves the desired sensitivity functions, which are obtained using the first order model with the PI controller by neglecting the inner-loop control system dynamics. The other involves the actual sensitivity functions, which are obtained using both first order model and the inner-loop closed-loop model together with the PI controller. The desired sensitivity functions reflect what we ask the system to do, whilst the actual sensitivity functions reflect the reality of the control system. They could be quite different in applications, causing the performance gaps between what we desired and what we actually achieved.

The desired complementary sensitivity function for the outer-loop system is calculated using the formula

$$T(s) = \frac{G(s)C(s)}{1 + G(s)C(s)}. \tag{5.69}$$

Substituting $C(s) = K_c(1 + \frac{1}{\tau_I s})$ and the first order design model $G(s) = \frac{b}{s+a}$ into (5.69), together with the calculations of the proportional gain K_c and the integral time constant τ_I, the desired complementary sensitivity function is expressed as

$$T(s) = \frac{(2\xi w_n - a)s + w_n^2}{s^2 + 2\xi w_n s + w_n^2}, \tag{5.70}$$

where the proportional controller gain K_c and integral time constant are chosen as

$$K_c = \frac{2\xi w_n - a}{b}; \quad \tau_I = \frac{2\xi w_n - a}{w_n^2}.$$

Here, we have used the pole-assignment controller design technique in choosing the proportional gain K_c and the integral time constant τ_I. The parameter ξ is the damping coefficient, which is chosen to be 0.707, and the parameter w_n then corresponds to the bandwidth of the desired closed-loop system. It is emphasized that if a proportional controller is used in the inner current control loop, the steady-state gain α should be included in the parameter b for the outer-loop controller design (see Chapter 3).

Based on the parameters of a PMSM, the desired complementary sensitivity function is

$$T(s) = \frac{(2\xi w_n - \frac{B_v}{J_m})s + w_n^2}{s^2 + 2\xi w_n s + w_n^2}. \tag{5.71}$$

Similarly, based on the complementary sensitivity function (5.71), we calculate the desired sensitivity function for the desired outer-loop system as

$$\begin{aligned} S(s) = 1 - T(s) &= \frac{(s + a)s}{s^2 + 2\xi w_n s + w_n^2} \\ &= \frac{(s + \frac{B_v}{J_m})s}{s^2 + 2\xi w_n s + w_n^2}. \end{aligned} \tag{5.72}$$

The actual sensitivity functions for the outer-loop system will take into consideration the neglected inner-loop closed-loop dynamics (see the cascade control system configurations in Chapter 3 illustrated in Figure 3.10). These will determine how the closed-loop control system responds to the set-point change, disturbance rejection and noise attenuation.

By letting $F(s)$ denote the the inner closed-loop transfer function that relates the desired current reference signal $I_q^*(s)$ to the output current $I_q(s)$ as

$$\frac{I_q(s)}{I_q^*(s)} = F(s), \tag{5.73}$$

the complementary sensitivity function for the actual outer-loop system is calculated as

$$\begin{aligned} T(s)^{actual} &= \frac{C(s)F(s)G(s)}{1 + C(s)F(s)G(s)} \\ &= \frac{C(s)F(s)G(s)}{1 + C(s)G(s) - C(s)G(s) + C(s)F(s)G(s)} \\ &= \frac{C(s)F(s)G(s)}{1 + C(s)G(s)(1 - \frac{C(s)G(s)(1-F(s))}{1+C(s)G(s)})}. \end{aligned} \tag{5.74}$$

Since the desired complementary sensitivity function is $T(s) = \frac{C(s)G(s)}{1+C(s)G(s)}$, the actual complementary sensitivity is simplified in notation as

$$T(s)^{actual} = T(s)F(s) \times \frac{1}{1 - T(s)(1 - F(s))}. \tag{5.75}$$

This basically says that if $|\frac{1}{1-T(j\omega)(1-F(j\omega))}|$ is approximately one for all $\omega \geq 0$, the actual complementary sensitivity function can be approximated using the desired complementary sensitivity and the transfer function $F(s)$:

$$T(s)^{actual} \approx T(s)F(s), \tag{5.76}$$

which gives a convenient estimate for the actual complementary sensitivity and the bandwidth of the closed-loop system.

From this relation, the actual sensitivity function is obtained:

$$S(s)^{actual} = 1 - T(s)^{actual} = (1 - T(s)) \times \frac{1}{1 - T(s)(1 - F(s))}, \tag{5.77}$$

which could be approximated if $|\frac{1}{1-T(j\omega)(1-F(j\omega))}|$ is approximately one for $\omega \geq 0$

$$S(s)^{actual} \approx 1 - T(s). \tag{5.78}$$

To measure how much performance loss could occur to the complementary sensitivity function when using the cascade PI control system, we will examine the magnitude of the weighting function $|\frac{1}{1-T(j\omega)(1-F(j\omega))}|$. In order to make it close to one over the low and medium frequency region, a quick look at the weighting function leads to two crude conclusions: either $|F(j\omega)|$ is close to one at the low and medium frequency region or $|T(j\omega)|$ is small over the same frequency band. With the smaller magnitude of the complementary sensitivity function, the closed-loop response speed to set-point following and disturbance rejection will be compromised. Thus, the following analysis is focused on how the bandwidth of the inner-loop control system affects the performance robustness of the outer-loop control system.

5.4.1.1 Actual Sensitivity Functions when Inner-loop Uses a P Controller for q-axis Current

If a proportional controller is used to control the q-axis current, then $F(s)$ is expressed as

$$F(s) = \frac{I_q(s)}{I_q^*(s)} = \frac{\frac{K_c^q}{L_q}}{s + \frac{R_s}{L_q} + \frac{K_c^q}{L_q}} = \frac{1}{(1-\alpha)\frac{L_q}{R_s}s + 1}, \tag{5.79}$$

where the proportional gain $K_c^q = \frac{\alpha}{1-\alpha}R$, $0 < \alpha < 1$.

The weighting function ($|\frac{1}{1-T(j\omega)(1-F(j\omega))}|$) that alters the behavior of the actual complementary sensitivity function is illustrated in Figure 5.21(a), where two α values are used in the calculation. It is seen from this figure that this weighting function has magnitude equal to one at both low and high frequency regions and close to one at the median frequency region, and its magnitude is larger when α is smaller. As a result, the difference between the desired and actual sensitivity functions becomes negligible when

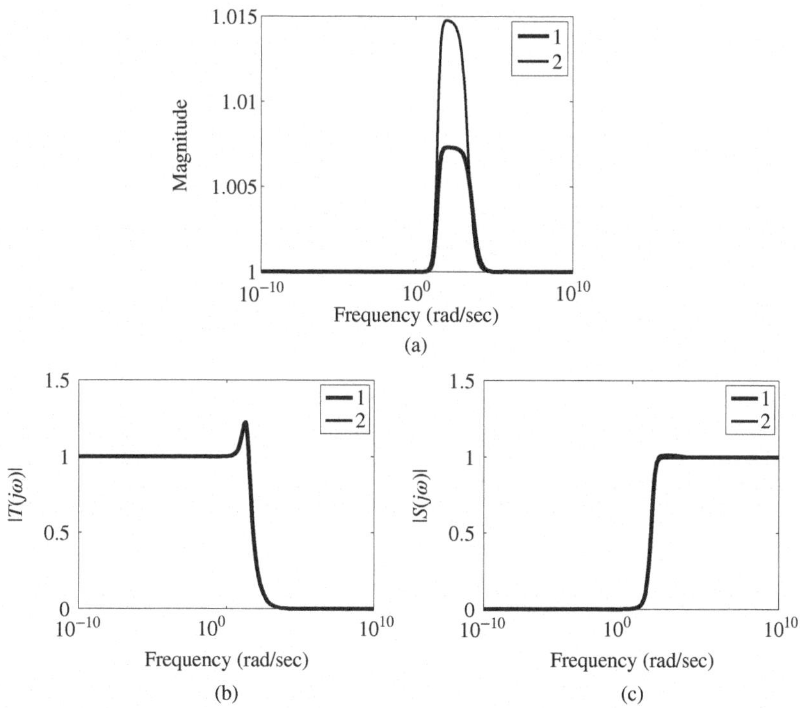

Figure 5.21 Desired and actual sensitivity functions (Key: (1) desired function, (2) actual function). Inner-loop current control specification: $\alpha = 0.9$, outer-loop velocity control specification: $\gamma = 0.9$. (a) Frequency weighting function ($|\frac{1}{1-T(j\omega)(1-F(j\omega))}|$) with varying inner-loop proportional current control specification (Key: line (1) $\alpha = 0.9$, line (2) $\alpha = 0.8$), (b) Complementary sensitivity function, and (c) Sensitivity function.

a large α is used in the inner-loop current control (see Figure 5.21(b)) for the complementary sensitivity function and 5.21(c) for the sensitivity function).

In summary, in order to reduce the closed-loop performance variation for the outer-loop control system, the closed-loop inner-loop current control system is required to have a large α value that leads to a large controller gain and wide closed-loop bandwidth.

5.4.1.2 Actual Sensitivity Functions when Inner-loop Uses a PI Controller

If a PI controller is used, the function $F(s)$ is expressed as

$$F(s) = \frac{(1-\gamma)(2\xi - 1 + \gamma)a^{-1}s + 1}{(1-\gamma)^2 a^{-2}s^2 + 2\xi(1-\gamma)a^{-1}s + 1}, \quad (5.80)$$

where $a = \frac{R_s}{L_q}$. With $F(s)$ given by (5.80), it is calculated that

$$1 - F(s) = \frac{(1-\gamma)^2 a^{-1}s(a^{-1}s + 1)}{(1-\gamma)^2 a^{-2}s^2 + 2\xi(1-\gamma)a^{-1}s + 1}. \quad (5.81)$$

From (5.81), it is seen that with the PI controller, $|(1 - F(j\omega))|$ is zero at $\omega = 0$ and $|(1 - F(j\omega))| \rightarrow 1$ as $\omega \rightarrow \infty$. Because the complementary function has the characteristic $|T(j\omega)| \rightarrow 0$ as $\omega \rightarrow \infty$,

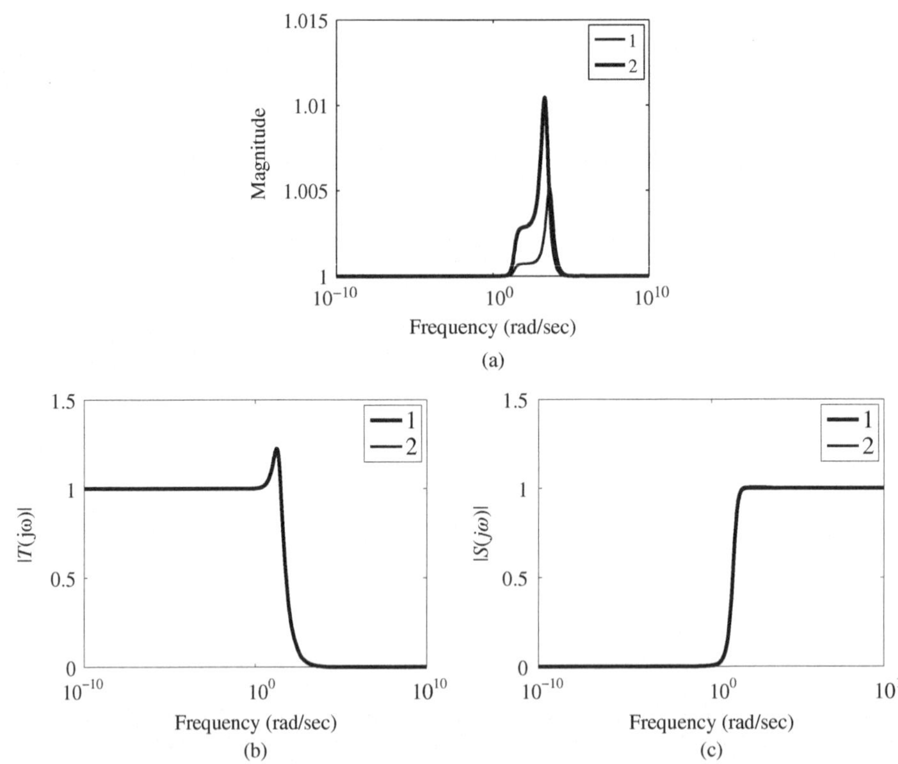

Figure 5.22 Desired and actual sensitivity functions (Key: (1) desired function, (2) actual function). Inner-loop current PI control specification: $\gamma = 0.9$, outer-loop velocity control specification: $\gamma = 0.9$. (a) Frequency weighting function ($|\frac{1}{1-T(j\omega)(1-F(j\omega))}|$) with varying inner-loop PI current control specification (Key: line (1) $\gamma = 0.9$, line (2) $\gamma = 0.8$), (b) Complementary sensitivity function, and (c) Sensitivity function.

$|T(j\omega)(1-F(j\omega))| \to 0$ as $\omega \to \infty$. The quantity $|T(j\omega)(1-F(j\omega))|$ is small at the medium frequency region for a large γ value for the inner-loop current control. Figure 5.22(a) shows the quantity $|\frac{1}{1-T(j\omega)(1-F(j\omega))}|$ for two γ values of 0.8 and 0.9 respectively. Since $|\frac{1}{1-T(j\omega)(1-F(j\omega))}|$ is one at both low and high frequency regions, the desired sensitivity functions are identical to the actual sensitivity functions. However, in the medium frequency region, this quantity is not one, indicating that there is a difference between the desired and actual sensitivity functions. When the inner-loop is controlled using a PI controller with specification of $\gamma = 0.9$, the difference is about 0.5 percent at the maximum magnitude. The results of the desired and actual complementary sensitivity functions are shown in Figure 5.22(b) whilst the desired and actual sensitivity functions are shown in Figure 5.22(c). The results confirm that when the inner-loop current control uses a PI controller with a large γ value that corresponds to a wide closed-loop bandwidth, the difference between the desired and actual sensitivity functions is negligible.

By comparing Figure 5.22(a) (PI controller used in the inner-loop system) with Figure 5.21(a) (P controller used in the inner-loop system), it is seen that there is a small difference in terms of the weighting function $|\frac{1}{1-T(j\omega)(1-F(j\omega))}|$, however, their magnitudes are about the same order. It is evident that the total area that the weighting function $|\frac{1}{1-T(j\omega)(1-F(j\omega))}|$ covers is smaller when the inner-loop current PI controller is used. This indicates that the combination of PI inner-loop control with outer-loop PI control provides some advantages in terms of performance robustness. This conclusion is obtained in the absence of parameter uncertainty. When there is parameter uncertainty in the inner-loop current control system, it is expected that the inner-loop current PI controller will provide a much better robust performance over the current P controller. This is because of the propagation of steady-state errors from inner-loop to outer-loop when a proportional controller is used for current control.

5.4.2 Input Sensitivity Functions for the Outer-loop System

The disturbances existing in the outer-loop velocity control system consist of the input disturbances coming from load torque variations, the harmonics from the PWM switching, and the current sensor bias error. The sources of noises in the outer-loop are from the acoustical noise of PWM, measurement noise of the encoder, which measures the angular position of the motor and measurement noise of the current sensors. To examine how the input disturbance affects the output velocity, the input sensitivity function is calculated using

$$S(s)^i = G(s)(1-T(s)) \times \frac{1}{1-T(s)(1-F(s))}, \tag{5.82}$$

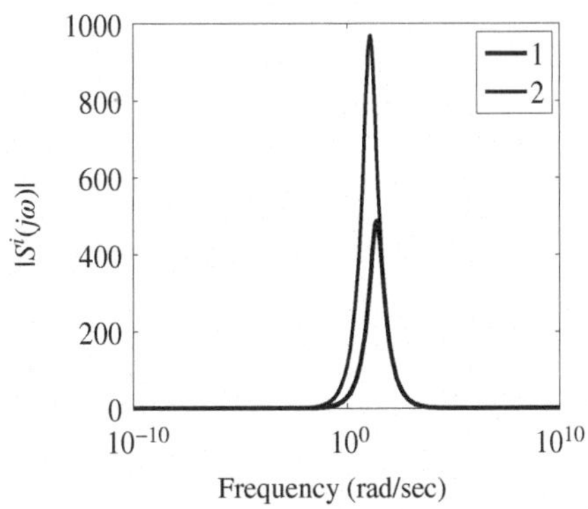

Figure 5.23 Magnitude of actual input sensitivity function. Inner-loop current PI control and outer-loop velocity control specification: $\gamma = 0.9$. Key: inner-loop current controller (1) $\gamma = 0.9$; (2) $\gamma = 0.8$.

where $G(s) = \frac{b}{s+a}$, and for the PMSM, $a = \frac{B}{J_m}$ and $b = \frac{3}{2}\frac{Z_p^2 \phi_{mg}}{J_m}$. Figure 5.23 shows the magnitude of the input sensitivity function where $a = 2.3404$ and $b = 1.5957 \times 10^4$. For this system, the magnitude of the input sensitivity function is quite large in the medium frequency region, because of the large parameter b, which indicates that if any input disturbances fall within this frequency band, they will be amplified. For a larger γ value in the inner-loop PI current control, the magnitude is smaller, which means the closed-loop system will have a better disturbance rejection property. The load torque disturbances are slow variations and have typically low frequency content. Therefore, according to the behavior of the input sensitivity function, the load torque disturbance will be completely rejected.

In summary, in order to maintain performance robustness when using cascade feedback control of velocity, the bandwidth of the inner-loop current control is required to be as large as possible. Namely, if the proportional current controller is used, a large α value is desired, or if a PI current controller is used, a large γ value is desired.

5.5 Analysis of Time-delay Effects

With the time-delay existing in the current-loop system, due to the PWM implementation and measurement delay, the actual transfer function for the q-axis current system is modelled as a first order plus time delay τ_d, that is

$$\tilde{G}(s) = \frac{\frac{1}{L_q}}{s + \frac{R_s}{L_q}} e^{-\tau_d s}, \tag{5.83}$$

where τ_d is a function of the PWM switching frequency and a function of the sampling interval of the current sensor. The time delay due to PWM implementation is commonly approximated by one sampling interval Δt ($\tau_d \approx \Delta t$). The maximum time delay in the current control system in the worst case scenario is $\tau_d \approx 2\Delta t$, which has taken into account of current sensor time delay. Because the time-delay τ_d is not accounted for in the design of either proportional current control or the PI current control, it causes modelling error between the actual plant $\tilde{G}(s)$ and the model $G(s)$ used for the design. With the model defined by $G(s) = \frac{\frac{1}{L_q}}{s+\frac{R_s}{L_q}}$, we can quantify the modeling error by either additive error:

$$\Delta G(s) = \tilde{G}(s) - G(s) = \frac{\frac{1}{L_q}}{s + \frac{R_s}{L_q}}(1 - e^{-\tau_d s}) \tag{5.84}$$

or the multiplicative error:

$$\Delta G_m(s) = \frac{\tilde{G}(s) - G(s)}{G(s)} = (1 - e^{-\tau_d s}). \tag{5.85}$$

The frequency response of multiplicative error is expressed by substituting $s = j\omega$ into (5.85) as

$$\Delta G_m(j\omega) = 1 - \cos(\tau_d \omega) + j\sin(\tau_d \omega) \tag{5.86}$$

which gives its magnitude as

$$|\Delta G_m(j\omega)| = \sqrt{(1 - \cos(\tau_d \omega))^2 + \sin(\tau_d \omega)^2}. \tag{5.87}$$

A calculation of the derivative of (5.87) yields the maximum magnitude occurring at $\tau_d \omega = \pi$, having the value of 2. Figure 5.24 shows the magnitude of the multiplicative modeling error with the time delay $\tau_d = 2\Delta t$ and $3\Delta t$, where $\Delta t = 100$ μs.

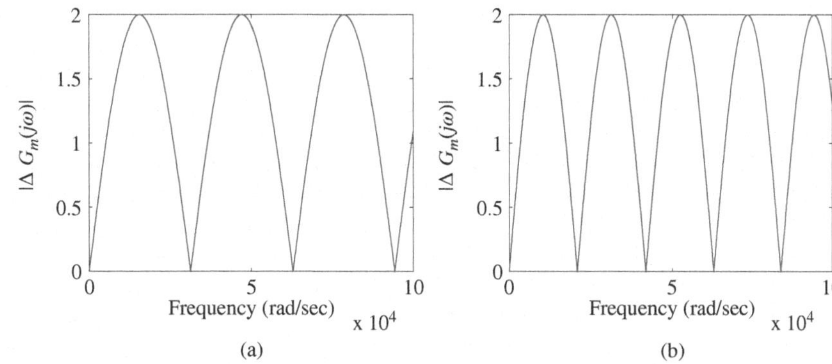

Figure 5.24 Magnitude of multiplicative error as function of the sampling interval Δt. (a) $|\Delta G_m(j\omega)|$ for $\tau_d = 2\Delta t$ and (b) $|\Delta G_m(j\omega)|$ for $\tau_d = 3\Delta t$.

In the case of $d = 2\Delta t$, it is seen that the first maximum value of the multiplicative error is 2 at $\omega = \frac{\pi}{\tau_d} = 15700$ rad/s, and in the case of $\tau_d = 3\Delta t$, the maximum occurs at $\omega = \frac{\pi}{\tau_d} = 10470$ rad/s. In order to guarantee closed-loop stability in the presence of unmodeled time-delay τ_d, the following condition is required to be satisfied:

$$|T(j\omega)||1 - e^{-j\tau_d\omega}| < 1 \tag{5.88}$$

for all $\omega \geq 0$.

5.5.1 PI Control of q-axis Current

If a PI controller is used in the current control, $|T(0)| = 1$ at $\omega = 0$ and is approximately equal to 1 in the low frequency region. It is known that at the desired closed-loop bandwidth $\omega = w_n = \frac{1}{1-\gamma}\frac{R_s}{L_q}$, $|T(j\omega)|$ is approximately $\frac{1}{\sqrt{2}} = 0.707$. A crude estimation of the desired closed-loop bandwidth w_n is that it should be much less than $\frac{\pi}{\tau_d}$ so that $|T(j\omega)| < 0.5$ when the first maximum of $|\Delta G_m(j\omega)|$ (= 2) occurs at $\omega = \frac{\pi}{\tau_d}$. In the example of $R_s = 2.98\ \Omega$ and $L_q = 0.07$ H, when $w_n = \frac{1}{1-\gamma}\frac{R_s}{L_q}$ rad/s, with $\gamma = 0.7$, $w_n = 1490$ rad/s, the closed-loop system will be robustly stable if the time delay is $2\Delta t$ because $|T(j\omega)||\Delta G_m(j\omega)| < 1$ which is confirmed by Figure 5.25(a). However, when γ is increased to 0.9, $|T(j\omega)||\Delta G_m(j\omega)| > 1$ (see Figure 5.25(b)), thus, the closed-loop current control system is no longer guaranteed to be stable with this choice of closed-loop bandwidth.

5.5.2 P Control of q-axis Current

Because the complementary sensitivity function has a magnitude less than 1 at the lower frequency region when the proportional controller is used in the current control, it is expected that the current control loop is more robust in the presence of unmodeled time-delay. Figure 5.26 compares the magnitude of the quantity $T(j\omega)||\Delta G_m(j\omega)$ for $\alpha = 0.7$ and $\alpha = 0.9$. It is seen that the current closed-loop system is robustly stable when $\tau_d = 2\Delta t$ as $|T(j\omega)||\Delta G_m(j\omega)| < 1$. It is apparent that the proportional current control system provides a larger stability margin in the presence of unmodeled time delay.

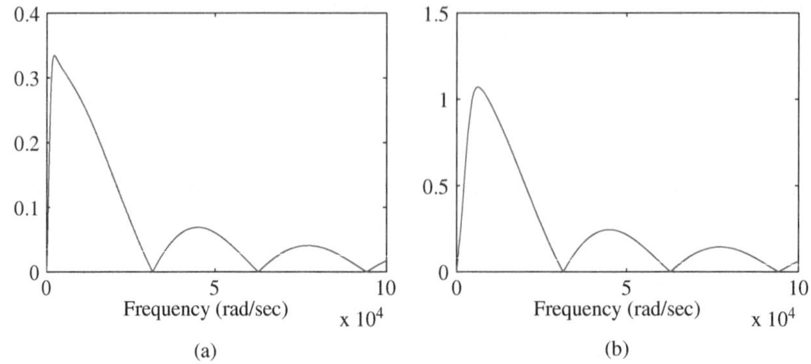

Figure 5.25 $|T(j\omega)||\Delta G_m(j\omega)|$, PI control of q-axis current. (a) $\tau_d = 2\Delta t$ and $\gamma = 0.7$, and (b) $\tau_d = 2\Delta t$ and $\gamma = 0.9$.

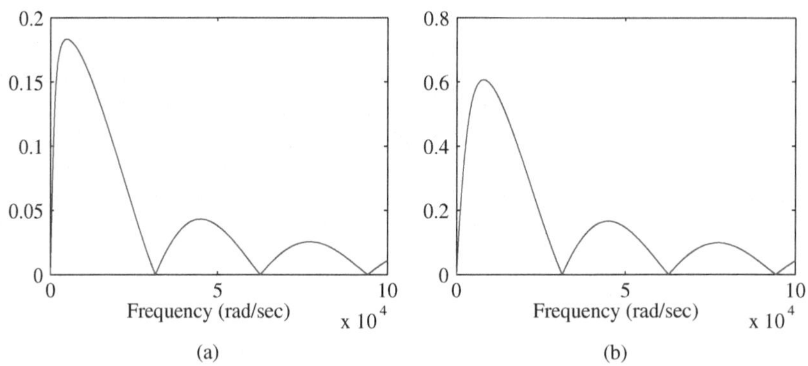

Figure 5.26 $|T(j\omega)||\Delta G_m(j\omega)|$, P control of q-axis current. (a) $\tau_d = 2\Delta t$ and $\gamma = 0.7$, and (b) $\tau_d = 2\Delta t$ and $\gamma = 0.9$.

5.6 Tuning Cascade PI Control Systems for Induction Motor

PI cascade control system design for induction motor is discussed in Chapter 3 (see Sections 3.6–3.7). This section is focused on tuning its closed-loop performance with experimental evaluations. For the induction motor test-bed, the system parameters for the current controller design are given as

$$a = \frac{1}{\tau'_\sigma} = 167.76; \quad b = \frac{1}{r_\sigma \tau'_\sigma} = 9.41.$$

With the specification of the closed-loop bandwidth for the current controller as

$$w_n = \frac{1}{1-\gamma} a$$

and $\xi = 0.707$, the PI controller gain K_c and τ_I for the current controller are:

$$K_c = \frac{2\xi w_n - a}{b}; \quad \tau_I = \frac{2\xi w_n - a}{w_n^2}.$$

The third section in Table 5.3 shows the parameters for the choices of $\gamma = 0.7, 0.8, 0.9, 0.95$.

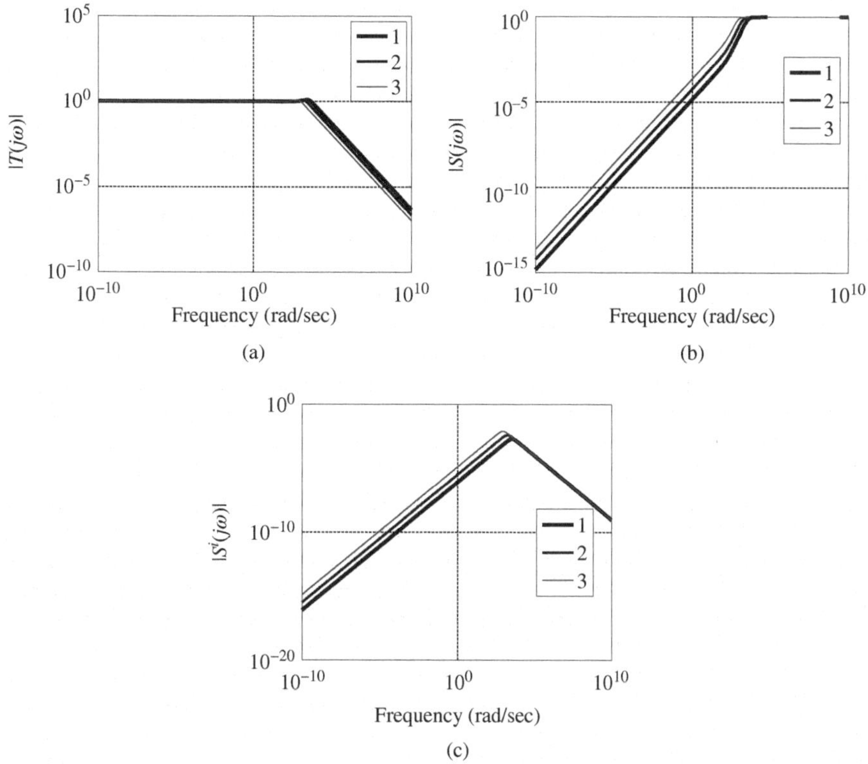

Figure 5.27 Sensitivity function and input sensitivity function for PI current control (induction motor). Key: line (1) $\gamma = 0.95$, line (2) $\gamma = 0.9$, line (3) $\gamma = 0.8$. (a) Complementary sensitivity function, (b) Sensitivity function, and (c) Input sensitivity function.

The complementary sensitivity, sensitivity and input sensitivity functions for the current control system are shown in Figure 5.27. There are six cases from experimental studies, presented in Table 5.3. In the first four case studies (see rows from A to D), the outer-loop velocity controller has the same closed-loop performance specification while the inner-loop current control systems have an increased bandwidth. In the experiments, the proportional control gain K_c is implemented on the output only. The

Table 5.3 Inner-loop PI control on both $d - q$ axes current and outer-loop PI control of velocity. $E_u = \sum(u_{sd}^2 + u_{sq}^2)/M$ and $E_\omega = \sum(\omega_m^* - \omega_m)^2/M$, where M is the data length

	Velocity controller			Current controller			Perform. measure	
	γ	K_c	τ_I	γ	K_c	τ_I	$E_u(\times 10^4)$	$E_\omega\ (\times 10^4)$
A	0.7	0.0095	0.1744	0.7	66.1902	0.0020	1.9664	1.8740
B	0.7	0.0095	0.1744	0.8	108.1978	0.0014	1.9745	1.9489
C	0.7	0.0095	0.1744	0.9	234.2206	7.8326×10^{-4}	2.1728	1.9840
D	0.7	0.0095	0.1744	0.95	486.2662	4.0653×10^{-4}	3.5607	2.1672
E	0.8	0.0155	0.1267	0.8	108.1978	0.0014	1.9725	1.3549
F	0.9	0.0335	0.0686	0.8	108.1978	0.0014	1.9804	0.8952

velocity overshoot response is due to the condition that the induction motor is running without load in order to test the reference following response. For this type of test, the steady-state i_{sq} current is very small, thus a large overshoot of the i_{sq} current occurs during the transient response. This free load test condition is deployed for the induction motor control in this chapter. In contrast, the experimental results will be obtained with appropriate load torque for induction motor control in Chapter 6, where this type of overshoot is avoided in general because of the implementation of the proportional control on the output only.

It is seen that with the increased current control loop bandwidth, the control signal variations increase. In particular, when the bandwidth increases from 10 times of the open-loop bandwidth ($\gamma = 0.9$) to 20 times of the open-loop bandwidth ($\gamma = 0.95$), the variation of the control signals $E_u = \sum(u_{sd}^2 + u_{sq}^2)/M$, where M is the data length, has increased from 2.1728×10^4 to 3.5607×10^4. In the last two case studies (see rows from E to F), the closed-loop bandwidth for the velocity control loop is increased from 5 times ($\gamma = 0.8$) of the open-loop bandwidth of the mechanical system to 10 times of that of the same system while the current loops have the same bandwidth for both cases. The numerical results indicate that the variations of the control signals are about the same; however, the mean squared error for the velocity is reduced when the closed-loop bandwidth of the velocity controller is larger. The experimental results are illustrated via the closed-loop control responses. In particular, comparative results are used to show the closed-loop performance differences when the bandwidth of the outer-loop system has a different performance specification. In Figure 5.28 where the outer-loop velocity control system has a relatively smaller bandwidth $\gamma = 0.8$, the velocity response is slower, but without oscillations. In contrast, In Figure 5.29, when the outer-loop velocity control system has a larger bandwidth, the velocity response is faster, but with oscillations. There are several factors that might cause the closed-loop oscillations, including the unmodeled dynamics from the inner-loop control system and a mismatched mechanical inertia parameter J_m.

5.6.1 *Robustness of Cascade PI Control System*

To test robustness of the cascade control system against physical parameter variations, experiments are performed with respect to changes in both inner-loop and outer-loop systems. Table 5.4 summarizes the numerical results about the robustness studies. For the inner-loop system, the resistance R_s is changed from the nominal value of 11.2 Ω to an increased value of 1.5×11.2 Ω, then to a decreased value of 0.5×11.2 Ω. Changes of resistance typically occur when the motor has been in operation for a long time and its temperature variation leads to this variation. For the outer-loop system, the inertial parameter J_m is changed from the nominal value of $0.00214\ \text{kg}\cdot\text{m}^2$ to an increased value of $3 \times 0.00214\ \text{kg}\cdot\text{m}^2$ to a decreased value of $0.5 \times 0.00214\ \text{kg}\cdot\text{m}^2$. Because the actual physical parameters are unknown, the robustness verification is performed by changing the model parameters used in the

Table 5.4 Robustness analysis on speed control. Inner-loop PI control on both $d-q$ axes current and outer-loop PI control of velocity. $E_u = \sum(u_{sd}^2 + u_{sq}^2)/M$ and $E_\omega = \sum(\omega_m^* - \omega_m)^2/M$, where M is the data length

Variation		Velocity controller			Current controller			Perform. measure	
R_s	J_m	γ	K_c	τ_I	γ	K_c	τ_I	$E_u(\times 10^4)$	$E_\omega(\times 10^4)$
×0.5		0.8	0.0155	0.1267	0.8	74.2058	0.0021	1.9478	1.3597
×1.5		0.8	0.0155	0.1267	0.8	142.1898	0.0011	2.0269	1.3562
	×0.5	0.8	0.0155	0.0634	0.8	108.1978	0.0014	1.9903	1.3229
	×3.0	0.8	0.0155	0.3802	0.8	108.1978	0.0014	2.0118	2.6659

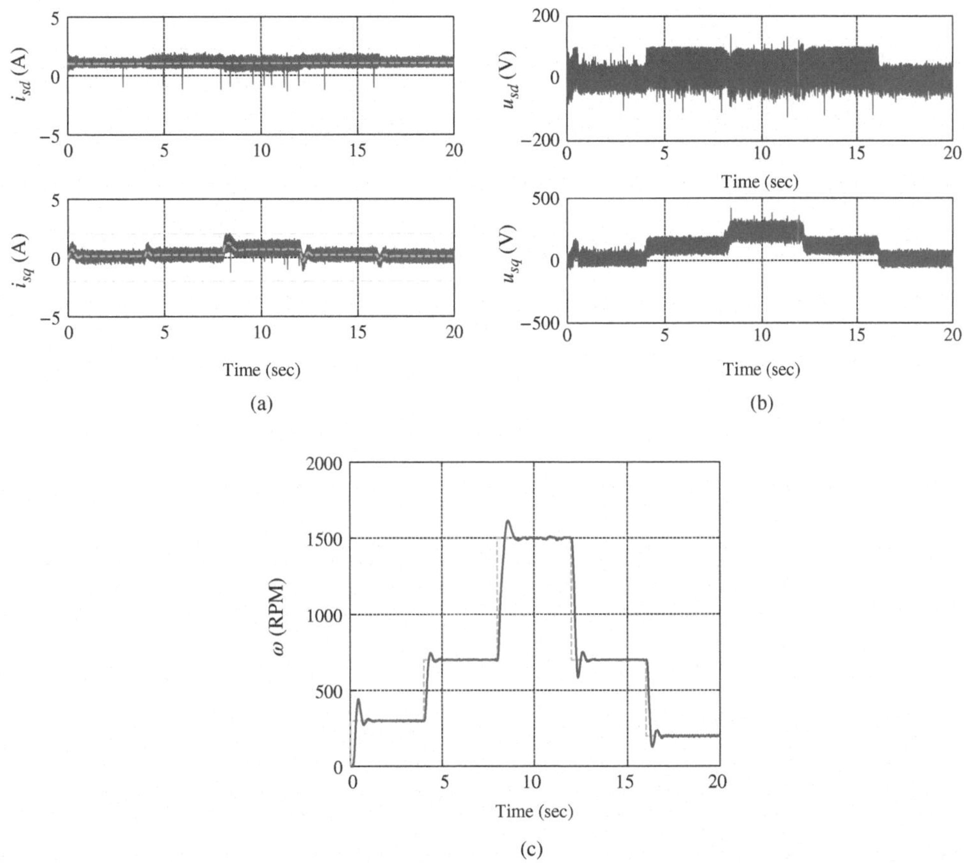

Figure 5.28 Induction motor control. PI current controller $\gamma = 0.8$, $\xi = 0.707$; outer-loop velocity control system specification: $\gamma = 0.8$ and $\xi = 0.707$. (a) Stator current (solid) and reference (dashed), (b) Stator voltage, and (c) Velocity (solid) and reference (dashed).

PI control system design. Therefore, when the model parameter changes, the controller parameters may also change.

5.6.1.1 Robustness of Current Control System

The first two cases presented in Table 5.4 are related to the variations of the resistance in the current system while the outer-loop velocity controller takes the nominal design. Because the closed-loop specification in the current control is related to the open-loop pole, the current controller has a lower proportional gain (74.2058) than the nominal proportional gain (108.1978) and a larger integral time constant (0.0021) than the nominal integral time constant (0.0014) by comparing the first row of the current controller with the third row of current controller. Therefore, we may conclude that with this current controller design, by using a smaller resistance value in the design, the resultant current control system will have a slower closed-loop response. An overestimated resistance value will lead to a larger proportional gain and a smaller integral time constant by comparing the second row of the current controller with the third row of current controller. We conclude that using a larger resistance value will lead to a faster closed-loop

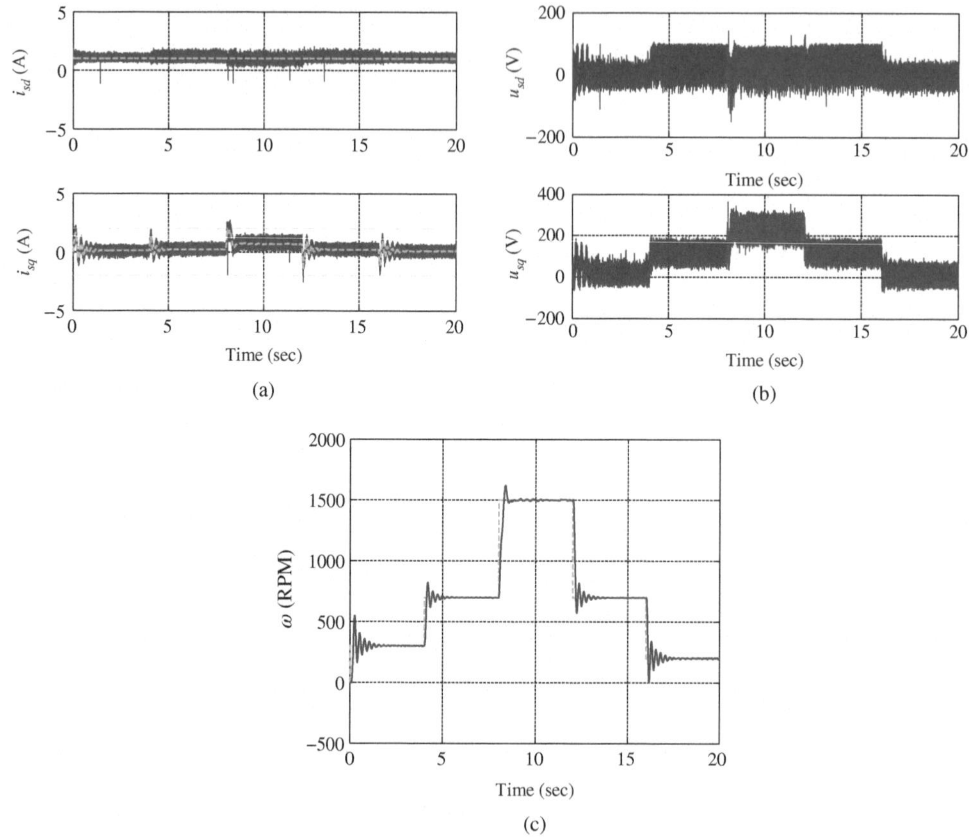

Figure 5.29 Induction motor control. PI current control. PI controller $\gamma = 0.8$, $\xi = 0.707$; outer-loop velocity control system specification: $\gamma = 0.9$ and $\xi = 0.707$. (a) Stator current (solid) and reference (dashed), (b) Stator voltage, and (c) Velocity (solid) and reference (dashed).

current response. For a conservative design, one may wish to consider the reduction of resistance if the motor is to be in operation for a long time by choosing a smaller resistance value for the current controller design. As we expected, the larger proportional gain and smaller integral time constant lead to a larger variation of the control signals (see E_u); however, it has a little impact on the variation of velocity (see E_ω).

Figure 5.30 shows the closed-loop responses when R_s is half of the nominal value. In comparison, Figure 5.31 shows the closed-loop responses when R_s is 1.5× the nominal value. By comparing these two figures, it is seen that the variations in current control loop has a little impact on the velocity response for this study; however, the variations of the control signals are larger when R_s is over-estimated.

5.6.1.2 Robustness of Velocity Control System

The last two cases presented in Table 5.4 are related to the variations of the inertia parameter J_m in the velocity control system while the inner-loop current controller takes the nominal design. Because the closed-loop bandwidth is specified in relation to the open-loop bandwidth, the change of the parameter J_m does not alter the value of the proportional gain K_c; however, it changes the integral time constant τ_I.

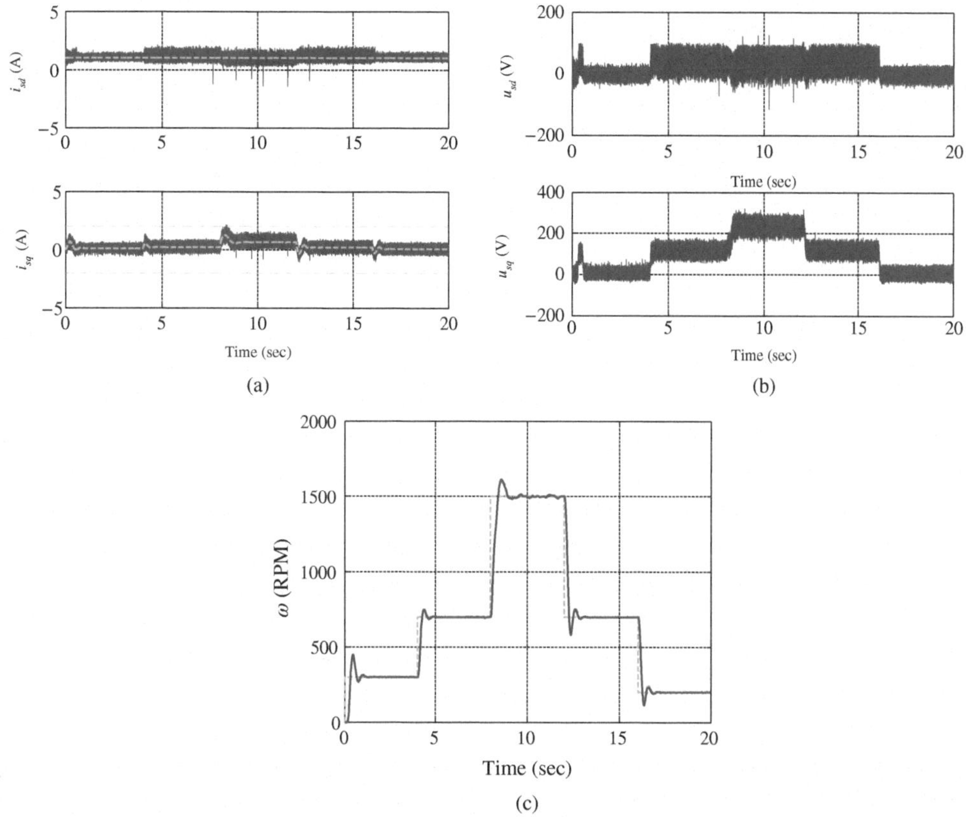

Figure 5.30 Induction motor control. Current control system specifications: PI controller $\gamma = 0.8$, $\xi = 0.707$; outer-loop velocity controller: $\gamma = 0.8$ and $\xi = 0.707$. The R_s used in the design is half of the nominal value. (a) Stator current (solid) and reference (dashed), (b) Stator voltage, and (c) Velocity (solid) and reference (dashed).

A smaller inertia J_m will lead to a smaller integral time constant and in contrast a larger J_m will lead to a larger integral time constant. Therefore, in the design, if a smaller value of J_m is selected, in general the closed-loop response speed is faster because of the smaller integral time constant. The variation of the velocity is smaller when using a smaller J_m as measured by the value of E_ω. In a conservative design, one wishes to use a larger J_m value to enhance the robustness property of the closed-loop control system.

Figure 5.32 shows the closed-loop responses when the inertia parameter J_m is smaller than the nominal value. In comparison, Figure 5.33 shows the responses when J_m is larger than the nominal value. When the inertia parameter J_m is smaller, it is seen that the closed-loop velocity response exhibits oscillatory behavior (see Figure 5.32); however, when the inertia parameter J_m is larger, the closed-loop velocity response is slower and without oscillation. The experimental results confirm that choosing a smaller J_m value will enhance the robustness property of the closed-loop velocity control system.

5.6.2 *Robustness Study Using Nyquist Plot*

Nyquist plots of the outer-loop control system provide an effective means to determine the bandwidth w_n to be used in the design. A good design should have reasonable gain and phase margins to ensure

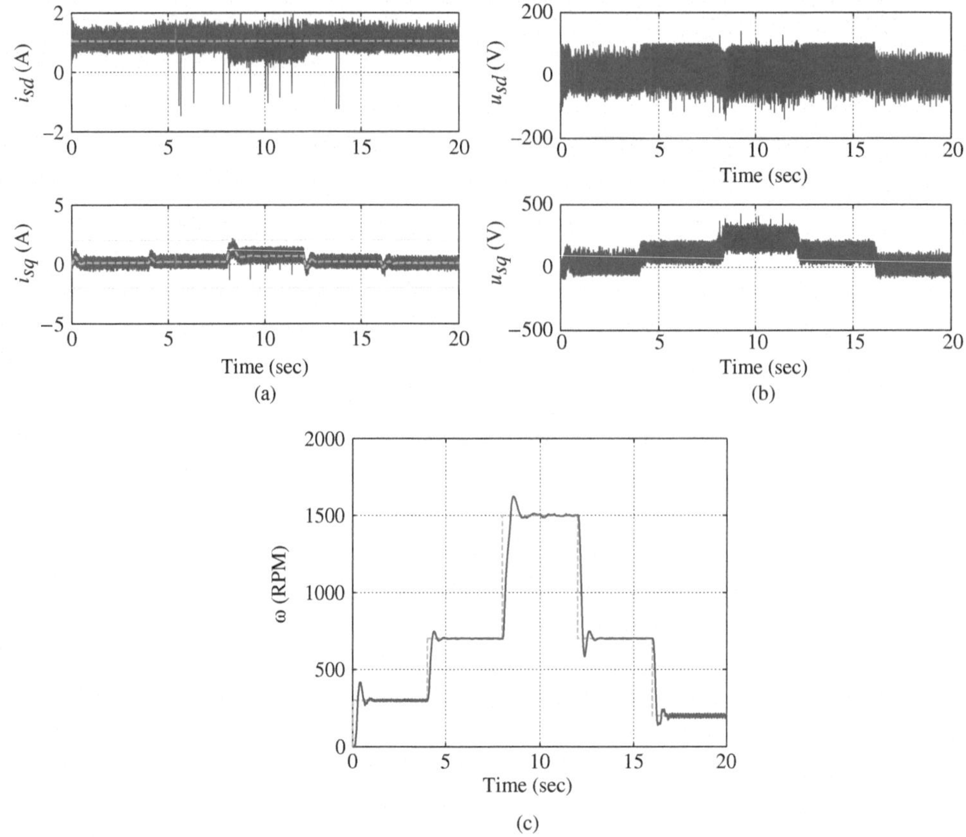

Figure 5.31 Induction motor control. Current control system specifications: PI controller $\gamma = 0.8$, $\xi = 0.707$; outer-loop velocity controller: $\gamma = 0.8$ and $\xi = 0.707$. The R_s used in the design is 1.5× the nominal value. (a) Stator current (solid) and reference (dashed), (b) Stator voltage, and (c) Velocity (solid) and reference (dashed).

that the closed-loop system is robustly stable in the presence of factors known and unknown. When using the Nyquist stability criterion, we examine the frequency response of the loop transfer function ($M(j\omega)$) that contains the transfer functions for the mechanical system, the outer-loop PI controller and the neglected inner-loop system. The criterion states that a feedback control system with single input and single output is stable if and only if, for the frequency response of the loop transfer function, number of counter clockwise encirclements of the $(-1, 0)$ point is equal to the number of poles of this loop transfer function with positive real parts. In the control problems of electrical drives and power converters, this loop transfer function does not contain any poles that have positive real parts; thus, for the closed-loop stability of the single-input and single-output system, the frequency response should not encircle $(-1, 0)$ point on the complex plane.

A good design should have reasonable gain and phase margins to ensure that the closed-loop system is robustly stable in the presence of factors known and unknown. Gain margin is defined as $GM = \frac{1}{k_g}$, where k_g is the distance between the origin of the complex plane and the point that $M(j\omega)$ intersects the real axis (see Figures 5.34 and 5.35). It means that if the loop gain were to exceed the reciprocal of k_g,

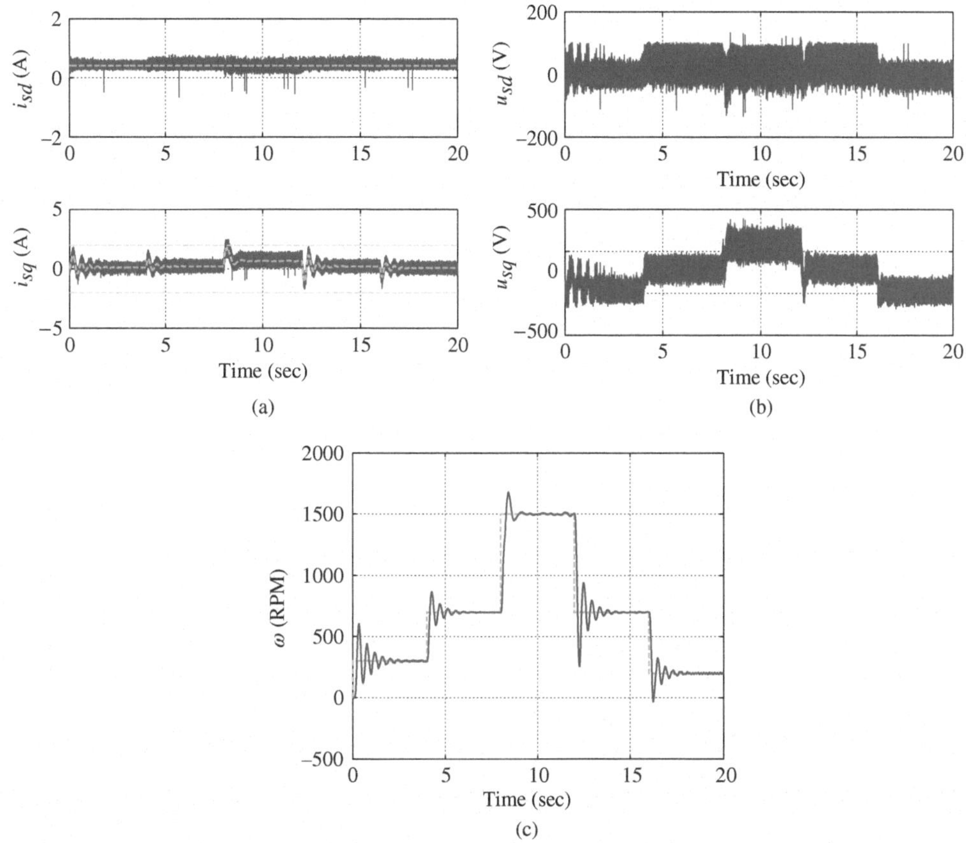

Figure 5.32 Induction motor control. Current control system specifications: PI controller $\gamma = 0.8$, $\xi = 0.707$; outer-loop velocity controller: $\gamma = 0.8$ and $\xi = 0.707$. The J_m used in the design is half of the nominal value. (a) Stator current (solid) and reference (dashed), (b) Stator voltage, and (c) Velocity (solid) and reference (dashed).

then the closed-loop system would become unstable. Clearly, there is no such intersection point for the Nyquist plots shown in both Figures. This indicates that the cascade control system has "infinite" large gain margin. Phase margin is defined as the additional phase lag that could be associated with $M(j\omega)$ before the closed-loop system became unstable. It is the angle between the negative real axis and the line that intersects the circle $|M(j\omega)| = 1$ (see Figure 5.35).

5.6.2.1 Robustness in Resistance Variations

Figure 5.34 shows the variations of the resistance have little effect on the Nyquist loci of the cascade control system, where the three Nyquist plots are not distinguishable. This is one of the advantages when using cascade control system to safeguard parameter variations in the inner-loop control system. The experimental validations also confirm that indeed the variations on R_s have very small effects on the closed-loop velocity responses (see Table 5.4).

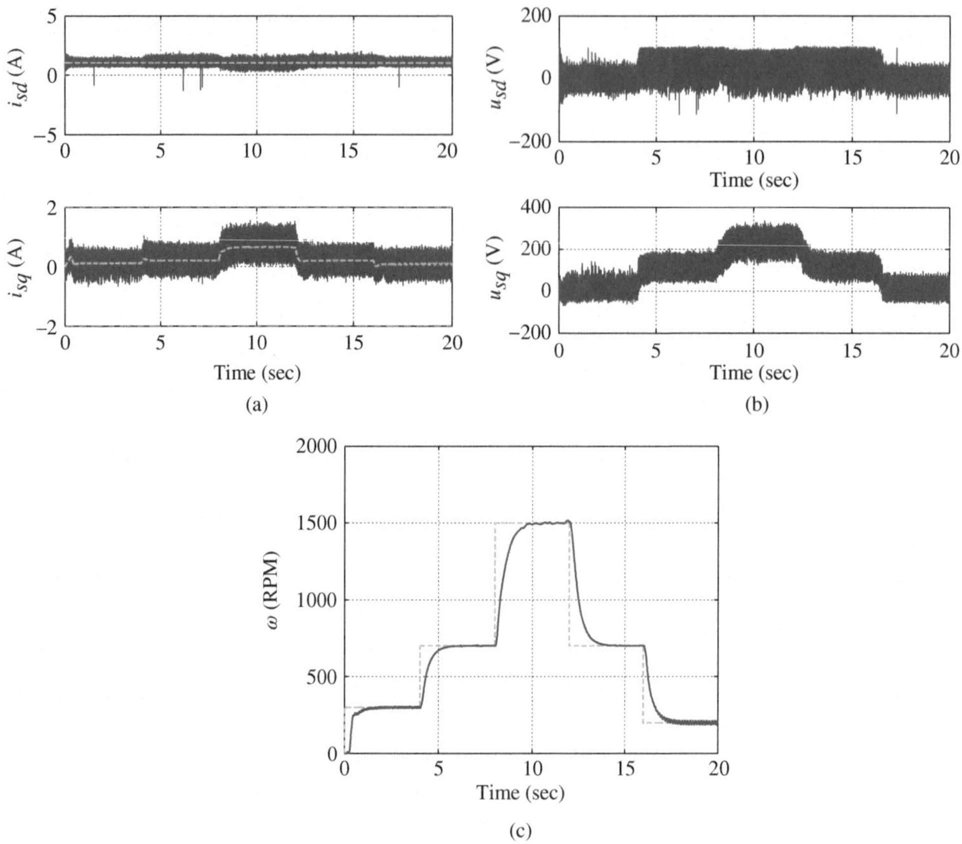

Figure 5.33 Induction motor control. Current control system specifications: PI controller $\gamma = 0.8$, $\xi = 0.707$; outer-loop velocity controller: $\gamma = 0.8$ and $\xi = 0.707$. The J_m used in the design is 3 times the original value. (a) Stator current (solid) and reference (dashed), (b) Stator voltage, and (c) Velocity (solid) and reference (dashed).

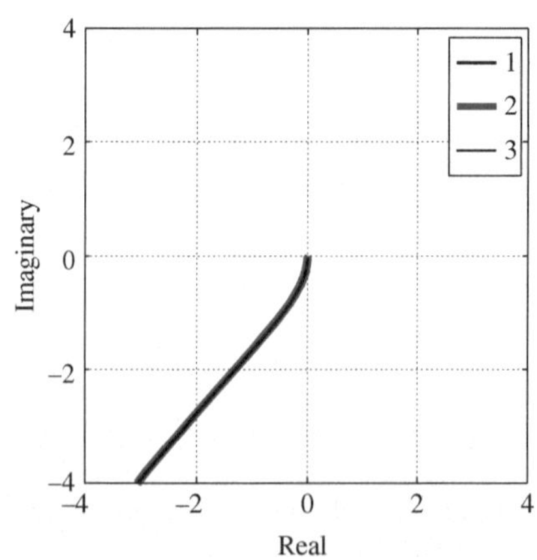

Figure 5.34 Nyquist plots with respect to changes in resistance with PI velocity control. Key: line (1) $R_s = 0.5R_s^0$, line (2) $R_s = 1.5R_s^0$ and line (3) $R_s = R_s^0$, where $R_s^0 = 11.2\ \Omega$.

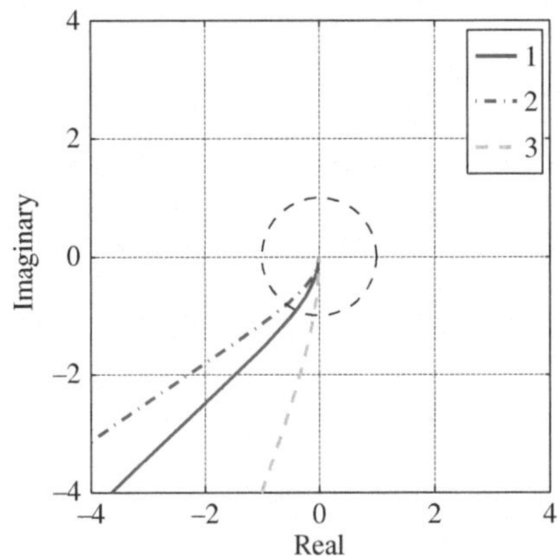

Figure 5.35 Nyquist plots with a unit circle. Key: line (1) with $J_m = J_m^0$, line (2) $J_m = 0.5J_m^0$, line (3) $J_m = 3J_m^0$ where $J_m^0 = 0.00214\ \text{kg}\cdot\text{m}^2$.

5.6.2.2 Robustness in Presence of Inertia Parameter Variation

To examine the changes of inertia parameter J_m on the robustness of the cascade control of velocity system, the parameter J_m was increased to 3 times of the nominal value (=0.0064) and then reduced to 0.5 of the nominal value (= 0.0013). With these two inertia values, two PI controllers are designed with $\tau_I^\omega = 0.2969$ for the larger inertia value and $\tau_I^\omega = 0.0594$ for the smaller inertia value, respectively. The proportional control gain K_c^ω is unchanged with respect to J_m in the proposed design. Figure 5.35 shows the Nyquist plots for the three cascade control systems.

A unit circle is superimposed on the Nyquist loci to help identify the phase margins for the three system, which is the angle between the left side of the real axis and the intersection point of the unit circle with the Nyquist loci. At the nominal design with $J_m = 0.0064$, the phase margin is 67°; with the increase of the inertia parameter J_m in the model, the phase margin is increased to 85°; but with the decrease of the inertia parameter in the model, the phase margin is decreased to 54°. The phase margin parameters indicate that the closed-loop response speed will become slower when the parameter J_m used in the design is larger than the actual inertia parameter, and the closed-loop response speed will become faster or oscillatory when the parameter J_m used in the design is smaller than the actual inertia parameter. The experimental results show that mismatch of the inertia parameter leads to sensitivity of the closed-loop control performance, particularly in the lower speed range (see Figure 5.32).

5.7 Tuning PI Control Systems for Power Converter

The design of the current controllers and the voltage controller has been discussed in detail in the previous chapter. The design equations are briefly listed here as a review, also in conjunction with the test-bed.

5.7.1 Overview of the Designs

The controller for the q-axis current loop is designed based on the first order transfer function model:

$$\frac{I_q(s)}{\hat{S}_q(s)} = \frac{-\frac{1}{2L_s}}{s + \frac{R_s}{L_s}}, \tag{5.89}$$

where $\hat{S}_q(s)$ is the Laplace transform of the auxiliary control variable, which is related to the actual control variable:

$$-\frac{1}{2L_s}\hat{S}_d(t) = -\frac{S_d(t)v_{dc}(t)}{2L_s} + \omega_g i_q(t) + \frac{E_d}{L_s}. \tag{5.90}$$

In the test-bed the resistance $R_s = 0.2\ \Omega$ and the inductance $L_s = 6.3 \times 10^{-3}$ H.

In order to maintain unity power factor in the steady-state, the set-point signal to the q-axis current is 0 and a PI controller is always used in the closed-loop control of the q-axis current. Therefore, with the pole-assignment controller design, the proportional controller gain K_c and the integral time constant are chosen to be

$$K_c = \frac{2\xi w_n - a}{b}; \quad \tau_I = \frac{2\xi w_n - a}{w_n^2}, \tag{5.91}$$

where $a = \frac{R_s}{L_s}$ and $b = -\frac{1}{2L_s}$. Specifically, with the damping coefficient $\xi = 0.707$, the closed-loop bandwidth w_n is selected as

$$w_n = \frac{1}{1-\gamma}\frac{R_s}{L_s}, \tag{5.92}$$

where the parameter $0 < \gamma < 1$ is assigned to achieve the desired closed-loop response speed.

For the outer-loop PI controller design, the following integrator model is used:

$$\frac{V_{dc}(s)}{I_d^*(s)} \approx \frac{3}{4}\frac{S_d^{ss}}{C_{dc}}\frac{\alpha}{s}. \tag{5.93}$$

Here, since a PI controller is used in the d-axis current control, the parameter α is assigned to 1, as the steady-state of the inner-loop current control gain is unity with PI control. The capacitor C_{dc} is 2.96×10^{-4} F. The steady-state value of control variable S_d is chosen as 0.7965.

In the outer-loop controller design, since the open-loop pole for the integrating system is at $s = 0$, there is no trivial benchmark for the selection of the desired closed-loop poles. However, what one could do is to select the desired closed-loop bandwidth for the outer-loop system in relation to the inner-loop system. The outer-loop bandwidth should be between 0.25 to 0.5 times the inner-loop bandwidth. The main reason why the outer-loop bandwidth is narrower than the inner-loop system bandwidth is because the dynamics from the inner-loop closed-loop system are neglected in the design of the outer-loop control system. In the specification of the inner-loop bandwidth, it is parameterized as $\frac{1}{1-\gamma}\frac{R_s}{L_s}$ for a PI controller. Say, if $\gamma = 0.9$, the inner-loop bandwidth is about 318 rad/s. In the case studies, the candidates of the parameter w_n for the outer-loop system are chosen to be 80 and 150. If a faster outer-loop voltage response is desired, then the inner-loop d-axis current control system should have a wider bandwidth to achieve this higher performance demand.

Table 5.5 lists the PI controller parameters for controlling i_d and i_q currents and the PI controller parameters for the outer-loop voltage control. There are nine cases presented in this table. The first three cases (A-C) show the variations of proportional gain and integral time constant for the current controllers where the bandwidth of outer-loop voltage controller is fixed as $w_n = 80$ rad/s, but the bandwidths of both current control systems increase. It is seen that as γ increases, the magnitude of the proportional gain increases, but the integral time constant decreases. All the proportional controller gains for the current controllers are negative because the steady-state gains of both d-axis and q-axis current systems are negative. The next three cases (D-F) show that as the outer-loop bandwidth w_n increases from 80 rad/s to 150 rad/s, the proportional controller gain of the voltage controller increases while the integral time constant decreases. The final three cases (G-I) show that as the outer-loop bandwidth w_n increases to 300 rad/s, the proportional controller gain of the voltage controller is further increased and the integral time constant is further decreased.

Table 5.5 Inner-loop PI control of both d-axis and q-axis current and outer-loop PI control of voltage. $E_s = \sum(S_d^2 + S_q^2)/M$, $E_{vdc} = \sum(V_{dc}^* - V_{dc})^2/M$ and $E_i = \sum((I_d^* - I_d)^2 + (I_q^* - I_q)^2)/M$, where M is the data length

	Voltage controller			i_q controller			i_d controller			Perform. measure		
	w_n	K_c	τ_I	γ	K_c	τ_I	γ	K_c	τ_I	E_s	E_{vdc}	E_i
A	80	0.0561	0.0177	0.8	−2.428	0.0076	0.8	−2.428	0.0076	0.5648	8.7256	0.4918
B	80	0.0561	0.0177	0.9	−5.256	0.0041	0.9	−5.256	0.0041	0.5706	8.4745	0.3742
C	80	0.0561	0.0177	0.95	−10.912	0.0021	0.95	−10.912	0.0021	0.5616	8.1558	0.3056
D	150	0.1051	0.0094	0.8	−2.428	0.0076	0.8	−2.428	0.0076	0.5642	3.6898	0.5014
E	150	0.1051	0.0094	0.9	−5.256	0.0041	0.9	−5.256	0.0041	0.5753	3.2023	0.3709
F	150	0.1051	0.0094	0.95	−10.912	0.0021	0.95	−10.912	0.0021	0.5616	3.0648	0.3050
G	300	0.2102	0.0047	0.8	−2.428	0.0076	0.8	−2.428	0.0076	0.5785	2.4530	0.5044
H	300	0.2102	0.0047	0.9	−5.256	0.0041	0.9	−5.256	0.0041	0.5733	1.918	0.3688
I	300	0.2102	0.0047	0.95	−10.912	0.0021	0.95	−10.912	0.0021	0.5656	1.7604	0.3049

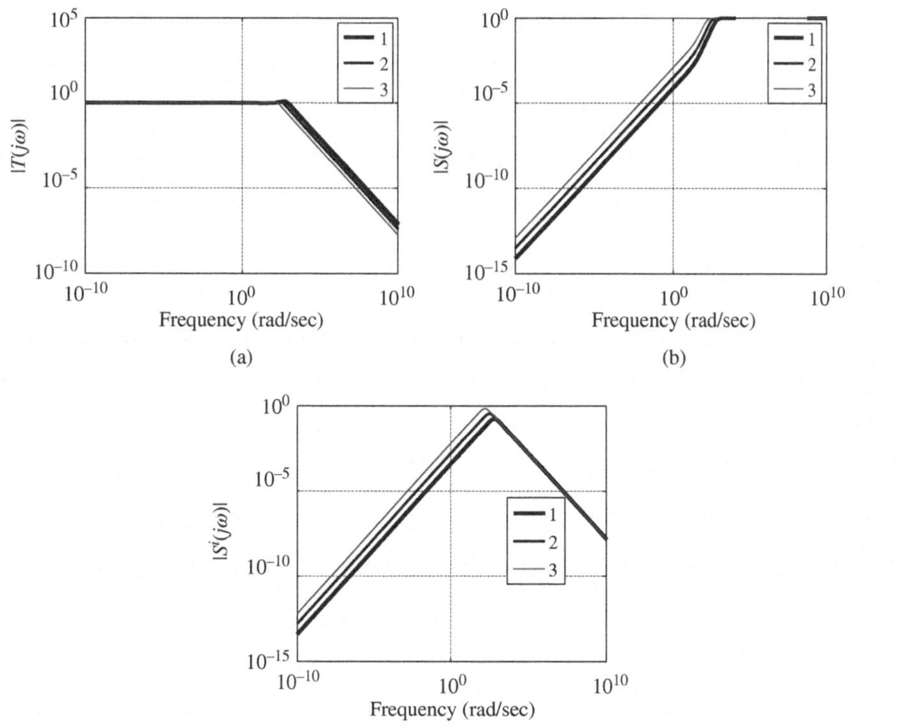

Figure 5.36 Sensitivity function and input sensitivity function for PI current control (power converter). Key: line (1) $\gamma = 0.95$, line (2) $\gamma = 0.9$, line (3) $\gamma = 0.8$. (a) Complementary sensitivity function, (b) Sensitivity function, and (c) Input sensitivity function.

5.7.2 Tuning the Current Controllers

Figure 5.36 shows the sensitivity functions for the PI current control for $\gamma = 0.95$, $\gamma = 0.9$ and $\gamma = 0.8$. The complementary sensitivity function shown in Figure 5.36(a) illustrates that its magnitude for all cases is unity at the low frequency region, indicating that the current output signal will track its reference signal without steady-state error. The bandwidth of the current control loop is calculated as $\frac{1}{1-\gamma}\frac{R_s}{L_s}$ rad/s, therefore, for $\gamma = 0.95$, the bandwidth of the current closed- loop system is about 20 times of that of the open-loop system. The sensitivity function shown in Figure 5.36(b) illustrates that at the zero frequency, its magnitude for all cases is zero, therefore, constant disturbances will be rejected without steady-state error; however, the magnitude at the high frequency is unity, indicating that high frequency noise will be amplified in the current control system. The input sensitivity function shown in Figure 5.36(c) illustrates that at both low and high frequency regions, its magnitude is small, indicating that disturbances occurring at the input variable, such as the error signal from the PWM switching, will be attenuated at both low and high frequency regions; however, in the medium frequency region, they will be amplified. Furthermore, the magnitude of the input sensitivity function is smaller when a larger γ value is used, namely a stronger effect of input disturbance rejection will be achieved. Because the PWM switching errors occur at the d-axis and q-axis voltages, which are the input variables to the current control systems, and these error signals contain undesired harmonics, it is essential to overcome their influence by using a large γ value to achieve a wide bandwidth for the current control system. On the other hand, the wide bandwidth in the current control system will amplify various noises existing in a power converter. Therefore, the ultimate choice of the parameter γ for the current control systems will be based on the balance between harmonic disturbance rejection and noise attenuation. It is envisaged that the closed-loop performance of a current control system will continue to improve when the parameter γ increases because a wider closed-loop bandwidth will result in a better rejection of current harmonics; however, there could be a critical point where further increasing the parameter γ leads to severe amplification of the noises in the current control system, and the benefit of harmonic reduction via wider closed-loop bandwidth diminishes.

5.7.3 Tuning Voltage Controller

Another benefit of a wider closed-loop bandwidth (or a larger γ value) can be explored for the design of the outer-loop voltage controller in terms of actual closed-loop performance and robust stability, which is discussed as follows.

The desired complementary sensitivity function for the PI controlled voltage in the outer-loop system power is

$$T(s) = \frac{2\xi w_n s + w_n^2}{s^2 + 2\xi w_n s + w_n^2}, \tag{5.94}$$

where the design model (5.93) is used in the calculation of the complementary sensitivity function. The PI controller parameters are shown in Table 5.5 where with $\xi = 0.707$, w_n is selected as 80 rad/s and 150 rad/s respectively. Similarly, based on the complementary sensitivity function (5.94), the desired sensitivity and input sensitivity functions for the outer-loop system are

$$S(s) = 1 - T(s) = \frac{s^2}{s^2 + 2\xi w_n s + w_n^2} \tag{5.95}$$

$$S^i(s) = \frac{3}{4}\frac{S_d^{ss}}{C_{dc}}\frac{1}{s}S(s) = \frac{3}{4}\frac{S_d^{ss}}{C_{dc}}\frac{s}{s^2 + 2\xi w_n s + w_n^2}. \tag{5.96}$$

Because in the design of the PI voltage controller the dynamics of the current closed-loop control system are neglected, the characteristics of the three sensitivity functions are not exactly realized in the voltage control system. There are discrepancies between the desired sensitivity functions which are specified by the designer and the actual sensitivity functions of the closed-loop system because of neglected dynamics in the inner-loop current control system. Let $F(s)$ represent the neglected dynamics in the inner-loop current control system. With $a = \frac{R_s}{L_s}$, more specifically, $F(s)$ is calculated as function of γ

$$F(s) = \frac{(1-\gamma)(2\xi - 1 + \gamma)a^{-1}s + 1}{(1-\gamma)^2 a^{-2}s^2 + 2\xi(1-\gamma)a^{-1}s + 1}. \tag{5.97}$$

For notational simplicity, we define

$$W(s) = \frac{1}{1 - T(s)(1 - F(s))}. \tag{5.98}$$

Following from the procedures used previously in the analysis of PMSM (see Equation (5.75)) the actual sensitivity functions in relation to the desired sensitivity functions are then expressed as

$$\begin{aligned} T(s)^{actual} &= T(s)F(s) \times \frac{1}{1 - T(s)(1 - F(s))} \\ &= \frac{2\xi w_n s + w_n^2}{s^2 + 2\xi w_n s + w_n^2} F(s) \times W(s) \end{aligned} \tag{5.99}$$

$$\begin{aligned} S(s)^{actual} &= S(s) \times \frac{1}{1 - T(s)(1 - F(s))} \\ &= \frac{s^2}{s^2 + 2\xi w_n s + w_n^2} W(s) \end{aligned} \tag{5.100}$$

$$\begin{aligned} S^i(s)^{actual} &= S^i(s) \times \frac{1}{1 - T(s)(1 - F(s))} \\ &= \frac{3}{4} \frac{S_d^{ss}}{C_{dc}} \frac{s}{s^2 + 2\xi w_n s + w_n^2} W(s). \end{aligned} \tag{5.101}$$

Equations (5.99 –5.101) indicate that if $|W(j\omega)|$ is approximately one for all $\omega \geq 0$, the actual sensitivity functions can be closely approximated using the desired sensitivity functions. Because $W(j\omega)$ is a function of the desired complementary sensitivity function $T(j\omega)$, the bandwidth of the voltage control system w_n has an effect on $|W(j\omega)|$.

5.7.3.1 Variation of Outer-loop Bandwidth, *Wn*

Figures 5.37–5.38 illustrate the effects of the outer-loop bandwidth w_n on the weighting function and the sensitivity functions. In this case, the bandwidth of the current closed-loop control system is fixed to 318 rad/s ($\gamma = 0.9$). If the outer-loop bandwidth is 30 (that is approximately 1/10 of the inner-loop bandwidth), as shown in Figure 5.37(a), the weighting function $|W(j\omega)|$ is close to one over the median frequency region (it has a maximum of 1.1) and equal to one for both low and high frequency regions, indicating that the discrepancies between the desired and actual sensitivity functions are small. In contrast, if w_n is 300 that is close to the same bandwidth of the inner-loop control system, then $|W(j\omega)|$ has a maximum of 6.5, which is very large and will significantly alter the characteristics of the actual sensitivity functions over the median frequency region and cause significant performance uncertainty. Figure 5.37(b) shows $|W(j\omega)|$ for w_n = 80 rad/s and 150 rad/s respectively, which again confirms that as the bandwidth of the outer-loop system increases with respect to a fixed bandwidth of the inner-loop system, the magnitude of $|W(j\omega)|$ increases over the medium frequency region, thus the degree of performance uncertainty increases. However, in comparison with the case where w_n = 300 rad/s, $|W(j\omega)|$ is three times smaller when w_n = 150 rad/s.

To investigate further, Figure 5.38 illustrates the comparison results for the actual sensitivity functions when using w_n = 80 rad/s and w_n = 150 rad/s respectively. For clarity, the linear scaled y-axis

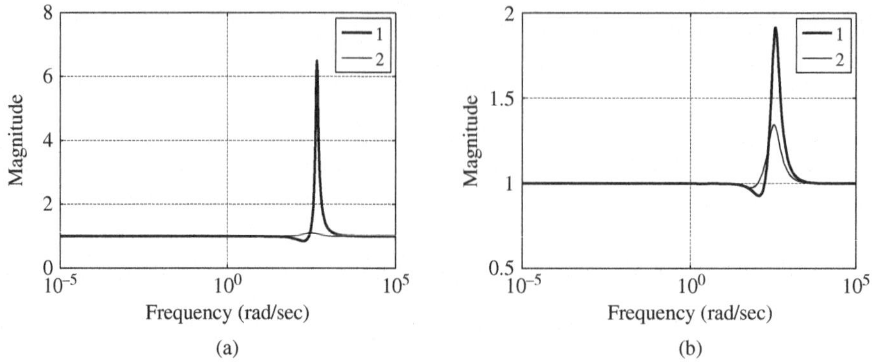

Figure 5.37 Weighting function ($|\frac{1}{1-T(j\omega)(1-F(j\omega)}|$) for outer-loop voltage control (power converter), where the bandwidth of the current inner-loop control system is fixed to 318 rad/s. (a) Key: line (1) $w_n = 300$ rad/s, line (2) $w_n = 30$ rad/s. (b) Key: line (1) $w_n = 150$ rad/s, line (2) $w_n = 80$ rad/s.

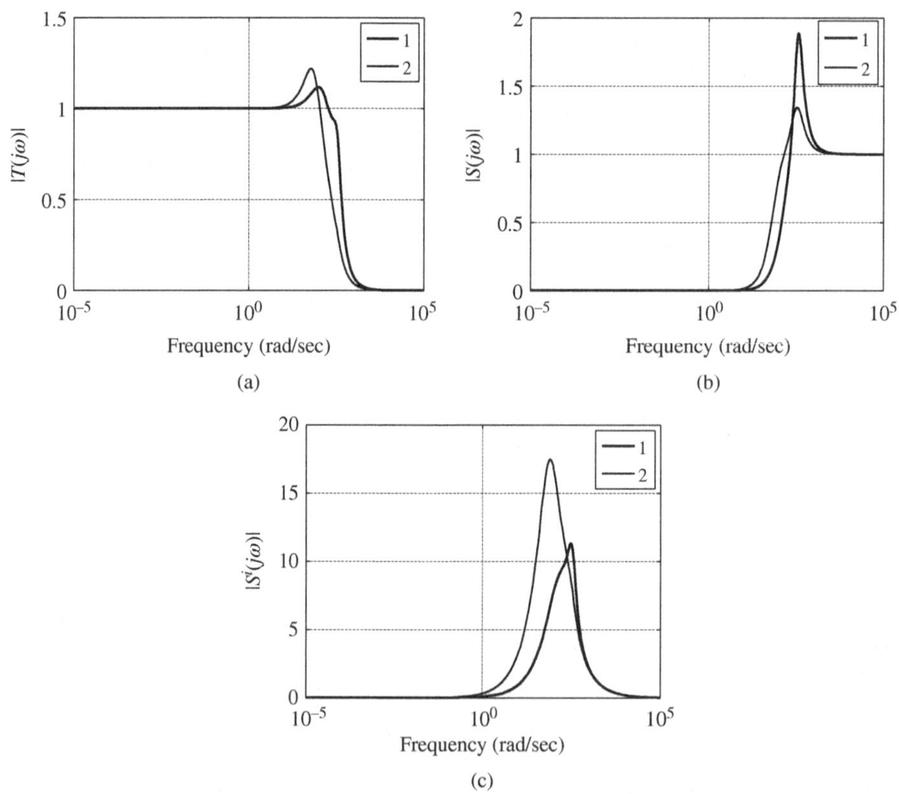

Figure 5.38 Sensitivity functions for the outer-loop voltage control (power converter) where PI current controllers are used. Inner-loop current control system has bandwidth 318 rad/s. Key: line (1) $w_n = 150$ rad/s, line (2) $w_n = 80$ rad/s. (a) Actual complementary sensitivity function, (b) Actual sensitivity function, and (c) Actual input sensitivity function.

is deployed in the plots. In Figure 5.38(a), the magnitude of the complementary sensitivity function shows that the bandwidth of the closed-loop system when using $w_n = 150$ is wider than the one when using $w_n = 80$ rad/s, meaning that the closed-loop response speed to a reference signal is faster, but the closed-loop system is more sensitive to measurement noise. From Figure 5.38(b), a much larger peak of the magnitude of the sensitivity is seen in the case of $w_n = 150$ rad/s. If a disturbance has a frequency content corresponding to the region where the peak occurs, then it will be amplified. The magnitude of the input sensitivity shown in Figure 5.38(c) leads to the conclusion that the property of input disturbance rejection when using $w_n = 150$ rad/s is much superior than the one with $w_n = 80$ rad/s because its magnitude is about half of the later case. Since disturbance rejection is often the main control objective for a power converter, the control system with the bandwidth $w_n = 150$ rad/s is preferable when the inner-loop current control bandwidth is fixed to 318 rad/s.

5.7.3.2 Relative Relationship of the Bandwidths between the Inner and Outer Loops

Figure 5.39 shows the results of the second investigation to reveal the relative relationship of the bandwidths between the inner and outer loops. In this case, the γ value for the current controllers is increased to 0.95 (see Table 5.5), leading to the closed-loop bandwidth of the current control system being 20 times of the open-loop system where $w_n = \frac{1}{1-\gamma}\frac{R_s}{L_s} = 634.9$ rad/s. With this increased closed-loop bandwidth

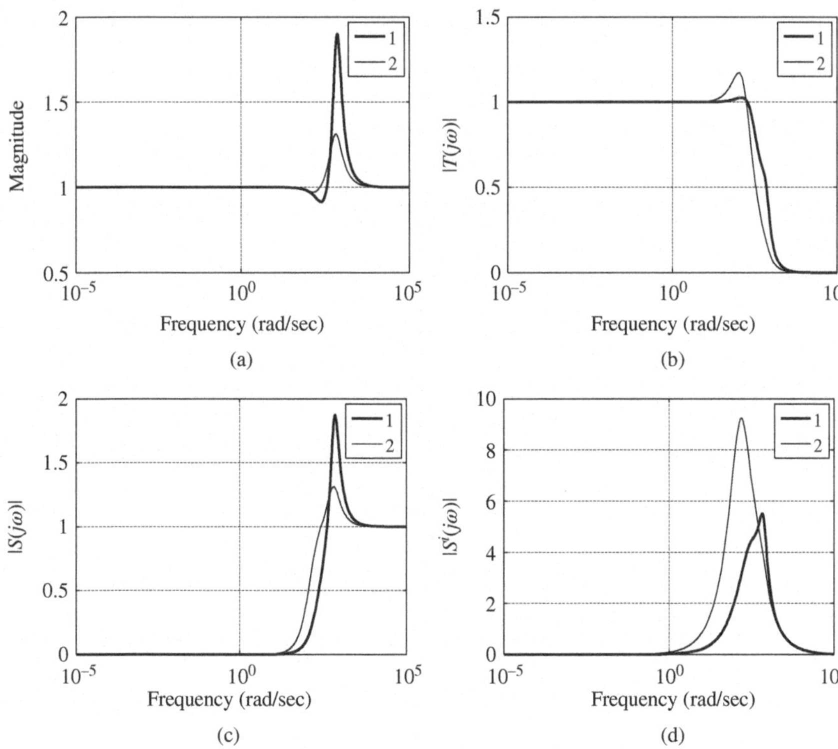

Figure 5.39 Sensitivity functions for outer-loop voltage control (power converter) where PI current controllers are used. Inner-loop current control system has bandwidth 634.9. Key: line (1) $w_n = 300$ rad/s, line (2) $w_n = 150$ rad/s. (a) Weighting function, (b) Actual complementary sensitivity function, (c) Actual sensitivity function, and (d) Actual input sensitivity function.

for the inner-loop current control, it is anticipated that the outer-loop bandwidth of the voltage control system could be increased. In this situation, if bandwidth of the outer-loop voltage control system is selected as 300 rad/s, it is less than half of the inner-loop bandwidth. The first difference is that with this increased inner-loop current controller bandwidth, the peak of the weighting function $|W(j\omega)|$ (see 5.98) has reduced, which is significant when comparing Figure 5.39(a) with Figure 5.37(a). In the previous case when the inner-loop current controller is designed using $\gamma = 0.9$, the peak value for the weighting function is 6.5 when the outer-loop bandwidth is 300 rad/s. In comparison, the peak value is 1.89 with the same outer-loop bandwidth, which results in much reduced performance uncertainty. The peak value is also reduced for the outer-loop controller with the bandwidth of 150 (see Figure 5.37(b)). The investigation of second case indicates that when the inner-loop current control system has a wider bandwidth, the bandwidth of the outer-loop system could be increased accordingly.

In summary, the sensitivity function analysis for the outer-loop voltage control system demonstrates that the bandwidth of the outer-loop control system is limited by the bandwidth of the inner-loop control system. The inner-loop control system should have a much wider bandwidth than the outer-loop control system in order to reduce performance uncertainty. For the power converter control, it is justifiable for the outer-loop bandwidth to be half of the inner-loop bandwidth.

5.7.4 Experimental Evaluations

Experiments are conducted to evaluate the nine control systems with their parameters listed in Table 5.5. The carrier frequency for the PWM signals is 1 kHz which is low due to the limitations of the experimental setup. Because of this low carrier frequency, it is expected that the problem with harmonics and noise will occur in the current control system. To determine the bandwidth of the d-axis and q-axis current controllers, three experiments are conducted in which $\gamma = 0.8$, $\gamma = 0.9$ and $\gamma = 0.95$ are used. These γ values correspond to the closed-loop bandwidth being 5, 10 and 20 times of the bandwidth of the open-loop current system. The γ value should be as large as possible to obtain a sufficiently large closed-loop bandwidth for disturbance rejection of the harmonics caused by the PWM switching and for the preparation of the design of outer-loop voltage controller upon which its bandwidth is dependent.

In the experiments, a voltage reference signal V_{dc}^* is set at 55 V when the current and voltage controllers are turned on, followed by a step change from 55 V to 60 V. At the operating condition of 60 V, a load disturbance with a positive amplitude and a load disturbance with a negative amplitude occur. The times of the step change and disturbance events are identical in the experiments. The disturbances are created using resistance as load that will be changed via a switch in the experiments. With M being the number of data point, the effort of the control signals is calculated as the sum of the mean squared error:

$$E_s = \frac{1}{M}\sum_{i=0}^{M-1}(S_d(t_i)^2 + S_q(t_i)^2), \tag{5.102}$$

where S_d and S_q are the control signals for d-axis and q-axis current systems. To measure the performance of the current controllers, the sum of the mean squared errors between the reference currents i_d^*, i_q^* and the measured currents i_d and i_q is calculated as

$$E_i = \frac{1}{M}\sum_{i=0}^{M-1}((i_d^*(t_i) - i_d(t_i))^2 + (i_q^*(t_i) - i_q(t_i))^2). \tag{5.103}$$

With the voltage controller fixed in the first three experiments, the mean squared error between the reference voltage and the measured voltage is computed as

$$E_{vdc} = \frac{1}{M}\sum_{i=0}^{M-1}(v_{dc}^*(t_i) - v_{dc}(t_i))^2. \tag{5.104}$$

Their values are listed in the final three columns of Table 5.5. It is interesting to note that the mean squared errors, E_s and E_i are strongly dependent on the bandwidth of the current control systems. For instance, when looking at Cases C, F and I where $\gamma = 0.95$ was used in the current control systems, the value of E_s reflecting the control signal variations is about 0.56, and the value of E_i reflecting the reference tracking performance is about 0.3. These values were obtained with three different bandwidths for the voltage control system. This means that the current controllers predominately determine the performance of current control systems and the voltage controller has little effect on the performance of the current control systems. As for the performance of voltage control system, it is seen from Table 5.5 that the value of E_{vdc} reduces drastically with the increase of outer-loop bandwidth w_n, in contrast, it has a very small dependence on the current controllers. For instance, for the first three cases (A, B, and C) with the same w_n being 80 rad/s, the value of E_{vdc} reflecting the capability of reference tracking and load disturbance rejection is between 8 and 8.8; and for the next three cases (D, E, and F) with the same w_n being about 150 rad/s, the value of E_{vdc} is between 3 and 3.7. For the final three cases (G, H, I), with the same w_n being 300 rad/s, the value of E_{vdc} is between 1.8 and 2.5. It is apparent that among the nine cases, when $\gamma = 0.95$, the current controllers with this large bandwidth produced superior performance in terms of the variations of the control signals (see the smaller E_s values) and the variations of the d-axis and q-axis currents (see the smaller E_i values), which is achieved regardless of the bandwidth of the voltage control system. This can be interpreted as a better reduction of harmonics in the current control when using a wider closed-loop bandwidth in the current control systems.

Figures 5.40–5.43 present the closed-loop responses based on the experimental results. In particular, Figure 5.40 illustrates control systems where the current controllers are designed using $\gamma = 0.9$ and the voltage control system has a bandwidth of 80 rad/s. It is seen that with this set of controller parameters, the closed-loop system has a slow set-point response and a slow response of disturbance rejection. In contrast, the responses become faster when the bandwidth of voltage control system is increased to 150 rad/s (see Figure 5.41). Because the current controllers remain the same for both cases, the plots also show that the d-axis current i_d and the q-axis current i_q have similar characteristics when Figure 5.40 is compared with Figure 5.41, except that the reference signal i_d^* has a faster dynamics response in Figure 5.41 due to the higher performance demand from the voltage controller. To illustrate how the current controllers affect the closed-loop performance, Figure 5.42 shows the

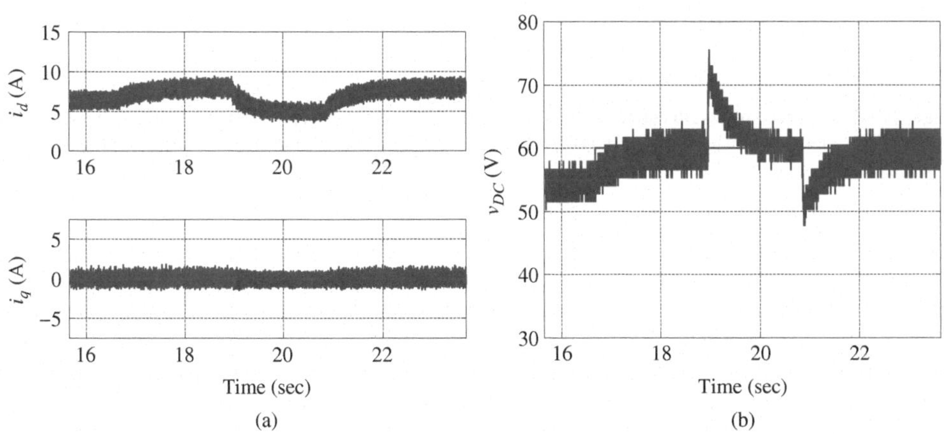

Figure 5.40 Power converter. PI current control of d-axis and q-axis currents, $\gamma = 0.90$, $\xi = 0.707$. Outer-loop voltage control system specification: $w_n = 80$ rad/s and $\xi = 0.707$. (a) Current and reference, and (b) Output voltage and reference.

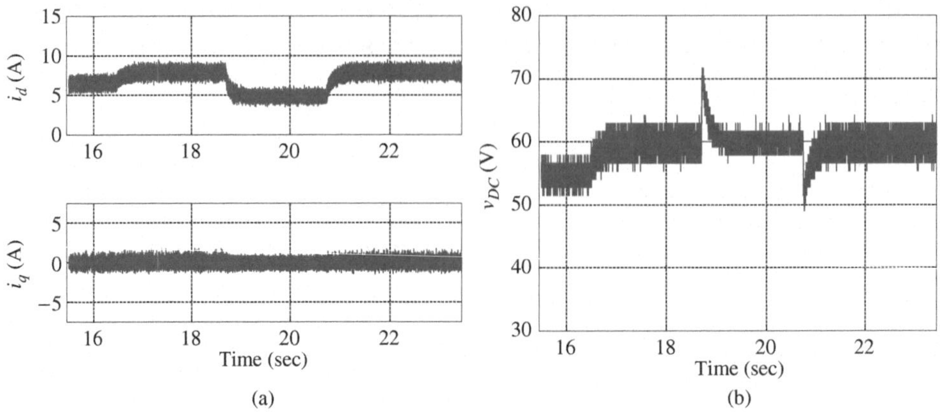

Figure 5.41 Power converter. PI current control of d-axis and q-axis currents, $\gamma = 0.9$, $\xi = 0.707$. Outer-loop voltage control system specification: $w_n = 150$ rad/s and $\xi = 0.707$. (a) Current and reference and (b) Output voltage and reference.

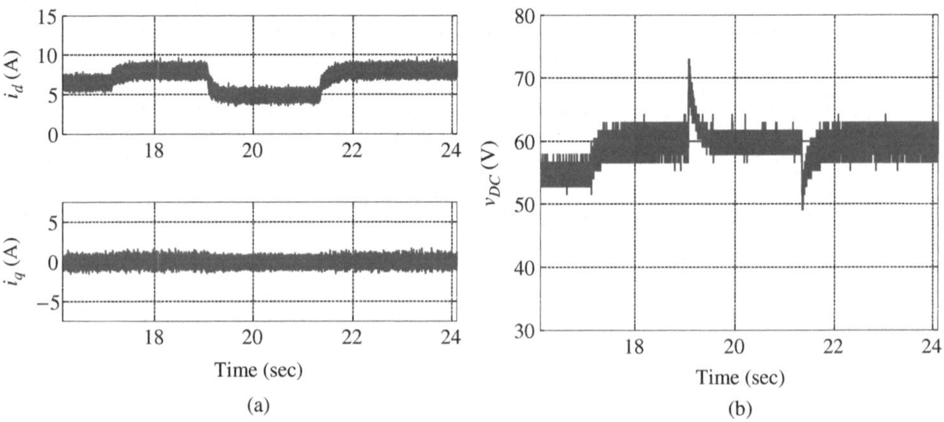

Figure 5.42 Power converter. PI current control of d-axis and q-axis currents, $\gamma = 0.95$, $\xi = 0.707$. Outer-loop voltage control system specification: $w_n = 150$ rad/s and $\xi = 0.707$. (a) Current and reference, and (b) Output voltage and reference.

case where the current controllers are designed using $\gamma = 0.95$ and the voltage control system has a bandwidth of 150 rad/s. In this case, because the voltage control system has the same bandwidth as the case presented in Figure 5.41, it is anticipated that both cases should have similar responses to set-point changes and disturbance rejections, and this is confirmed as the outcome of the comparison. A further increase of bandwidth w_n for the voltage control system results in a significant reduction in the response speed to reference following and disturbance rejection as shown in Figure 5.43. Among the nine cases, this is the best case in terms of closed-loop response speed of reference following and disturbance rejection.

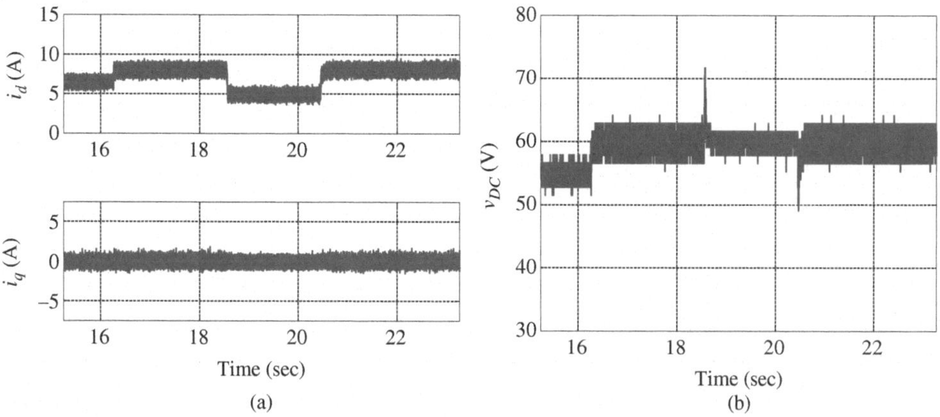

Figure 5.43 Power converter. PI current control of d-axis and q-axis currents, $\gamma = 0.95$, $\xi = 0.707$. Outer-loop voltage control system specification: $w_n = 300$ rad/s and $\xi = 0.707$. (a) Current and reference, and (b) Output voltage and reference.

5.8 Tuning P Plus PI Controllers for Power Converter

5.8.1 Design and Sensitivity Functions

There are two types of controllers that may be used to control the d-axis current. If the controller is a PI controller, then the proportional controller gain and integral time constant are identical to the controller parameters calculated for the q-axis current. However, if one chooses to use a P controller for the d-axis, the proportional gain for the d-axis current is selected as

$$K_c = -\frac{2\alpha}{1-\alpha}R_s, \tag{5.105}$$

where $0 < \alpha < 1$ is the performance parameter selected by the user to achieve the desired closed-loop response speed.

For the outer-loop PI controller design, the following integrator model is used:

$$\frac{V_{dc}(s)}{I_d^*(s)} \approx \frac{3}{4}\frac{S_d^{ss}}{C_{dc}}\frac{\alpha}{s}, \tag{5.106}$$

where α is the value used in the proportional control of d-axis current (see Equation (5.105)).

The tuning procedure is identical to the case studies presented in Section 5.7. Here, the results are presented in summary form. Again, nine cases are investigated in this study. Table 5.6 presents the current and voltage controller parameters and the performance measurements in terms of E_s, E_{vdc} and E_i which are defined in the previous section.

Figure 5.44 shows the complementary sensitivity, sensitivity and input sensitivity functions for the inner-loop proportional control of d-axis current. From Figure 5.44(a), it is seen that with the increase of α, the magnitude of the complementary sensitivity function increases, in particular, the $|T(j\omega)|$ approaches unity as α approaches 1. With a proportional controller, $|T(j\omega)|$ will not be equal 1 at $\omega = 0$. In comparison with a PI control of d-axis current that ensures constant value of $T(j\omega) = 1$, this is one of the disadvantages of using a proportional current controller. In order for the current control system to

Table 5.6 Inner-loop P control on d-axis current and outer-loop PI control of voltage. $E_s = \sum(S_d^2 + S_q^2)/M$, $E_{vdc} = \sum(V_{dc}^* - V_{dc})^2/M$ and $E_i = \sum((I_d^* - I_d)^2 + (I_q^* - I_q)^2)/M$, where M is the data length

	Voltage controller			i_q controller			i_d controller		Perform. measure		
	w_n	K_c	τ_I	γ	K_c	τ_I	α	K_c	E_s	E_{vdc}	E_i
A	80	0.0701	0.0177	0.8	−2.428	0.0076	0.8	−1.6	0.5712	7.3716	4.1316
B	80	0.0623	0.0177	0.9	−5.256	0.0041	0.9	−3.6	0.5757	7.9901	1.0976
C	80	0.059	0.0177	0.95	−10.912	0.0021	0.95	−7.6	0.571	8.3825	0.4648
D	150	0.1314	0.0094	0.8	−2.428	0.0076	0.8	−1.6	0.5696	3.3802	3.7977
E	150	0.1168	0.0094	0.9	−5.256	0.0041	0.9	−3.6	0.5717	3.1963	1.024
F	150	0.1106	0.0094	0.95	−10.912	0.0021	0.95	−7.6	0.5765	3.2665	0.4702
G	300	0.2627	0.0047	0.8	−2.428	0.0076	0.8	−1.6	0.5765	3.1442	3.4188
H	300	0.2335	0.0047	0.9	−5.256	0.0041	0.9	−3.6	0.5734	2.0089	1.0278
I	300	0.2213	0.0047	0.95	−10.912	0.0021	0.95	−7.6	0.5752	1.8477	0.458

track a reference signal, a larger value of α is required. Perhaps, the problem with disturbance rejection should demand more investigation here. From Figure 5.44(b), it is seen that at the lower frequency, the magnitude of the sensitivity function is larger when α is smaller, while at the higher frequency, the magnitude of the sensitivity is smaller when α is smaller. In comparison, the magnitude of input sensitivity (see Figure 5.44(c)) behaves similarly to that of sensitivity in the lower frequency region, but in the higher frequency region, the magnitude of the input sensitivity becomes very small. The behavior of the sensitivity and input sensitivity functions means that when α is small, the proportional current control system will have a poor disturbance rejection property. This again reinforces that in order for the current control system to work well, a larger α value is required for disturbance rejection and reference following. In the control of power converter, the existence of harmonics due to PWM switching is a problem, the characteristics of the sensitivity and input sensitivity indicate that the proportional control system is less effective in terms of disturbance rejection when compared with the use of a PI controller (see the characteristics of the sensitivity functions in Figure 5.36).

The complementary sensitivity, sensitivity and input sensitivity functions for the outer-loop voltage control system when using proportional control of d-axis current are shown in Figure 5.45. Because the outer-loop voltage control uses a PI controller, the characteristics of these functions are similar to those when the d-axis current control uses a PI controller. Therefore, it is anticipated that the properties of set-point following and load disturbance rejection are similar in both cases. Indeed, when the mean squared errors E_{vdc} listed in Table 5.6 are compared with those in Table 5.5, it is found that they are about the same.

5.8.2 Experimental Results

Figures 5.46–5.49 show the experimental results for the control system configuration where a PI controller is used to control the q-axis current and a P controller is for controlling the d-axis current, as well as a PI controller is used to control the voltage in the outer-loop system. There are four cases presented in these figures for comparison purposes. In Figure 5.46, both α and γ are set to be 0.9 for the current control systems that lead to modest current control bandwidths, while the outer-loop bandwidth is selected to be $w_n = 80$ rad/s which is considered to be a narrow bandwidth. In contrast, Figure 5.47 shows the same current control systems; however with a wider bandwidth applied to the outer-loop voltage control system. The comparison between the two sets of experimental results leads to the conclusions that because of the choice of identical current controllers, the variations of the control signals S_d and S_q, i_d and i_q currents

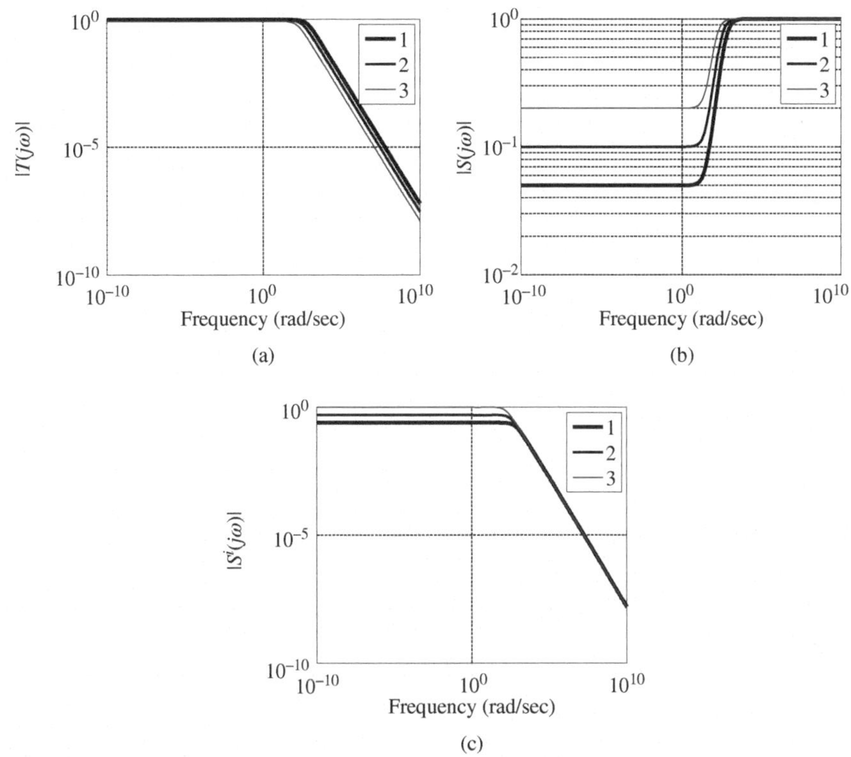

Figure 5.44 Power converter. Sensitivity function and input sensitivity function for inner-loop P control of d-axis current. Key: line (1) $\alpha = 0.95$ ($\gamma = 0.95$), line (2) $\alpha = 0.9$ ($\gamma = 0.9$), line (3) $\alpha = 0.8$ ($\gamma = 0.8$). (a) Complementary sensitivity function, (b) Sensitivity function, and (c) Input sensitivity function.

are about the same level, except that the set-point signal to the i_d current (i_d^*) has a faster dynamic response due to the higher performance specification from the outer-loop voltage controller. Because of the wider bandwidth used in the control results presented in Figure 5.46, the voltage response to set-point change and disturbance rejection in Figure 5.47 is much faster than that presented in Figure 5.46. The control results presented in Figure 5.48 and 5.49 have used a higher closed-loop performance specification for the current controllers, where both α and γ are chosen to be 0.95 (20 times of the open-loop bandwidth). It is noted that with this higher performance demand for the current controllers, the variations of the control signal S_d and S_q are larger than the cases presented in Figures 5.46 and 5.47. Furthermore, when the bandwidth for the voltage controller w_n is increased from 150 rad/s to 300 rad/s, the dynamic response to set-point change and disturbance rejection becomes extremely faster (see Figures 5.48 and 5.49 for comparison).

5.9 Robustness of Power Converter Control System Using PI Current Controllers

It is difficult to obtain the exact value for the inductance in a power converter since this parameter may change under the influence of the power network. The PI control system designed must be robust against

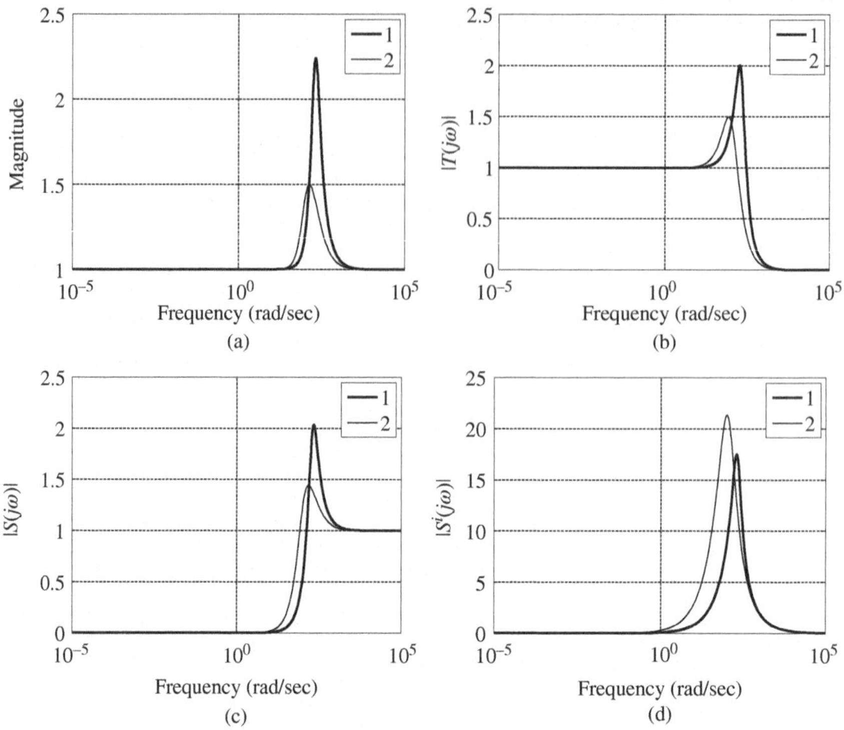

Figure 5.45 Sensitivity functions for outer-loop voltage control (power converter) where the inner-loop d-axis P current controller is used. Key: line (1) $w_n = 150$ rad/s, line (2) $w_n = 80$ rad/s. (a) Weighting function, (b) Actual complementary sensitivity function, (c) Actual sensitivity function, and (d) Actual input sensitivity function.

the variations of the physical parameters. An investigation is performed here to evaluate the robustness of the cascade PI control system.

5.9.1 Variation of Inductance Using PI Current Controllers

In the design of the PI current controller, the parameters K_c and τ_I are calculated using the following equations:

$$K_c = \frac{2\xi w_n - a}{b}; \quad \tau_I = \frac{2\xi w_n - a}{w_n^2}, \tag{5.107}$$

where $a = \frac{R_s}{L_s}$, $b = -\frac{1}{2L_s}$ and $w_n = \frac{1}{1-\gamma}\frac{R_s}{L_s}$ rad/s, which gives

$$K_c = -\frac{R_s}{2}\left(\frac{2\xi}{1-\gamma} - 1\right); \quad \tau_I = (1-\gamma)^2 \frac{L_s}{R_s}\left(\frac{2\xi}{1-\gamma} - 1\right).$$

These calculations imply that in the PI current controller design, the proportional gain K_c is not affected by the uncertainty of the inductance, but the integral time constant is affected if the value of L_s changes when maintaining the same desired closed-loop bandwidth. The value of τ_I increases if the inductance increases. In other words, an underestimated inductance will lead to a smaller integral time constant while an overestimated inductance will lead to a larger integral time constant.

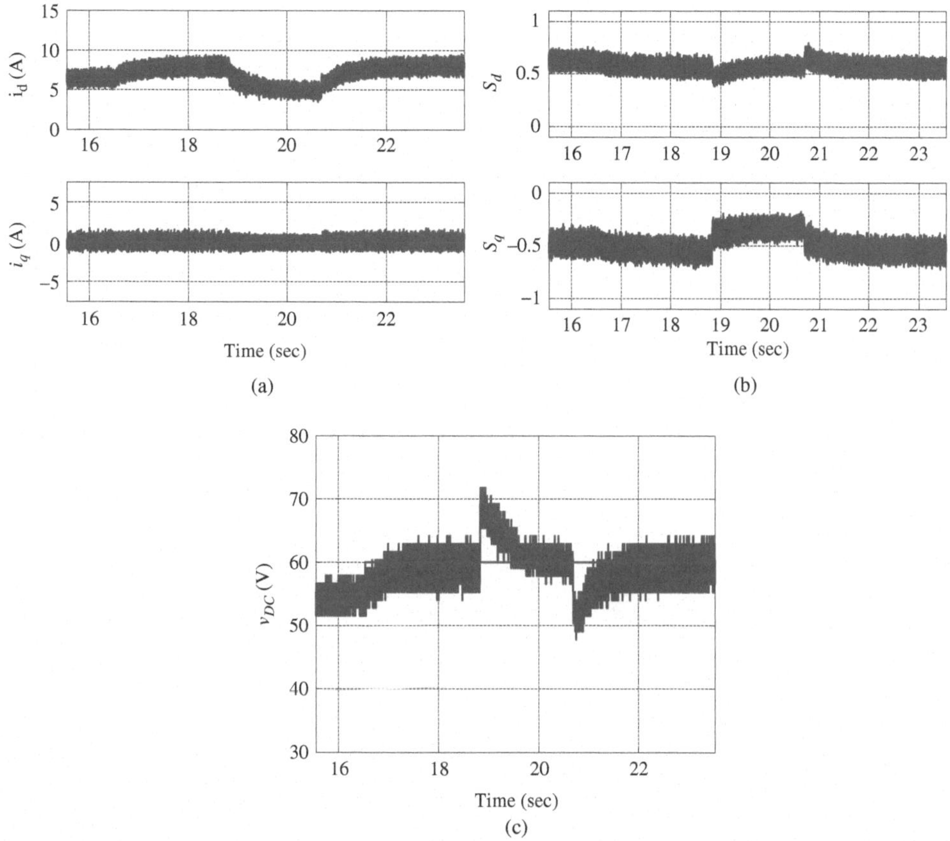

Figure 5.46 Power converter. PI control of q-axis current and P control of d-axis current, $\alpha, \gamma = 0.90$, $\xi = 0.707$. Outer-loop voltage control system specification: $w_n = 80$ rad/s and $\xi = 0.707$. (a) Current and reference, (b) S_d and S_q, and (c) Output voltage and reference.

Because it is difficult to change the actual physical system, in the assessment of robustness of the closed-loop control system, a set of inductances are used in the design of the current controllers, which practically speaking is an attempt to create the scenario where inductance variation occurs without changing the actual physical system. With this test-bed the nominal inductance was measured as $L_s^0 = 6.3 \times 10^{-3}$ H. There are six inductance values used in the design of current controllers with variations ranging from $0.1L_s^0$ to $3L_s^0$. The parameters for the six current controllers are listed in Table 5.7, where it is seen that the proportional controller gain K_c remains unchanged as the inductance increases; however, the integral time constant τ_I increases.

To assess how the variation of induction affects the closed-loop stability and performance, the family of Nyquist plots is generated where the nominal L_s is taken as the "true" system's inductance and the current PI controllers are applied to this nominal system in order to obtain the inner-loop closed-loop control systems. In the assessment, the outer-loop voltage controller is designed with a bandwidth $w_n = 150$ rad/s. Figure 5.50 shows the family of the Nyquist plots for the seven cases, which indicate that all the closed-loop systems are stable for the variations of inductance. Additionally, all gain margins and phase margins appear to be sufficient to guarantee a satisfactory closed-loop performance.

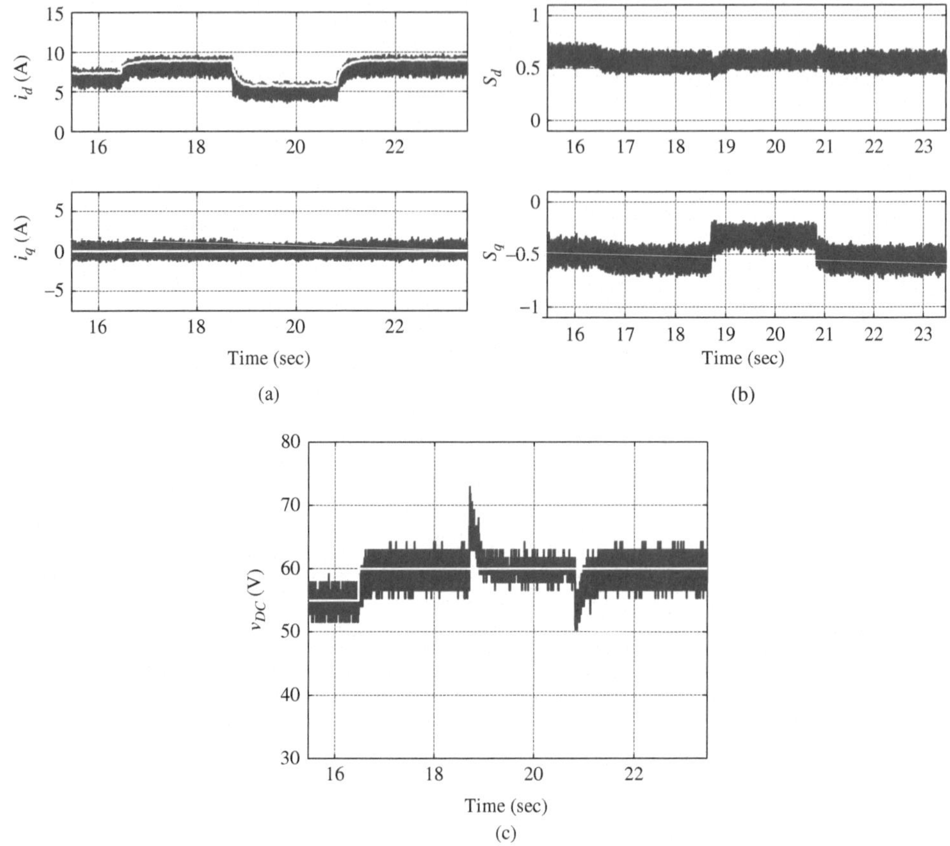

Figure 5.47 Power converter. PI control of q-axis current and P control of d-axis current, $\alpha, \gamma = 0.90$, $\xi = 0.707$. Outer-loop voltage control system specification: $w_n = 150$ rad/s and $\xi = 0.707$. (a) Current and reference, (b) S_d and S_q, and (c) Output voltage and reference.

However, there are small variations on the gain and phase margins when the inductance changes. In particular, when the inductance is overestimated ($3L_s^0$), it is seen that the closed-loop control system has the least phase margin (see the thickest line in Figure 5.50), but largest gain margin. In comparison when the inductance is underestimated ($0.1L_s^0$), then closed-loop control system has the least gain margin (see the thicker line in Figure 5.50), but largest phase margin. The Nyquist plots for the remaining cases lie between these two extreme cases. The assessment using Nyquist plots indicates that when the L_s value is overestimated, the closed-loop control system will be more tolerant to gain variation, but less tolerant to phase variation. The opposite is true when L_s is underestimated.

Experiments are conducted to evaluate the robustness of the PI control systems, where the set-point signal for V_{dc} takes a step change from 55 V to 60 V, followed by two disturbance rejections. The experimental results show that the mean squared errors E_s and E_{vdc} have small variations for the range of inductances, whilst the variation of the mean squared error E_i is relatively larger (see Table 5.7).

Figure 5.51 shows the closed-loop control results when the inductance used in the design is $0.1L_s^0$ whilst Figure 5.52 shows the control results when the inductance is $3L_s^0$. The experimental results confirm that the PI control system is robust against the variations of inductance.

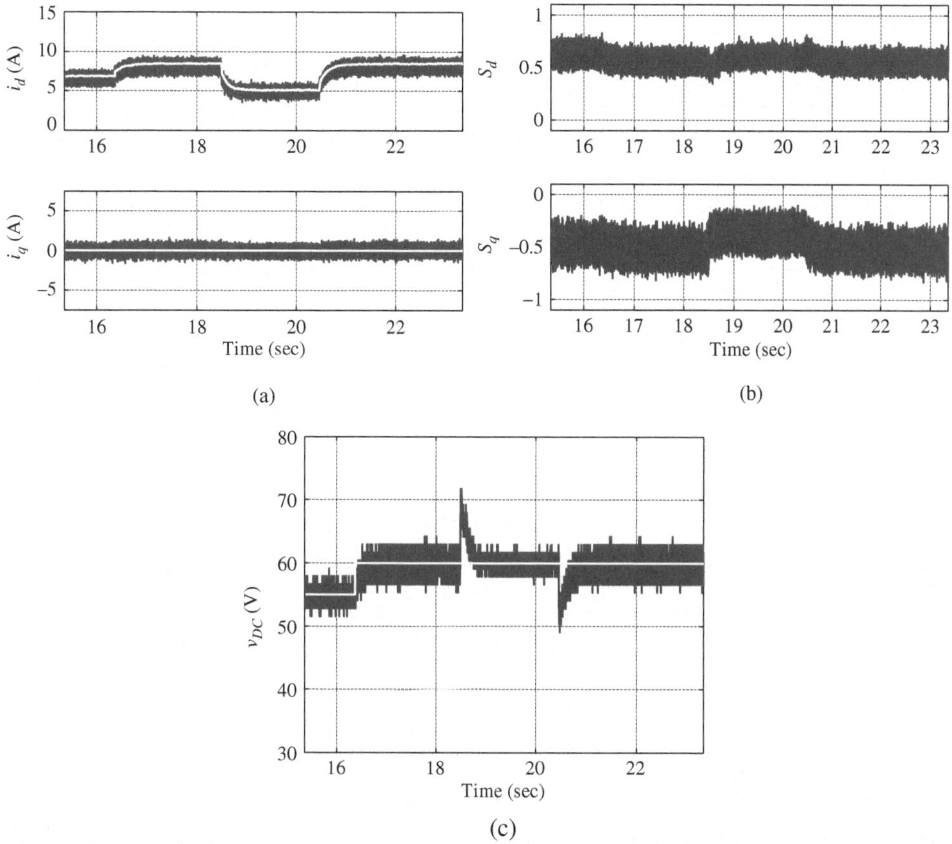

Figure 5.48 Power converter. PI control of q-axis current and P control of d-axis current, $\alpha, \gamma = 0.95$, $\xi = 0.707$. Outer-loop voltage control system specification: $w_n = 150$ rad/s and $\xi = 0.707$. (a) Current (solid) and reference (dashed), (b) S_d and S_q, and (c) Output voltage (solid) and reference (dashed).

5.9.2 *Variation of Capacitance on Closed-loop Performance*

The proportional controller gain K_c and τ_I for the voltage controller are calculated using

$$K_c = \frac{2\xi w_n - a}{b}; \quad \tau_I = \frac{2\xi w_n - a}{w_n^2}.$$

In the design, the model for the voltage control is an integrator, thus $a = 0$, and b is calculated using (5.93) that is $b = \frac{3}{4}\frac{S_d^{ss}}{C_{dc}}$. Thus, the variation of capacitance C_{dc} in the design only affects the proportional controller gain K_c for the outer-loop voltage control. If the value of C_{dc} used in the design is larger than the actual capacitance, then the proportional control gain K_c calculated is larger. This implies that an overestimated capacitance value will result in a higher gain and faster closed-loop response speed than the desired specification. Conversely, an underestimated capacitance will lead to a smaller K_c.

In the assessment of the robustness of the cascade closed-loop system in the presence of capacitance uncertainty, a set of capacitances are used in the design of voltage controller. With this test-bed the nominal capacitance was estimated as $C_{dc}^0 = 2.96 \times 10^{-4}$ F. There are six capacitance values, in addition

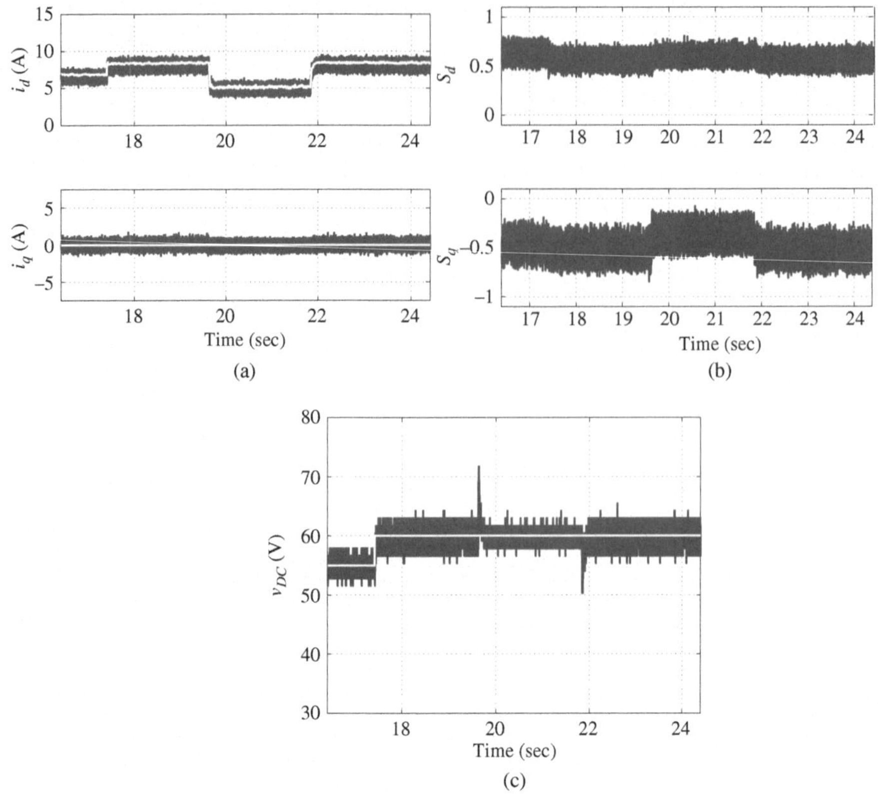

Figure 5.49 Power converter. PI control of q-axis current and P control of d-axis current, $\alpha, \gamma = 0.95$, $\xi = 0.707$. Outer-loop voltage control system specification: $w_n = 300$ rad/s and $\xi = 0.707$. (a) Current (solid) and reference (dashed), (b) S_d and S_q, and (c) Output voltage (solid) and reference (dashed).

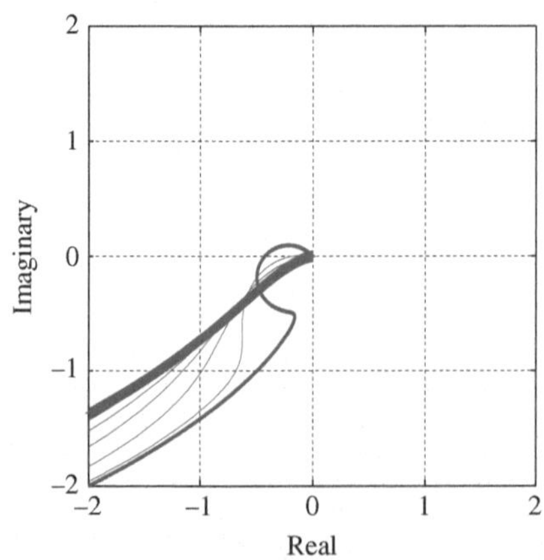

Figure 5.50 Nyquist plots for the family of inner-loop d-axis PI current controllers together with the outer-loop voltage controller as in Table 5.7. Key: thickest line $L_s = 3L_s^0$, thicker line $L_s = 0.5L_s^0$. The remaining lines are for the other five cases listed in Table 5.7.

Table 5.7 Robustness on inner-loop inductance variation, where the nominal inductance is $L_s^0 = 6.3 \times 10^{-3}$ H. Inner-loop PI control of both d-axis and q-axis current and outer-loop PI control of voltage. Voltage controller w_n is 150 rad/s and current controller γ is 0.9. $E_s = \sum(S_d^2 + S_q^2)/M$, $E_{vdc} = \sum(V_{dc}^* - V_{dc})^2/M$ and $E_i = \sum((I_d^* - I_d)^2 + (I_q^* - I_q)^2)/M$, where M is the data length

	Inductance	Voltage controller		i_q controller		i_d controller		Perform. measure		
	L_s	K_c	τ_I	K_c	τ_I	K_c	τ_I	E_s	E_{vdc}	E_i
A	$0.1L_s^0$	0.1051	0.0094	−5.256	0.0004	−5.256	0.0004	0.5575	3.6958	0.608
B	$0.5L_s^0$	0.1051	0.0094	−5.256	0.0021	−5.256	0.0021	0.5703	3.3675	0.3699
C	L_s^0	0.1051	0.0094	−5.256	0.0041	−5.256	0.0041	0.5695	3.3902	0.3739
D	$1.5L_s^0$	0.1051	0.0094	−5.256	0.0062	−5.256	0.0062	0.5651	3.1933	0.3811
E	$2L_s^0$	0.1051	0.0094	−5.256	0.0083	−5.256	0.0083	0.5559	3.2586	0.4118
F	$2.5L_s^0$	0.1051	0.0094	−5.256	0.0103	−5.256	0.0103	0.5593	3.5586	0.5079
G	$3L_s^0$	0.1051	0.0094	−5.256	0.0124	−5.256	0.0124	0.5569	3.9859	0.6975

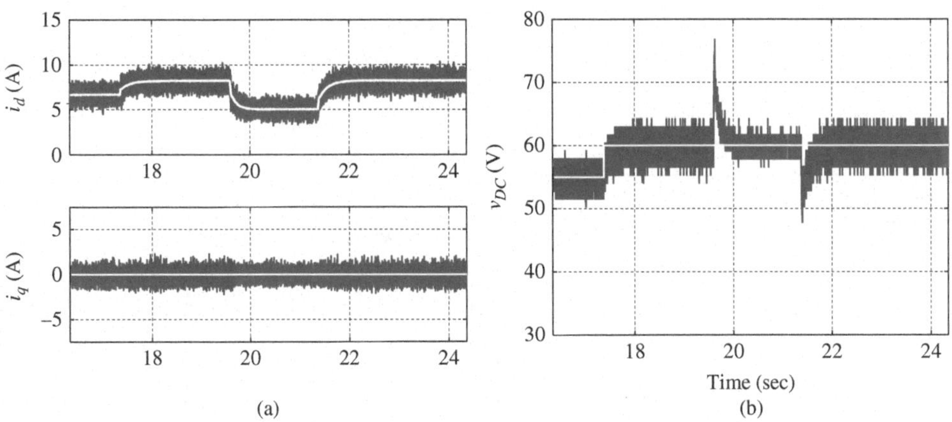

Figure 5.51 Power converter. Assessment of robustness of the PI control system with $L_s = 0.1L_s^0$. Outer-loop voltage control system specification: $w_n = 150$ rad/s and $\xi = 0.707$. (a) Current and reference, and (b) Output voltage and reference.

to the nominal value, used in the design of voltage controllers with variations ranging from $0.1C_{dc}^0$ to $3C_{dc}^0$. With the desired closed-loop bandwidth w_n is fixed to 150 rad/s, the parameters for the seven voltage controllers are calculated and listed in Table 5.8, where it is seen that the integral time constant remains unchanged while K_c increases as the capacitance increases.

To assess how the variation of capacitance affects the cascade closed-loop system stability and performance, the family of Nyquist plots is generated for the seven cases. In the plots, the nominal C_{dc}^0 is taken as the converter's true capacitance, and the PI controllers take their values from Table 5.8. PI controllers are used to control the d and q axes currents, where the parameter $\gamma = 0.9$, meaning that the current control systems have a bandwidth 10 times that of its open-loop bandwidth. Figure 5.53 shows the family of the Nyquist plots for the seven cases with different capacitance values. It is seen that all the closed-loop systems are stable for the variations of capacitance. It appears that the gain margin for all the control

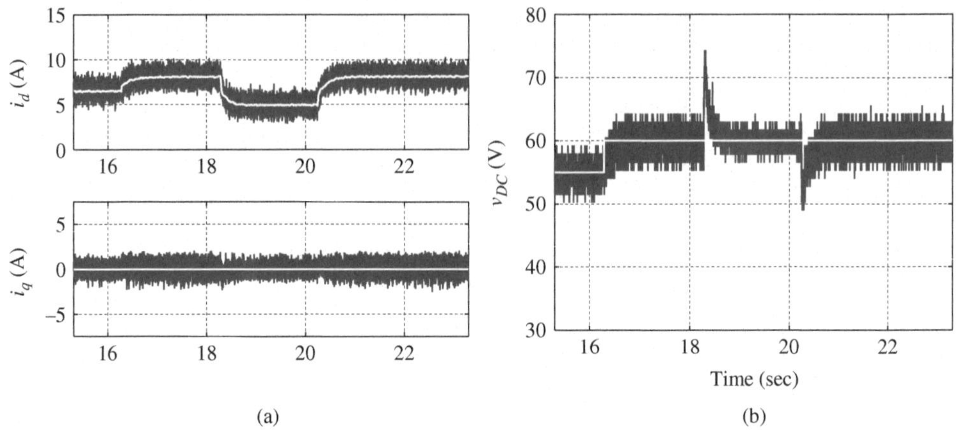

Figure 5.52 Power converter. Assessment of robustness of the PI control system with $L_s = 3L_s^0$. Outer-loop voltage control system specification: $w_n = 150$ rad/s and $\xi = 0.707$. (a) Current and reference, and (b) Output voltage and reference.

Table 5.8 Robustness on outer-loop capacitance variation, where the nominal capacitance C_{dc}^0 is 2.96×10^{-4} F. PI control of both d-axis and q-axis current and outer-loop PI control of voltage. Voltage control system bandwidth w_n is 150 rad/s and current controller with γ is 0.9. $E_s = \sum(S_d^2 + S_q^2)/M$, $E_{vdc} = \sum(V_{dc}^* - V_{dc})^2/M$ and $E_i = \sum((I_d^* - I_d)^2 + (I_q^* - I_q)^2)/M$, where M is the data length

	Capacitance	Voltage controller		i_q controller		i_d controller		Perform. measure		
	C_{dc}	K_c	τ_I	K_c	τ_I	K_c	τ_I	E_s	E_{vdc}	E_i
A	$0.1C_{dc}^0$	0.0105	0.0094	−5.256	0.0041	−5.256	0.0041	0.5440	22.6975	0.3665
B	$0.5C_{dc}^0$	0.0525	0.0094	−5.256	0.0041	−5.256	0.0041	0.5649	5.6644	0.3686
C	C_{dc}^0	0.1051	0.0094	−5.256	0.0041	−5.256	0.0041	0.5769	3.33047	0.3834
D	$1.5C_{dc}^0$	0.1576	0.0094	−5.256	0.0041	−5.256	0.0041	0.5626	2.4917	0.3655
E	$2C_{dc}^0$	0.2102	0.0094	−5.256	0.0041	−5.256	0.0041	0.5775	2.1956	0.3839
F	$2.5C_{dc}^0$	0.2627	0.0094	−5.256	0.0041	−5.256	0.0041	0.5755	2.0425	0.3779
G	$3C_{dc}^0$	0.3153	0.0094	−5.256	0.0041	−5.256	0.0041	0.5678	1.9347	0.3791

systems is quite large and less critical to the closed-loop performance. On the other hand, phase margin becomes more critical for the design, as it becomes smaller when a larger C_{dc} value is used in the design. In particular, when the capacitance is overestimated ($3C_{dc}^0$), it is seen that the closed-loop control system has the least phase margin (see the thickest line in Figure 5.53).

Figure 5.54 shows the closed-loop responses for the case when the capacitance is underestimated ($C_{dc} = 0.1C_{dc}^0$). From this figure, it is seen that the underestimated capacitance leads to a slower closed-loop voltage response in both reference following and disturbance rejection. In contrast, Figure 5.55 shows the closed-loop responses for the case when the capacitance is overestimated ($C_{dc} = 3C_{dc}^0$). It appears that the closed-loop response speed is much faster for both reference following and disturbance rejection when a larger capacitance is used in the design of PI controller that leads to a higher proportional control gain.

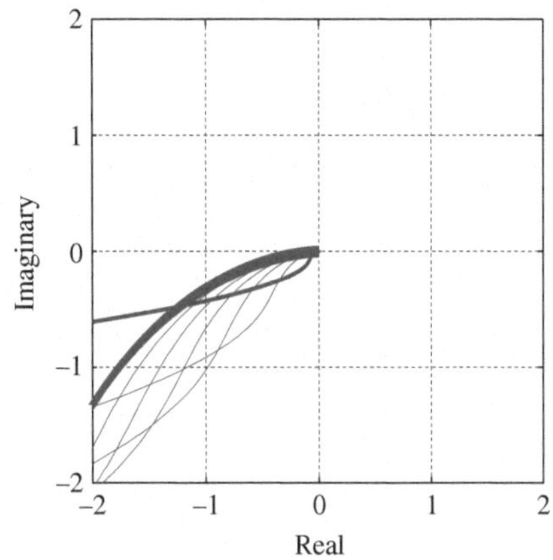

Figure 5.53 Nyquist plots for the family of outer-loop PI voltage controllers together with the inner-loop PI current controller as in Table 5.8. Key: thickest line $C_{dc} = 3C^0_{dc}$, thicker line $C_{dc} = 0.1C^0_{dc}$. The remaining lines are for the other five cases listed in Table 5.8.

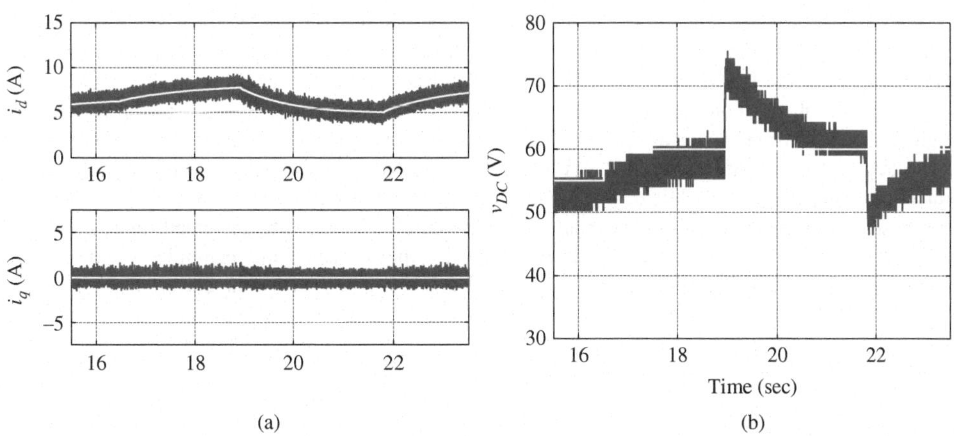

Figure 5.54 Power converter. Assessment of robustness of the PI control system with $C_{dc} = 0.1C^0_{dc}$. Outer-loop voltage control system specification: w_n = 150 rad/s and $\xi = 0.707$. (a) Current and reference, and (b) Output voltage and reference.

5.10 Summary

This chapter has investigated the tuning of the PID controllers for electrical drives and power converters. The results are discussed and summarized as follows.

5.10.1 Current Controllers

In the current control system design, a high gain controller is preferable for three important reasons. The first reason is that a high gain current controller will ensure a fast dynamic response to reference

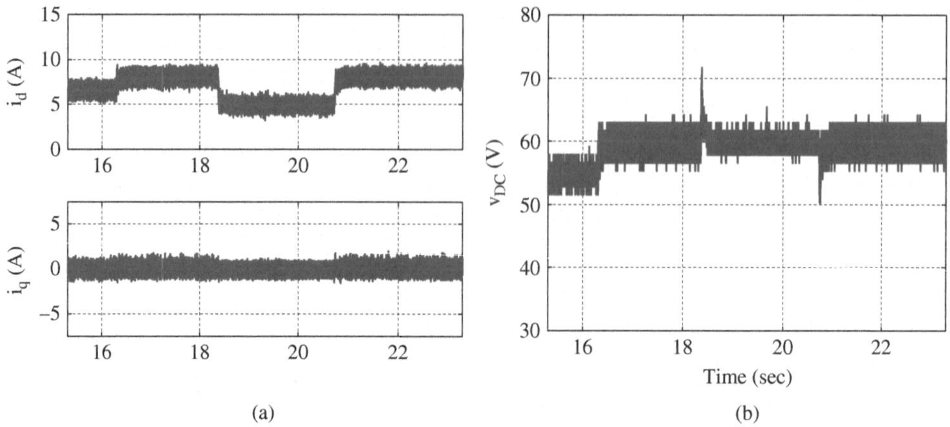

Figure 5.55 Power converter. Assessment of robustness of the PI control system with $C_{dc} = 3C_{dc}^0$. Outer-loop voltage control system specification: $w_n = 150$ rad/s and $\xi = 0.707$. (a) Current and reference, and (b) Output voltage and reference.

changes and disturbance rejections. The second reason is to reduce the PWM errors which are modeled as an input disturbance where this disturbance has the characteristics of periodic signals with a central frequency at the PWM switching frequency. Therefore, to reduce the effect of modulation errors, the amplitude of the input sensitivity function needs to be small at the low and medium frequency regions. The third reason for a high gain current controller is that for velocity and voltage control, a cascade control structure is needed. With this cascade control structure, the design is based on the outer-loop dynamic model while neglecting the inner-loop feedback dynamics. With the high gain current control, the closed-loop time constants from the current control become small, thus, the effects of the neglected dynamics on the cascade control system performance are negligible as demonstrated by the sensitivity function analysis. In short, if a PI controller is used for the current control and the band width of the current control system is specified as $w_n = \frac{1}{1-\gamma}a$ rad/s where $s = -a$ is the open-loop pole of the current system, then γ is recommended in the range of 0.9 to 0.95, which is 10 to 20 times open-loop bandwidth.

5.10.2 Velocity, Position and Voltage Controllers

The velocity and voltage controllers are in the outer-loop of the control system structure. The primary concern for these outer-loops is the robustness of the closed-loop system against unmodeled dynamics neglected from the inner-loop system and the modeling errors of the mechanical part of the system. In the outer-loop control system design, PI controllers are used for velocity and voltage control and PID controller is used for position control because of the requirement of steady-state performance for load disturbance rejection and reference following. For electrical drive control, when $w_n = \frac{1}{1-\gamma}a$ rad/s where $s = -a$ is the open-loop pole of the mechanical system, γ is recommended to be in the range of 0.6 to 0.9. Smaller γ will lead to a slower closed-loop response, but will have a larger tolerance to the unmodeled dynamics from the inner-loop system and from the potential mismatch of the inertial parameter in the mechanical system. For the voltage control of power converters, because the outer-loop system is modeled as an integrator system, there is no physical parameter to be used as a benchmark for the open-loop dynamics. Thus, the bandwidth of the outer-loop controller is benchmarked against the bandwidth of the inner-loop current control system. It is recommended to choose the bandwidth of the outer-loop system to be in the range of 5% to 10% of the bandwidth of the inner-loop current control system. For position control, a smaller w_n is needed because of the pure integrator in the position model.

5.10.3 Choice between P Current Control and PI Current Control

Although the proportional current control is simpler in its structure, it has limited applications in current control. If the control objective is for current regulation only, then PI controller is strongly recommended because the integral action is necessary for eliminating steady-state errors to ensure satisfactory operations of the electrical systems. When there is a cascade control for the voltage and velocity, then proportional controller can be used for the inner-loop current control of q-axis current (electrical drives) or d-axis current current (power converter) systems. However, the simpler controller structure when using a P controller does not necessarily provide the advantages over a slightly more complicated PI controller structure, because the the inner-loop PI controller has a better performance in disturbance rejection, also a better performance against performance losses in the outer-loop control system. But if there is a time delay in the current control system, the proportional controller could provide better robustness against the neglected time delay.

5.11 Further Reading

Books on control system design and analysis include Goodwin *et al.* (2001), Astrom and Murray (2008). Direct torque control for PMSM drives was analyzed in Zhong *et al.* (1997). Zmood *et al.* (2001) presented a frequency-domain analysis for three-phase linear current regulators.

Torque ripple minimization in PM synchronous motors using iterative learning control was presented in Qian *et al.* (2004), Xu *et al.* (2004) and using repetitive control Mattavelli *et al.* (2005), Escobar *et al.* (2007). Torque ripple minimization techniques for permanent magnet AC motor drives were discussed and reviewed in Jahns and Soong (1996), Mariethoz *et al.* (2009), Chai *et al.* (2013). Current measurement error in vector-controlled *AC* motor drives was compensated in Chung and Sul (1998).

References

Astrom KJ and Murray RM 2008 *Feedback Systems: An Introduction for Scientists and Engineers*. Princeton University Press.

Chai S, Wang L and Rogers E 2013 A cascade MPC control structure for a PMSM with speed ripple minimization. *IEEE Transactions on Industrial Electronics* **60**(8), 2978–2987.

Chung DW and Sul SK 1998 Analysis and compensation of current measurement error in vector-controlled AC motor drives. *IEEE Transactions on Industry Applications* **34**(2), 340–345.

Escobar G, Valdez AA, Leyva-Ramos J and Mattavelli P 2007 Repetitive-based controller for a UPS inverter to compensate unbalance and harmonic distortion. *IEEE Transactions on Industrial Electronics* **54**(1), 504–510.

Gan WC and Qiu L 2004 Torque and velocity ripple elimination of AC permanent magnet motor control systems using the internal model principle. *IEEE/ASME Transactions on Mechatronics* **9**(2), 436–447.

Goodwin GC, Graebe SF and Salgado ME 2001 *Control System Design*. Prentice Hall, New Jersey.

Jahns T and Soong W 1996 Pulsating torque minimization techniques for permanent magnet AC motor drives-a review. *IEEE Transactions on Industrial Electronics* **43**(2), 321–330.

Mariethoz S, Domahidi A and Morari M 2009 A model predictive control scheme with torque ripple mitigation for permanent magnet motors *Industrial Electronics, 2009. IECON '09. 35th Annual Conference of IEEE*, pp. 2943–2948.

Mattavelli P, Tubiana L and Zigliotto M 2005 Torque-ripple reduction in PM synchronous motor drives using repetitive current control. *IEEE Transactions on Power Electronics* **20**(6), 1423–1431.

Qian W, Panda S and Xu J 2004 Torque ripple minimization in PM synchronous motors using iterative learning control. *IEEE Transactions on Power Electronics* **19**(2), 272 – 279.

Xu J, Panda S, Pan Y, Lee T and Lam B 2004 A modular control scheme for pmsm speed control with pulsating torque minimization. *IEEE Transactions on Industrial Electronics* **51**(3), 526 – 536.

Zhong L, Rahman M, Hu WY and Lim KW 1997 Analysis of direct torque control in permanent magnet synchronous motor drives. *IEEE Transactions on Power Electronics* **12**(3), 528–536.

Zmood D, Holmes D and Bode G 2001 Frequency-domain analysis of three-phase linear current regulators. *IEEE Transaction on Industry Applications* **37**(2), 601–610.

6

FCS Predictive Control in $d - q$ Reference Frame

In essence, the core technique used in the design of PID controllers is to assign the actual closed-loop poles to some desired locations specified by the user (see Chapters 3–5). With the PID controllers, the implementation of the control system is achieved using the PWM technologies to control the semiconductor switches of a voltage source inverter (VSI), where approximation of the three phase voltage control signals took place (see Chapter 2).

This chapter departs from the approaches used in the traditional PID control system design and implementation by directly optimizing the inverter states. As a result, the PWM is not required in the implementation of the control systems, which simplifies the implementation procedure. Since for a two level VSI, there are eight combinations of inverter states, the terminology of finite control set (FCS) is given. Furthermore, the optimization of the inverter states is performed using the receding horizon control principle, which is the core of model predictive control. By combining the finite control set with model predictive control technology, FCS-MPC is termed. The original FCS-MPC did not have integral action, and relied on high gain feedback control to reduce the steady-state errors, resulting in performance losses at its steady-state operation. How to add integrator to this class of control system is not obvious because the original control problem is solved via numerical optimization. A reverse engineering practice takes place in this chapter. Starting from the numerical optimization, it is shown in this chapter that, without constraints, the original FCS-MPC system is a deadbeat feedback control system with a time-varying feedback control gain where the discrete-time closed-loop poles are at the origin of the complex plane. Subsequently, an integral controller is added to the original FCS-model predictive controller via a cascade control structure in which the characteristics of the deadbeat control system are considered. In the presence of the inverter states constraints, the original objective function is shown to be equivalent to an objective function expressed in terms of the differences between the optimal inverter state calculated from using the deadbeat controller and the candidate inverter states. This new FCS-MPC contains an integrator to overcome steady-state errors, which is called an I-FCS model predictive controller (I-FCS-MPC). A key outcome of the proposed approach is that the new objective function has an explicit relationship with the sampling interval Δt, which acts as a weight coefficient towards the squared errors. This naturally puts the sampling interval Δt as the parameter for determining the desired closed-loop performance. Not only does it simplify the design, but also this performance parameter is similar to other classical control strategies (such as hysteresis used control) in electrical drives and power converters, where their minimum sampling frequency is restricted.

PID and Predictive Control of Electrical Drives and Power Converters using MATLAB®/Simulink®, First Edition.
Liuping Wang, Shan Chai, Dae Yoo, Lu Gan and Ki Ng.

Companion Website: www.wiley.com/go/wang/pid

The remainder of this chapter is outlined as follows. In Section 6.1, the states of IGBT inverter are discussed in relation to its operational constraints, leading to the candidate variables for the control signals. The next six sections will use the current control of a PMSM as an example to illustrate the FCS-MPC algorithm and its modified version (I-FCS-MPC) that has embedded integrators in the control system. In Section 6.2, the original FCS-MPC algorithm is used to control a PMSM, followed by a MATLAB tutorial on real-time implementation of FCS-MPC (see Section 6.3). This FCS-MPC method is analyzed to yield the feedback control gain and the closed-loop system eigenvalues, showing that they are at the origin of the complex plane. Since the original FCS-MPC algorithm did not have integral action, which is required to overcome steady-state errors and protect against performance uncertainty, a modified I-FCS-MPC is proposed in a framework of cascade feedback control (see Section 6.5). Furthermore, for convenience of digital implementation, in Section 6.6 the I-FCS-MPC algorithm is derived using the principle of receding horizon to take advantages of the incremental variables. MATLAB tutorial is given in Section 6.7 for real-time implementation of the I-FCS-MPC algorithm. In Sections 6.8–6.9, this I-FCS-MPC is applied to the current control of an induction motor and a power converter with experimental validations. Robustness of the closed-loop system performance for the original FCS-MPC and the modified I-FCS-MPC is compared by using Monte-Carlo simulations with random parameter variations (see Section 6.10). The final two sections of this chapter are devoted to design and implementation of velocity and position control of a PMSM and an induction motor via the cascade feedback control structures (see Sections 6.11 and 6.12).

6.1 States of IGBT Inverter and the Operational Constraints

In the derivation of operational constraints, it is assumed that a three-phase 2L-VSI inverter is used in the implementation of the current control. The operation of the three-phase 2L-VSI inverter and its states were illustrated in Figure 2.1 and Table 6.1. Corresponding to the switching states in Table 6.1, the resulting output voltage v_i are summarized in Table 6.2. The transformation

Table 6.1 Switching states of inverter

	$\vec{V_0}$	$\vec{V_1}$	$\vec{V_2}$	$\vec{V_3}$	$\vec{V_4}$	$\vec{V_5}$	$\vec{V_6}$	$\vec{V_7}$
S_a	0	1	1	0	0	0	1	1
S_b	0	0	1	1	1	0	0	1
S_c	0	0	0	0	1	1	1	1

Table 6.2 Output voltage of inverter

	$\vec{V_0}$	$\vec{V_1}$	$\vec{V_2}$	$\vec{V_3}$	$\vec{V_4}$	$\vec{V_5}$	$\vec{V_6}$	$\vec{V_7}$
v_a	$-\frac{V_{dc}}{2}$	$\frac{V_{dc}}{2}$	$\frac{V_{dc}}{2}$	$-\frac{V_{dc}}{2}$	$-\frac{V_{dc}}{2}$	$-\frac{V_{dc}}{2}$	$\frac{V_{dc}}{2}$	$\frac{V_{dc}}{2}$
v_b	$-\frac{V_{dc}}{2}$	$-\frac{V_{dc}}{2}$	$\frac{V_{dc}}{2}$	$\frac{V_{dc}}{2}$	$\frac{V_{dc}}{2}$	$-\frac{V_{dc}}{2}$	$-\frac{V_{dc}}{2}$	$\frac{V_{dc}}{2}$
v_c	$-\frac{V_{dc}}{2}$	$-\frac{V_{dc}}{2}$	$-\frac{V_{dc}}{2}$	$-\frac{V_{dc}}{2}$	$\frac{V_{dc}}{2}$	$\frac{V_{dc}}{2}$	$\frac{V_{dc}}{2}$	$\frac{V_{dc}}{2}$

of the three-phase voltages to their components in α-β frame is achieved by the Clarke transformation,

$$\begin{bmatrix} v_\alpha \\ v_\beta \end{bmatrix} = \frac{2}{3}\begin{bmatrix} 1 & -\frac{1}{2} & -\frac{1}{2} \\ 0 & \frac{\sqrt{3}}{2} & -\frac{\sqrt{3}}{2} \end{bmatrix}\begin{bmatrix} v_{an} \\ v_{bn} \\ v_{cn} \end{bmatrix}. \tag{6.1}$$

Using (2.1) and (2.2), the α-β representation of three-phase output voltages can be expressed in terms of the switching states of three upper leg switches,

$$\begin{aligned}\begin{bmatrix} v_\alpha \\ v_\beta \end{bmatrix} &= \frac{2}{3}\begin{bmatrix} 1 & -\frac{1}{2} & -\frac{1}{2} \\ 0 & \frac{\sqrt{3}}{2} & -\frac{\sqrt{3}}{2} \end{bmatrix}\left(\begin{bmatrix} v_a \\ v_b \\ v_c \end{bmatrix} - \begin{bmatrix} v_n \\ v_n \\ v_n \end{bmatrix}\right) \\ &= \frac{2}{3}V_{dc}\begin{bmatrix} 1 & -\frac{1}{2} & -\frac{1}{2} \\ 0 & \frac{\sqrt{3}}{2} & -\frac{\sqrt{3}}{2} \end{bmatrix}\begin{bmatrix} S_a - \frac{1}{2} \\ S_b - \frac{1}{2} \\ S_c - \frac{1}{2} \end{bmatrix} \\ &= \frac{2}{3}V_{dc}\begin{bmatrix} 1 & -\frac{1}{2} & -\frac{1}{2} \\ 0 & \frac{\sqrt{3}}{2} & -\frac{\sqrt{3}}{2} \end{bmatrix}\begin{bmatrix} S_a \\ S_b \\ S_c \end{bmatrix},\end{aligned} \tag{6.2}$$

where the property that the Clarke transformation of any constant vector leads to a zero vector has been utilized.

Let the matrix U be defined by the switching states:

$$U = \begin{bmatrix} 0 & 1 & 1 & 0 & 0 & 0 & 1 & 1 \\ 0 & 0 & 1 & 1 & 1 & 0 & 0 & 1 \\ 0 & 0 & 0 & 0 & 1 & 1 & 1 & 1 \end{bmatrix} \tag{6.3}$$

and matrix D be the Clarke transformation from the three phase voltage to α-β frame:

$$D = \begin{bmatrix} 1 & -\frac{1}{2} & -\frac{1}{2} \\ 0 & \frac{\sqrt{3}}{2} & -\frac{\sqrt{3}}{2} \end{bmatrix}. \tag{6.4}$$

The multiplication of D and U matrices leads to

$$DU = \begin{bmatrix} 0 & 1 & \frac{1}{2} & -\frac{1}{2} & -1 & -\frac{1}{2} & \frac{1}{2} & 0 \\ 0 & 0 & \frac{\sqrt{3}}{2} & \frac{\sqrt{3}}{2} & 0 & -\frac{\sqrt{3}}{2} & -\frac{\sqrt{3}}{2} & 0 \end{bmatrix}. \tag{6.5}$$

With the transformation matrix (6.5), the operational constraints due to the voltage source inverter are expressed in the $\alpha-\beta$ frame as the equality constraints, which are characterized by the values in the matrix

$$\begin{bmatrix} 0 & 1 & \frac{1}{2} & -\frac{1}{2} & -1 & -\frac{1}{2} & \frac{1}{2} & 0 \\ 0 & 0 & \frac{\sqrt{3}}{2} & \frac{\sqrt{3}}{2} & 0 & -\frac{\sqrt{3}}{2} & -\frac{\sqrt{3}}{2} & 0 \end{bmatrix}\frac{2}{3}V_{dc}, \tag{6.6}$$

where V_{dc} is the voltage for the DC power supply. More precisely, the control variable v_α will only take the values defined by the first row of the matrix (6.6) while the control variable v_β will only take the values of the second row, with further constraints that their values must form the exactly paired relations as in (6.6). These constraints result in the situation where the control movements are restricted to a finite control set of parameters. For instance, with the *DC* bus voltage V_{dc} given, the v_α and v_β voltage values are chosen among the following set of parameters in pairs:

$$v_\alpha^0 = 0; \quad v_\alpha^1 = V_{dc}; \quad v_\alpha^2 = \frac{1}{2}V_{dc}; \quad v_\alpha^3 = -\frac{1}{2}V_{dc};$$

$$v_\beta^0 = 0; \quad v_\beta^1 = 0; \quad v_\beta^2 = \frac{\sqrt{3}}{2}V_{dc}; \quad v_\beta^3 = \frac{\sqrt{3}}{2}V_{dc};$$

$$v_\alpha^4 = -V_{dc}; \quad v_\alpha^5 = -\frac{1}{2}V_{dc}; \quad v_\alpha^6 = \frac{1}{2}V_{dc}; \quad v_\alpha^7 = 0;$$

$$v_\beta^4 = 0; \quad v_\beta^5 = -\frac{\sqrt{3}}{2}V_{dc}; \quad v_\beta^6 = -\frac{\sqrt{3}}{2}V_{dc}; \quad v_\beta^7 = 0.$$

The superscripts of v_α and v_β correspond to the indices of the IGBT's switching states. Once one of the indices is identified, the control action is determined and is implemented via the VSI inverter with the actual voltage values shown in Table 6.2. Although the pair v_α^0, v_β^0 are identical to v_α^7, v_β^7, what action should the inverter take is different in the sense that one corresponds to all off-states while the other to all on-states. To avoid excessive switching actions from the inverter, when $v_\alpha = 0$ and $v_\beta = 0$, the action that the inverter takes will depend on the inverter's previous action. For instance, if the previous action is two states on and one state off, then the inverter's present action corresponding to $v_\alpha = 0$ and $v_\beta = 0$ should be all states on.

The constraints on the control variables (v_d, v_q) in $d - q$ frame are functions of electrical angle θ_e due to the deployment of the Park-transformation. Namely, the operational constraints in the $d - q$ frame due to the VSI inverter are also expressed as equality constraints with the following form:

$$\begin{bmatrix} \cos\theta_e & \sin\theta_e \\ -\sin\theta_e & \cos\theta_e \end{bmatrix} \begin{bmatrix} 0 & 1 & \frac{1}{2} & -\frac{1}{2} & -1 & -\frac{1}{2} & \frac{1}{2} & 0 \\ 0 & 0 & \frac{\sqrt{3}}{2} & \frac{\sqrt{3}}{2} & 0 & -\frac{\sqrt{3}}{2} & -\frac{\sqrt{3}}{2} & 0 \end{bmatrix} \frac{2}{3}V_{dc}, \tag{6.7}$$

where θ_e is the electrical angle. A more compact form of (6.7) is given as

$$\begin{bmatrix} 0 & \cos\theta_e & \cos\left(\theta_e - \frac{2\pi}{3}\right) & \cos\left(\theta_e - \frac{4\pi}{3}\right) & -\cos\theta_e & -\cos\left(\theta_e - \frac{2\pi}{3}\right) & -\cos\left(\theta_e - \frac{4\pi}{3}\right) & 0 \\ 0 & -\sin\theta_e & -\sin\left(\theta_e - \frac{2\pi}{3}\right) & -\sin\left(\theta_e - \frac{4\pi}{3}\right) & \sin\theta_e & \sin\left(\theta_e - \frac{2\pi}{3}\right) & \sin\left(\theta_e - \frac{4\pi}{3}\right) & 0 \end{bmatrix} \frac{2}{3}V_{dc}.$$

This basically says that with a given electrical angle θ_e, there are only seven pairs of v_d and v_q values that can be exactly realized by the VSI inverter. In the $\alpha - \beta$ frame the constraints are the finite set of constant parameters defined by (6.6) once the *DC* bus voltage V_{dc} is given, however, in the $d - q$ frame, this set of constant parameters becomes functions of the electrical angle θ_e.

It is emphasized that with a given θ_e value and a sampling time t, in the $d - q$ frame, the set of v_d and v_q values are constant. For instance, with the given θ_e, the v_d and v_q are chosen among the following parameters:

$$v_d^0 = 0; \quad v_d^1 = \frac{2}{3}V_{dc}\cos\theta_e; \quad v_d^2 = \frac{2}{3}V_{dc}\cos\left(\theta_e - \frac{2\pi}{3}\right); \quad v_d^3 = \frac{2}{3}V_{dc}\cos\left(\theta_e - \frac{4\pi}{3}\right);$$

$$v_q^0 = 0; \quad v_q^1 = -\frac{2}{3}V_{dc}\sin\theta_e; \quad v_q^2 = -\frac{2}{3}V_{dc}\sin\left(\theta_e - \frac{2\pi}{3}\right); \quad v_q^3 = -\frac{2}{3}V_{dc}\sin\left(\theta_e - \frac{4\pi}{3}\right);$$

$$v_d^4 = -\frac{2}{3}V_{dc}\cos\theta_e; \quad v_d^5 = -\frac{2}{3}V_{dc}\cos\left(\theta_e - \frac{2\pi}{3}\right); \quad v_d^6 = -\frac{2}{3}V_{dc}\cos\left(\theta_e - \frac{4\pi}{3}\right); \quad v_d^7 = 0;$$

$$v_q^4 = \frac{2}{3}V_{dc}\sin\theta_e; \quad v_q^5 = \frac{2}{3}V_{dc}\sin\left(\theta_e - \frac{2\pi}{3}\right); \quad v_q^6 = \frac{2}{3}V_{dc}\sin\left(\theta_e - \frac{4\pi}{3}\right); \quad v_q^7 = 0.$$

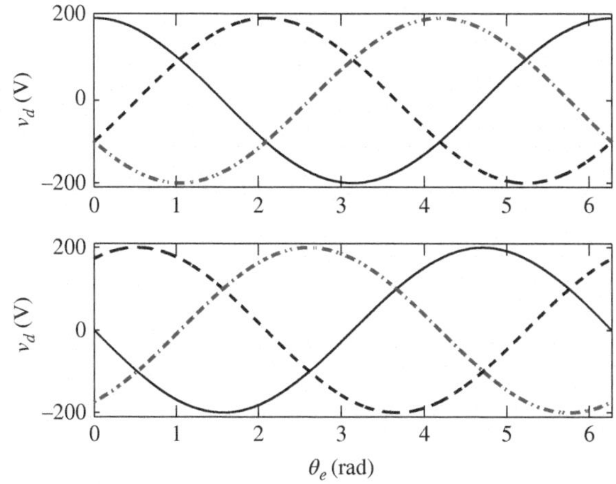

Figure 6.1 The trajectories of v_d and v_q, where $V_{dc} = 300$ V. Solid-line: v_d^1 and v_q^1; dashed-line v_d^2 and v_q^2; dash-dotted line v_d^3 and v_q^3.

Assuming that $V_{dc} = 300$ V, among them, the six sets of v_d and v_q (see the indices from 1 to 6) are sinusoidal functions with an amplitude of 200, the rest being zero. Because the first candidate set (v_d^1, v_q^1) and the last candidate set (v_d^7, v_q^7) have identical value of zero, there are only seven sets of candidate variables to be evaluated. Figure 6.1 shows the three sets of trajectories of v_d and v_q with respect to the variation of electrical angle θ_e ($0 \leq \theta_e \leq 2\pi$).

6.2 FCS Predictive Control of PMSM

After establishing the trajectories of the control signal in the $d - q$ frame with respect to the electrical angle θ_e, without using PWM in the implementation of the control signal, the selection of the control signal v_d and v_q is confined to the seven possible choices listed in the previous section. The seven sets of v_d and v_q values, at a given sampling time and θ_e, form the candidate sets for finite control set.

In order to determine which pair of control signals should be used for controlling the drives, an objective function is required to reflect the purpose of controller design. Current control is often considered in the majority of the applications when using the FCS-MPC approach. For current control, at the sampling time t_i, the objective function is chosen as sum of the square errors between the desired and predicted signals:

$$J = (i_d^*(t_i) - i_d(t_{i+1}))^2 + (i_q^*(t_i) - i_q(t_{i+1}))^2, \tag{6.8}$$

where $i_d(t_{i+1})$ and $i_q(t_{i+1})$ are one-step-ahead predictions of $i_d(t_i)$ and $i_q(t_i)$, respectively.

In the calculation of the current prediction, a physical model in the $d - q$ frame is needed, which describes the relationship between the inputs v_d, v_q and the outputs i_d and i_q. In the case of PMSM control, the following differential equations are used in the calculation:

$$\frac{di_d(t)}{dt} = \frac{1}{L_d}(v_d(t) - Ri_d(t) + \omega_e(t)L_q i_q(t)) \tag{6.9}$$

$$\frac{di_q(t)}{dt} = \frac{1}{L_q}(v_q(t) - Ri_q(t) - \omega_e(t)L_d i_d(t) - \omega_e(t)\phi_{mg}). \tag{6.10}$$

At sampling time t_i, assuming that a sampling interval Δt is used, let

$$\frac{di_d(t)}{dt} \approx \frac{i_d(t_{i+1}) - i_d(t_i)}{\Delta t}; \quad \frac{di_q(t)}{dt} \approx \frac{i_q(t_{i+1}) - i_q(t_i)}{\Delta t}.$$

Then, the discretized differential equations become the difference equations:

$$i_d(t_{i+1}) = i_d(t_i) + \frac{\Delta t}{L_d}(v_d(t_i) - Ri_d(t_i) + \omega_e(t_i)L_q i_q(t_i)) \tag{6.11}$$

$$i_q(t_{i+1}) = i_q(t_i) + \frac{\Delta t}{L_q}(v_q(t_i) - Ri_q(t_i) - \omega_e(t_i)L_d i_d(t_i) - \omega_e(t_i)\phi_{mg}). \tag{6.12}$$

By substituting (6.11) and (6.12) into (6.8), the objective function J now contains the variables that are measured at the sampling time t_i and the manipulated variables $v_d(t_i)$ and $v_q(t_i)$, that is

$$J = \left(i_d^*(t_i) - i_d(t_i) - \frac{\Delta t}{L_d}(v_d(t_i) - Ri_d(t_i) + \omega_e(t_i)L_q i_q(t_i)) \right)^2 + \left(i_q^*(t_i) - i_q(t_i) - \frac{\Delta t}{L_q}(v_q(t_i) - Ri_q(t_i) - \omega_e(t_i)L_d i_d(t_i) - \omega_e(t_i)\phi_{mg}) \right)^2. \tag{6.13}$$

Since at the sampling instant t_i, from Section 6.1, there are seven pairs of $v_d(t_i)$ and $v_q(t_i)$ available as candidates, the next step in the FCS-MPC design is to find the pair of manipulated variables that will minimize the objective function J (6.13). For this purpose, the seven values of the objective function J are calculated with respect to the candidate pairs of $v_d(t_i)$ and $v_q(t_i)$ and denoted as $J_0, J_1, J_2, \ldots, J_7$. A simple search function is used to find the minimal value of J and its associated index k. Once this index is found, the control signal at time t_i to the VSI is determined through Table 6.1 and the corresponding voltage is obtained through Table 6.2. However, in order to reduce unnecessary switchings, if the index is found to be 0, then the previous states of the VS inverter are required to determine whether the index 0 or 7 should be used in the control action.

When the sampling time progresses to $t = t_i + \Delta t$, the new measurements of $i_d(t_{i+1})$, $i_q(t_{i+1})$ currents and velocity measurement $\omega_e(t_{i+1})$ are obtained, the seven new pairs of candidates $v_d(t_{i+1})$ and $v_q(t_{i+1})$ are computed due to the new electrical angle $\theta_e(t_{i+1})$. With all the variables in the objective function (6.13) being updated, a minimization is performed to find the new minimal value of J_k and its index k at sampling time $t = t_i + \Delta t$, leading to the control signals for the voltage source inverter.

The essence of the finite control set method is based on the receding horizon control principle, which uses one-step-ahead prediction and on-line optimization to solve the constrained optimal control problem. The closed-loop feedback mechanism is generated when using the updated $i_d(t_i)$ and $i_q(t_i)$ current measurements in the prediction. Therefore, the corresponding algorithm is called FCS-MPC algorithm.

For convenience of programming, the difference equations (6.11) and (6.12) may also be expressed in matrix and vector forms:

$$\begin{bmatrix} i_d(t_{i+1}) \\ i_q(t_{i+1}) \end{bmatrix} = (I + \Delta t A_m(t_i)) \begin{bmatrix} i_d(t_i) \\ i_q(t_i) \end{bmatrix} - \begin{bmatrix} 0 \\ \frac{\omega_e(t_i)\phi_{mg}\Delta t}{L_q} \end{bmatrix} + \Delta t B_m \begin{bmatrix} v_d(t_i) \\ v_q(t_i) \end{bmatrix}, \tag{6.14}$$

where I is the identity matrix with dimension 2×2 and the system matrices $A_m(t_i)$ and B_m are defined as

$$A_m(t_i) = \begin{bmatrix} -\frac{R_s}{L_d} & \frac{\omega_e(t_i)L_q}{L_d} \\ -\frac{\omega_e(t_i)L_d}{L_q} & -\frac{R_s}{L_q} \end{bmatrix}; B_m = \begin{bmatrix} \frac{1}{L_d} & 0 \\ 0 & \frac{1}{L_q} \end{bmatrix}.$$

The finite control set method discussed here has found many applications in electrical drive control and power converter control due to its simplicity in both concept and computation.

6.3 MATLAB Tutorial on Real-time Implementation of FCS-MPC

Tutorial 2 *The objective of this tutorial is to learn how to implement finite control set algorithm in real-time. The core of this activity is to produce a MATLAB embedded function that can be used in a Simulink simulation as well as in a xPC Target based implementation. This embedded function is based on the computational algorithm given in Section 6.2. The entire embedded MATLAB completes one cycle of computation for the control signals. For every sampling period, it will repeat the same computation procedure.*

Step by Step

1. *Create a new Simulink file called FCS.mdl*
2. *In Simulink's directory of User-Defined Functions, find the icon of embedded MATLAB function and copy it to the FCS model.*
3. *Define the input and output variables to the FCS model so that the embedded function has the following form:*

```
function [Ua,Ub,Uc,Ind]=
FCS(idRef,iqRef,id,iq,we,theta,Ld,Lq,Rs,Phi_mg,Tin,D,U)
```

where the first six elements among the input variables are the reference and measurement signals of i_d and i_q, electrical velocity ω and angle θ; the next four elements are the parameters for the PMSM, Tin is the sampling interval (Δt); D is the data matrix (see (6.4)) defined as

$$D = \frac{2}{3}V_{dc}\begin{bmatrix} 1 & -\frac{1}{2} & -\frac{1}{2} \\ 0 & \frac{\sqrt{3}}{2} & -\frac{\sqrt{3}}{2} \end{bmatrix}.$$

U is the data matrix (see (6.3)) defined as

$$U = \begin{bmatrix} 0 & 1 & 1 & 0 & 0 & 0 & 1 & 1 \\ 0 & 0 & 1 & 1 & 1 & 0 & 0 & 1 \\ 0 & 0 & 0 & 0 & 1 & 1 & 1 & 1 \end{bmatrix}.$$

4. *At the top of the embedded function, find "Model Explorer" among the "Tools". When opening the Model Explorer, select "discrete" for the "update method" and input "Tin" into the "sample time"; select "Support variable-size arrays"; select "Saturate on integer overflow"; select "Fixed point". Click "Apply" to save the changes.*
5. *We need to edit the input and output data ports in order to let the embedded function know which input ports are the real-time variables and which are the parameters. This editing task is performed using Model Explorer.*
   - *click on "idRef", on Scope, select "input", assign port "1" and size "-1", complexity "Inherited", type "Inherit: Same as Simulink". Repeat the same editing procedure for the remaining input variables, "iqRef", "id", "iq", "we", "theta".*
   - *The remaining 7 inputs to the embedded function are the parameters required in the computation. Click on "Ld", on Scope, select "Parameter" and click "Tunable" and click "Apply" to save the changes. Repeat the same editing procedure for the remaining parameters, "Lq", "Rs", "Phi_mg", "Tin", "D", "U".*
   - *To edit the output ports from the embedded function, click on "Ua", on Scope, select "Output", Port "1", Size "-1", Sampling Model "Sample based", Type "Inherit: Same as Simulink", and click on "Apply" to save the changes. Repeat the same editing procedure for Ub, Uc and Ind.*

6. *In the following, the program will declare those variables that are stored in the embedded function during the repetition for their dimensions and initial values. "Xe" is the error vector between the predicted and measured state vectors; "err" is the norm of the error vector and "ind_old" is the previous ind. Enter the following program into the file:*

```
persistent Xe
if isempty(Xe)
 Xe=zeros(2,7);
end
persistent err
if isempty(err)
     err=zeros(1,7)
end
persistent ind_old
if isempty(ind_old)
     ind_old=0;
end
```

7. *Form the system matrices $F = (I + A_m(t_i)\Delta t)$, $G = B_m\Delta t$ and disturbance vector H. Enter the following program into the file:*

```
F=[1-Rs*Tin/Ld Tin*we;
     -Tin*we 1-Rs*Tin/Lq];
G=[Tin/Ld 0;
     0    Tin/Lq];
H=[0; -Tin*we*Phi_mg/Lq];
```

8. *Form the rotation matrix M with the real-time θ value from the input of the embedded function. Enter the following program into the file:*

```
M=[cos(theta) sin(theta);-sin(theta) cos(theta)];
```

9. *Form the state vectors. Enter the following program into the file:*

```
xRef=[idRef;iqRef];
x=[id;iq];
```

10. *Note that the multiplication of the rotation matrix M with the D and U matrices will produce the 8 pairs of v_d and v_q values for a given θ at the sampling time t_i. In a recursive manner, calculate the one-step-ahead prediction of i_d and i_q currents using (6.14) based on the 7 pairs of v_d and v_q values (the 8th pair is a zero vector identical to the first pair, thus neglected). The state errors between the desired and the measured i_d and i_q currents are calculated using a vector form and its norm is evaluated.*

```
for i=1:7
xn=F*x+G*M*D*U(:,i)+H;
Xe(:,i)=xRef-xn;
err(:,i)=Xe(:,i)'*Xe(:,i);
end
```

11. *In order to determine which pair of control signals should be used, the evaluation of the norm of the state error vector is performed to find the index number that corresponds to the minimum*

sum of square errors (the objective function J given by (6.13)). If this index corresponds to the first vector, then in order to reduce unnecessary control movement (or switches of the VSI), the information of previous control switches will be needed to determine whether the all states-off or all states-on control sequence should be used (see Table 6.1). Enter the following program into the file:

```
[val ind]=min(err);
ind=ind-1;
if (ind==0)
    if (ind_old==1 || ind_old==3 || ind_old==5)
        ind=0;
    else
        ind=7;
    end
end
```

12. *On obtaining the index number that minimizes the objective function, the control sequence is found. This index is translated into the switching states of the IGBT inverter. Enter the following program into the file:*

```
ind_old=ind;
Ua=U(1,ind+1);Ub=U(2,ind+1);Uc=U(3,ind+1);
```

13. *The embedded function takes the values of Ua, Ub and Uc as its outputs, which is connected with the next components of this finite control set system.*

This embedded function is now interfaced with the simulator of PMSM to produce the Simulink simulator called "simulator4FCS.mdl".

6.3.1 Simulation Results

The finite control set method is attractive for the applications because of its simplicity in implementation, where the PWMs are no longer required. The tuning parameter for the closed-loop performance is the sampling interval Δt. In order to achieve satisfactory closed-loop performance, the sampling interval Δt is selected as small as possible. The choice of Δt is critical in the design and implementation of finite set control method. For a large Δt, the finite control set method fails to produce satisfactory results.

A PMSM model used in the simulation of FCS-MPC has the following parameters. The pair of poles is 2; the resistance is $R_s = 2.98\ \Omega$; inductance $L_d = L_q = 7 \times 10^{-3}$ H; $\phi_{mg} = 0.125\ Wb$; inertia $J_m = 1 \times 10^{-5}\ \text{kg} \cdot \text{m}^2$; $B_v = 11 \times 10^{-5}\ \text{Nm} \cdot \text{s}$. The reference signal to the i_d current is $i_d^* = 0$ A and i_q^* is a step signal with amplitude of ± 2 A. The torque load disturbance is also a step signal with unit amplitude at the start of the simulation, followed by a step change to -1 at half of the simulation time. The *DC* power supply has voltage $V_{dc} = 300$ V.

There are two sampling intervals used in the simulation studies. In the first case, the sampling interval is selected as $\Delta t = 40 \times 10^{-6}$ sec and in the second case, Δt is reduced to 10×10^{-6} sec.

6.3.1.1 Case A

Figure 6.2(a) shows the i_d and i_q closed-loop current control results, and Figure 6.2(b) shows the normalized three phase control signals, where the sampling interval Δt is 40×10^{-6} sec. It is seen that with

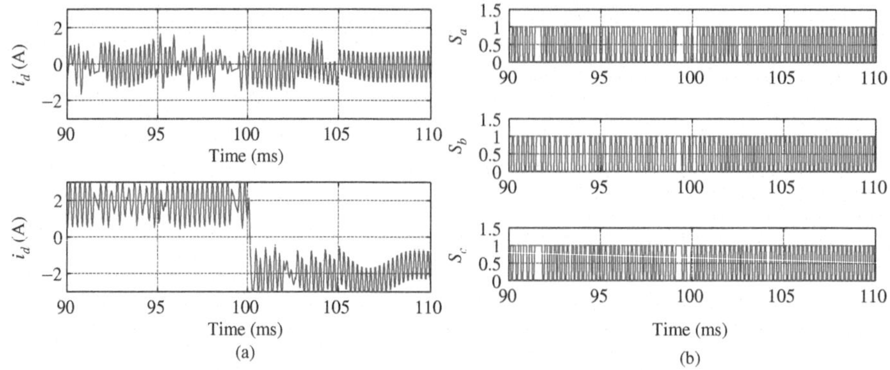

Figure 6.2 Case A. Sampling interval $\Delta t = 40 \times 10^{-6}$ sec. (a) i_d and i_q currents and (b) The normalized control signals.

this selection of Δt, the closed-loop control results are very poor because of the large variations in both i_d and i_q currents.

6.3.1.2 Case B

In the second case study, the sampling interval is reduced to $\Delta t = 10 \times 10^{-6}$ sec. It is clearly seen from Figure 6.3 that the closed-loop performance is significantly improved over the slower sampling case. The improvement is evident as the variations of i_d and i_q currents become very small.

Since the reference signal to the i_d current is 0 A, we can easily compute the mean value of i_d signal to determine its offset error. When $\Delta t = 40 \times 10^{-6}$ sec, the mean value for the i_d current is −0.1947 A, and $\Delta t = 10 \times 10^{-6}$ sec, the mean value is 0.0023 A. In an ideal case, the mean value of the i_d current should be 0 A. Thus, it is seen that the FCS-MPC has a large offset error when the sampling interval is not sufficiently small.

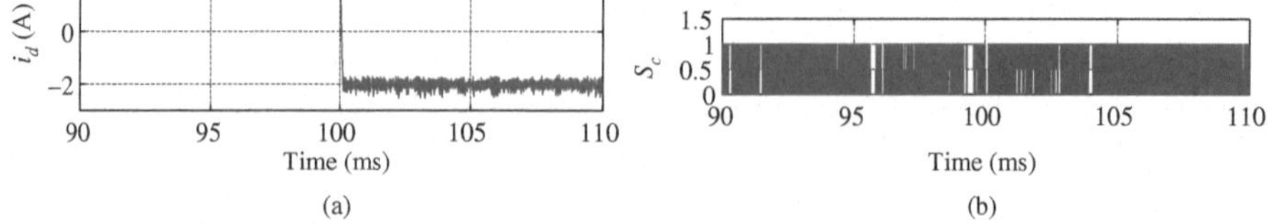

Figure 6.3 Case B. Sampling interval $\Delta t = 10 \times 10^{-6}$ sec. (a) i_d and i_q currents and (b) The normalized control signals.

This simulation results demonstrate that choice of sampling interval is critical in the implementation of FCS-MPC method. In order for this method to produce satisfactory results, the sampling interval Δt should be sufficiently small for practical implementations.

6.3.2 Experimental Results of FCS Control

The FCS-MPC algorithm is implemented on an experimental test-bed (see Figure 10.6 for illustration of the test-bed) equipped with an eZdsp-F28335 module, a two-level IGBT inverter, a driving PMSM, a rigid coupling, a second PMSM, a three-phase rectifier and a pure resistor load. The driving PMSM is coupled with the second PMSM through the rigid coupling. An unknown load torque disturbance is generated by varying the speed of the second PMSM. Furthermore, the second PMSM is connected to the resistor load through the three-phase rectifier. With this setup, the kinetic energy is converted to the electric energy dissipated by the resistor loads finally.

The algorithm is implemented with the per-unit model of the PMSM with its parameters given in the simulation results (see Section 6.3.1). With the per-unit model, all the parameters and variables in the $d - q$ model are scaled by their respective unit values. The benefit of such an approach is that the influence of the different units on the calculation, such as ampere, volts and rad/s, is eliminated. Since all calculations are computed on a per-unit basis, it is more suitable for the implementation using a Digital Signal Processor (DSP), particularly for the fixed-point implementation. Very critically, using the DSP in the implementation of the FCS-MPC algorithm, a much faster sampling rate can be used in the experiments.

The control objective is to maintain d-axis current zero and q-axis current to its desired set-point for the purpose of optimal torque to current ratio. Thus, in the experiment, the d-axis current reference is kept at zero for the control with optimal torque current ratio, while the q-axis reference signal is chosen to be a square-wave signal as shown in Figure 6.4(b). In the experiment, the sampling interval Δt is 10×10^{-6} sec. With this fast sampling rate, the FCS-MPC algorithm successfully tracked the reference i_d^* and i_q^* signals and rejected the unknown load torque disturbance, as shown by the experimental data illustrated in Figure 6.4. It is interesting to note that the dynamic response to a step change in the reference i_q^* is immediate and the variations of both i_d and i_q currents are very small; however, there are small offset errors in both i_d and i_q currents.

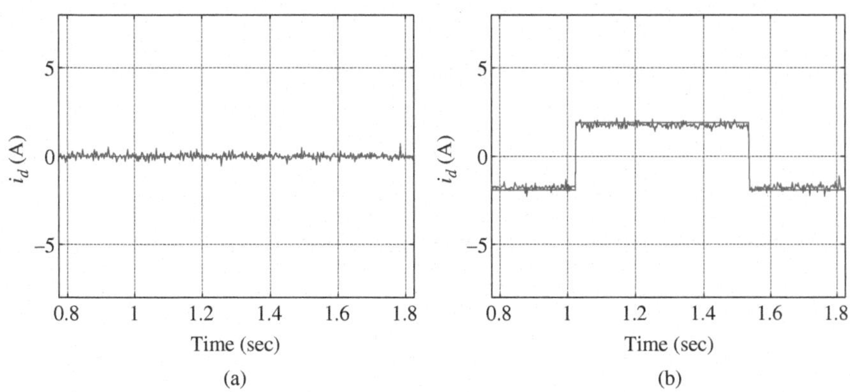

Figure 6.4 Experimental results for FCS-MPC system. (a) i_d and (b) i_q and i_q^*.

6.4 Analysis of FCS-MPC System

This section will examine the FCS predictive control algorithm and derive the corresponding closed-loop feedback controller gain in which the locations of the closed-loop poles are identified.

6.4.1 Optimal Control System

In order to analyze the closed-loop performance via feedback control, the objective function J is rewritten in vector form:

$$J = \begin{bmatrix} i_d^*(t_i) - i_d(t_{i+1}) & i_q^*(t_i) - i_q(t_{i+1}) \end{bmatrix} \begin{bmatrix} i_d^*(t_i) - i_d(t_{i+1}) \\ i_q^*(t_i) - i_q(t_{i+1}) \end{bmatrix}. \tag{6.15}$$

Furthermore, the difference equations (6.11) and (6.12) are expressed in matrix and vector forms given by (6.14). For notational simplicity, let the vector $\begin{bmatrix} f_d(t_i) & f_q(t_i) \end{bmatrix}^T$ be defined as

$$\begin{bmatrix} f_d(t_i) \\ f_q(t_i) \end{bmatrix} = \begin{bmatrix} i_d^*(t_i) \\ i_q^*(t_i) \end{bmatrix} - (I + \Delta t A_m(t_i)) \begin{bmatrix} i_d(t_i) \\ i_q(t_i) \end{bmatrix} + \begin{bmatrix} 0 \\ \frac{\omega_e(t_i)\phi_{mg}\Delta t}{L_q} \end{bmatrix}. \tag{6.16}$$

Then it can be verified by combining (6.16) with (6.14) that the objective function (6.15) has the compact expression:

$$J = \left(\begin{bmatrix} f_d(t_i) & f_q(t_i) \end{bmatrix} - \Delta t \begin{bmatrix} v_d(t_i) & v_q(t_i) \end{bmatrix} B_m^T \right) \left(\begin{bmatrix} f_d(t_i) \\ f_q(t_i) \end{bmatrix} - \Delta t B_m \begin{bmatrix} v_d(t_i) \\ v_q(t_i) \end{bmatrix} \right)$$

which is in the quadratic objective function form:

$$J = \begin{bmatrix} f_d(t_i) & f_q(t_i) \end{bmatrix} \begin{bmatrix} f_d(t_i) \\ f_q(t_i) \end{bmatrix} - 2 \begin{bmatrix} v_d(t_i) & v_q(t_i) \end{bmatrix} \Delta t B_m^T \begin{bmatrix} f_d(t_i) \\ f_q(t_i) \end{bmatrix}$$

$$+ \begin{bmatrix} v_d(t_i) & v_q(t_i) \end{bmatrix} \Delta t^2 B_m^T B_m \begin{bmatrix} v_d(t_i) \\ v_q(t_i) \end{bmatrix}. \tag{6.17}$$

Optimizing J becomes a least square minimization problem. Defining vector $v_{dq}(t_i) = \begin{bmatrix} v_d(t_i) & v_q(t_i) \end{bmatrix}^T$, from the first derivative of the objective function J:

$$\frac{\partial J}{\partial v_{dq}(t_i)} = -2\Delta t B_m^T \begin{bmatrix} f_d(t_i) \\ f_q(t_i) \end{bmatrix} + 2\Delta t^2 B_m^T B_m \begin{bmatrix} v_d(t_i) \\ v_q(t_i) \end{bmatrix}, \tag{6.18}$$

the necessary condition of an extremum J is obtained as,

$$\frac{\partial J}{\partial v_{dq}(t_i)} = 0. \tag{6.19}$$

To determine whether this extremum is a minimum, the second derivative is calculated as

$$\frac{\partial^2 J}{\partial^2 v_{dq}(t_i)} = \Delta t^2 B_m^T B_m. \tag{6.20}$$

Note that since $L_d > 0$ and $L_q > 0$, the matrix $B_m^T B_m$ is positive definite where

$$B_m^T B_m = \begin{bmatrix} \frac{1}{L_d^2} & 0 \\ 0 & \frac{1}{L_q^2} \end{bmatrix} > 0,$$

which is the sufficient condition to guarantee that the solution of (6.19) is a minimum of the objective function J. From (6.19), we find the control signal vector $\upsilon_{dq}(t_i) = [\upsilon_d(t_i)\upsilon_q(t_i)]^T$ that minimizes the objective function J as:

$$\begin{bmatrix} \upsilon_d(t_i) \\ \upsilon_q(t_i) \end{bmatrix} = (\Delta t^2 B_m^T B_m)^{-1} \Delta t B_m^T \begin{bmatrix} f_d(t_i) \\ f_q(t_i) \end{bmatrix} = \frac{1}{\Delta t} \begin{bmatrix} L_d & 0 \\ 0 & L_q \end{bmatrix} \begin{bmatrix} f_d(t_i) \\ f_q(t_i) \end{bmatrix}. \tag{6.21}$$

By substituting (6.16) into (6.21), with the definition of the $\upsilon_{dq}(t_i)$, the optimal solution of the control signals is obtained as

$$\begin{bmatrix} \upsilon_d(t_i) \\ \upsilon_q(t_i) \end{bmatrix} = \begin{bmatrix} \frac{L_d}{\Delta t} & 0 \\ 0 & \frac{L_q}{\Delta t} \end{bmatrix} \left(\begin{bmatrix} i_d^*(t_i) \\ i_q^*(t_i) \end{bmatrix} - (I + \Delta t A_m(t_i)) \begin{bmatrix} i_d(t_i) \\ i_q(t_i) \end{bmatrix} + \begin{bmatrix} 0 \\ \frac{\omega_e(t_i)\phi_{mg}\Delta t}{L_q} \end{bmatrix} \right). \tag{6.22}$$

Without considering the restriction of the control signals, this is the unconstrained optimal solution of the predictive control system with one-step-ahead prediction. Because the actual electrical velocity $\omega_e(t_i)$ is used in the computation of the prediction, the control law is linear time-varying.

There is an alternative way to find the minimum of the objective function via the technique of completing squares. This completing squares approach will lead to a different method to evaluate the objective function for finding the control signals among the candidate variables.

From the quadratic objective function (6.17), by adding and subtracting the term

$$\begin{bmatrix} f_d(t_i) & f_q(t_i) \end{bmatrix} \Delta t B_m (\Delta t^2 B_m^T B_m)^{-1} B_m^T \Delta t \begin{bmatrix} f_d(t_i) \\ f_q(t_i) \end{bmatrix}$$

to the original objective function J, its value remains unchanged. With this term added, the following three terms lead to completed squares, which is denoted by J_0,

$$\begin{aligned} J_0 &= \begin{bmatrix} \upsilon_d(t_i) & \upsilon_q(t_i) \end{bmatrix} \Delta t^2 B_m^T B_m \begin{bmatrix} \upsilon_d(t_i) \\ \upsilon_q(t_i) \end{bmatrix} - 2 \begin{bmatrix} \upsilon_d(t_i) & \upsilon_q(t_i) \end{bmatrix} \Delta t B_m^T \begin{bmatrix} f_d(t_i) \\ f_q(t_i) \end{bmatrix} \\ &+ \begin{bmatrix} f_d(t_i) & f_q(t_i) \end{bmatrix} \Delta t B_m (\Delta t^2 B_m^T B_m)^{-1} B_m^T \Delta t \begin{bmatrix} f_d(t_i) \\ f_q(t_i) \end{bmatrix} \\ &= \left(\begin{bmatrix} \upsilon_d(t_i) \\ \upsilon_q(t_i) \end{bmatrix} - (\Delta t^2 B_m^T B_m)^{-1} B_m^T \Delta t \begin{bmatrix} f_d(t_i) \\ f_q(t_i) \end{bmatrix} \right)^T (\Delta t^2 B_m^T B_m) \\ &\times \left(\begin{bmatrix} \upsilon_d(t_i) \\ \upsilon_q(t_i) \end{bmatrix} - (\Delta t^2 B_m^T B_m)^{-1} B_m^T \Delta t \begin{bmatrix} f_d(t_i) \\ f_q(t_i) \end{bmatrix} \right). \end{aligned} \tag{6.23}$$

Now, with J_0 given by the completed squares (6.23), the original objective function J becomes

$$J = J_0 + J_{min}, \tag{6.24}$$

where J_{min} is

$$J_{min} = - \begin{bmatrix} f_d(t_i) f_q(t_i) \end{bmatrix} B_m (B_m^T B_m)^{-1} B_m^T \begin{bmatrix} f_d(t_i) \\ f_q(t_i) \end{bmatrix} + \begin{bmatrix} f_d(t_i) f_q(t_i) \end{bmatrix} \begin{bmatrix} f_d(t_i) \\ f_q(t_i) \end{bmatrix}.$$

Note that the weighting matrix $\Delta t^2 B_m^T B_m$ in J_0 (see (6.23)–(6.24)) is positive definite and J_{min} is independent of the variables $\upsilon_d(t_i)$ and $\upsilon_q(t_i)$. Thus, the minimum of the original objective function J is achieved if J_0 is minimized. Furthermore, it is seen that the minimum of J_0 is zero, from (6.23), if variables $v_d(t_i)$ and $v_q(t_i)$ are chosen to be

$$\begin{bmatrix} v_d(t_i) \\ v_q(t_i) \end{bmatrix} = (\Delta t^2 B_m^T B_m)^{-1} B_m^T \Delta t \begin{bmatrix} f_d(t_i) \\ f_q(t_i) \end{bmatrix}. \tag{6.25}$$

The solution obtained via completing the squares is identical to that obtained before (see (6.21)).

With the completing squares approach, the constant term J_{min} can be easily examined via

$$J_{min} = \begin{bmatrix} f_d(t_i) f_q(t_i) \end{bmatrix} (I - B_m(B_m^T B_m)^{-1} B_m^T) \begin{bmatrix} f_d(t_i) \\ f_q(t_i) \end{bmatrix}.$$

Since

$$B_m = \begin{bmatrix} \frac{1}{L_d} & 0 \\ 0 & \frac{1}{L_q} \end{bmatrix}$$

it is easy to verify that the matrix $I - B_m(B_m^T B_m)^{-1} B_m^T$ is a zero matrix, which leads to

$$J_{min} = \begin{bmatrix} f_d(t_i) f_q(t_i) \end{bmatrix} (I - B_m(B_m^T B_m)^{-1} B_m^T) \begin{bmatrix} f_d(t_i) \\ f_q(t_i) \end{bmatrix} = 0, \tag{6.26}$$

hence $J = J_0$ from (6.24). This is an interesting conclusion, which basically says that the sum of squared error between the predicted and the reference signals is zero if the control signals are chosen according to (6.25).

6.4.2 *Feedback Controller Gain*

From (6.22), the feedback control gain in the one-step-ahead predictive control system at sampling instant t_i is identified

$$K_{fcs}(t_i) = \begin{bmatrix} \frac{L_d}{\Delta t} & 0 \\ 0 & \frac{L_q}{\Delta t} \end{bmatrix} (I + \Delta t A_m(t_i)), \tag{6.27}$$

which is obtained by examining the relationship between $\begin{bmatrix} v_d(t_i) & v_q(t_i) \end{bmatrix}^T$ and $\begin{bmatrix} i_d(t_i) & i_q(t_i) \end{bmatrix}^T$. Immediately, (6.27) reveals that the feedback controller gain K_{fcs} increases as the sampling interval Δt decreases. As $\Delta t \rightarrow 0$, the feedback controller gain $K_{fcs} \rightarrow \infty$. Furthermore, for sufficiently small Δt, the controller gain could be approximated by

$$K_{fcs}(t_i) \approx \begin{bmatrix} \frac{L_d}{\Delta t} & 0 \\ 0 & \frac{L_q}{\Delta t} \end{bmatrix}. \tag{6.28}$$

To determine the internal closed-loop stability of the one-step-ahead predictive control system, consider the discretized system model (6.14), which has the form,

$$\begin{bmatrix} i_d(t_{i+1}) \\ i_q(t_{i+1}) \end{bmatrix} = (I + \Delta t A_m(t_i)) \begin{bmatrix} i_d(t_i) \\ i_q(t_i) \end{bmatrix} - \begin{bmatrix} 0 \\ \frac{\omega_e(t_i)\phi_{mg}\Delta t}{L_q} \end{bmatrix} + \Delta t B_m \begin{bmatrix} v_d(t_i) \\ v_q(t_i) \end{bmatrix}.$$

By substituting the feedback control signal (see (6.22)),

$$\begin{bmatrix} v_d(t_i) \\ v_q(t_i) \end{bmatrix} = -K_{fcs}(t_i) \begin{bmatrix} i_d(t_i) \\ i_q(t_i) \end{bmatrix} + \begin{bmatrix} \frac{L_d}{\Delta t} & 0 \\ 0 & \frac{L_q}{\Delta t} \end{bmatrix} \begin{bmatrix} 0 \\ \frac{\omega_e(t_i)\phi_{mg}\Delta t}{L_q} \end{bmatrix} \tag{6.29}$$

into this equation where the reference signals are considered to be 0 in the original control law (see (6.22)), it can be readily verified that the closed-loop system has the following form:

$$\begin{bmatrix} i_d(t_{i+1}) \\ i_q(t_{i+1}) \end{bmatrix} = \begin{bmatrix} 0 & 0 \\ 0 & 0 \end{bmatrix} \begin{bmatrix} i_d(t_i) \\ i_q(t_i) \end{bmatrix}. \tag{6.30}$$

The two eigenvalues of the closed-loop system (6.30) are at the origin of the complex plane. Because this is a discrete system, the location of the closed-loop eigenvalues at zero guarantees its stability at sampling time t_i. However, in order to guarantee the internal closed-loop stability for $0 \leq i < \infty$, an additional condition on the slow variation of the system matrix is required because the system is time varying and the control law is time varying. This is based on the classical work in the area of linear system theory Desoer (1969) which stated that for a linear time varying system, it is stable if it has all eigenvalues lying strictly inside the unit circle and if it is slowly time varying. A question arises from the fact that in this design, the closed-loop system matrix is zero in (6.30), but where does the time variation of the control system come from? The answer to the question lies in the derivation of (6.30) where the assumption that all the electrical parameters in the PMSM are exactly known values is implicitly used. Under this assumption, the control law will result in a cancellation of the dynamics. Therefore, in reality with some degree of parameter mismatch between the model and the PMSM, there would not be the perfect cancellation, leading to the time-varying nature of the system matrix. Hence, the slow time-variation of the controller gain is needed as part of the closed-loop stability condition. However, if the sampling interval Δt is sufficiently small, then the controller gain is close to a constant gain matrix (see (6.28)).

Because the closed-loop eigenvalues of the optimal control system are located at the origin of the complex plane, the FCS-MPC design method is identical to the deadbeat control technique, which then has a long history in the applications of electrical drives and power converters (see, for example, Gokhale *et al.* (1987), Kawabata *et al.* (1990) and Malesani *et al.* (1999)).

6.4.3 Constrained Optimal Control

The one-step-ahead prediction of the current control system presented is an optimal control system without constraints. As there are only seven sets of candidates of $v_d(t_i)$ and $v_q(t_i)$ for the implementation of the one-step-ahead predictive control law, the optimal control signals computed using (6.22) are not necessarily equal to one of the seven pair values. As discussed before, a search procedure was needed in the determination of the actual control signals $v_d(t_i)$ and $v_q(t_i)$ among the candidates.

In the optimal control without constraints, the solution that minimizes the objective function is given by (6.25), which virtually leads to the zero value of the objective function J. Letting the optimal control signals be denoted by

$$\begin{bmatrix} v_d(t_i)^{opt} \\ v_q(t_i)^{opt} \end{bmatrix} = (\Delta t^2 B_m^T B_m)^{-1} B_m^T \Delta t \begin{bmatrix} f_d(t_i) \\ f_q(t_i) \end{bmatrix} \tag{6.31}$$

and replacing the corresponding terms in the objective function J_0 (see (6.23)–(6.24)) with $\left[v_d(t_i)^{opt} \;\; v_q(t_i)^{opt}\right]^T$, we obtain the objective function for the constrained control problem:

$$J = \left(\begin{bmatrix} v_d(t_i) \\ v_q(t_i) \end{bmatrix} - \begin{bmatrix} v_d(t_i)^{opt} \\ v_q(t_i)^{opt} \end{bmatrix} \right)^T (\Delta t^2 B_m^T B_m) \left(\begin{bmatrix} v_d(t_i) \\ v_q(t_i) \end{bmatrix} - \begin{bmatrix} v_d(t_i)^{opt} \\ v_q(t_i)^{opt} \end{bmatrix} \right), \tag{6.32}$$

where $J = J_0$ because $J_{min} = 0$. Since the weighting matrix $\Delta t^2 B_m^T B_m$ is

$$\Delta t^2 B_m^T B_m = \begin{bmatrix} \frac{\Delta t^2}{L_d^2} & 0 \\ 0 & \frac{\Delta t^2}{L_q^2} \end{bmatrix},$$

the objective function J can also be written as

$$J = \frac{\Delta t^2}{L_d^2}(v_d(t_i) - v_d(t_i)^{opt})^2 + \frac{\Delta t^2}{L_q^2}(v_q(t_i) - v_q(t_i)^{opt})^2. \tag{6.33}$$

An immediate comment follows from (6.33). Note that the minimum value of the objective function when $v_d(t_i) \neq v_d(t_i)^{opt}$ and $v_q(t_i) \neq v_q(t_i)^{opt}$ is weighted by Δt^2, where Δt is the sampling interval that is also the design parameter in this control scheme. With the same system parameters, a reduction of the sampling interval Δt will reduce the minimum of the objective function J_{min}. Equation (6.33) also provides an insight into the choice of sampling interval Δt, where it is seen that the selection of Δt should be made in relation to the size of inductance.

To seek the optimal solution that will minimize the objective function J with the limited choices of $v_d(t_i)$ and $v_q(t_i)$, namely the seven pairs of $v_d(t_i)$ and $v_q(t_i)$, the seven values of the objective function J (6.33) are calculated with respect to the candidate pairs of $v_d(t_i)$ and $v_q(t_i)$ and denoted as J_0, J_1, J_2, … ,J_7. A simple search function is used to find the minimal value of J_{k*} and its associated index $k*$. This search procedure is identical to that stated in Section 6.2. The control implementation procedure is also identical. There is a geometric interpretation for the minimization of the objective function (6.33) subject to the finite control set. When $L_d = L_q$, the variations of J form a family of circles centered at $(v_d(t_i)^{opt}, v_q(t_i)^{opt})$. The optimal solution is the pair of $v_d(t_i)^k$ and $v_q(t_i)^k$ values that form a line to touch the circle in the shortest distance. This geometric interpretation is illustrated in Figure 6.5.

Although the original objective function (6.13) is identical to the objective function (6.33) after the analysis, the latter case offers an insight into the design problem, and is also more convenient in the computation of the control law. For the objective function (6.33), we can firstly calculate the feedback control gain K_{fcs} and the optimal control signal without constraints. Then we evaluate the cost function with the actual seven pairs of voltage variables against the optimal solution. The pair that yields a smallest cost function is the solution of the control signal.

The control law is summarized as follows.

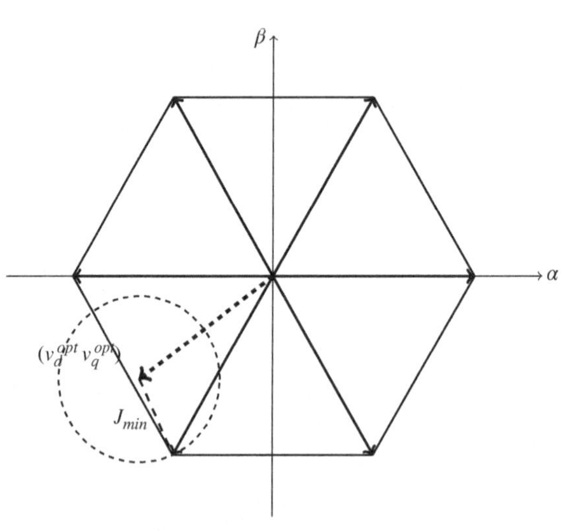

Figure 6.5 Illustration of FCS-MPC solution with $\theta_e = 0$.

Algorithm 3

1. *At sampling time t_i, with the measured currents $i_d(t_i)$, $i_q(t_i)$, and the reference currents $i_d^*(t_i)$ and $i_q^*(t_i)$, compute the optimal control signals $v_d(t_i)^{opt}$ and $v_q(t_i)^{opt}$ via*

$$\begin{bmatrix} v_d(t_i)^{opt} \\ v_q(t_i)^{opt} \end{bmatrix} = \begin{bmatrix} \frac{L_d}{\Delta t} & 0 \\ 0 & \frac{L_q}{\Delta t} \end{bmatrix} \left(\begin{bmatrix} i_d^*(t_i) \\ i_q^*(t_i) \end{bmatrix} - (I + \Delta t A_m(t_i)) \begin{bmatrix} i_d(t_i) \\ i_q(t_i) \end{bmatrix} + \begin{bmatrix} 0 \\ \omega_e(t_i)\phi_{mg} \end{bmatrix} \right).$$

2. *Compute the value of the objective function for $k = 0, 1, 2 \ldots, 6$*[1]

$$J_k = \frac{\Delta t^2}{L_d^2}(v_d(t_i)^k - v_d(t_i)^{opt})^2 + \frac{\Delta t^2}{L_q^2}(v_q(t_i)^k - v_q(t_i)^{opt})^2.$$

3. *Find the minimum of the objective function J_k and its corresponding index number.*
4. *From this index number, construct the three phase voltage control signals.*

Note that in the case that $L_d = L_q$, the weighting factors on the errors of v_d and v_q are identical, thus it is sufficient to evaluate the objective function using

$$J_k = (v_d(t_i)^k - v_d(t_i)^{opt})^2 + (v_q(t_i)^k - v_q(t_i)^{opt})^2.$$

However, the factors $\frac{\Delta t^2}{L_d^2}$ and $\frac{\Delta t^2}{L_q^2}$ could help resolve the scaling problem that may arise from large errors between these variables.

6.5 Overview of FCS-MPC with Integral Action

The FCS-MPC method discussed in the previous sections has found many applications in electrical drives and power converters. As from the previous analysis, the design method adopted a high gain feedback control with constraints in which the feedback control gain is designed using least squares optimization and the seven candidate pairs of the electrical voltage values, $v_d(t_i)$ and $v_q(t_i)$, form the finite control set for the constrained control. The finite control set controller does not have integral action. As a result, from the simulation and experimental evaluation, it is seen that there are steady-state errors in the closed-loop current responses. Similar to the other linear control systems, the steady-state errors will be reduced if the feedback controller gain K_{fcs} is increased. Here, in order to increase the feedback controller gain K_{fcs}, the sampling interval Δt should be reduced as shown in (6.28).

Without integral action, the finite control set method faces two challenges. One is the existence of steady-state errors between the desired current reference signals $i_d^*(t)$, $i_q^*(t)$, and the actual current reference signals $i_d(t)$, $i_q(t)$. These steady-state errors could occur when the closed-loop control system changes its operating conditions. Their existence affects the performance of the closed-loop control system in the steady-state operations. The other challenge is the quality of disturbance rejection. Without integral action, as shown before in the PID controller chapters, the low frequency disturbance could not be completely compensated with the feedback controller because the controller gain is not

[1] Although the index k changes from 0 to 7, the first pair and the last pair of the candidate variables have the identical values of zero (see Section 6.1). Thus, the actual evaluation neglects the last pair of candidate values.

sufficiently high at the low frequency regions. This disturbance rejection problem particularly needs attention in the control of power converters. Another important factor is that without integral action, when the system parameters vary as it happens frequently in drives and power converters (for instance the inductance and resistance change with the variation of load and temperature), the closed-loop control system performance will face uncertainty, particularly in steady-state operations. Thus, from the application point of view, it will be particularly important to incorporate integral action into the FCS-MPC algorithm.

This section modifies the existing finite control set method to include integral action in the controller. There are two mainstream approaches to design a control system with integral action. The first approach is to embed the integrator into the controller and the second approach is to estimate a constant input disturbance using an observer followed by substraction of this constant disturbance from the control signal. The details of the latter approach can be found in Goodwin *et al.* (2000). The first approach is the most widely used method in the applications. Among them are the PID controllers that have embedded the integral function in the controller and the model predictive controllers with incremental models (see Wang (2009)). Given that the FCS-MPC methods have been accepted with proven record in the electrical drives and power electronics community, in this section, we will propose some modifications to the original schemes so that integrators are embedded into the scheme.

The essence of a simplified finite control set scheme is an optimal output feedback control with the gain matrix K_{fcs} where the optimal control signals $v_d(t_i)$ and $v_q(t_i)$ are expressed in the feedback control framework:

$$\begin{bmatrix} v_d(t_i)^{opt} \\ v_q(t_i)^{opt} \end{bmatrix} = K_{fcs}\left(\begin{bmatrix} i_d^*(t_i) \\ i_q^*(t_i) \end{bmatrix} - \begin{bmatrix} i_d(t_i) \\ i_q(t_i) \end{bmatrix}\right), \tag{6.34}$$

where the K_{fcs} is used for the reference signals as well as the measured current signals (a small modification from the original scheme), also the feedforward compensation is neglected in this simpler expression. This finite control set controller has proportional control, but does not include integral action which is evident from (6.34). In order to generate integral action in the feedback controller, the integrated error signals between the current reference signals $i_d^*(t_i)$, $i_q^*(t_i)$ and $i_d(t_i)$, $i_q(t_i)$ will need to be included in the controller. Since the FCS predictive controller is a discrete time controller, discrete-time control system design is better suited for the proposed solution.

The operator for integrator in the discrete-time system is expressed as $\frac{1}{1-q^{-1}}$ where q^{-1} is the backward shift operator defined as $q^{-1}x(t_i) = x(t_{i-1})$. In a similar expression to a PI controller, an additional term that has the functionality of an integrator is added to the original finite control set scheme, leading to the new FCS predictive controller:

$$\begin{bmatrix} v_d(t_i)^{opt} \\ v_q(t_i)^{opt} \end{bmatrix} = K_{fcs}\begin{bmatrix} \frac{k_d}{1-q^{-1}}(i_d^*(t_i) - i_d(t_i)) \\ \frac{k_q}{1-q^{-1}}(i_q^*(t_i) - i_q(t_i)) \end{bmatrix} - K_{fcs}\begin{bmatrix} i_d(t_i) \\ i_q(t_i) \end{bmatrix}, \tag{6.35}$$

where k_d and k_q are the integral gains for the i_d and i_q currents where $0 < k_d \leq 1$ and $0 < k_q \leq 1$. It could be interpreted that the FCS-MPC gain K_{fcs} acts similarly in a role to the proportional controller gain K_c while the parameters k_d and k_q are integral gains in the discrete PI controller. Figure 6.6 shows the configuration of the new FCS predictive controller with integral action, where the control signals calculated are the v_d^{opt} and v_q^{opt}.

In the presence of constraints, there are seven pairs of candidate variables for the v_d and v_q voltages. As before, upon obtaining the signals $v_d(t_i)^{opt}$ and $v_q(t_i)^{opt}$ at the sampling time t_i, the actual control signals $v_d(t_i)$ and $v_q(t_i)$ are determined by computing the value of the objective function for $k = 0, 1, 2, \ldots, 6$

$$J_k = \frac{\Delta t^2}{L_d^2}(v_d(t_i)^k - v_d(t_i)^{opt})^2 + \frac{\Delta t^2}{L_q^2}(v_q(t_i)^k - v_q(t_i)^{opt})^2$$

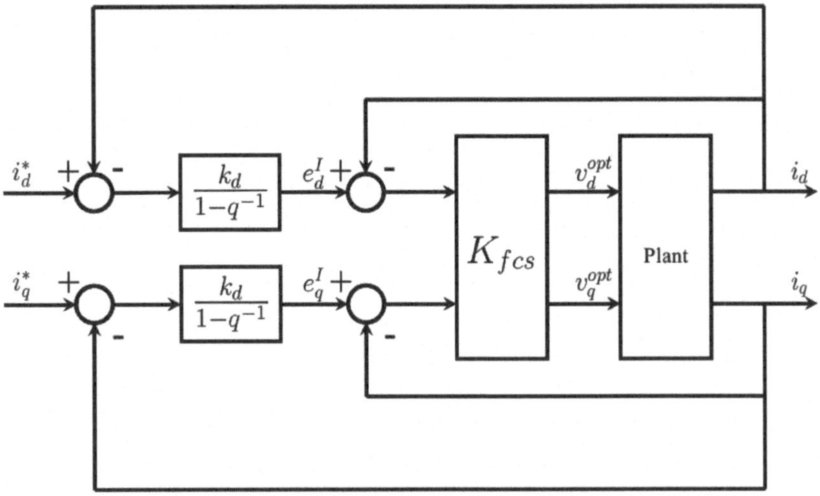

Figure 6.6 Feedback current control using I-FCS predictive control without constraints.

and finding the minimum of the objective function J_k and its corresponding index number. We call this modified controller I-FCS-MPC.

The question remains as how to choose the integral gains k_d and k_q in the framework of I-FCS controller. It is seen from Figure 6.6 that the I-FCS controller has two feedback loops. One is the inner-loop proportional control system with the controller gain K_{fcs} while the outer-loop controllers are the integral controllers. So from this design topology, we evaluate the closed-loop transfer function for the inner-loop system based on which the outer-loop integral controllers will be designed.

Since the inner-loop system has the discrete-time state space model (from (6.14)):

$$\begin{bmatrix} i_d(t_{i+1}) \\ i_q(t_{i+1}) \end{bmatrix} = (I + \Delta t A_m(t_i)) \begin{bmatrix} i_d(t_i) \\ i_q(t_i) \end{bmatrix} - \begin{bmatrix} 0 \\ \frac{\omega_e(t_i)\phi_{mg}\Delta t}{L_q} \end{bmatrix} + \Delta t B_m \begin{bmatrix} v_d(t_i) \\ v_q(t_i) \end{bmatrix}, \tag{6.36}$$

with the feedback control signals $v_d(t_i)$ and $v_q(t_i)$ calculated using the feedback gain matrix K_{fcs} as illustrated in the inner-loop control system:

$$\begin{bmatrix} v_d(t_i) \\ v_q(t_i) \end{bmatrix} = K_{fcs} \begin{bmatrix} e_d(t_i)^I - i_d(t_i) \\ e_q(t_i)^I - i_q(t_i) \end{bmatrix}, \tag{6.37}$$

where $e_d(t_i)^I$ and $e_q(t_i)^I$ are the reference signals to the inner-loop, also the manipulated signals from the outer-loop system. The superscript I means integrated error. By substituting the control signals into the state-space model (6.36), the inner-loop closed-loop system is formed:

$$\begin{bmatrix} i_d(t_{i+1}) \\ i_q(t_{i+1}) \end{bmatrix} = (I + \Delta t A_m(t_i) - \Delta t B_m K_{fcs}) \begin{bmatrix} i_d(t_i) \\ i_q(t_i) \end{bmatrix} + \Delta t B_m K_{fcs} \begin{bmatrix} e_d(t_i)^I \\ e_q(t_i)^I \end{bmatrix} - \begin{bmatrix} 0 \\ \frac{\omega_e(t_i)\phi_{mg}\Delta t}{L_q} \end{bmatrix}. \tag{6.38}$$

Define the z-transforms of the reference signals and the feedback signals as $E_d(z)^I$, $E_q(z)^I$, $I_d(z)$ and $I_q(z)$. Note that the current vector on the left-hand side of (6.38) has the z-transform of $[zI_d(z) \; zI_q(z)]^T$. Thus the closed-loop relationship between the z-transform of the reference signals and the feedback signals is obtained as

$$\begin{bmatrix} I_d(z) \\ I_q(z) \end{bmatrix} = (zI - (I + A_m(t_i)\Delta t) + \Delta t B_m K_{fcs})^{-1} \Delta t B_m K_{fcs} \begin{bmatrix} E_d(z)^I \\ E_q(z)^I \end{bmatrix}. \tag{6.39}$$

Since the feedback control gain K_{fcs} is

$$K_{fcs}(t_i) = \begin{bmatrix} \frac{L_d}{\Delta t} & 0 \\ 0 & \frac{L_q}{\Delta t} \end{bmatrix} (I + \Delta t A_m(t_i)),$$

it can easily be verified that the matrix $(zI - (I + A_m(t_i)\Delta t) + \Delta t B_m K_{fcs})^{-1}$ is a diagonal matrix, having the following form:

$$(zI - (I + A_m(t_i)\Delta t) + \Delta t B_m K_{fcs})^{-1} = \begin{bmatrix} \frac{1}{z} & 0 \\ 0 & \frac{1}{z} \end{bmatrix},$$

where the matrix $\Delta t B_m K_{fcs} = I + \Delta t A_m(t_i)$. Therefore, the closed-loop transfer function given by (6.39) becomes:

$$\begin{bmatrix} I_d(z) \\ I_q(z) \end{bmatrix} = \left(\begin{bmatrix} \frac{1}{z} & 0 \\ 0 & \frac{1}{z} \end{bmatrix} + \Delta t A_m(t_i) z^{-1} \right) \begin{bmatrix} E_d(z)^I \\ E_q(z)^I \end{bmatrix}. \tag{6.40}$$

For sufficiently small Δt, the quantity $\Delta t A_m(t_i) z^{-1}$ could be neglected. So, in terms of approximation, the inner-loop dynamics are expressed as

$$\frac{I_d(z)}{E_d(z)^I} \approx z^{-1} \tag{6.41}$$

$$\frac{I_q(z)}{E_q(z)^I} \approx z^{-1}. \tag{6.42}$$

By taking this approximation justified by using a small sampling interval Δt, the closed-loop transfer functions here have some degree of mismatch with the steady-state gain. Also, the interactions in the $A_m(t_i)\Delta t$ matrix are neglected.

Upon simplifying the inner-loop system, the design of the outer-loop integral controller becomes straightforward. It is apparent that considering the d-axis current, the open-loop transfer function for the outer-loop system includes the integral controller $\frac{k_d}{1-z^{-1}}$ together with the time delay z^{-1} from the inner closed-loop system. Hence, the outer closed-loop has the transfer function:

$$\frac{I_d(z)}{I_d^*(z)} = \frac{k_d z^{-1}}{1 - z^{-1} + k_d z^{-1}}, \tag{6.43}$$

where the closed-loop pole for this first order system is $z = 1 - k_d$. By choosing a desired closed-loop pole as $0 \le p_{cl} < 1$, the integral controller gain is determined as $k_d = 1 - p_{cl}$. Similarly, the outer closed-loop system for the q-axis current has the transfer function

$$\frac{I_q(z)}{I_q^*(z)} = \frac{k_q z^{-1}}{1 - z^{-1} + k_q z^{-1}}, \tag{6.44}$$

if the same desired closed-loop pole is used for the q-axis current controller design, then the integral controller gain $k_q = 1 - p_{cl}$.

The desired closed-loop pole p_{cl} is a pole in the discrete-time system and this parameter is the design parameter selected by the user. We can choose its value relative its continuous-time counterpart. For instance, if the continuous-time counterpart is $s = -a_{cl}$, the pole in discrete-time is $z = p_{cl} = e^{-a_{cl}\Delta t}$ where Δt is the sampling interval. Thus, we can determine the value of p_{cl} according to the desired response time in the continuous-time. For instance, if we wish the desired closed-loop current control system to have a time constant of 100×10^{-6} second ($a_{cl} = \frac{1}{100\times 10^{-6}}$), then with a sampling interval $\Delta t = 10 \times 10^{-6}$ second, $p_{cl} = e^{-a_{cl}\times\Delta t} = e^{-0.1} = 0.9048$. If we reduce the desired closed-loop time constant to 50×10^{-6} second, then $p_{cl} = e^{-0.2} = 0.8187$.

In the selection of the desired closed-loop pole, one should also take into consideration the small mismatch of the steady-state gain and the existence of the interaction terms in the $A_m(t_i)$ matrix, which may affect the closed-loop performance. In general, the effects of the mismatch will be smaller if smaller integral gains k_d and k_q are used.

6.6 Derivation of I-FCS Predictive Control Algorithm

The I-FCS predictive control algorithm outlined in Section 6.5 is derived in this section. The derivation process gives the justification for the algorithm and furthermore leads to the actual implementation algorithm for the I-FCS predictive controller.

6.6.1 Optimal Control without Constraints

Consider the discretized linear model for the PMSM:

$$\begin{bmatrix} i_d(t_{i+1}) \\ i_q(t_{i+1}) \end{bmatrix} = (I + \Delta t A_m(t_i)) \begin{bmatrix} i_d(t_i) \\ i_q(t_i) \end{bmatrix} + \Delta t B_m \begin{bmatrix} v_d(t_i) \\ v_q(t_i) \end{bmatrix} - \begin{bmatrix} 0 \\ \frac{\omega_e(t_i)\phi_{mg}\Delta t}{L_q} \end{bmatrix}. \tag{6.45}$$

This approximation of the continuous-time differential equation model also holds at the sampling time $t_i - \Delta t$, which has the form:

$$\begin{bmatrix} i_d(t_i) \\ i_q(t_i) \end{bmatrix} = (I + \Delta t A_m(t_{i-1})) \begin{bmatrix} i_d(t_{i-1}) \\ i_q(t_{i-1}) \end{bmatrix} + \Delta t B_m \begin{bmatrix} v_d(t_{i-1}) \\ v_q(t_{i-1}) \end{bmatrix} - \begin{bmatrix} 0 \\ \frac{\omega_e(t_{i-1})\phi_{mg}\Delta t}{L_q} \end{bmatrix}. \tag{6.46}$$

Subtracting (6.46) from (6.45) leads to the difference model between the two sampling instants:

$$\begin{aligned} \begin{bmatrix} i_d(t_{i+1}) - i_d(t_i) \\ i_q(t_{i+1}) - i_q(t_i) \end{bmatrix} &= (I + \Delta t A_m(t_i)) \begin{bmatrix} i_d(t_i) - i_d(t_{i-1}) \\ i_q(t_i) - i_q(t_{i-1}) \end{bmatrix} \\ &+ \Delta t B_m \begin{bmatrix} v_d(t_i) - v_d(t_{i-1}) \\ v_q(t_i) - v_q(t_{i-1}) \end{bmatrix} - \begin{bmatrix} 0 \\ \frac{(\omega_e(t_i)-\omega_e(t_{i-1}))\phi_{mg}\Delta t}{L_q} \end{bmatrix} \\ &+ (A_m(t_i) - A_m(t_{i-1}))\Delta t \begin{bmatrix} i_d(t_{i-1}) \\ i_q(t_{i-1}) \end{bmatrix}, \end{aligned} \tag{6.47}$$

where in the process of derivation the following term is both added and subtracted to (6.47):

$$A_m(t_i)\Delta t \begin{bmatrix} i_d(t_{i-1}) \\ i_q(t_{i-1}) \end{bmatrix}.$$

Note that matrix $(A_m(t_i) - A_m(t_{i-1}))\Delta t$ contained in the final term of (6.47) is expressed for the PMSM case as

$$(A_m(t_i) - A_m(t_{i-1}))\Delta t = \begin{bmatrix} 0 & a_{12} \\ a_{21} & 0 \end{bmatrix},$$

where $a_{12} = \Delta t \frac{L_q}{L_d}(\omega_e(t_i) - \omega_e(t_{i-1}))$ and $a_{21} = -\Delta t \frac{L_d}{L_q}(\omega_e(t_i) - \omega_e(t_{i-1}))$.

Because the quantity $\Delta t(\omega_e(t_i) - \omega_e(t_{i-1}))$ is sufficiently small for a small sampling interval Δt (say, 50×10^{-6} sec), the matrix $(A_m(t_i) - A_m(t_{i-1}))\Delta t$ is approximated by a zero matrix. Here, the incremental change of the velocity $\omega_e(t_i) - \omega_e(t_{i-1})$ is also relatively small because of physically limited acceleration of a motor, and in steady-state operation, this term is assumed to be zero. Thus the final term of (6.47) is neglected. Approximation using zero for the second last term of (6.47) is also performed with the same reasoning.

The following incremental variables are defined for notational simplicity:

$$\Delta i_d(t_{i+1}) = i_d(t_{i+1}) - i_d(t_i) \tag{6.48}$$

$$\Delta i_q(t_{i+1}) = i_q(t_{i+1}) - i_q(t_i) \tag{6.49}$$

$$\Delta i_d(t_i) = i_d(t_i) - i_d(t_{i-1}) \tag{6.50}$$

$$\Delta i_q(t_i) = i_q(t_i) - i_q(t_{i-1}) \tag{6.51}$$

$$\Delta v_d(t_i) = v_d(t_i) - v_d(t_{i-1}) \tag{6.52}$$

$$\Delta v_q(t_i) = v_q(t_i) - v_q(t_{i-1}). \tag{6.53}$$

With these incremental variables defined and the approximations taken, the incremental model of a PMSM (6.47) becomes

$$\begin{bmatrix} \Delta i_d(t_{i+1}) \\ \Delta i_q(t_{i+1}) \end{bmatrix} = (I + \Delta t A_m(t_i)) \begin{bmatrix} \Delta i_d(t_i) \\ \Delta i_q(t_i) \end{bmatrix} + \Delta t B_m \begin{bmatrix} \Delta v_d(t_i) \\ \Delta v_q(t_i) \end{bmatrix}. \tag{6.54}$$

To include the integral action into the controller, choose the weighted current errors

$$e_d(t_i) = k_d(i_d^*(t_i) - i_d(t_i)); \; e_q(t_i) = k_q(i_q^*(t_i) - i_q(t_i))$$

as the steady-states of the $\Delta i_d(t_i)$ and $\Delta i_q(t_i)$, where $0 < k_d < 1$, $0 < k_q < 1$. By subtracting the steady-states from the incremental model (6.54), we obtain:

$$\begin{bmatrix} \Delta i_d(t_{i+1}) - e_d(t_i) \\ \Delta i_q(t_{i+1}) - e_q(t_i) \end{bmatrix} = (I + \Delta t A_m(t_i)) \begin{bmatrix} \Delta i_d(t_i) - e_d(t_i) \\ \Delta i_q(t_i) - e_q(t_i) \end{bmatrix} + \Delta t B_m \begin{bmatrix} \Delta v_d(t_i) \\ \Delta v_q(t_i) \end{bmatrix}. \tag{6.55}$$

The control objective is to minimize the error function J, where

$$J = \begin{bmatrix} \Delta i_d(t_{i+1}) - e_d(t_i) \\ \Delta i_q(t_{i+1}) - e_q(t_i) \end{bmatrix}^T \begin{bmatrix} \Delta i_d(t_{i+1}) - e_d(t_i) \\ \Delta i_q(t_{i+1}) - e_q(t_i) \end{bmatrix}, \tag{6.56}$$

which is to regulate the incremental current signals $\Delta i_d(t_{i+1})$, $\Delta i_q(t_{i+1})$ to be as close as possible to $e_d(t_i)$ and $e_q(t_i)$.

For notational simplicity, define the vector:

$$\begin{aligned} \begin{bmatrix} g_d(t_i) \\ g_q(t_i) \end{bmatrix} &= -(I + \Delta t A_m(t_i)) \begin{bmatrix} \Delta i_d(t_i) - e_d(t_i) \\ \Delta i_q(t_i) - e_q(t_i) \end{bmatrix} \\ &= (I + \Delta t A_m(t_i)) \begin{bmatrix} e_d(t_i) - \Delta i_d(t_i) \\ e_q(t_i) - \Delta i_q(t_i) \end{bmatrix}. \end{aligned} \tag{6.57}$$

Then by substituting (6.55) with the simplified notation into the objective function (6.56), it can be readily verified that the objective function has the expression

$$J = \left(\begin{bmatrix} g_d(t_i) \\ g_q(t_i) \end{bmatrix} - \Delta t B_m \begin{bmatrix} \Delta v_d(t_i) \\ \Delta v_q(t_i) \end{bmatrix} \right)^T \left(\begin{bmatrix} g_d(t_i) \\ g_q(t_i) \end{bmatrix} - \Delta t B_m \begin{bmatrix} \Delta v_d(t_i) \\ \Delta v_q(t_i) \end{bmatrix} \right), \tag{6.58}$$

which is

$$J = \begin{bmatrix} g_d(t_i) & g_q(t_i) \end{bmatrix} \begin{bmatrix} g_d(t_i) \\ g_q(t_i) \end{bmatrix} - 2 \begin{bmatrix} \Delta v_d(t_i) & \Delta v_q(t_i) \end{bmatrix} \Delta t B_m^T \begin{bmatrix} g_d(t_i) \\ g_q(t_i) \end{bmatrix}$$
$$+ \begin{bmatrix} \Delta v_d(t_i) & \Delta v_q(t_i) \end{bmatrix} \Delta t^2 B_m^T B_m \begin{bmatrix} \Delta v_d(t_i) \\ \Delta v_q(t_i) \end{bmatrix}. \tag{6.59}$$

The objective function (6.59) is a quadratic function similar to the objective function (6.17) in Section 6.4. Following the same least squares minimization procedure as outlined in Section 6.4, the optimal incremental control signals $\Delta v_d(t_i)$ and $\Delta v_q(t_i)$ that minimize the objective function (6.59) are found as

$$\begin{aligned} \begin{bmatrix} \Delta v_d(t_i) \\ \Delta v_q(t_i) \end{bmatrix} &= (\Delta t^2 B_m^T B_m)^{-1} \Delta t B_m^T \begin{bmatrix} g_d(t_i) \\ g_q(t_i) \end{bmatrix} \\ &= \frac{1}{\Delta t} \begin{bmatrix} L_d & 0 \\ 0 & L_q \end{bmatrix} \begin{bmatrix} g_d(t_i) \\ g_q(t_i) \end{bmatrix}, \end{aligned} \tag{6.60}$$

where the matrix $B_m^T B_m$ is positive definite, and given by

$$B_m^T B_m = \begin{bmatrix} \frac{1}{L_d^2} & 0 \\ 0 & \frac{1}{L_q^2} \end{bmatrix}.$$

By substituting the variables $g_d(t_i)$ and $g_q(t_i)$ (see (6.57)) into the optimal solution (6.60), we obtain the expression of the incremental control signals

$$\begin{aligned} \begin{bmatrix} \Delta v_d(t_i) \\ \Delta v_q(t_i) \end{bmatrix} &= \frac{1}{\Delta t} \begin{bmatrix} L_d & 0 \\ 0 & L_q \end{bmatrix} (I + \Delta t A_m(t_i)) \begin{bmatrix} e_d(t_i) - \Delta i_d(t_i) \\ e_q(t_i) - \Delta i_q(t_i) \end{bmatrix} \\ &= K_{fcs} \begin{bmatrix} e_d(t_i) - \Delta i_d(t_i) \\ e_q(t_i) - \Delta i_q(t_i) \end{bmatrix}, \end{aligned} \tag{6.61}$$

where the feedback control gain K_{fcs} is defined by

$$K_{fcs} = \frac{1}{\Delta t} \begin{bmatrix} L_d & 0 \\ 0 & L_q \end{bmatrix} (I + \Delta t A_m(t_i)),$$

which is identical to the case without integrator in Section 6.4.

Furthermore, since the weighted error signals are $e_d(t_i) = k_d(i_d^*(t_i) - i_d(t_i))$ and $e_q(t_i) = k_q(i_q^*(t_i) - i_q(t_i))$, the optimal incremental control signals finally have the following form:

$$\begin{bmatrix} \Delta v_d(t_i) \\ \Delta v_q(t_i) \end{bmatrix} = K_{fcs} \left(\begin{bmatrix} k_d(i_d^*(t_i) - i_d(t_i)) \\ k_q(i_q^*(t_i) - i_q(t_i)) \end{bmatrix} - \begin{bmatrix} \Delta i_d(t_i) \\ \Delta i_q(t_i) \end{bmatrix} \right). \tag{6.62}$$

The important point here is that the computation is presented in terms of $\Delta v_d(t_i)$ and $\Delta v_q(t_i)$ that are the incremental variables. Also, the incremental feedback variables $\Delta i_d(t_i)$ and $\Delta i_q(t_i)$, the error signals $i_d^*(t_i) - i_d(t_i)$ and $i_q^*(t_i) - i_q(t_i))$ are used in the control law. Therefore, the steady-state information of the i_d and i_q currents is no longer required. The control signals are calculated based on the incremental signals and their past values as

$$v_d(t_i) = \Delta v_d(t_i) + v_d(t_{i-1}) \tag{6.63}$$

$$v_q(t_i) = \Delta v_q(t_i) + v_q(t_{i-1}). \tag{6.64}$$

Again, there is no need to consider the steady-state values of the voltage signals in the implementation.

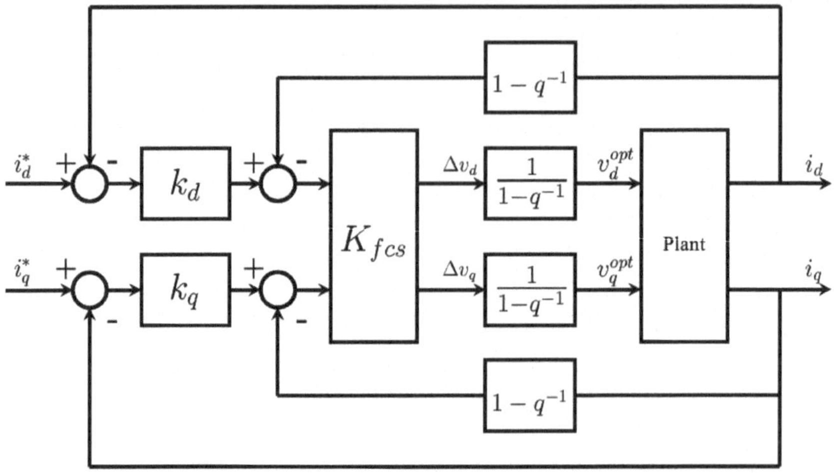

Figure 6.7 I-FCS feedback current control using incremental variables.

Since in terms of the q^{-1} shift operator, the incremental control signal $\Delta v_d(t_i) = v_d(t_i) - v_d(t_{i-1})$ is expressed as $(1 - q^{-1})v_d(t_i)$ and $\Delta v_q(t_i) = v_q(t_i) - v_q(t_{i-1})$ as $(1 - q^{-1})v_q(t_i)$, dividing (6.62) with the factor $1 - q^{-1}$ leads to

$$\begin{bmatrix} v_d(t_i) \\ v_q(t_i) \end{bmatrix} = K_{fcs} \begin{bmatrix} \frac{k_d}{1-q^{-1}}(i_d^*(t_i) - i_d(t_i)) \\ \frac{k_q}{1-q^{-1}}(i_q^*(t_i) - i_q(t_i)) \end{bmatrix} - K_{fcs} \begin{bmatrix} i_d(t_i) \\ i_q(t_i) \end{bmatrix}. \tag{6.65}$$

Here, we have also replaced the incremental current signals $\Delta i_d(t_i) = i_d(t_i) - i_d(t_{i-1})$ by $\Delta i_d(t_i) = (1 - q^{-1})i_d(t_i)$ and $\Delta i_q(t_i) = i_q(t_i) - i_q(t_{i-1})$ by $\Delta i_q(t_i) = (1 - q^{-1})i_q(t_i)$. The optimal control signals $v_d(t_i)$ and $v_q(t_i)$ given by (6.65) are identical to the proposed control signals in (6.35). This completes the derivation of the optimal control signal with integral action without constraints.

An equivalent expression of the cascade feedback system based on (6.62) is shown in Figure 6.7, which can also be seen by moving the integrators in Figure 6.6 from outer-loops to the inner-loop.

6.6.2 I-FCS Predictive Controller with Constraints

In the previous section, the optimal controller with integral action was derived without constraints. Considerable efforts were devoted to show the link between the cascade PI controller structure and the optimal controller designed using an objective function. It becomes clear in this section that the objective function J used in the previous section provides the key to solution of the I-FCS controller. The derivation procedure uses some of the results obtained from the analysis of FCS controller (see Section 6.4).

Consider the objective function (6.59). By adding and subtracting the term

$$\begin{bmatrix} g_d(t_i) & g_q(t_i) \end{bmatrix} \Delta t B_m (\Delta t^2 B_m^T B_m)^{-1} B_m^T \Delta t \begin{bmatrix} g_d(t_i) \\ g_q(t_i) \end{bmatrix}$$

to the original objective function J given by (6.59), its value remains unchanged. With this term added, the following three terms lead to completed squares:

$$\begin{aligned} J_0 &= \begin{bmatrix} \Delta v_d(t_i) & \Delta v_q(t_i) \end{bmatrix} \Delta t^2 B_m^T B_m \begin{bmatrix} \Delta v_d(t_i) \\ \Delta v_q(t_i) \end{bmatrix} \\ &- 2 \begin{bmatrix} \Delta v_d(t_i) & \Delta v_q(t_i) \end{bmatrix} \Delta t B_m^T \begin{bmatrix} g_d(t_i) \\ g_q(t_i) \end{bmatrix} \\ &+ \begin{bmatrix} g_d(t_i) & g_q(t_i) \end{bmatrix} \Delta t B_m (\Delta t^2 B_m^T B_m)^{-1} B_m^T \Delta t \begin{bmatrix} g_d(t_i) \\ g_q(t_i) \end{bmatrix} \\ &= \left(\begin{bmatrix} \Delta v_d(t_i) \\ \Delta v_q(t_i) \end{bmatrix} - (\Delta t^2 B_m^T B_m)^{-1} B_m^T \Delta t \begin{bmatrix} g_d(t_i) \\ g_q(t_i) \end{bmatrix} \right)^T (\Delta t^2 B_m^T B_m) \\ &\times \left(\begin{bmatrix} \Delta v_d(t_i) \\ \Delta v_q(t_i) \end{bmatrix} - (\Delta t^2 B_m^T B_m)^{-1} B_m^T \Delta t \begin{bmatrix} g_d(t_i) \\ g_q(t_i) \end{bmatrix} \right). \end{aligned} \tag{6.66}$$

Now, with J_0 given by the completed squares (6.66), the original objective function J becomes

$$J = J_0 + J_{min} \tag{6.67}$$

where J_{min} is

$$J_{min} = \begin{bmatrix} g_d(t_i) & g_q(t_i) \end{bmatrix} (I - B_m (B_m^T B_m)^{-1} B_m^T) \begin{bmatrix} g_d(t_i) \\ g_q(t_i) \end{bmatrix} = 0,$$

because of the special form of the B_m matrix, where

$$B_m = \begin{bmatrix} \frac{1}{L_d} & 0 \\ 0 & \frac{1}{L_q} \end{bmatrix}.$$

Thus, the minimum of the original objective function J is achieved if J_0 is minimized with the optimal control:

$$\begin{bmatrix} \Delta v_d(t_i) \\ \Delta v_q(t_i) \end{bmatrix} = (\Delta t^2 B_m^T B_m)^{-1} B_m^T \Delta t \begin{bmatrix} g_d(t_i) \\ g_q(t_i) \end{bmatrix}. \tag{6.68}$$

This solution is identical to the solution presented in (6.60) that has been derived without constraints.

We define the unconstrained solution as the optimal solution denoted by $\Delta v_d(t_i)^{opt}$ and $\Delta v_q(t_i)^{opt}$, yielding

$$\begin{bmatrix} \Delta v_d(t_i)^{opt} \\ \Delta v_q(t_i)^{opt} \end{bmatrix} = (\Delta t^2 B_m^T B_m)^{-1} B_m^T \Delta t \begin{bmatrix} g_d(t_i) \\ g_q(t_i) \end{bmatrix}. \tag{6.69}$$

By substituting (6.69) into the objective function (6.67), we obtain

$$J = \left(\begin{bmatrix} \Delta v_d(t_i) \\ \Delta v_q(t_i) \end{bmatrix} - \begin{bmatrix} \Delta v_d(t_i)^{opt} \\ \Delta v_q(t_i)^{opt} \end{bmatrix} \right)^T (\Delta t^2 B_m^T B_m) \left(\begin{bmatrix} \Delta v_d(t_i) \\ \Delta v_q(t_i) \end{bmatrix} - \begin{bmatrix} \Delta v_d(t_i)^{opt} \\ \Delta v_q(t_i)^{opt} \end{bmatrix} \right).$$

Also, by definition of the incremental control signals, the following relationship is true:

$$\begin{bmatrix} \Delta v_d(t_i)^{opt} \\ \Delta v_q(t_i)^{opt} \end{bmatrix} = \begin{bmatrix} v_d(t_i)^{opt} \\ v_q(t_i)^{opt} \end{bmatrix} - \begin{bmatrix} \Delta v_d(t_{i-1})^{opt} \\ \Delta v_q(t_{i-1})^{opt} \end{bmatrix}.$$

Thus, by calculating the actual incremental control signals using the same past control signal states, that is,

$$\begin{bmatrix} \Delta v_d(t_i) \\ \Delta v_q(t_i) \end{bmatrix} = \begin{bmatrix} v_d(t_i) \\ v_q(t_i) \end{bmatrix} - \begin{bmatrix} \Delta v_d(t_{i-1})^{opt} \\ \Delta v_q(t_{i-1})^{opt} \end{bmatrix},$$

we arrive at the objective function that can be used in the I-FCS MPC design as

$$J = \left(\begin{bmatrix} v_d(t_i) \\ v_q(t_i) \end{bmatrix} - \begin{bmatrix} v_d(t_i)^{opt} \\ v_q(t_i)^{opt} \end{bmatrix} \right)^T (\Delta t^2 B_m^T B_m) \left(\begin{bmatrix} v_d(t_i) \\ v_q(t_i) \end{bmatrix} - \begin{bmatrix} v_d(t_i)^{opt} \\ v_q(t_i)^{opt} \end{bmatrix} \right)$$
$$= \frac{\Delta t^2}{L_d^2}(v_d(t_i) - v_d(t_i)^{opt})^2 + \frac{\Delta t^2}{L_q^2}(v_q(t_i) - v_q(t_i)^{opt})^2. \quad (6.70)$$

In the presence of constraints, there are seven pairs of candidate variables for the v_d and v_q voltages. When having the integrators in the I-FCS predictive controller, upon obtaining the signals $v_d(t_i)^{opt}$ and $v_q(t_i)^{opt}$ with integral action at the sampling time t_i, the actual control signals $v_d(t_i)$ and $v_q(t_i)$ are determined by computing the value of the objective function for $k = 0, 1, 2 \ldots, 6$

$$J_k = \frac{\Delta t^2}{L_d^2}(v_d(t_i)^k - v_d(t_i)^{opt})^2 + \frac{\Delta t^2}{L_q^2}(v_q(t_i)^k - v_q(t_i)^{opt})^2. \quad (6.71)$$

The pair of constrained control signals $v_d(t_i)^k$ and $v_q(t_i)^k$ is found to minimize the objective function J_k subject to the index number k.

6.6.3 Implementation of I-FCS-MPC Algorithm

The design of the I-FCS predictive control system is shown in Figure 6.6 as the configuration of a cascade feedback control system, which clearly indicated that integrators have been embedded in the outer-loop systems. The equivalent control system structure is shown in Figure 6.7. Based on the latter structure, the implementation algorithm for the I-FCS-MPC scheme is presented using the difference of the control signals with recursive computation, offering convenience for real-time computation.

The I-FCS-MPC algorithm is summarized for the convenience of implementation.

Algorithm 4

1. *Assume that the closed-loop control action begins at the sampling instant t_0. It is reasonable to take the initial values $\Delta i_d(t_0) = 0$ and $\Delta i_q(t_0) = 0$. At the sampling time $t_0 - \Delta t$, take the measurements of $v_d(t_0 - \Delta t)$, $v_q(t_0 - \Delta t)$, $i_d(t_0 - \Delta t)$, $i_q(t_0 - \Delta t)$. Initialize the optimal control signals at the sampling time $t_i - \Delta t$ as*
$$v_d(t_{i-1})^{opt} = v_d(t_0 - \Delta t);\; v_q(t_{i-1})^{opt} = v_q(t_{i-1})$$
$$\Delta i_d(t_i) = 0;\; \Delta i_q(t_i) = 0.$$
2. *Perform the following computation at the sampling time t_i with current measurements $i_d(t_i)$, $i_q(t_i)$ and their reference signals $i_d^*(t_i)$, $i_q^*(t_i)$. Calculate the optimal incremental control signals using the following equation:*
$$\begin{bmatrix} \Delta v_d(t_i)^{opt} \\ \Delta v_q(t_i)^{opt} \end{bmatrix} = K_{fcs} \begin{bmatrix} k_d(i_d^*(t_i) - i_d(t_i)) \\ k_q(i_q^*(t_i) - i_q(t_i)) \end{bmatrix} - K_{fcs} \begin{bmatrix} \Delta i_d(t_i) \\ \Delta i_q(t_i) \end{bmatrix}.$$
3. *Calculate the optimal control signals using the past optimal control states:*
$$\begin{bmatrix} v_d(t_i)^{opt} \\ v_q(t_i)^{opt} \end{bmatrix} = \begin{bmatrix} v_d(t_{i-1})^{opt} \\ v_q(t_{i-1})^{opt} \end{bmatrix} + \begin{bmatrix} \Delta v_d(t_i)^{opt} \\ \Delta v_q(t_i)^{opt} \end{bmatrix}.$$
4. *Calculate the value of the objective function J with respect to the finite control set for $k = 0, 1, 2, \ldots, 6$*
$$J_k = \frac{\Delta t^2}{L_d^2}(v_d(t_i)^k - v_d(t_i)^{opt})^2 + \frac{\Delta t^2}{L_q^2}(v_q(t_i)^k - v_q(t_i)^{opt})^2,$$

where the values of $v_d(t_i)^k$ and $v_q(t_i)^k$ are selected among the pairs of the given array:

$$\begin{bmatrix} 0 & \cos\theta_e & \cos\left(\theta_e - \frac{2\pi}{3}\right) & \cos\left(\theta_e - \frac{4\pi}{3}\right) & -\cos\theta_e & -\cos\left(\theta_e - \frac{2\pi}{3}\right) & -\cos\left(\theta_e - \frac{4\pi}{3}\right) \\ 0 & -\sin\theta_e & -\sin\left(\theta_e - \frac{2\pi}{3}\right) & -\sin\left(\theta_e - \frac{4\pi}{3}\right) & \sin\theta_e & \sin\left(\theta_e - \frac{2\pi}{3}\right) & \sin\left(\theta_e - \frac{4\pi}{3}\right) \end{bmatrix} \frac{2}{3}V_{dc}.$$

5. *Find the minimum of J_k and its corresponding index, which leads to the control signals to be implemented.*
6. *Go to Step 2 in the computation as the sampling time progresses to $t_{i+1} = t_i + \Delta t$.*

It is seen that the I-FCS-MPC algorithm is straightforward for implementation.

6.7 MATLAB Tutorial on Implementation of I-FCS Predictive Controller

Tutorial 3 *The real-time implementation of the I-FCS-MPC for PMSM is developed on the basis of the original FCS-MPC program from Tutorial 2. Most of this MATLAB embedded function is identical to the embedded MATLAB program called FCS.mdl, except that the optimal voltage control signals $v_d(t_i)^{opt}$ and $v_q(t_i)^{opt}$ are computed explicitly.*

Step by Step

1. *Edit the Simulink file called FCS.mdl and save it as FCSInt.mdl. This is the new MATLAB embedded function that performs the computation of I-FCS-MPC algorithm in real-time.*
2. *Change the embedded function name from "FCS" to "FCSInt" and double check the input and output variables that appear in the following order together with the new file name.*

```
[Ua,Ub,Uc,Ind]=
FCSInt(idRef,iqRef,id,iq,we,theta,Ld,Lq,Rs,Phi_mg,Tin,D,U)
```

3. *The stages from Step 2 to Step 9 in Tutorial 2 remain unchanged. However, delete matrices G and H from Step 7 in Tutorial 2 because they are no longer needed.*
4. *Delete the embedded program in Step 10 of the Tutorial 2.*
5. *Define the feedback controller gain $k_d = k_q = 1 - p_{cl}$, where p_{cl} is the location of the closed-loop pole for the outer-loop current control system. In this example, p_{cl} is selected to be 0.8. Enter the following program into the file.*

```
pcl=0.8;
kd=1-pcl;
```

6. *Calculate the controller gain from FCS-MPC, where*

$$K_{fcs} = \frac{1}{\Delta t}\begin{bmatrix} L_d & 0 \\ 0 & L_q \end{bmatrix}(I + \Delta t A_m(t_i)).$$

Enter the following program into the file.

```
Kfcs=[Ld/Tin 0; 0 Lq/Tin]*F;
```

7. *Calculate the optimal voltage control signals $v_d(t_i)^{opt}$ and $v_q(t_i)^{opt}$ using the updating equation:*

$$\begin{bmatrix} v_d(t_i)^{opt} \\ v_q(t_i)^{opt} \end{bmatrix} = \begin{bmatrix} v_d(t_{i-1})^{opt} \\ v_q(t_{i-1})^{opt} \end{bmatrix} + K_{fcs}\begin{bmatrix} k_d(i_d^*(t_i) - i_d(t_i)) \\ k_d(i_q^*(t_i) - i_q(t_i)) \end{bmatrix} - K_{fcs}\begin{bmatrix} \Delta i_d(t_i) \\ \Delta i_q(t_i) \end{bmatrix},$$

where $\left[v_d(t_{i-1})^{opt} \; v_q(t_{i-1})^{opt}\right]^T$ *is called the "upast" vector and* $\Delta i_d(t_i)$ *and* $\Delta i_q(t_i)$ *are generated using the difference between current and past vectors. Enter the following program into the file.*

```
vopt=upast-Kfcs*(x-xpast)+Kfcs*kd*(xRef-x);
```

8. *Calculate the norm of the error vector between the seven* v_d *and* v_q *candidate variables and the optimal control signals* $v_d(t_i)^{opt}$ *and* $v_q(t_i)^{opt}$*. Enter the following program into the file.*

```
for i=1:7
Xe(:,i)=vopt-M*D*U(:,i);
err(:,i)=Xe(:,i)'*Xe(:,i)*(Tin)^2/(Ld)^2;
end
```

9. *Step 11, Step 12 and Step 13 in Tutorial 2 remain unchanged.*
10. *Update the past optimal control signal and past state variables. Enter the following program into the file.*

```
upast=vopt;
xpast=x;
```

This embedded function is now interfaced with the simulator of PMSM to produce the Simulink simulator called "simulator4PIFCS.mdl".

6.7.1 Simulation Results

In comparison with the simulation results presented in Section 6.3.1, the simulation setup conditions in this section are identical to those used previously. Namely, the pair of poles is 2; the resistance is R_s = 2.98c Ω; inductance $L_d = L_q = 7 \times 10^{-3}$ H; $\phi_{mg} = 0.125$ *Wb*; inertia $J_m = 0.01 \times 10^{-3}$ kg · m²; $Bv = 11 \times 10^{-5}$ Nm · s. The reference signal to the i_d current is $i_d^* = 0$ A and i_q^* is a step signal with amplitude of ±2 A. The torque load disturbance is also a step signal with unit amplitude at the start of the simulation, followed by a step change to −1 at half of the simulation time. The *DC* power supply has voltage V_{dc} = 300 V.

There four simulation case studies presented in this section. In the first two cases, we will choose the closed-loop pole for the outer-loop integral system as $p_{cl} = 0.8$, hence leading to $k_d = k_q = 1 - 0.8 = 0.2$. In the second two cases, the closed-loop pole is increased to $p_{cl} = 0.9$, and $k_d = k_q = 1 - 0.9 = 0.1$. To investigate how the choice of sampling rate affects the closed-loop control performance, we will use two sampling intervals for in the simulation studies: $\Delta t = 40 \times 10^{-6}$ sec and $\Delta t = 10 \times 10^{-6}$ sec.

The feedback control gain K_{fcs} is a function of the electrical velocity ω_e and is calculated using

$$K_{fcs}(t_i) = \begin{bmatrix} \frac{L_d}{\Delta t} & 0 \\ 0 & \frac{L_q}{\Delta t} \end{bmatrix} (I + \Delta t A_m(t_i)),$$

where

$$A_m(t_i) = \begin{bmatrix} -\frac{R_s}{L_d} & \omega_e(t_i) \\ -\omega_e(t_i) & -\frac{R_s}{L_q} \end{bmatrix}.$$

6.7.1.1 Case A

The sampling interval is selected as $\Delta t = 40 \times 10^{-6}$ sec, which gives the time-varying FCS-MPC gain as

$$K_{fcs}(t_i) = \begin{bmatrix} 172.025 & 40 \times 10^{-6}\omega_e(t_i) \\ -40 \times 10^{-6}\omega_e(t_i) & 172.025 \end{bmatrix}.$$

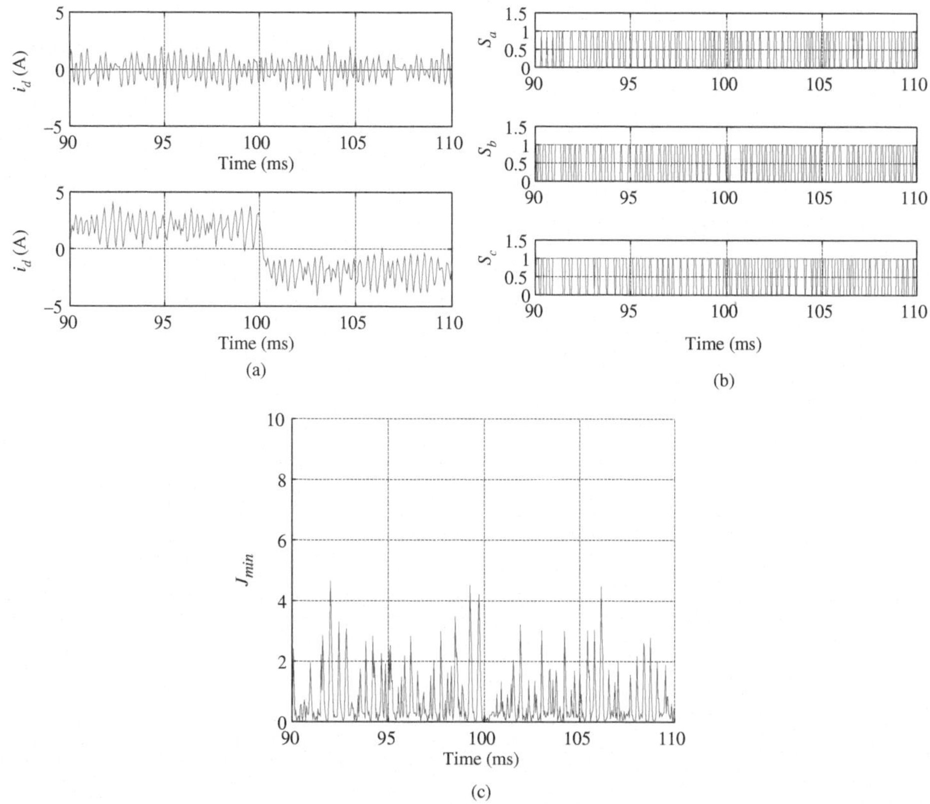

Figure 6.8 Simulation results for I-FCS-MPC of PMSM. Case A. Sampling interval $\Delta t = 40 \times 10^{-6}$ sec, $k_d = k_q = 0.2$. (a) i_d and i_q currents, (b) The normalized control signals, and (c) Minimum of the objective function.

Figures 6.8(a)–(c) show the i_d and i_q closed-loop current control results, the normalized three phase control signals and the minimum of the objective function $J_{min} = \frac{\Delta t^2}{L_d^2}(v_d(t_i)^{act} - v_d(t_i)^{opt})^2 + \frac{\Delta t^2}{L_q^2}(v_q(t_i)^{act} - v_q(t_i)^{opt})^2$ where $v_d(t_i)^{act}$ and $v_q(t_i)^{act}$ are the actual voltage variables implemented on the PMSM. With this larger sampling interval Δt, it is seen from Figure 6.8(a) that the steady-state errors in both i_d and i_q become very small. In general, when Δt is large, the overall performance is poor because the variations of the i_d and i_q currents are large. This variation is also reflected by the minimum of the objective function J_{min} that has a relatively large amplitude.

6.7.1.2 Case B

In the second case, the sampling interval Δt is reduced to 10×10^{-6} sec. The FCS-MPC gain is

$$K_{fcs}(t_i) = \begin{bmatrix} 696.99 & 10 \times 10^{-6}\omega_e(t_i) \\ -10 \times 10^{-6}\omega_e(t_i) & 696.99 \end{bmatrix}.$$

Figures 6.9(a)–(c) show that the closed-loop control performance is much improved when the sampling interval is reduced, where it is seen that the variations of i_d and i_q currents become very small. However,

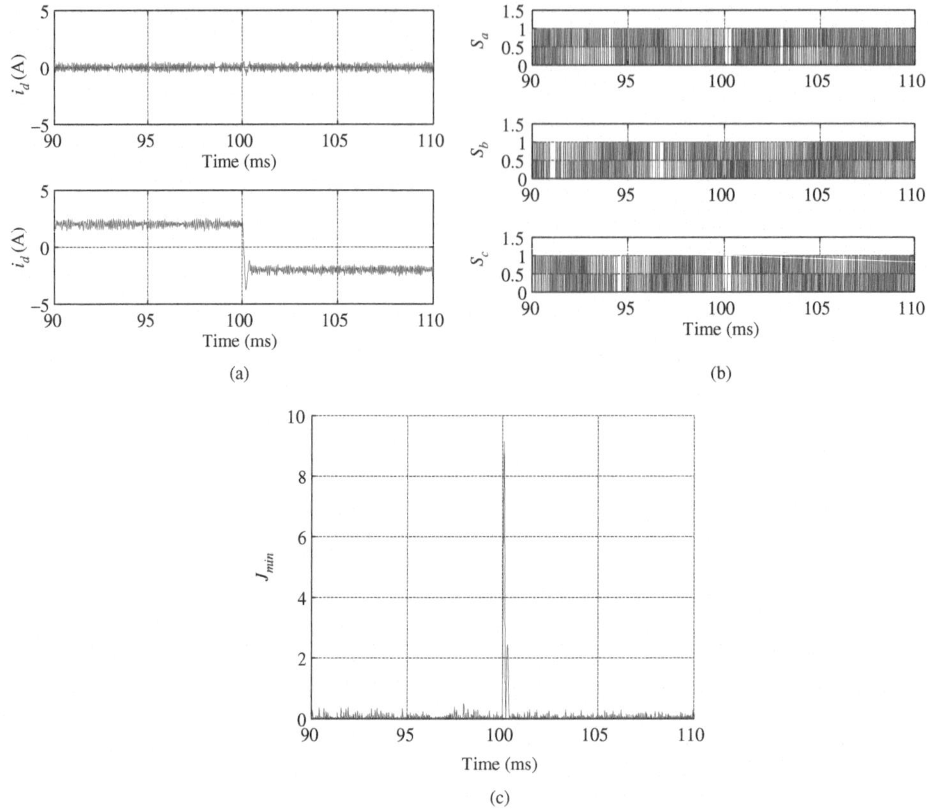

Figure 6.9 Simulation results for I-FCS-MPC of PMSM. Case B. Sampling interval $\Delta t = 10 \times 10^{-6}$ sec, $k_d = k_q = 0.2$. (a) i_d and i_q currents, (b) The normalized control signals, and (c) Minimum of the objective function.

at $t = 100$ msec when the torque disturbance changes to -1 and the reference signal i_q^* changes from $+2$ to -2, there is an undershoot in the i_q current response, which is also seen by a large spike in the minimum of the objective function (see Figure 6.9(c)). The simulation results in Case D where the integral gain is reduced confirm that the undershoot and spike are caused by the large integral gain.

6.7.1.3 Case C

In the third case, we reduce the integral controller gain to $k_d = k_q = 0.1$, but with the sampling interval $\Delta t = 40 \times 10^{-6}$ sec. Figures 6.10(a)–(c) show that the variations of the i_d and i_q currents are slightly reduced, which can also been seen from the minimum of the objective function (see Figure 6.10(c)).

6.7.1.4 Case D

In the fourth case, we will use the integral controller gain $k_d = k_q = 0.1$, together with the faster sampling interval $\Delta t = 10 \times 10^{-6}$ sec. Figures 6.11(a)–(c) show the closed-loop control system performance. It is seen that with the reduction of the integral controller gain, there is a slight reduction of the variations of the i_d and i_q currents. More importantly, at $t = 100$ μs when the torque disturbance and the reference

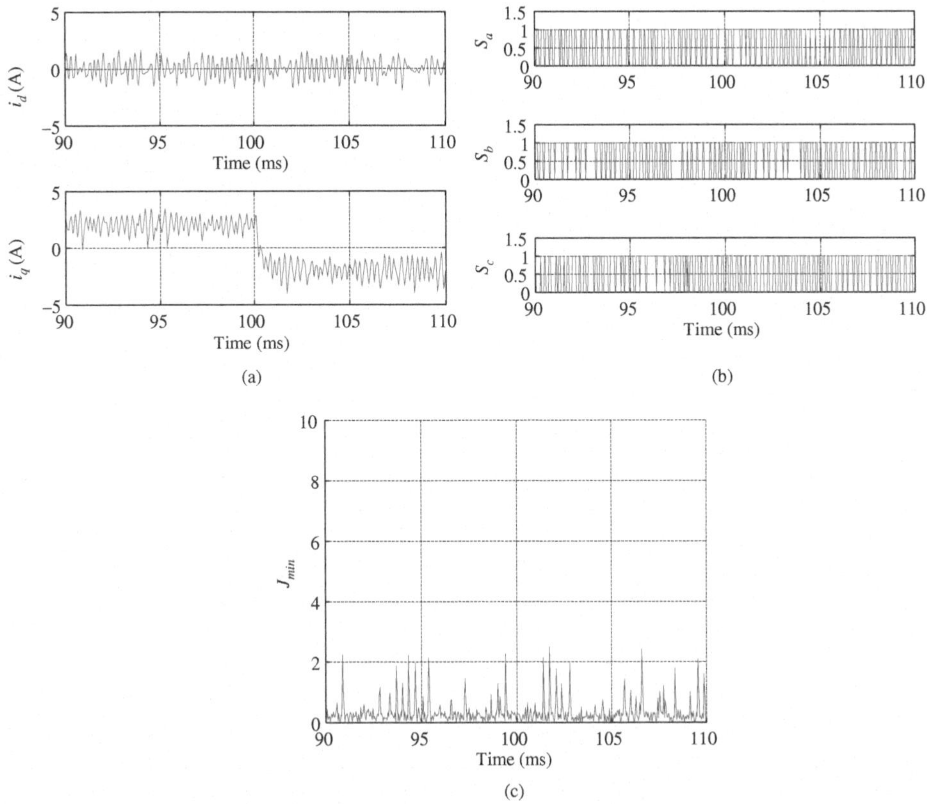

Figure 6.10 Simulation results for I-FCS-MPC of PMSM. Case C. Sampling interval $\Delta t = 40 \times 10^{-6}$ sec, $k_d = k_q = 0.1$. (a) i_d and i_q currents, (b) The normalized control signals, and (c) Minimum of the objective function.

signal change, neither the undershoot occurs in the i_q current response, nor is there a large spike in the minimum of the objective function (see Figure 6.11(c)).

It is interesting to note that with the I-FCS-MPC, the offset errors of the i_d and i_q currents are significantly reduced. For instance, in the slower sampling rate case ($\Delta t = 40 \times 10^{-6}$ sec, $k_d = k_q = 0.2$), the mean value of i_d current is 0.0011 A. In comparison with the case using the original FCS controller, where its mean value was -0.1947 A, the offset error is significantly reduced. In the faster sampling rate case ($\Delta t = 10 \times 10^{-6}$ sec, $k_d = k_q = 0.2$), the mean value of the i_d current is -2.26×10^{-5} A, which is also much smaller than the case when using the original FCS predictive controller where its mean value is 0.0023. When $k_d = k_q = 0.1$, and using the I-FCS predictive controller, the offset error for the slower sampling rate is 0.0021 A and for the faster sampling rate is 4.2475×10^{-5} A. Because these numbers are very small, their increase has little effect on the closed-loop performance.

6.8 I-FCS Predictive Control of Induction Motor

The design of I-FCS-MPC for induction motor follows the exact steps outlined in the previous sections for the PMSM. In particular, the embedded function for I-FCS-MPC function written in Tutorial 3 is used to control the currents of induction motor with appropriate changes of motor parameters and input

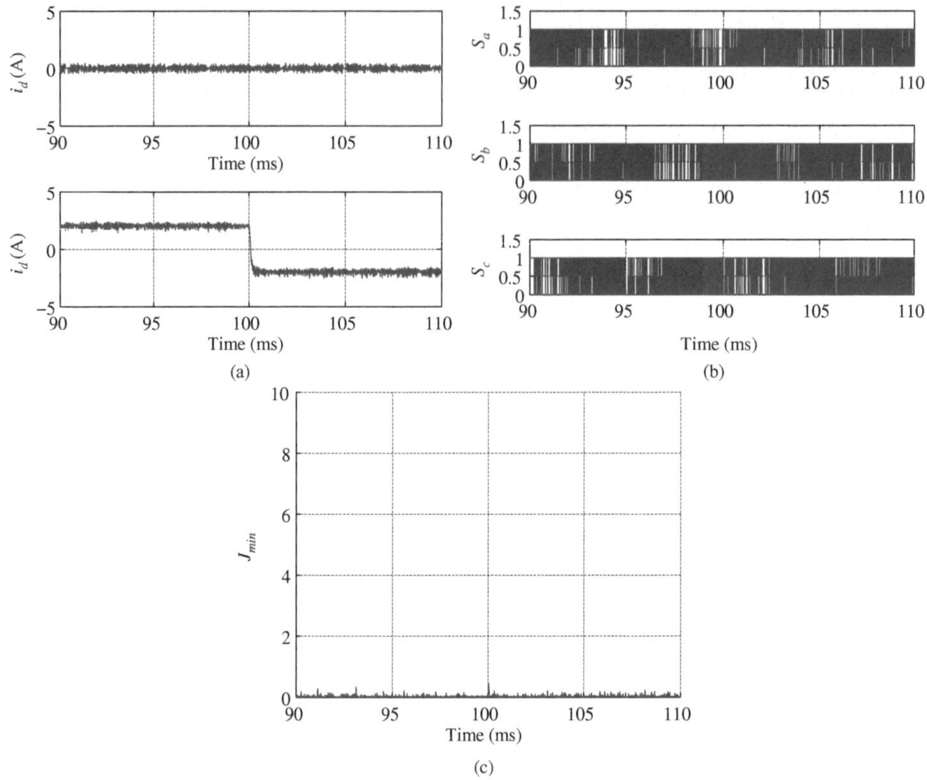

Figure 6.11 Simulation results for I-FCS-MPC of PMSM. Case D. Sampling interval $\Delta t = 10 \times 10^{-6}$ sec, $k_d = k_q = 0.1$. (a) i_d and i_q currents, (b) The normalized control signals, and (c) Minimum of the objective function.

and output variables. We assume that the reader has already progressed through the sections that have detailed the I-FCS-MPC of a PMSM. Therefore, in this section, we will only outline the differences between the two controllers.

6.8.1 The Control Algorithm for an Induction Motor

The continuous-time model that describes the dynamics of an induction motor in $d-q$ reference frame are given by the differential equations:

$$i_{sd}(t) + \tau_\sigma \frac{di_{sd}(t)}{dt} = \omega_s(t)\tau_\sigma i_{sq}(t) + \frac{k_r}{r_\sigma \tau_r}\psi_{rd}(t) + \frac{1}{r_\sigma}u_{sd}(t) \tag{6.72}$$

$$i_{sq}(t) + \tau_\sigma \frac{di_{sq}(t)}{dt} = -\omega_s(t)\tau_\sigma i_{sd}(t) - \frac{k_r}{r_\sigma}\omega_e(t)\psi_{rd}(t) + \frac{1}{r_\sigma}u_{sq}(t). \tag{6.73}$$

From (6.72) and (6.73), it can be easily verified that the system matrices $A_m(t)$ and B_m are written in the following forms:

$$A_m(t) = \begin{bmatrix} -\frac{1}{\tau_\sigma} & \omega_s(t) \\ -\omega_s(t) & -\frac{1}{\tau_\sigma} \end{bmatrix}; \; B_m = \begin{bmatrix} \frac{1}{\tau_\sigma r_\sigma} & 0 \\ 0 & \frac{1}{\tau_\sigma r_\sigma} \end{bmatrix}.$$

The discretized prediction equations corresponding to (6.72) and (6.73) are in matrix form:

$$\begin{bmatrix} i_{sd}(t_{i+1}) \\ i_{sq}(t_{i+1}) \end{bmatrix} = (I + \Delta t A_m(t_i)) \begin{bmatrix} i_{sd}(t_i) \\ i_{sq}(t_i) \end{bmatrix} + \Delta t B_m \begin{bmatrix} u_{sd}(t_i) \\ u_{sq}(t_i) \end{bmatrix} + \begin{bmatrix} \frac{k_r \Delta t}{\tau_\sigma r_\sigma \tau_r} \psi_{rd}(t_i) \\ -\frac{k_r \Delta t}{\tau_\sigma r_\sigma} \omega_e(t_i) \psi_{rd}(t_i) \end{bmatrix}, \tag{6.74}$$

where I is the identity matrix with dimension 2×2.

At sampling time t_i, the gain matrix for the FCS controller is calculated based on A_m and B_m matrices using a sampling interval Δt:

$$\begin{aligned} K_{fcs}(t_i) &= (\Delta t^2 B_m^T B_m)^{-1} B_m^T \Delta t (I + \Delta t A_m(t_i)) \\ &= \begin{bmatrix} \frac{\tau_\sigma r_\sigma}{\Delta t} & 0 \\ 0 & \frac{\tau_\sigma r_\sigma}{\Delta t} \end{bmatrix} \begin{bmatrix} 1 - \frac{\Delta t}{\tau_\sigma} & \omega_s(t_i)\Delta t \\ -\omega_s(t_i)\Delta t & 1 - \frac{\Delta t}{\tau_\sigma} \end{bmatrix} \\ &= \begin{bmatrix} \frac{\tau_\sigma r_\sigma}{\Delta t}\left(1 - \frac{\Delta t}{\tau_\sigma}\right) & \omega_s(t_i)\tau_\sigma r_\sigma \\ -\tau_\sigma r_\sigma \omega_s(t_i) & \frac{\tau_\sigma r_\sigma}{\Delta t}\left(1 - \frac{\Delta t}{\tau_\sigma}\right) \end{bmatrix}. \end{aligned} \tag{6.75}$$

Then the optimal control signals $u_{sd}(t_i)^{opt}$ and $u_{sq}(t_i)^{opt}$ are calculated at sampling time t_i,

$$\begin{aligned} \begin{bmatrix} u_{sd}(t_i)^{opt} \\ u_{sq}(t_i)^{opt} \end{bmatrix} &= \begin{bmatrix} u_{sd}(t_{i-1})^{opt} \\ u_{sq}(t_{i-1})^{opt} \end{bmatrix} \\ &+ K_{fcs}(t_i) \begin{bmatrix} k_d(i_{sd}^*(t_i) - i_{sd}(t_i)) \\ k_q(i_{sq}^*(t_i) - i_{sq}(t_i)) \end{bmatrix} - K_{fcs}(t_i) \begin{bmatrix} \Delta i_{sd}(t_i) \\ \Delta i_{sq}(t_i) \end{bmatrix}, \end{aligned}$$

where $0 < k_d < 1$ and $0 < k_q < 1$ are the integral controller gain, and $\Delta i_{sd}(t_i) = i_{sd}(t_i) - i_{sd}(t_{i-1})$, $\Delta i_{sq}(t_i) = i_{sq}(t_i) - i_{sq}(t_{i-1})$.

After calculating the value of the objective function J with respect to the finite control set for $k = 0, 1, 2, \dots, 6$

$$J_k = \frac{\Delta t^2}{(\tau_\sigma r_\sigma)^2}(u_{sd}(t_i)^k - u_{sd}(t_i)^{opt})^2 + \frac{\Delta t^2}{(\tau_\sigma r_\sigma)^2}(u_{sq}(t_i)^k - u_{sq}(t_i)^{opt})^2 \tag{6.76}$$

and finding the minimum of J_k and its corresponding index $k*$, the pair of control signals $u_{sd}(t_i)^{k*}$ and $u_{sq}(t_i)^{k*}$ is chosen for implementation.

One key difference between the FCS control systems of a PMSM and an induction motor is the computation of the seven pairs of $u_{sd}(t_i)$ and $u_{sq}(t_i)$ values used in the evaluation of the objective function. In the induction motor control, $\theta_s(t)$ is used to generate these voltage values, where θ_s is computed using the following relationship:

$$\begin{aligned} \theta_s(t) &= \theta_e(t) + \frac{1}{\tau_r} \int_0^t \frac{i_{sq}(\tau)}{i_{sd}(\tau)} d\tau \\ &\approx Z_p \theta_m(t) + \frac{1}{\tau_r} \int_0^t \frac{i_{sq}^*(\tau)}{i_{sd}^*(\tau)} d\tau, \end{aligned} \tag{6.77}$$

where $\theta_m(t)$ is the measured rotor position from the encoder, Z_p is the number of pole pair, and $i_{sd}^*(t)$ and $i_{sq}^*(t)$ are the current reference signals used to approximate the current feedback signals i_{sd} and i_{sq} because of their measurement noise. With $\theta_s(t_i)$ determined at the sampling time t_i, the seven pairs of candidate voltage variables are formed:

$$u_{sd}^0 = 0; \quad u_{sd}^1 = \frac{2}{3}V_{dc}\cos\theta_s(t_i);$$

$$u_{sq}^0 = 0; \quad u_{sq}^1 = -\frac{2}{3}V_{dc}\sin\theta_s(t_i);$$

$$u_{sd}^2 = \frac{2}{3}V_{dc}\cos\left(\theta_s(t_i) - \frac{2\pi}{3}\right); \quad u_{sd}^3 = \frac{2}{3}V_{dc}\cos\left(\theta_s(t_i) - \frac{4\pi}{3}\right);$$

$$u_{sq}^2 = -\frac{2}{3}V_{dc}\sin\left(\theta_s(t_i) - \frac{2\pi}{3}\right); \quad u_{sq}^3 = -\frac{2}{3}V_{dc}\sin\left(\theta_s(t_i) - \frac{4\pi}{3}\right);$$

$$u_{sd}^4 = -\frac{2}{3}V_{dc}\cos\theta_s(t_i); \quad u_{sd}^5 = -\frac{2}{3}V_{dc}\cos\left(\theta_s(t_i) - \frac{2\pi}{3}\right);$$

$$u_{sq}^4 = \frac{2}{3}V_{dc}\sin\theta_s(t_i); \quad u_{sq}^5 = \frac{2}{3}V_{dc}\sin\left(\theta_s(t_i) - \frac{2\pi}{3}\right);$$

$$u_{sd}^6 = -\frac{2}{3}V_{dc}\cos\left(\theta_s(t_i) - \frac{4\pi}{3}\right);$$

$$u_{sq}^6 = \frac{2}{3}V_{dc}\sin\left(\theta_s(t_i) - \frac{4\pi}{3}\right).$$

6.8.2 Simulation Results

In the simulation example, the embedded MATLAB function FCSInt.mdl written in Tutorial 3 is used to control the induction motor with small modifications. For instance, the input and output variables are defined using the following form:

```
function [Ua,Ub,Uc,val]= FCSInt(isdRef,isqRef,ws,thetas,isM,
Tin,U,r_sigma,D,tau_sigmat)
```

Once the parameters are determined, the embedded function can be changed accordingly.

In the simulation of current control, the reference signal to i_{sd} is 0.8 A and the reference signal to i_{sq} is 3 A in the beginning of the simulation and changes to 1 A after 0.1 second. The *DC* power supply is $V_{dc} = 520$ V. The parameters $r_\sigma = 17.8250$, $\tau_\sigma = 0.0060$ and $\tau_r = 0.0769$. The load torque is $T_L = 2$ Nm in the simulation. In applications, k_d and k_q are often chosen to be either 0.2 or 0.1. It seems that the simulation results show that when $k_d = k_q = 0.2$, there is a large overshoot on the i_{sq} current when it performs a set-point change, which again confirmed the results shown in the Case A and B in the PMSM control studies. Therefore, they are preferred to be $k_d = k_q = 0.1$ so to avoid the large spike when the i_{sq} current makes a set-point change. There are two cases to be presented here: Case A is for $\Delta t = 80 \times 10^{-6}$ sec and Case B is for $\Delta t = 50 \times 10^{-6}$ sec.

6.8.2.1 Case A

The sampling interval is $\Delta t = 80 \times 10^{-6}$ sec, and $k_d = k_q = 0.1$. Figure 6.12 shows the closed-loop control results. Although the sampling interval is quite large, the variations of the i_{sd} and i_{sq} currents are relatively small. The offset error is negligible. Overall, the performance is acceptable with this sampling rate.

6.8.2.2 Case B

The sampling interval is reduced to $\Delta t = 50 \times 10^{-6}$ sec while keeping the integral gains $k_d = k_q = 0.1$. Figure 6.13 shows the closed-loop control results. With the reduction of the sampling interval Δt, the variations of the i_{sd} and i_{sq} currents are further reduced. This is consistent with the results obtained from I-FCS predictive control of the PMSM (see Figures 6.8–6.11).

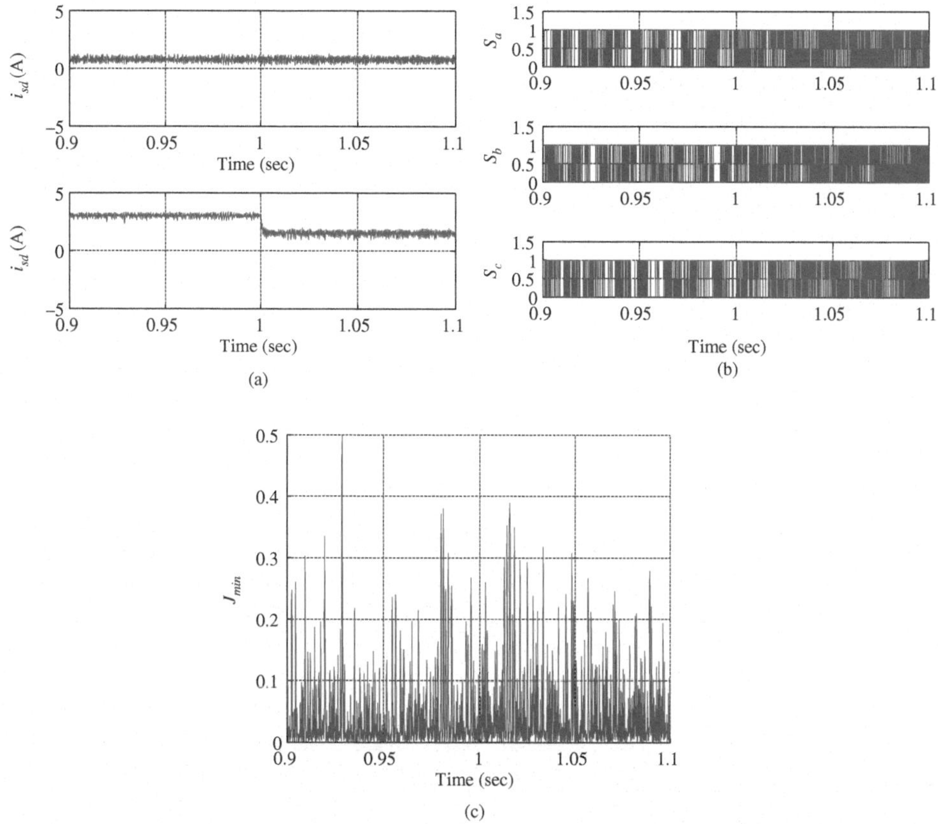

Figure 6.12 Simulation results for I-FCS-MPC of induction motor. Case A. Sampling interval $\Delta t = 80 \times 10^{-6}$ sec, $k_d = k_q = 0.1$. (a) i_{sd} and i_{sq} currents, (b) The normalized control signals, and (c) Minimum of the objective function.

6.8.3 *Experimental Results*

The experimental results are obtained from the induction motor control test-bed where Simulink xPC Target is used for control algorithm implementation. The induction motor is coupled with a servo *DC* motor as its load (see Chapter 10). The supply voltage at *DC*-link is 520V.

In the current control scheme, the sampling interval is $\Delta t = 80\,\mu s$ that is the lowest limit restricted by the equipment. The closed-loop reference signals for the d-axis stator current is $i_{sd}^* = 0.877$ A and the q-axis stator current is $i_{sq}^* = 1.5$ A. Current control experiments have been conducted using both the original FCS predictive control scheme and the I-FCS predictive control scheme.

6.8.3.1 Current Response

Figure 6.14 shows the experimental results obtained from using the original FCS-MPC scheme and Figure 6.15 shows the corresponding results obtained using the I-FCS-MPC scheme where the integral gains are selected as $k_d = k_q = 0.15$. Comparing Figure 6.14(a)–(b) with Figure 6.15(a)–(b), it is seen that the noise levels of both d-axis stator and q-axis stator currents are about the same. However,

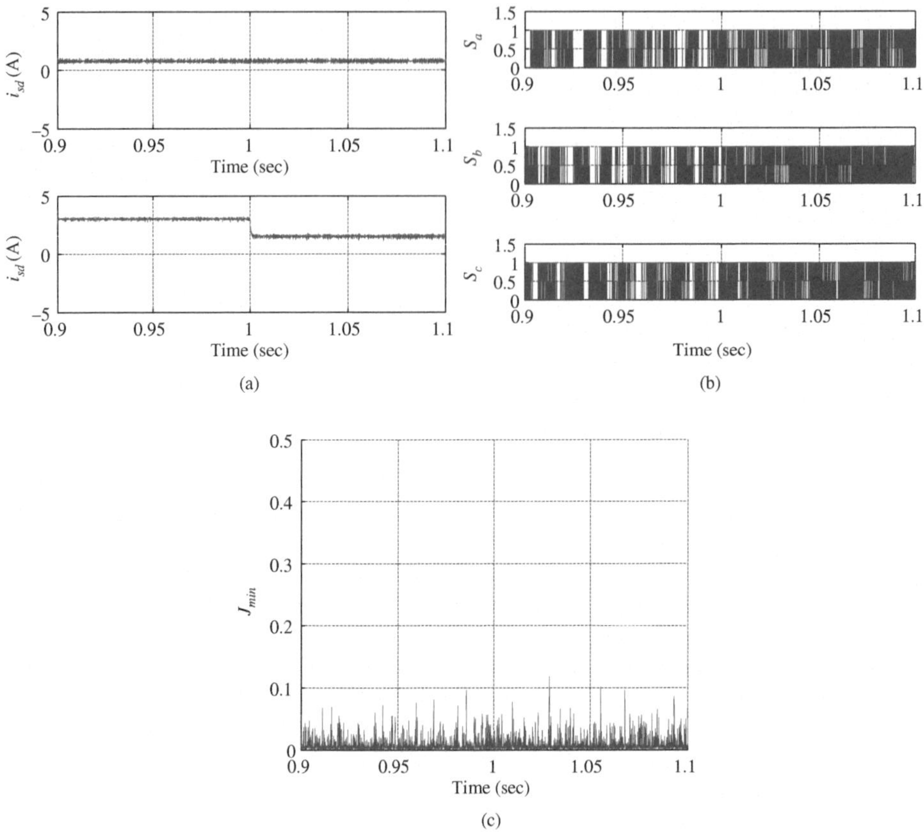

Figure 6.13 Simulation results for I-FCS-MPC of induction motor. Case B. Sampling interval $\Delta t = 50 \times 10^{-6}$ sec, $k_d = k_q = 0.1$. (a) i_{sd} and i_{sq} currents, (b) The normalized control signals, and (c) Minimum of the objective function.

calculations show that when using the original FCS predictive controller, the mean value of the error $i_{sd}^* - i_{sd}$ is 0.0164 A and the same error for the q-axis current is 0.036 A. In comparison, when using the I-FCS predictive controller, the mean value of the error $i_{sd}^* - i_{sd}$ is reduced to -2.5397×10^{-5} A and the same error for the q-axis current is reduced to -3.6636×10^{-4} A. It means that when using the original FCS-MPC scheme, the magnitudes of the off-set errors are about 648 and 98 times larger with respect to d-axis and q-axis currents. It is worthwhile to note that the steady-state response of the original FCS predictive control system is dependent on the selection of the system physical parameters. However, with the integral FCS predictive controller, this performance uncertainty in steady-state operation is removed.

6.8.3.2 Frequency Response of Phase Current

Comparing Figure 6.14(c) with 6.15(c), no significant difference between the amplitude of the frequency response of the phase current is observed. It is worthwhile to emphasize that because there is no modulation used in the implementation, a key difference from PI controller implementation, the amplitude of the phase current is quite widely spread for both cases.

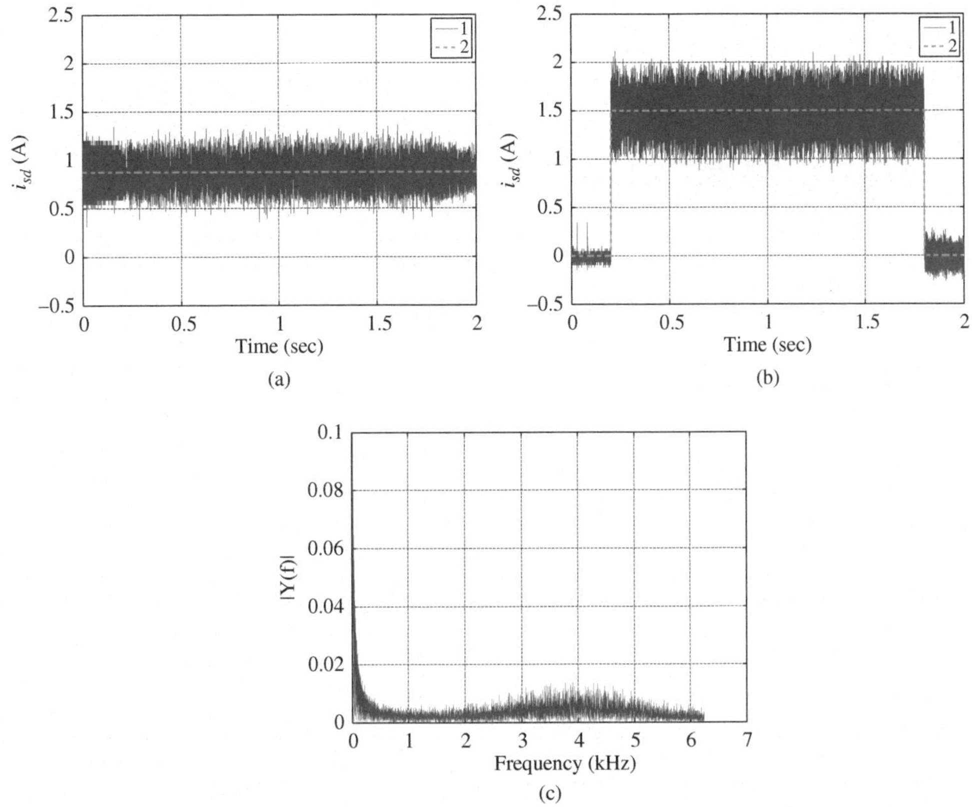

Figure 6.14 Experimental results of FCS-MPC of induction motor. $\Delta t = 80$ μs. (a) d-axis stator current (1) and reference (2), (b) q-axis stator current (1) and reference (2), and (c) Amplitude of the frequency response of A phase current.

6.8.3.3 Velocity Response

Although the steady-state errors when using the original FCS predictive controller are quite small for this induction motor with this set of physical parameters, the dynamic response of the velocity is affected. Figure 6.16 compares the open-loop responses of the speed of the motor with the original FCS predictive controller and the I-FCS predictive controller. It is interesting to see that, under the identical load condition of the coupled DC motor, the motor runs a bit faster during steady-state when using the I-FCS method. The reason for the difference could be that with the I-FCS predictive control the mean values of currents produced by the induction motor are larger and closer to the reference signals, hence leading to a faster steady-state speed. This could also imply that the motor runs more efficiently when using the I-FCS predictive controller which removes the steady-state current errors.

6.8.3.4 Selection of the integral Gain

The integral gains k_d and k_q are the performance parameters for the closed-loop current responses that affect the dynamic response of the currents. In general, the larger the gains, the faster the current dynamic response will be. Figure 6.17 shows the transient response of i_{sq} during the step change from 0 to 1.5 A

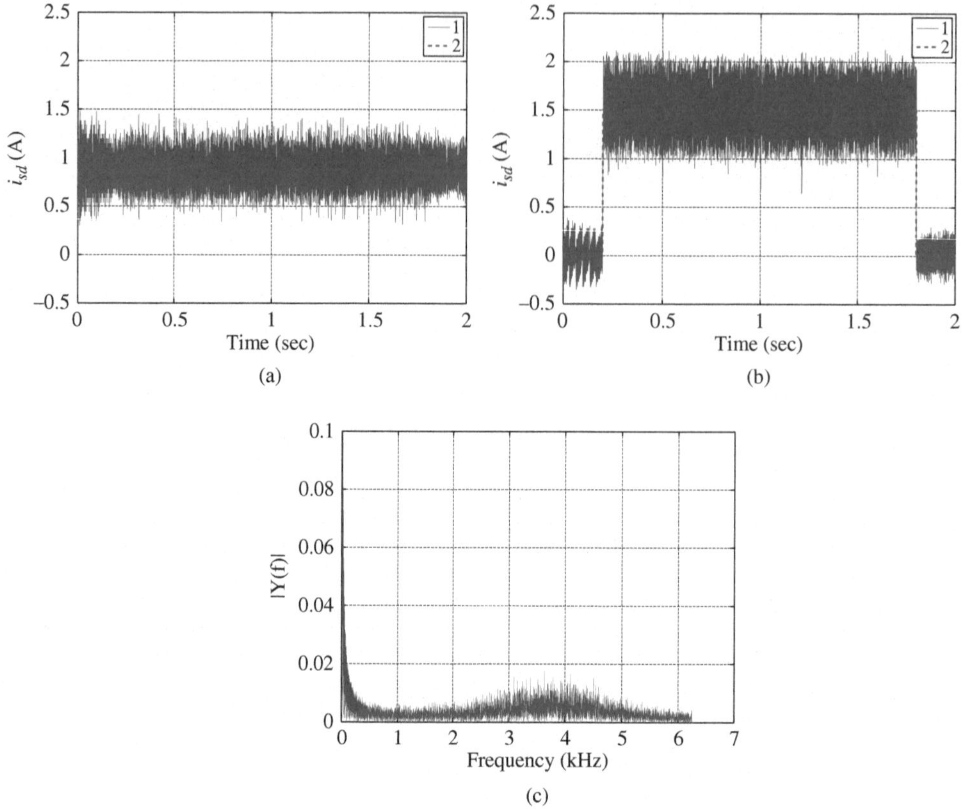

Figure 6.15 Experimental results of I-FCS-MPC of induction motor. $\Delta t = 80$ μs, and $k_d = k_q = 0.15$. (a) d-axis stator current (1) and reference (2), (b) q-axis stator current (1) and reference (2), and (c) Amplitude of the frequency response of A phase current.

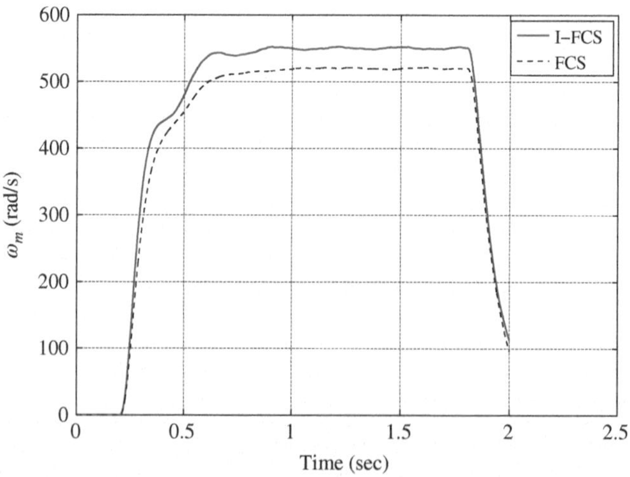

Figure 6.16 Comparison of open-loop responses of the motor speed between I-FCS-MPC and FCS-MPC schemes.

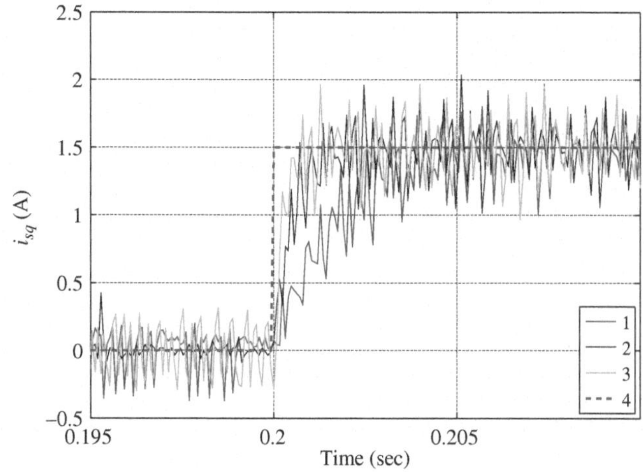

Figure 6.17 Step responses during transient with different k_d, k_q values. Key: line (1) $k_d = k_q = 0.05$; line (2) $k_d = k_q = 0.15$; line (3) $k_d = k_q = 0.25$; line (4) set-point signal.

where three different sets of integral gains are used. When $k_d = k_q = 0.05$, the response time is slowest as the closed-loop pole is closer to the unit circle. The fastest response is obtained with $k_d = k_q = 0.25$. Note that, the transient responses are very close between $k_d = k_q = 0.25$ and $k_d = k_q = 0.15$, perhaps due to the limitations of the inverter operation.

6.9 I-FCS Predictive Control of Power Converter

The design of I-FCS predictive control for power converter is identical to the controller design outlined in the previous sections for the PMSM. We assume that the reader has already progessed through the sections that have detailed the I-FCS predictive control of PMSM.

6.9.1 *I-FCS Predictive Control of a Power Converter*

The dynamic model of a power converter for current control, in $d - q$ reference frame, is expressed as

$$L_s \frac{di_d}{dt} = -R_s i_d + \omega_g L_s i_q + E_d - v_d \tag{6.78}$$

$$L_s \frac{di_q}{dt} = -v_q - R_s i_q - \omega_g L_s i_d \tag{6.79}$$

$$C_{dc} \frac{dv_{dc}}{dt} = \frac{3}{4}(S_d i_d + S_q i_q) - i_L, \tag{6.80}$$

where ω_g is the grid frequency, E_d is a grid source voltage that is a constant or a slow varying variable, i_d and i_q are the output variables, and v_d, v_q are the control variables. S_d and S_q are related to the control variables v_d and v_q through the following relations:

$$v_d = S_d \times (v_{dc}/2) \tag{6.81}$$

$$v_q = S_q \times (v_{dc}/2). \tag{6.82}$$

Equation (6.80) will not be used in the design of current controller, however, it describes the effect of the load current disturbance i_L, which is the disturbance to the current control system.

The system matrices A_m and B_m for the converter are written in the following forms:

$$A_m = \begin{bmatrix} -\frac{R_s}{L_s} & \omega_g \\ -\omega_g & -\frac{R_s}{L_s} \end{bmatrix}; \; B_m = \begin{bmatrix} -\frac{1}{L_s} & 0 \\ 0 & -\frac{1}{L_s} \end{bmatrix}.$$

The discrete prediction equations corresponding to (6.78) and (6.79) are

$$\begin{bmatrix} i_d(t_{i+1}) \\ i_q(t_{i+1}) \end{bmatrix} = (I + \Delta t A_m) \begin{bmatrix} i_d(t_i) \\ i_q(t_i) \end{bmatrix} + \begin{bmatrix} \frac{E_d}{L_s}\Delta t \\ 0 \end{bmatrix} + \Delta t B_m \begin{bmatrix} v_d(t_i) \\ v_q(t_i) \end{bmatrix}, \tag{6.83}$$

where I is the identity matrix with dimension 2×2. Note that in the application of power converter control, the system matrix A_m is a constant matrix.

The gain matrix for the FCS controller is calculated based on A_m and B_m matrices, independent of sample time t_i:

$$\begin{aligned} K_{fcs} &= (\Delta t^2 B_m^T B_m)^{-1} B_m^T \Delta t (I + \Delta t A_m) \\ &= -\frac{L_s}{\Delta t} \begin{bmatrix} 1 - \frac{R_s}{L_s}\Delta t & \omega_g \Delta t \\ -\omega_g \Delta t & 1 - \frac{R_s}{L_s}\Delta t \end{bmatrix}. \end{aligned} \tag{6.84}$$

Note that the FCS controller gain K_{fcs} is time invariant because of the invariance of the grid frequency ω_g, thus it can be calculated off-line.

Then the optimal control signals $v_d(t_i)^{opt}$ and $v_q(t_i)^{opt}$ are calculated at sampling time t_i,

$$\begin{aligned} \begin{bmatrix} v_d(t_i)^{opt} \\ v_q(t_i)^{opt} \end{bmatrix} &= \begin{bmatrix} v_d(t_{i-1})^{opt} \\ v_q(t_{i-1})^{opt} \end{bmatrix} \\ &+ K_{fcs} \begin{bmatrix} k_d(i_d^*(t_i) - i_d(t_i)) \\ k_q(i_q^*(t_i) - i_q(t_i)) \end{bmatrix} - K_{fcs} \begin{bmatrix} \Delta i_d(t_i) \\ \Delta i_q(t_i) \end{bmatrix}, \end{aligned}$$

where $0 < k_d < 1$ and $0 < k_q < 1$ are the integral controller gains, and $\Delta i_d(t_i) = i_d(t_i) - i_d(t_{i-1})$, $\Delta i_q(t_i) = i_q(t_i) - i_q(t_{i-1})$.

The objective function J with respect to the finite control set for $k = 0, 1, 2, \ldots, 6$ is defined using the optimal control signals, $v_d(t_i)^{opt}$ and $v_q(t_i)^{opt}$, as

$$J_k = \frac{\Delta t^2}{L_s^2}(v_d(t_i)^k - v_d(t_i)^{opt})^2 + \frac{\Delta t^2}{L_s^2}(v_q(t_i)^k - v_q(t_i)^{opt})^2. \tag{6.85}$$

The next step is to find the minimum of J_k and its corresponding index $k*$, with which the pair of control signals $v_d(t_i)^{k*}$ and $v_q(t_i)^{k*}$ is chosen for implementation.

A key difference between the FCS control systems of a power converter and the *AC* drives is the computation of the seven pairs of $v_d(t_i)$ and $v_q(t_i)$ values used in the evaluation of the objective function. In the power converter control, $\theta_g(t)$ is used to generate these voltage values, where $\theta_g(t)$ is the grid electrical angle estimated by the phase locked loop discussed in Section 3.9.5. With $\theta_g(t_i)$ determined at the sampling time t_i, the seven pairs of candidate voltage variables are formed:

$$v_d^0 = 0; \quad v_d^1 = \frac{2}{3} v_{dc} \cos\theta_g(t_i);$$

$$v_q^0 = 0; \quad v_q^1 = -\frac{2}{3} v_{dc} \sin\theta_g(t_i);$$

$$v_d^2 = \frac{2}{3}v_{dc}\cos\left(\theta_g(t_i) - \frac{2\pi}{3}\right);\quad v_d^3 = \frac{2}{3}v_{dc}\cos\left(\theta_g(t_i) - \frac{4\pi}{3}\right);$$
$$v_q^2 = -\frac{2}{3}v_{dc}\sin\left(\theta_g(t_i) - \frac{2\pi}{3}\right);\quad v_q^3 = -\frac{2}{3}v_{dc}\sin\left(\theta_g(t_i) - \frac{4\pi}{3}\right);$$
$$v_d^4 = -\frac{2}{3}v_{dc}\cos\theta_g(t_i);\quad v_d^5 = -\frac{2}{3}v_{dc}\cos\left(\theta_g(t_i) - \frac{2\pi}{3}\right);$$
$$v_q^4 = \frac{2}{3}v_{dc}\sin\theta_g(t_i);\quad v_q^5 = \frac{2}{3}v_{dc}\sin\left(\theta_g(t_i) - \frac{2\pi}{3}\right);$$
$$v_d^6 = -\frac{2}{3}v_{dc}\cos\left(\theta_g(t_i) - \frac{4\pi}{3}\right);$$
$$v_q^6 = \frac{2}{3}v_{dc}\sin\left(\theta_g(t_i) - \frac{4\pi}{3}\right).$$

Note that the v_{dc} is the output of the power converter. Therefore this variable is measured and is required to be updated when forming the candidate variables.

6.9.2 Simulation Results

In the simulation example, the embedded MATLAB function FCSInt.mdl written in Tutorial 3 is used to control the power converter with small modifications. For instance, the input and output variables are defined using the following form:

```
[Ua,Ub,Uc,val]=FCSInt(idqRef,idqMea,Vdc,Kfcs,U,kd,thetag)
```

where *idqRef* is the vector of the reference signals i_d^*, i_q^* and *idqMea* is the vector of the measured current signals i_d and i_q, *Vdc* is the *DC* voltage, which becomes a variable in the power converter case as an input to the embedded function, *Kfcs* is the gain of the FCS controller, *kdq* is the integral gain set to be 0.05, *thetag* is the grid electrical angle estimated by the phase locked loop, and *U* is the data matrix defined by

$$U = \begin{bmatrix} 0 & 1 & 1 & 0 & 0 & 0 & 1 & 1 \\ 0 & 0 & 1 & 1 & 1 & 0 & 0 & 1 \\ 0 & 0 & 0 & 0 & 1 & 1 & 1 & 1 \end{bmatrix}.$$

Because the gain of FCS can be calculated off-line, the embedded function "FCSInt" does not require the values of sampling interval and the inductance L_s, which means that the minimum of the objective function *val* is not weighted by $\frac{\Delta t^2}{L_s^2}$. There are two sampling intervals used in the simulation studies. In Case A, the sampling interval $\Delta t = 80 \times 10^{-6}$ sec and in Case B, $\Delta t = 40 \times 10^{-6}$ sec. The model parameters used in the computation of gain matrix K_{fcs} are converter side resistance $R_s = 0.2\ \Omega$, inductance $L_s = 6.3 \times 10^{-3}$ H and $\omega_g = 100\pi$ where the grid frequency is assumed to 50 Hz.

For $\Delta t = 80 \times 10^{-6}$, the gain matrix is

$$K_{fcs} = \begin{bmatrix} -78.55 & -1.9792 \\ 1.9792 & -78.55 \end{bmatrix}$$

and for $\Delta t = 40 \times 10^{-6}$,

$$K_{fcs} = \begin{bmatrix} -157.3 & -1.9792 \\ 1.9792 & -157.3 \end{bmatrix}.$$

This simulation is going to be used to illustrate various operational scenarios. In Scenario A, for the time interval $0 \leq t \leq 0.1$ sec, the power converter starts up. In Scenario B, for the time interval $0.1 \leq$

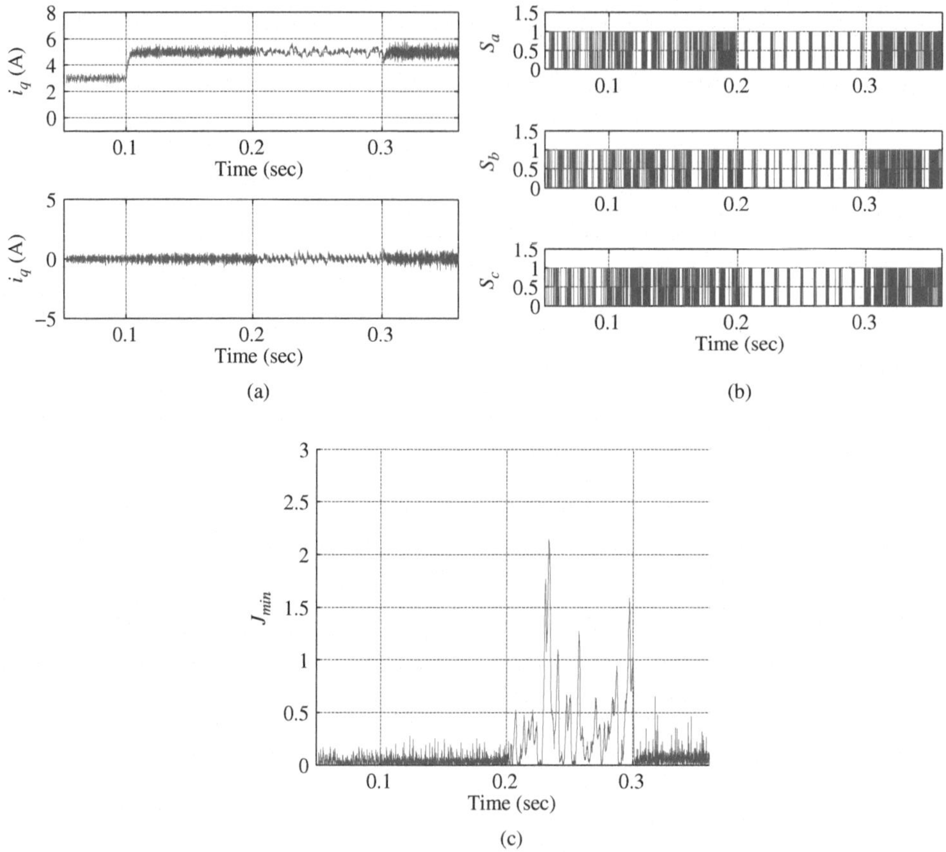

Figure 6.18 Simulation results for I-FCS-MPC of power converter. Case A. Sampling interval $\Delta t = 80 \times 10^{-6}$ sec, $k_d = k_q = 0.05$. (a) i_d and i_q currents, (b) The normalized control signals, and (c) Minimum of the objective function.

$t \leq 0.2$ sec, the d-axis current makes a set-point change. In Scenario C, for the time interval $0.2 \leq t \leq 0.3$ sec, current load disturbance $i_L = 2$ A is added to the system. In Scenario D, for the time interval $0.3 \leq t \leq 0.38$ sec, a negative current load disturbance $i_L = -2$ A is added to the system. The simulation results with respect to the four scenarios are discussed. Figure 6.18 shows the simulation results for the I-FCS-MPC control system with sampling interval $\Delta t = 80 \times 10^{-6}$ sec, and in comparison. Figure 6.19 shows the results for the control system with sampling interval $\Delta t = 40 \times 10^{-6}$ sec.

6.9.2.1 Scenario A

In the simulation for the time period $0 \leq t \leq 0.1$ sec, the I-FCS controller maintains the steady-state operation of the power converter with the defined reference signals, $i_d^* = 3$ A, and the $i_q^* = 0$. In this initial stage, the load current $i_L = 0$. This scenario represents the situation when the converter starts up its initial operation. Figures 6.18(a)–(c) illustrate the closed-loop control system responses with sampling interval $\Delta t = 80 \times 10^{-6}$ sec. It is seen that in this initial stage, the variations of the i_d and i_q currents are very

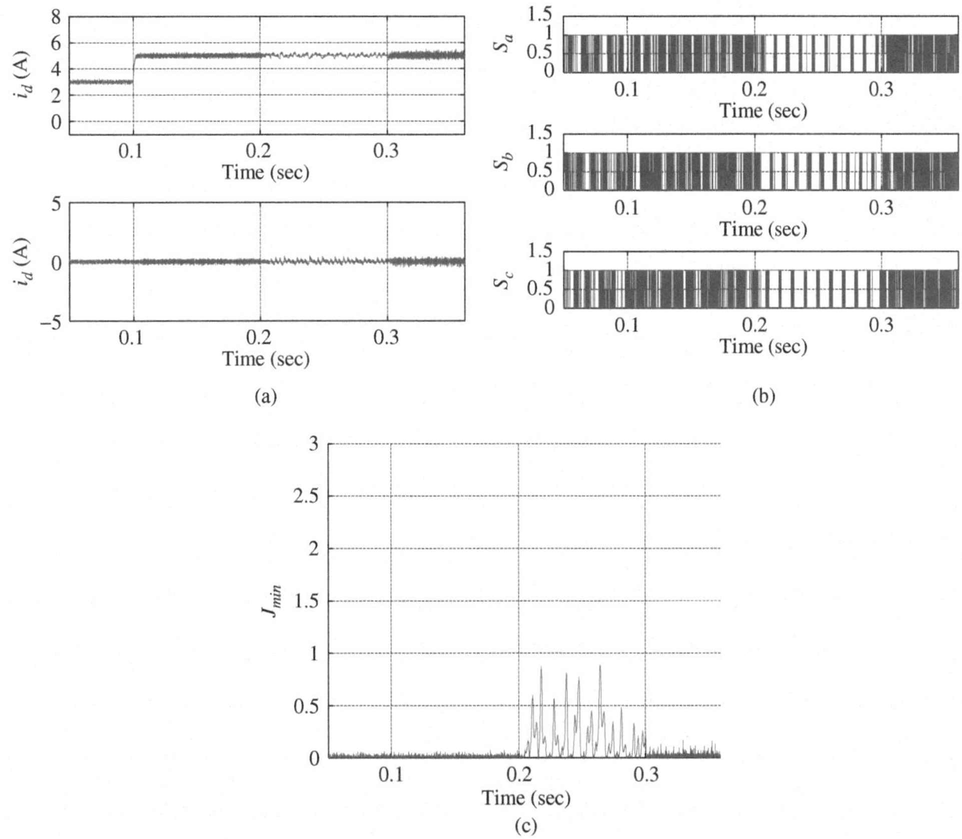

Figure 6.19 Simulation results for I-FCS-MPC of power converter. Case B. Sampling interval $\Delta t = 40 \times 10^{-6}$ sec, $k_d = k_q = 0.05$. (a) i_d and i_q currents, (b) The normalized control signals, and (c) Minimum of the objective function.

small. Improved responses can be seen in Figures 6.19(a)–(c) arising from the reduction of sampling interval to $\Delta t = 40 \times 10^{-6}$ sec.

6.9.2.2 Scenario B

The second scenario for the time period $0.1 \leq t \leq 0.2$ sec represents the situation of set-point change on i_d current with i_q^* being 0 and the load current i_L being 0. Here, i_d^* is a step signal with its amplitude increased from 3 A to 5 A. It is seen from Figures 6.18(a) and 6.19(a) that the output i_d has a fast response to the step change without oscillation and overshoot. The steady-state error is zero. The faster sampling rate produced better control results.

6.9.2.3 Scenario C

In the third scenario simulation for the time period $0.2 \leq t \leq 0.3$ sec, a positive current load disturbance $i_L = 2$ A is added to the system. Note that with a positive i_L disturbance, from (6.80), it is seen that the effect is to decrease v_{dc}. This represents the situation of drawing current from the DC bus. The simulation results show that the closed-loop control system is stable in the operation. However, the variations of the control signals are much larger as seen from the minimum of the objective function (see Figures 6.18(c) and 6.19(c)), which may be caused by the variations in the uncontrolled v_{dc}. In addition, there are low frequency oscillations in the i_d and i_q currents. Because the magnitudes of the oscillations are small, they are hardly noticed.

6.9.2.4 Scenario D

In the fourth scenario simulation for the time period $0.3 \leq t \leq 0.38$ sec, a negative current load disturbance $i_L = -2$ A is added to the system. This simulates the situation where current load is added to the system. Because the reference signal i_d current is positive and there is no feedback control of voltage, the load disturbance will charge the capacitor. If voltage control were used, this would correspond to the situation of putting current to the electrical grid. During this operation, the low frequency oscillations in Scenario C vanished and the variations of currents are small. Note that the current load disturbance i_L had a step change of magnitude of 4 when it changes from 2 A (see Scenario C) to -2 A for this scenario, however, this large change did not cause performance degradation in the i_d and i_q currents.

6.9.3 *Experimental Results*

In the experiments, the reference current i_q^* is selected to be zero. A reference signal i_d^* is set at 3 A when the current controllers are turned on, followed by a step change from 3 A to 5 A. The inductance is $L_s^0 = 6.3 \times 10^{-3}$ H and the resistance is $R_s = 0.2\ \Omega$. Sampling interval $\Delta t = 80 \times 10^{-6}$ sec. There is a unknown load current in the system. In the first case, the original FCS predictive controller is used for the current control. The experimental results are shown in Figure 6.20. Because there is no integral action, there is a steady-state error between the reference current signal and the actual measured current signal. It is seen that the steady-state error is larger for the i_d current. In the second case, the I-FCS predictive controller is used in the experiment. The experimental results are shown in Figure 6.21. The original FCS predictive controller had mean values of 3.3129 A and 5.2666 A for the set-points 3 A and 5 A in the d-axis respectively, while I-FCS predictive controller had mean values of 3.0008 A and 4.9611 A, which are significantly smaller.

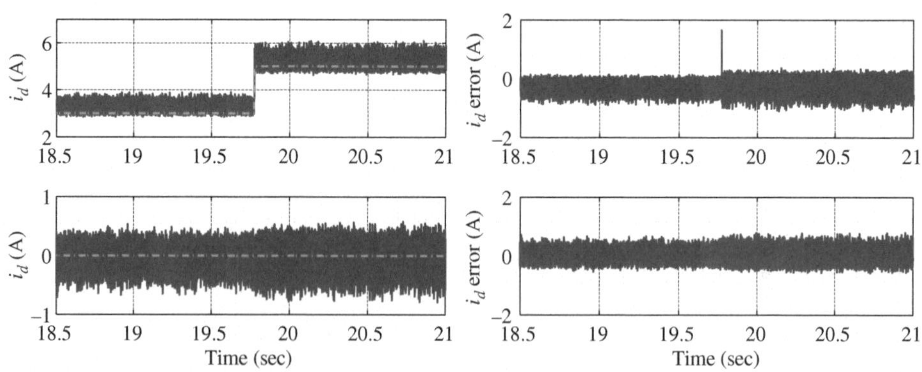

(a) (b)

Figure 6.20 Experimental results for FCS-MPC of power converter. $\Delta t = 80 \times 10^{-6}$ sec. (a) i_d and i_q currents and (b) i_d and i_q current errors.

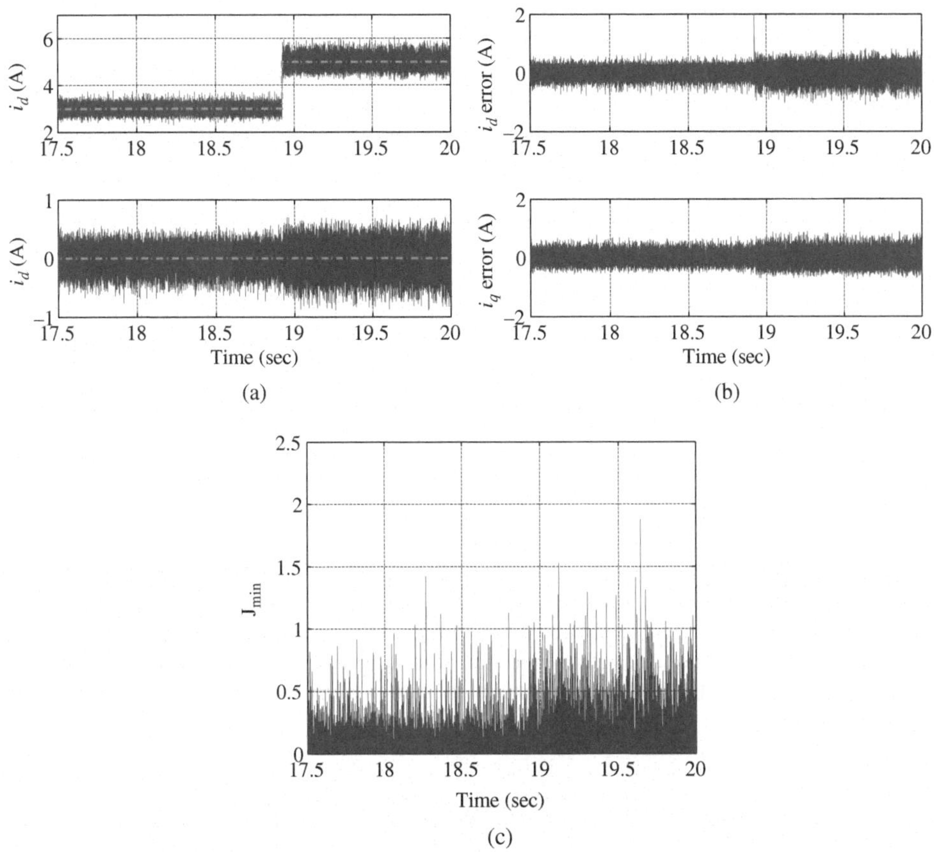

Figure 6.21 Experimental results for I-FCS-MPC of power converter. $\Delta t = 80 \times 10^{-6}$ sec, $k_d = k_q = 0.05$. (a) i_d and i_q currents, (b) i_d and i_q current errors, and (c) Minimum of the objective function.

6.10 Evaluation of Robustness of I-FCS-MPC via Monte-Carlo Simulations

This section will evaluate the robustness of the original FCS-MPC and the I-FCS-MPC systems with respect to parameter variations in the current control system. More specifically, we will use the I-FCS predictive control of power converter as the benchmark for the case studies.

In the I-FCS predictive control of a power converter, the main parameter uncertainty comes from the variations of the inductance L_s in the converter. In the Monte-Carlo simulation, the inductance L_s is taken as a random number within the range $(0, 2L_s^0)$ H. There are 500 simulated experiments conducted with the inductance parameter L_s calculated with the equation

$$L_s = L_s^0 \pm \epsilon L_s^0,$$

where $0 < \epsilon < 1$ is a random number with uniform distribution generated using the MATLAB random number generator. Among the 500 experiments, 250 of them are calculated with $-\epsilon$ and 250 with $+\epsilon$. Here, the nominal inductance $L_s^0 = 6.3 \times 10^{-3}$ H.

The sampling interval in the Monte-Carlo simulation is chosen to be $\Delta t = 80 \times 10^{-6}$ sec. The integral gain is set to be $k_d = k_q = 0.05$ for all cases. The reference signals and load current disturbances vary in an identical manner to the four scenarios described in Section 6.9.2, where $i_q^* = 0$ and i_d^* is a step signal with magnitude of 2 A, and the load current i_L varying between ± 2 A.

The model parameters used in the computation of gain matrix K_{fcs} are converter side resistance $R_s = 0.2\ \Omega$ and inductance $L_s = 6.3 \times 10^{-3}(1 \pm \epsilon)$ H, which leads to

$$K_{fcs} = \begin{bmatrix} -\frac{L_s}{\Delta t} + R_s & -\omega_g L_s \\ \omega_g L_s & -\frac{L_s}{\Delta t} + R_s \end{bmatrix} = \begin{bmatrix} -78.75(1 \pm \epsilon) + R_s & -1.9792(1 \pm \epsilon) \\ 1.9792(1 \pm \epsilon) & -78.75(1 \pm \epsilon) + R_s \end{bmatrix}, \tag{6.86}$$

where $\omega_g = 2\pi \times 50$ and $0 < \epsilon < 1$ is a uniformly distributed random number.

To evaluate the robust closed-loop performance of the control systems, the mean square error is defined as

$$MSE = \frac{1}{M} \sum_{i=0}^{M-1} ((i_d^*(t_i) - i_d(t_i))^2 + (i_q^*(t_i) - i_q(t_i))^2),$$

where M is the number of samples used in the evaluation.

The original FCS predictive control algorithm discussed in Sections 6.2 and 6.4 is implemented through the embedded MATLAB function given in Tutorial 2 to control the power converter where the K_{fcs} gain varies with the L_s parameter generated through the Monte-Carlo experiments. Similarly evaluated is the I-FCS predictive control algorithm with the identical simulation environment and parameter variations. All physical parameters in the Simulink simulator of the converter remain unchanged in the Monte-Carlo simulation experiments.

6.10.1 Discussion on Mean Square Errors

The mean square error results for the 500 simulations are shown in Figures 6.22–6.23 for the original FCS predictive control system and the I-FCS predictive control system, respectively. From the larger scale

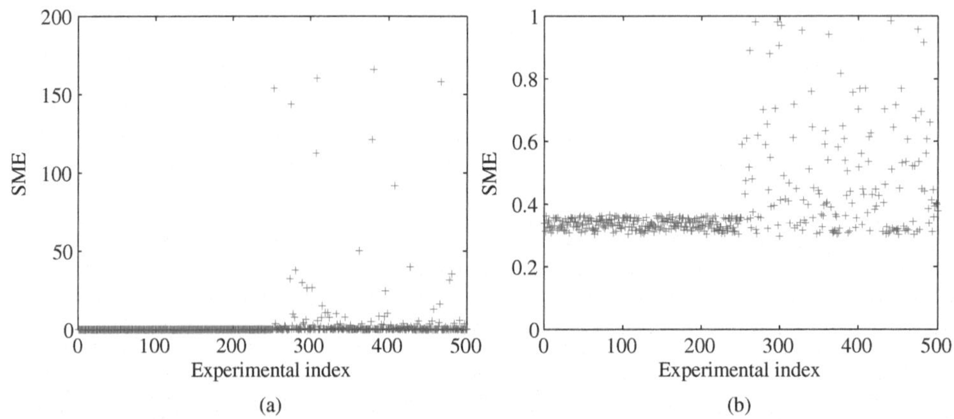

Figure 6.22 Monte-Carlo simulation results for original FCS-MPC of power converter. Mean square error for variations of L_s. (a) Larger scale and (b) Smaller scale.

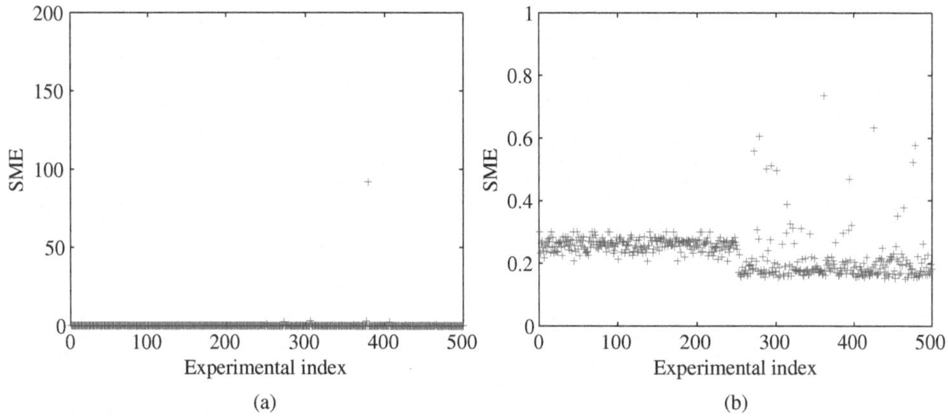

Figure 6.23 Monte-Carlo simulation results for I-FCS-MPC of power converter. Mean square error for variations of L_s. (a) Larger scale and (b) Smaller scale.

presentations of the mean square error, the failures of the closed-loop control results can be immediately identified. There are 25 data points visible in Figure 6.22(a) and 1 data point visible in Figure 6.23(a). If we count these large MSE numbers as the failed control systems because of the unacceptable deviations of i_d and i_q from their reference signals, then there are 25 FCS predictive control system failures and 1 I-FCS predictive control system failure. In the smaller scale presentations of the mean square errors (see Figures 6.22(b) and 6.23(b)), there are far more data points in Figure 6.22(b) than those in Figure 6.23(b), deviating from its targeted value occurring between the index of 250 to 500. It is clearly seen from the mean square error plots that the I-FCS predictive controller has far better robust performance in the presence of inductance variations.

To understand how the inductance variations affect the closed-loop performance, the mean square errors are plotted against the variation of the inductance L_s. Figures 6.24 and 6.25 show the mean square errors (<5), respectively. By comparing these two figures, several comments follow. In each figure, there is a window to zoom in on the region where the smallest MSE values are. It is seen that the I-FCS

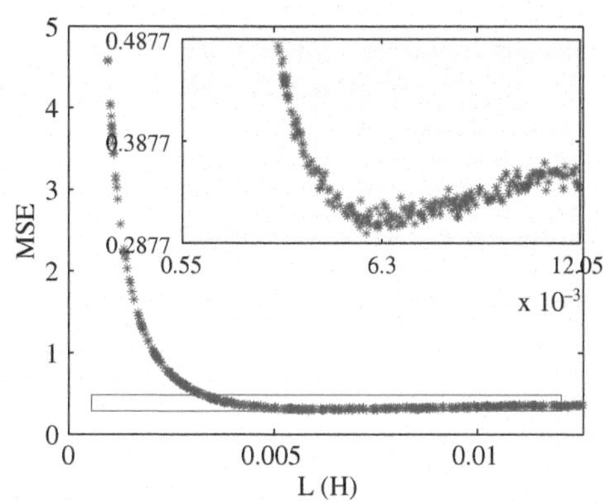

Figure 6.24 Monte-Carlo simulation results of FCS predictive control of power converter. Mean square error for variations of L_s.

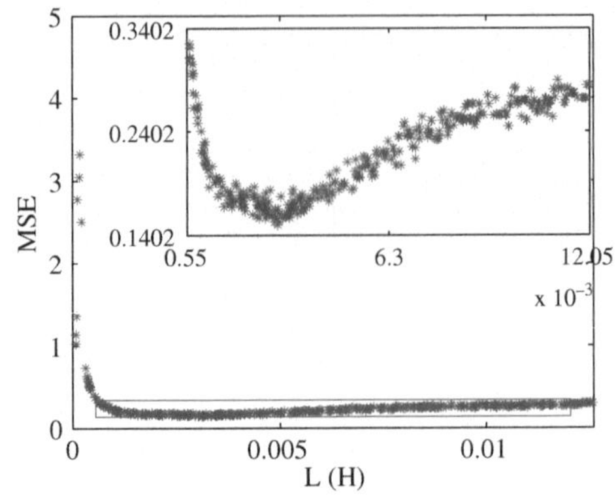

Figure 6.25 Monte-Carlo simulation results of I-FCS predictive control of power converter. Mean square error for variations of L_s.

predictive control system has the smallest MSE, which is half of that value arising from the original FCS predictive control system. We also notice that in the region where $L_s = 6.3 \times 10^{-3}(1 + \epsilon)$ H $(0 < \epsilon < 1)$, namely when L_s value used in the model is larger than the inductance of the physical system, both FCS and I-FCS control systems do not have significant performance deterioration, as demonstrated by the MSE values. However, in the region where $L_s = 6.3 \times 10^{-3}(1 - \epsilon)$ H, namely when the L_s value used in the model is smaller than the inductance of the physical system, the closed-loop control performance using the original FCS-MPC system deteriorates rapidly as illustrated by the increase of MSE values (see Figure 6.24). In contrast, the I-FCS-MPC system maintains its closed-loop control performance over a much wider variation of the inductance parameter L_s, as illustrated by the almost constant value of MSEs in Figure 6.25 with a much smaller L_s value.

The selected closed-loop control system responses for using FCS predictive controller and I-FCS predictive controller are illustrated in Figure 6.26 and Figure 6.27, respectively, where the failed control responses are excluded. It is clearly seen that the FCS predictive control system has a steady-state error in tracking the current references and its performance deteriorates severely when the model inductance is less than the inductance in the physical system. In summary, the roles of the integrators used in current control are to eliminate the steady-state errors, and equally importantly to safe-guard the closed-loop performance in the presence of parameter variations. In the case of converter control, the under-estimated inductance in the design could cause severe performance degrade without using integral action in the controller. However by incorporating integrators in the current control, the closed-loop performance is more robust against the variations of the parameters.

6.11 Velocity and Position Control of PMSM Using I-FCS-MPC

In this section, we will discuss the velocity and position control of a PMSM using I-FCS-MPC. These control systems are configured in cascade feedback control structures and are identical to those used in the design of PI current control systems as discussed in Chapters 3 to 5.

Figure 6.28 illustrates the control system configuration of velocity control of a PMSM, where the inner-loop current controllers are the I-FCS controllers and the outer-loop control system uses a PI for the velocity control. A similar figure is presented in Figure 6.29 to illustrate the position control of a PMSM, in which the outer-loop control system uses a PID controller for position control. Although the cascade control structures are identical to those introduced in the previous chapters (see Chapters 3 to 5), there are several differences that will be emphasized in the following discussions.

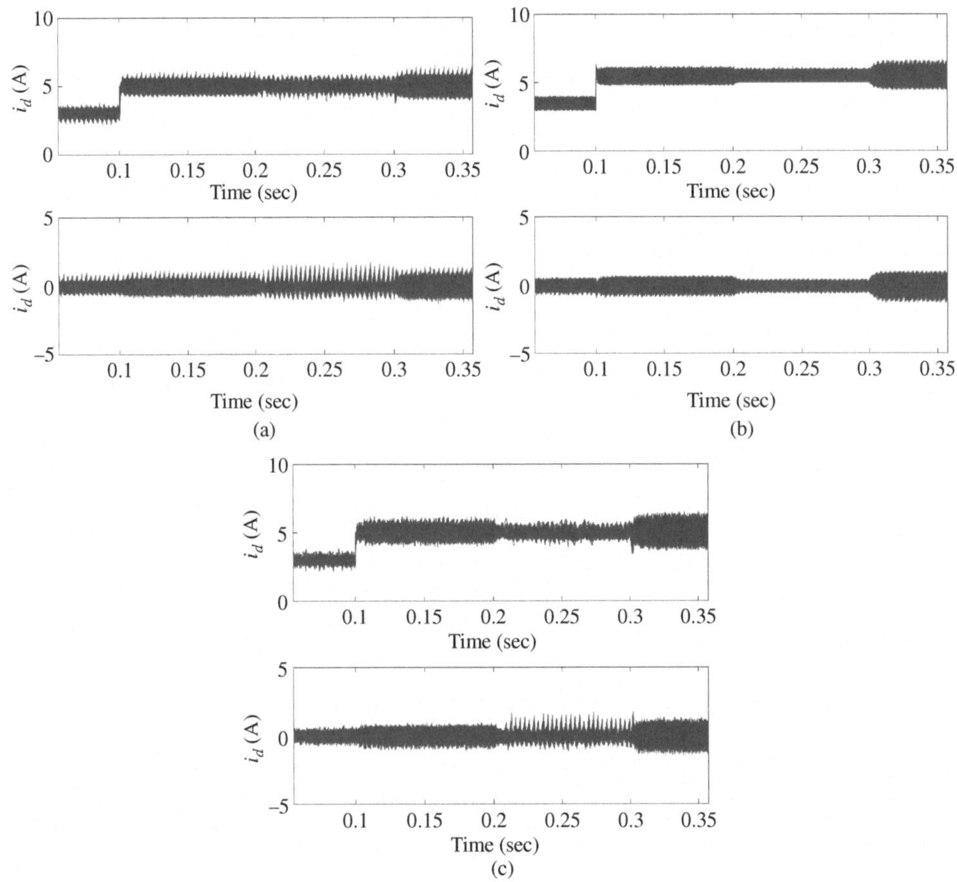

Figure 6.26 Monte-Carlo simulation results of original FCS-MPC system when L_s changes. Variations of i_d and i_q currents. (a) $L_s < 6.3 \times 10^{-3}(1-0.5\epsilon)$ H, (b) $6.3 \times 10^{-3}(1-0.5\epsilon)$ H $< L_s < 6.3 \times 10^{-3}(1+0.5\epsilon)$ H, and (c) $L_s > 6.3 \times 10^{-3}(1-0.5\epsilon)$ H.

6.11.1 Choice of Sampling Rate for the Outer-loop Control System

As we noticed from the diagrams, the inner-loop current control systems are discrete-time control systems, in contrast to the previously PI or P control of the currents. Although one considers that the PI controllers of the currents are also implemented using discretized algorithms (see Chapter 4), the design procedures between the continuous-time and discrete-time approaches are somewhat different to a certain degree. As we recall, the I-FCS-MPC is designed using a discrete model, and in order to obtain a satisfactory control performance, the sampling interval Δt is chosen to be as small as possible to reduce the variations of the i_d and i_q currents. This, without any doubt, will result in the use of dual sampling rate in the cascade control of velocity and position because the computational load of the single sampling rate becomes too large if both inner-loop and outer-loop deploy the same fast sampling rate. For the outer-loop velocity, the choice of sampling rate will take into consideration two factors: the neglected dynamics from the current control loop and the computational cost without compromising closed-loop performance.

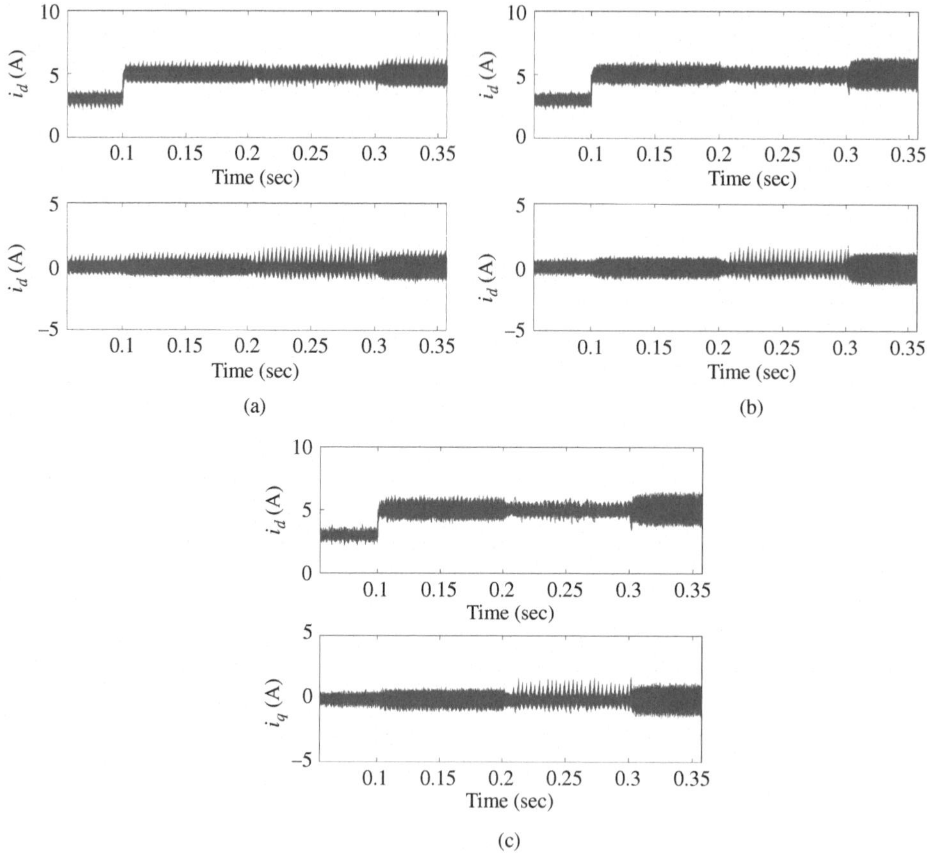

Figure 6.27 Monte-Carlo simulation results of I-FCS-MPC system when L_s changes. Variations of i_d and i_q currents. (a) $L_s < 6.3 \times 10^{-3}(1 - 0.5\epsilon)$ H, (b) $6.3 \times 10^{-3}(1 - 0.5\epsilon)$ H $< L_s < 6.3 \times 10^{-3}(1 + 0.5\epsilon)$ H, and (c) $L_s > 6.3 \times 10^{-3}(1 - 0.5\epsilon)$ H.

6.11.1.1 Neglected Dynamics from I-FCS Current Control

Note that with the I-FCS controlling the currents, the closed-loop pole for the i_d current is $z = p_{cl} = 1 - k_d$ and the i_q current $z = p_{cl} = 1 - k_q$. Because the integral gain k_d and k_q are quite small in the applications, effectively, the closed-loop pole is quite large. For example, assuming the $k_d = k_q = 0.1$, the closed-loop pole for the current control system is $z = p_{cl} = 0.9$. Note that this pole is calculated in relation to the very fast sampling rate used in the current controller. Now, if the outer-loop velocity control used the same fast sampling rate as the inner-loop current control, a problem would arise from neglecting such dynamics from the inner-loop system because of the relatively large closed-loop pole (for instance, 0.9). However, if a much slower sampling rate is used to control the velocity in the outer-loop system, the large closed-loop pole from the current control loop will be converted to the discrete-time pole location using the slower sampling rate, which then becomes relatively small. For example, at the fast sampling rate, the discrete pole is $p_{cl} = e^{-a\Delta t}$ and at the slow sampling rate, the discrete-time pole is $e^{-a\Delta T}$ where $\Delta T = n\Delta t$. With this relationship, the original pole becomes $(p_{cl})^n$ with respect to slow sampling rate

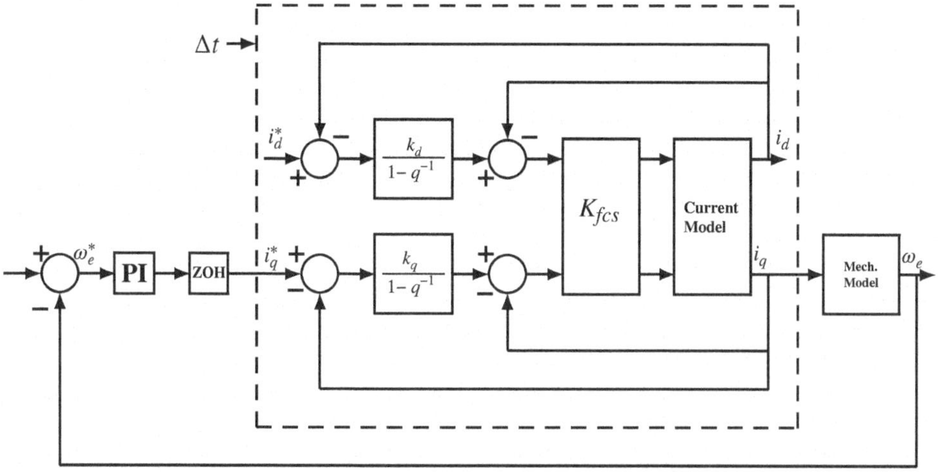

Figure 6.28 I-FCS velocity control of PMSM with dual sampling rate. Fast sampling rate for the dashed box and slow sampling rate for the rest of the system.

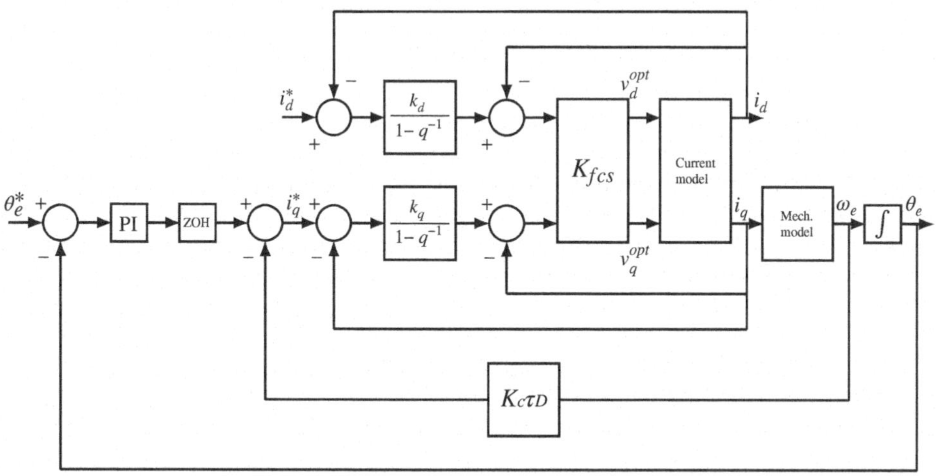

Figure 6.29 I-FCS-MPC position control of PMSM.

(or large ΔT). Supposing that the sampling interval for the outer-loop velocity control is 20 times larger than the inner-loop current control, the closed-loop pole of the current control system becomes $0.9^{20} = 0.1216$, which is quite small.

With respect to the slow sampling rate in the velocity control, the closed-loop z-transfer function between the desired current reference signal i_q^* and the measured current signal i_q is the first order transfer function,

$$\frac{I_q(z)}{I_q^*(z)} = \frac{(1-(p_{cl})^n)z^{-1}}{1-(p_{cl})^n z^{-1}}. \tag{6.87}$$

For a small $(p_{cl})^n$, this transfer function is approximated by a unit delay

$$\frac{I_q(z)}{I_q^*(z)} \approx z^{-1}, \tag{6.88}$$

where the integrator used in the current controller ensures that the steady-state gain. and the magnitude of the frequency response of the transfer function (6.87) at the low frequency region is unity. This leads to the neglected dynamics from the current control system, which may be beneficial from the discrete-time design of the velocity control system.

Another way to examine the neglected dynamics from the current control system is to convert the discrete-time pole p_{cl} into a continuous-time pole. The corresponding continuous-time pole to the discrete-time pole p_{cl} with the fast sampling rate is to calculate the parameter a from the relationship $p_{cl} = e^{-a_{cl}\Delta t}$, which gives

$$a_{cl} = -\frac{\ln(p_{cl})}{\Delta t}.$$

Because of the integrator used in the current control, the steady-state gain of the current control system is unity. Thus, in the continuous-time, the closed-loop transfer function between the desired reference current signal $I_q^*(s)$ and the measured current control signal $I_q(s)$ is described by

$$\frac{I_q(s)}{I_q^*(s)} = \frac{a_{cl}}{s + a_{cl}} = \frac{1}{\frac{1}{a_{cl}}s + 1}. \tag{6.89}$$

For a large value of a_{cl}, the closed-loop transfer function is approximated by

$$\frac{I_q(s)}{I_q^*(s)} \approx 1. \tag{6.90}$$

Again, the dynamics from the inner-loop current control system are neglected if the value of a_{cl} is large, or the sampling interval Δt used in the inner-loop current control is small. When using a continuous-time controller such as a PI controller for velocity control, the sampling rate for the outer-loop system is less restricted than its counter-part when using a discrete-time design technique. Because in the continuous-time case, we have reconstructed the neglected dynamics in the continuous-time domain. Thus, as long as we select the bandwidth of the velocity control system to be much narrower than the bandwidth of the neglected current control system (a_{cl}), robust closed-loop stability and performance can be established (see the analysis presented in Chapter 5).

6.11.1.2 Reducing Computational Cost

In general, a faster sampling rate demands a faster computational speed within one sampling interval, and as a consequence, the computational cost increases as the sampling interval Δt reduces, because of the requirement for more expensive digital signal processors. When the I-FCS predictive controllers are used for the current control, a fast sampling rate is necessary to guarantee a satisfactory performance, which will increase the computational cost. However, the simplified implementation without PWMs may lead to computational savings. A further study is needed to clarify whether there is a saving when using the I-FCS controllers.

Because the dynamics from the mechanical system of the motor (mainly its inertia), are much slower than the dynamics in the current control system, suitably increasing the sampling interval Δt (or reducing the sampling rate) for the outer-loop control system will not significantly affect the closed-loop control performance.

Typically, the sampling rate $\frac{1}{\Delta t}$ is chosen to be 5 to 10 times the closed-loop bandwidth. In the PI velocity control system, this bandwidth is the parameter w_n. In reality, this range of sampling rates will result in more or less performance degradation from the desired closed-loop performance we aimed at. To make the sampling effect truly negligible, the sampling rate should be 20 times the closed-loop bandwidth.

6.11.2 Velocity and Position Controller Design

Since the inner-loop current control systems are designed and implemented using a discrete-time model, the option becomes available for the outer-loop velocity controller and position controller to be designed using a discrete-time model. The advantages of using a discrete-time model for the design are that the sampling rate is selected at the design stage and there are no approximation errors at the implementation stage. The disadvantages of discrete-time design include the lack of flexibility with the sampling rate when implementing the controller and the controller parameters are not directly related to the physical parameters of the system. The discrete-time controller design techniques can be found in Goodwin *et al.* (2001). We present our design the continuous-time domain in the next section.

6.11.2.1 Design of PI Velocity Controller

The design of PI velocity controller entirely follows the design procedure outlined in Section 3.4.4 of Chapter 3. We briefly list the design equations for convenience.

For the PMSM, substituting the Laplace transform of the measured current $I_q(s)$ by the the reference current $I_q^*(s)$ based on the approximation of the inner-loop current control system (see (6.90)), we obtain the design model for controlling the velocity,

$$\frac{\Omega_e(s)}{I_q^*(s)} \approx \frac{\frac{3}{2}\frac{Z_p^2\phi_{mg}}{J_m}}{s+\frac{B_v}{J_m}}. \tag{6.91}$$

Here, we let

$$a=\frac{B_v}{J_m};\ b=\frac{3}{2}\frac{Z_p^2\phi_{mg}}{J_m}.$$

By choosing a pair of desired closed-loop poles $s_{1,2}=-\xi w_n \pm jw_n\sqrt{1-\xi^2}$, where the damping coefficient ξ is chosen to be 0.707, the proportional gain K_c is calculated as

$$K_c=\frac{2\xi w_n - a}{b} \tag{6.92}$$

and the integral time constant is calculated as

$$\tau_I=\frac{2\xi w_n - a}{w_n^2}. \tag{6.93}$$

The specification of the closed-loop bandwidth for the velocity controller can also be selected in a relationship to the open-loop pole a (see Chapter 5) as

$$w_n=\frac{1}{1-\gamma}a, \tag{6.94}$$

where γ is preferably chosen between 0.8 and 0.95.

6.11.2.2 Design of PID Position Controller

The design of a PID position controller follows from the procedure listed in Section 3.5 of Chapter 3. Begin from

$$\frac{\Omega_e(s)}{I_q^*(s)} \approx \frac{\frac{3}{2}\frac{Z_p^2\phi_{mg}}{J_m}}{s\left(s+\frac{B_v}{J_m}\right)}. \tag{6.95}$$

By defining the model parameters a and b as

$$a = \frac{B}{J_m};\ b = \frac{3}{2}\frac{Z_p^2\phi_{mg}}{J_m},$$

with the pole-assignment controller design, the PID controller parameters are calculated using the following equations,

$$K_c = \frac{(2\xi n+1)w_n^2}{b} \tag{6.96}$$

$$\tau_I = \frac{(2\xi n+1)w_n^2}{nw_n^3} = \frac{(2\xi n+1)}{nw_n} \tag{6.97}$$

$$\tau_D = \frac{(2\xi+n)w_n - a}{(2\xi n+1)w_n^2}, \tag{6.98}$$

where the parameters $n \geq 1$ and w_n are used in the specification of desired closed-loop polynomial $A_{cl}^d(s)$,

$$A_{cl}^d(s) = (s^2 + 2\xi w_n s + w_n^2)(s + n \times w_n).$$

The derivative term is implemented with a first order filter that has a time constant of $\beta\tau_D$ (see Figure 6.29).

In the selection of w_n for the outer-loop position control, the parameter a in the model could be used as a benchmark parameter, for example, $w_n = \frac{1}{1-\gamma}a$. However, the open-loop system contains an integrator that is the dominant pole of the system and thus this benchmark parameter may not be as useful for the velocity control case. Instead, one could consider the bandwidth of the outer-loop position control in relation to the inner-loop current control system based on the performance analysis presented in Chapter 5. For instance, the I-FCS-MPC current control leads to the closed-loop pole at $-a_{cl}$ where a_{cl} is calculated using the discrete-time closed-loop pole p_{cl},

$$a_{cl} = -\frac{\ln(p_{cl})}{\Delta t}.$$

Then, the closed-loop bandwidth parameter w_n for the position control could be selected as a fraction of the parameter a_{cl}, say $w_n = 0.1a_{cl}$, or using the ratio of the sampling rates in the relationship:

$$w_n = \epsilon\frac{\Delta t}{\Delta T}a_{cl}, \tag{6.99}$$

where Δt is the sampling interval for the inner-loop current control and ΔT is the sampling interval for the outer-loop position control, and $0 < \epsilon < 1$ is a tuning parameter. However, because the parameter a_{cl} is rather large in the I-FCS-MPC current control system, a small parameter ϵ may be needed.

6.12 Velocity and Position Control of Induction Motor Using I-FCS-MPC

In this section, we will apply the I-FCS cascade control systems discussed in Section 6.11 to control the velocity and position of an induction motor.

6.12.1 I-FCS Cascade Velocity Control of Induction Motor

To demonstrate PI velocity control algorithm using the I-FCS-MPC inner-loop current control, two simulation examples with different inner-loop sampling intervals are presented in this section. The outer-loop PI controller called PIV.mdl programmed in Tutorial 1 is used in the simulation.

The induction motor model that describes the relationship between i_{sq} and the velocity $\omega_m(t)$ is given by (3.107) in Chapter 3. Its linearized form is expressed as

$$\begin{aligned}\frac{d\omega_m(t)}{dt} &= -\frac{f_d}{J_m}\omega_m(t) + \frac{3Z_pL_h}{2L_rJ_m}\psi_{rd}^{ss}i_{sq}(t) + \frac{3Z_pL_h}{2L_rJ_m}i_{sq}^{ss}\psi_{rd}(t) - \frac{3Z_pL_h}{2L_rJ_m}i_{sq}^{ss}\psi_{rd}^{ss} - \frac{T_L}{J_m}\\ &= -a\omega_m(t) + bi_{sq}(t) + d(t),\end{aligned} \tag{6.100}$$

where $d(t)$ represents the disturbances including the effects of $\psi_{rd}(t)$ and the load torque T_L as well as the steady-state parameters.

In the simulation, the parameters for the PI controller design are $a = 0.4423$ and $b = 257.7164$. Choose the damping coefficient $\xi = 0.707$ and the bandwidth $w_n = \frac{1}{1-\gamma}a = 8.846$, where $\gamma = 0.95$. Thus, the bandwidth of the velocity control loop is 20 times the plant bandwidth.

To determine whether the choice of bandwidth for the outer-loop system is appropriate with regard to unmodeled dynamics from the inner-loop system, the bandwidth of the current control system needs to be examined. The inner-loop current control with the integral gain being 0.2 has the discrete-time closed-loop pole located at 0.8. With the larger sampling interval $\Delta t = 80 \times 10^{-6}$ sec, the parameter that determines the continuous-time pole location is computed as

$$a_{cl} = -\frac{\ln(p_{cl})}{\Delta t} = -\frac{\ln(0.8)}{80 \times 10^{-6}} = 2789. \tag{6.101}$$

The current control loop in an ideal situation, where there was no constraint on the voltage, would have its closed-loop pole located at $s = -2789$. Thus, the selection of the velocity control bandwidth is far less than the current control bandwidth so that the inner-loop dynamics can be safely neglected in the design.

With these choices of desired closed-loop performance parameters, the parameters for the PI controller are calculated as

$$K_c = \frac{2\xi w_n - a}{b} = 0.0468 \tag{6.102}$$

and

$$\tau_I = \frac{2\xi w_n - a}{w_n^2} = 0.1542. \tag{6.103}$$

In the implementation of the velocity controller, the sampling interval is selected as $\Delta T = 400 \times 10^{-6}$ sec and the constraints on the i_{sq} are imposed in the implementation with

$$-3\ \text{A} \leq i_{sq} \leq 3\ \text{A}.$$

The reference signal for i_{sd} is selected as 0.8772 A to reflect the operating condition of the induction motor. A constant torque disturbance $T_L = 1$ Nm is added to the simulation.

Since the sampling interval is adjustable and is the performance parameter in the I-FCS-MPC current controller, two sampling intervals $\Delta t = 80 \times 10^{-6}$ sec and 50×10^{-6} sec are used, respectively, in the simulation studies. Figures 6.30(a)–(c) illustrate the closed-loop control results for sampling interval $\Delta t = 80 \times 10^{-6}$ sec in the inner-loop current control. In comparison, Figures 6.31(a)–(c) show the closed-loop control results for sampling interval $\Delta t = 50 \times 10^{-6}$ sec. By comparing Figure 6.30(a) with Figure 6.31(a), it is seen that in both cases, the current controllers track the reference currents without steady-state errors; however, the current variations in both i_{sd} and i_{sq} are larger when the sampling interval Δt is larger. Also, the current variations slightly increased when the output velocity took a step change from 700 RPM to 1400 RPM (see Figure 6.30(a)–(b)). By contrast, when the sampling interval is

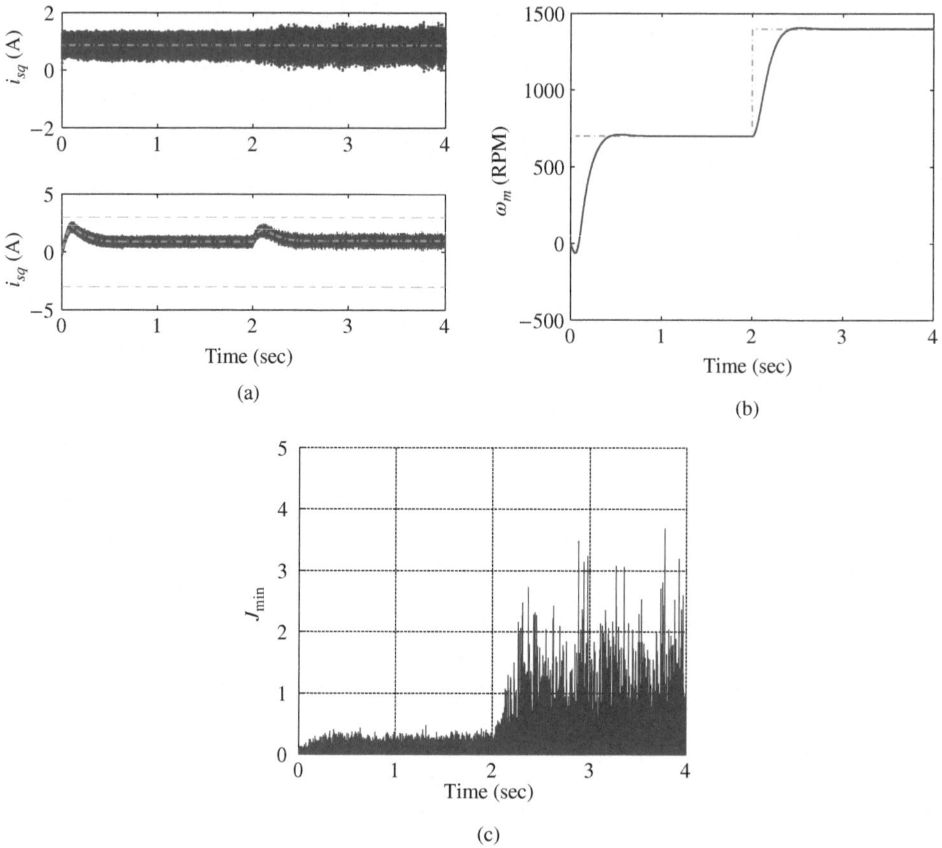

Figure 6.30 Simulation results for I-FCS-MPC of induction motor. Case A. Sampling interval $\Delta t = 80 \times 10^{-6}$ sec, $k_d = k_q = 0.2$. (a) i_{sd} and i_{sq} currents, (b) Velocity. Dashed-line: ω_m^*, solid-line: ω_m, and (c) Minimum of the objective function.

smaller, the current variations were much less affected when the step change in output velocity occurred (see Figure 6.31(a)–(b)). This discrepancy in variations can also be noticed from the magnitudes of the minimum of the objective functions (see Figures 6.30(c) and 6.31(c)) in which it is clearly seen that the magnitude of J_{min} has increased more significantly when Δt is larger. This indicates that if the *AC* motor is to operate at a high speed, a faster sampling rate is required to reduce the variations of the currents.

It is interesting to note that with the two different sampling rates for the inner-loop current control, despite the differences in current variations, the output velocity is almost identical to each other as seen from Figures 6.30(b) and 6.31(b). This means that the dynamics of the current control systems are so fast that they make little difference to the velocity control system.

6.12.2 I-FCS-MPC Cascade Position Control of Induction Motor

In the I-FCS-MPC cascade position control of an induction motor, the outer-loop position controller is a PID controller. The design of the PID controller follows the same procedure used for position controlling a PMSM that was discussed in Section 6.11.2. Therefore, the algorithm is not repeated here.

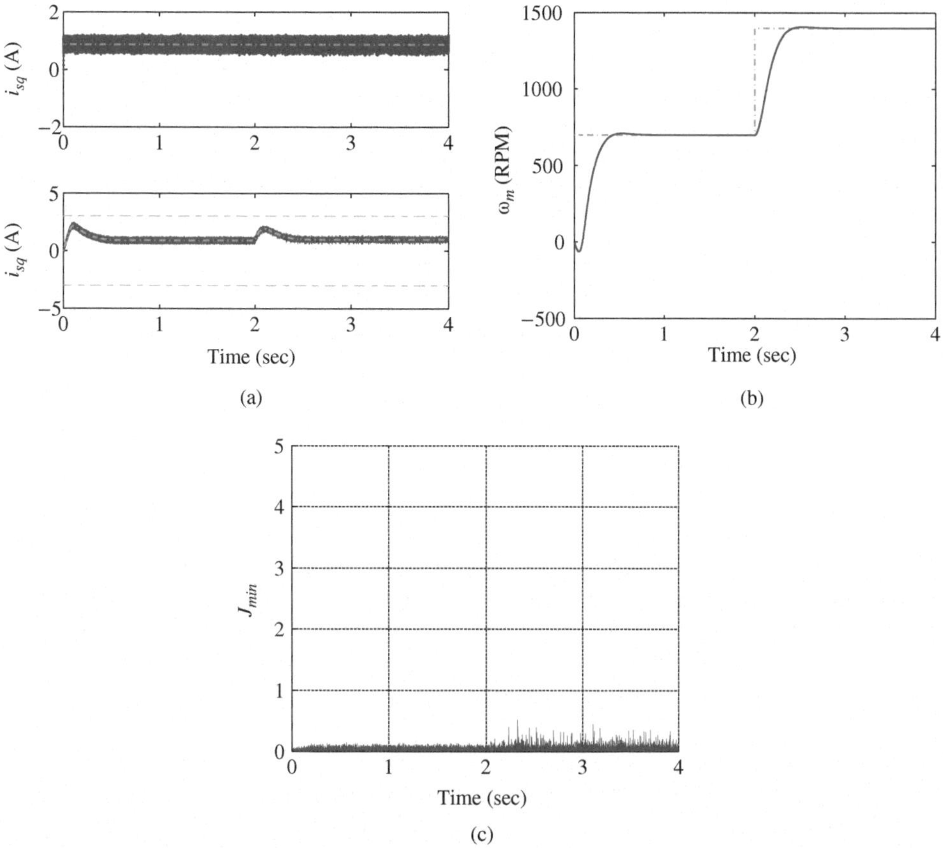

Figure 6.31 Simulation results for I-FCS-MPC of induction motor. Case B. Sampling interval $\Delta t = 50 \times 10^{-6}$ sec, $k_d = k_q = 0.2$. (a) i_{sd} and i_{sq} currents, (b) Velocity. Dashed-line: ω_m^*, solid-line: ω_m, and (c) Minimum of the objective function.

Similar to the position control of a PMSM, the mathematical model used in the design of a PID controller for an induction motor is given as

$$\frac{\Theta_r(s)}{I_{sq}^*(s)} = \frac{b}{s(s+a)}, \tag{6.104}$$

where $\Theta_r(s)$ and $I_{sq}^*(s)$ are the Laplace transforms of the rotor's angular position and the reference signal for the stator q-axis current. Here, the inner-loop current dynamics are approximated a unity gain because the I-FCS-MPC system has a unity gain and a relatively small time constant (see the discussions in Section 6.11).

In the simulation studies, the model parameters are identical to those used in velocity control where $a = 0.4423$ and $b = 257.7164$ (see the previous simulation). The sampling interval for the inner-loop current control is chosen to be $\Delta t = 80 \times 10^{-6}$ sec and $\Delta t = 50 \times 10^{-6}$ sec, respectively in order to study the effect of sampling interval of current control on the closed-loop performance. The integral gain for the current control is chosen as $k_d = k_q = 0.1$. For the outer-loop position control, the sampling interval ΔT is selected as 400×10^{-6} sec. for both cases.

The PID controller parameters are calculated using (6.96)–(6.98) where the bandwidth parameter w_n is the performance tuning parameter. Because the model for position control contains an integrator, the parameter w_n is often chosen in relation to bandwidth of the inner-loop control system, which is, in this case, $a_{cl} = 2789$ (see (6.101)). On the first attempt to design of a PID position controller, the bandwidth parameter w_n is selected as $w_n = \epsilon \frac{\Delta t}{\Delta T} a_{cl} = 0.125 \times 2789 = 348$ rad/s ($\epsilon = 0.625$) (see (6.99)). The closed-loop system simulation results indicate that this choice of closed-loop bandwidth leads to an unstable system. Clearly, this bandwidth parameter w_n is too large because of the very large a_{cl} parameter from the current control system. This indicates that the parameter ϵ in (6.99) should be very small for the large a_{cl} parameter. On the second attempt, the bandwidth w_n is selected in relation to the dynamics of the velocity response, namely the a parameter in the model, which gives a much better choice for w_n. By choosing the damping coefficient $\xi = 0.707$ and the bandwidth $w_n = \frac{1}{1-\gamma} a = 22$ rad/s, where $\gamma = 0.98$, and $n = 1$, the parameters for the PID position controller are calculated as

$$K_c = \frac{(2\xi n + 1)w_n^2}{b} = 4.5813 \tag{6.105}$$

$$\tau_I = \frac{(2\xi n + 1)}{n w_n} = 0.1092 \tag{6.106}$$

$$\tau_D = \frac{(2\xi + n)w_n - a}{(2\xi n + 1)w_n^2} = 0.0448. \tag{6.107}$$

The derivative term is implemented with a first order filter that has a time constant of $\beta \tau_D$ with $\beta = 0.1$(see Figure 6.29). Note that the w_n parameter used in the second attempt is 15 times smaller than the one used in the first attempt. In the implementation of the position controller, the constraints on the i_{sq} are imposed via the constraints on the current reference signal i_{sq}^* with

$$-3 \text{ A} \leq i_{sq}^* \leq 3 \text{ A}.$$

The reference signal for i_{sd} is selected as 0.8772 A to reflect the operating condition of the induction motor. A constant torque disturbance $T_L = 1$ is added to the simulation.

The reference signal to the angular position is a step signal that simulates the situation of stopping the motor to a given position. Figures 6.32–6.33 show the closed-loop control system responses for the two cases where the sampling interval for the I-FCS current control is $\Delta t = 80 \times 10^{-6}$ sec and $\Delta t = 50 \times 10^{-6}$ sec, respectively. By comparing these two figures, it is seen that the larger sampling interval used in the I-FCS-MPC system leads to larger current variations, however, there is little difference in the position responses.

6.12.3 Experimental Evaluation of Velocity Control

The cascaded PI velocity control of induction motor using I-FCS current controller for the inner-loop system is evaluated experimentally. The sampling interval for the I-FCS current controller is selected as 80×10^{-6} sec, and the integral gain is selected as $k_d = k_q = 0.15$. The closed-loop current control system has a discrete-pole located at 0.85 for both d-axis and q-axis currents. This inner-loop q-axis current control system is approximated in discrete-time using the transfer function

$$\frac{I_{sq}(z)}{I_{sq}^*(z)} = \frac{0.15z^{-1}}{1 - 0.85z^{-1}}, \tag{6.108}$$

which has a unit steady-state gain and discrete-time pole at $z = 0.85$. With sampling interval of 80×10^{-6} second, this pole can also be converted to its continuous-time counterpart equivalent to $s = -2031$. This gives the closed-loop time constant for the q-axis stator current control as approximately 4.92×10^{-4}.

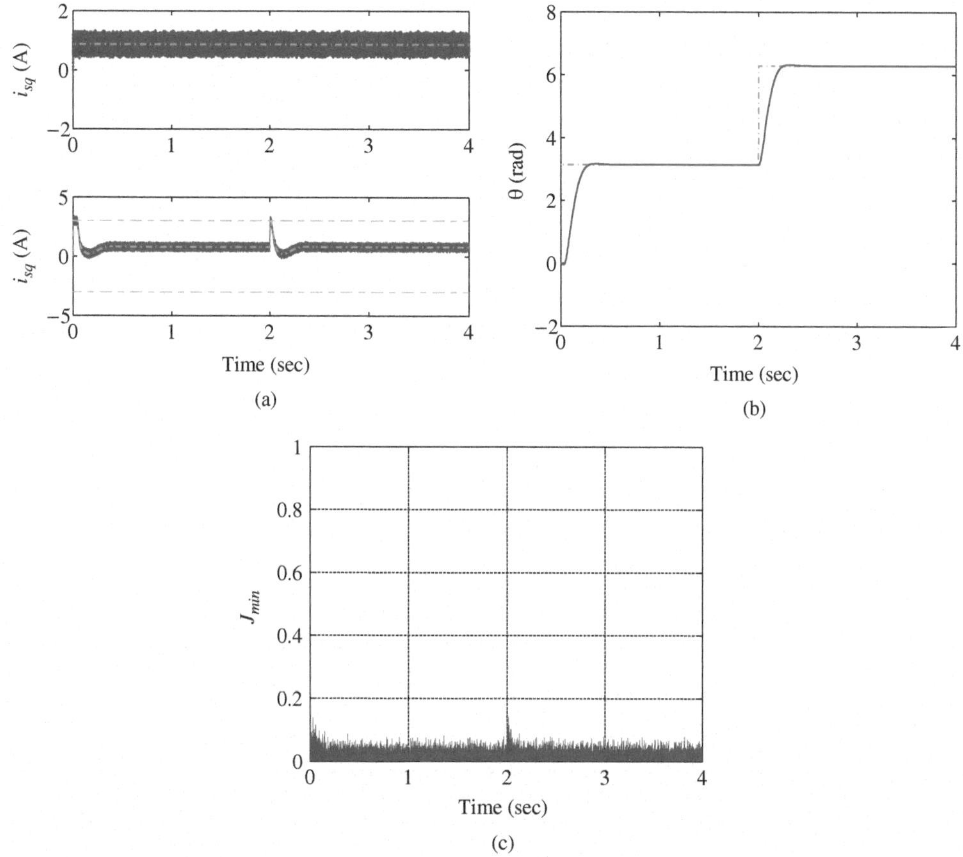

Figure 6.32 Induction motor position control, Case A. Sampling interval $\Delta t = 80 \times 10^{-6}$ sec, $k_d = k_q = 0.1$. Bandwidth for the position control system is $w_n = 22$ rad/s. (a) i_{sd} and i_{sq} currents, (b) Angular position (rad), and (c) Minimum of the objective function.

In the design of the outer-loop velocity PI controller, sampling interval is selected as 400×10^{-6} second. The bandwidth tuning parameter $\gamma = 0.981$ is selected for a fast velocity response, yielding $w_n = \frac{1}{1-\gamma}a = 23.28$ rad/s where $a = 0.4423$. Comparing the bandwidths of both the inner-loop current control system (approximately 2031 rad/s) and the velocity control system (23.28 rad/s), which is about 87 times wider, it is concluded that the fast dynamics from the inner-loop current control system can be safely neglected. With the parameter $b = 257.7164$ and $\xi = 0.707$, the PI controller parameters used in the experiments are

$$K_c = \frac{2\xi w_n - a}{b} = 0.126 \ \tau_I = \frac{2\xi w_n - a}{w_n^2} = 0.0599.$$

In the experiments, constraints are imposed on the i_{sq} current to ensure safe operation, where $|i_{sq}| \leq 3$ A. These constraints are implemented using the PI controller with anti-windup mechanism. To avoid overshoot in the velocity response, the proportional control is implemented on the output only (see Section 5.1). Two experiments are conducted with two different sets of velocity reference signals and the corresponding results are discussed below.

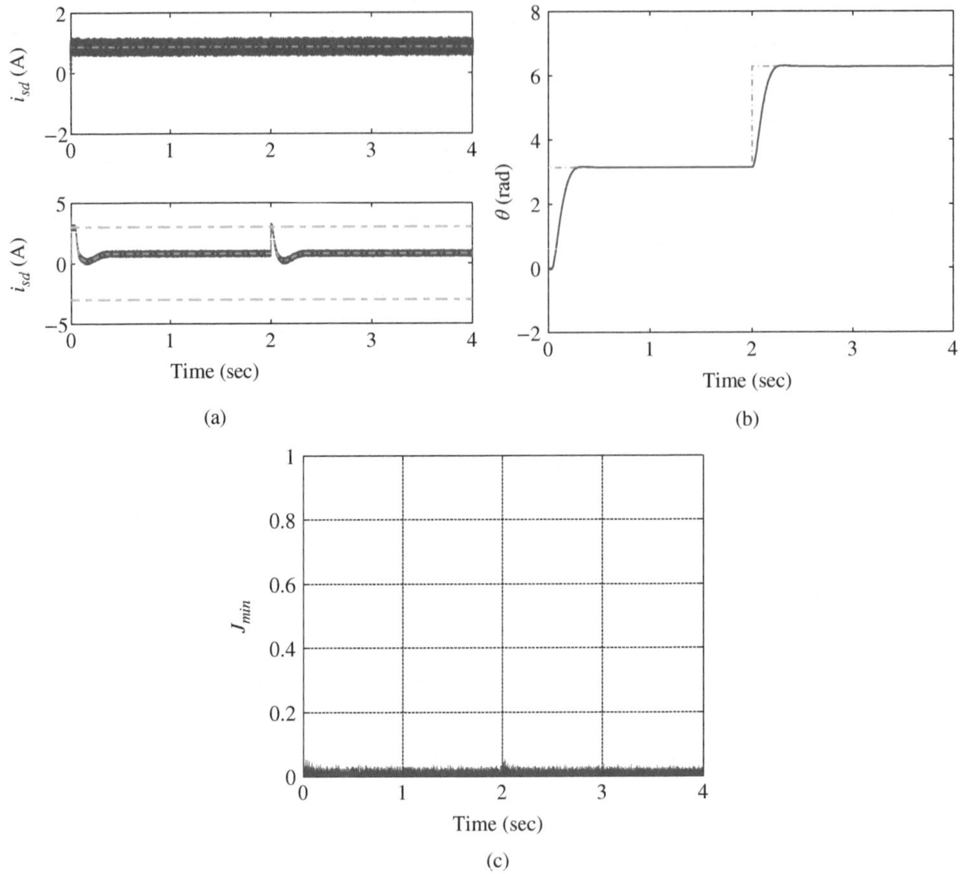

Figure 6.33 Induction motor position control, Case B. Sampling interval $\Delta t = 50 \times 10^{-6}$ sec, $k_d = k_q = 0.1$. Bandwidth for the position control system is $w_n = 22$ rad/s. (a) i_{sd} and i_{sq} currents, (b) Angular position (rad), and (c) Minimum of the objective function.

6.12.3.1 Step Reference Signal

Figure 6.34(a) shows the induction motor velocity control results with a series of step signals which illustrate the behavior of the velocity control system over different operating conditions. The induction motor starts from zero speed followed by a step change to 700 RPM and to 1400 RPM. Then the operation reverses following a series of step down to −1400 RPM. From Figure 6.34(a), it is seen that despite the existing nonlinearity in the induction motor, the velocity response has no oscillation and no overshoot over the entire operating regions. Also the velocity response speed is about the same over the entire operating region. This means that with the I-FCS predictive current control, the nonlinearities existing in the induction motor are overcome and it behaves like a linear velocity control system. From Figure 6.34(b), it is seen that there is no steady-state error for the velocity control system over the entire operating region; also the transient behavior of the error signal has similar characteristics over the entire operating region, which further confirms the characteristics of linear velocity control when the inner-loop current control uses the I-FCS predictive controller. There is also no overshoot in the velocity response. Figure 6.34(c)

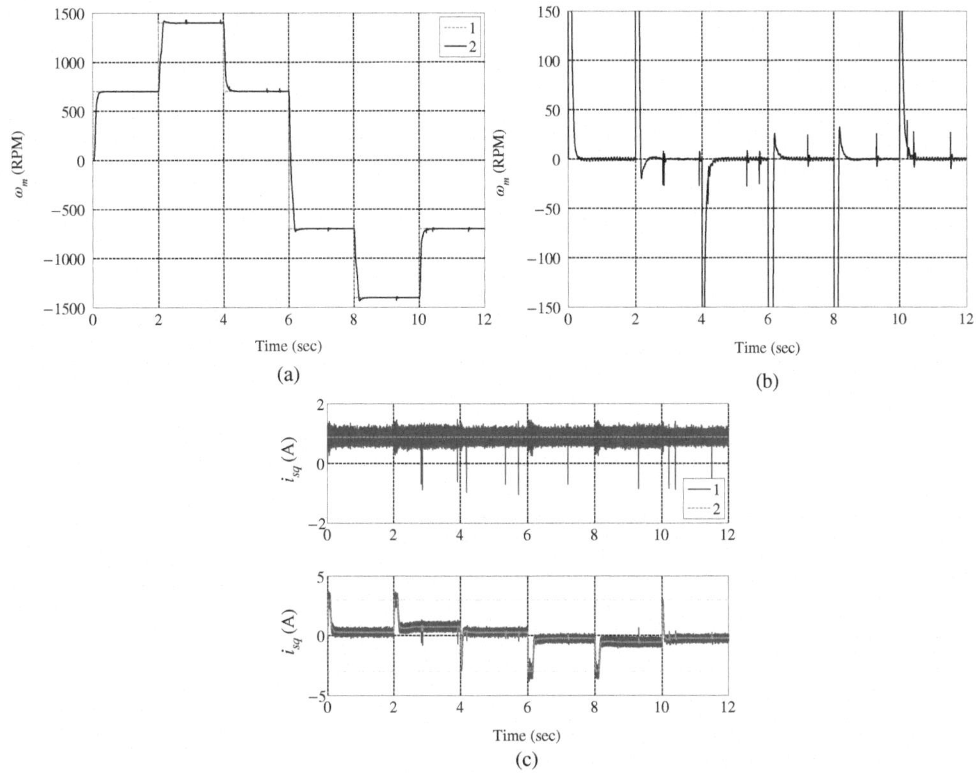

Figure 6.34 Experimental results for induction motor speed control using cascade I-FCS controller with step reference. Key: line (1): actual measurement; line (2): set-point signal. Sampling interval $\Delta t = 80 \times 10^{-6}$ sec, $k_d = k_q = 0.15$. (a) Velocity response, (b) Velocity error, and (c) i_{sd} and i_{sq} currents.

illustrates the responses of i_{sd} and i_{sq} currents. It is seen from this figure that the i_{sd} current maintains the steady-state operation and there are small perturbations when the step changes in velocity occur. Current i_{sq} closely follows the current reference signal i_{sq}^* which is the control signal from the outer-loop velocity PI controller. It is seen that the constraints on i_{sq}^* become active during the step changes, which means that the maximum and minimum currents are deployed to achieve the closed-loop response.

6.12.3.2 Ramp Reference Signal

This set of test signals is intended to demonstrate the acceleration, de-acceleration, braking and zero speed operations. Figure 6.35(a) shows the closed-loop velocity response that follows a ramp reference signal for acceleration, steady-state operation, de-acceleration, braking and zero speed operations. It is seen from this figure that the velocity response is smooth without oscillations. The error signal given in Figure 6.35(b) indicates that there is steady-state error for the ramp reference signal, but no steady-state error at the steady-state operation at 1400 RPM and at the zero speed operation. Because there is only one integrator in the PI velocity controller, the steady-state error for the ramp reference signal is expected. Figure 6.35(c) shows the i_{sd} and i_{sq} current responses. It is seen that the i_{sd} current maintains the steady-state operation and the i_{sq} current follows the reference current signal i_{sq}^*. It is interesting to

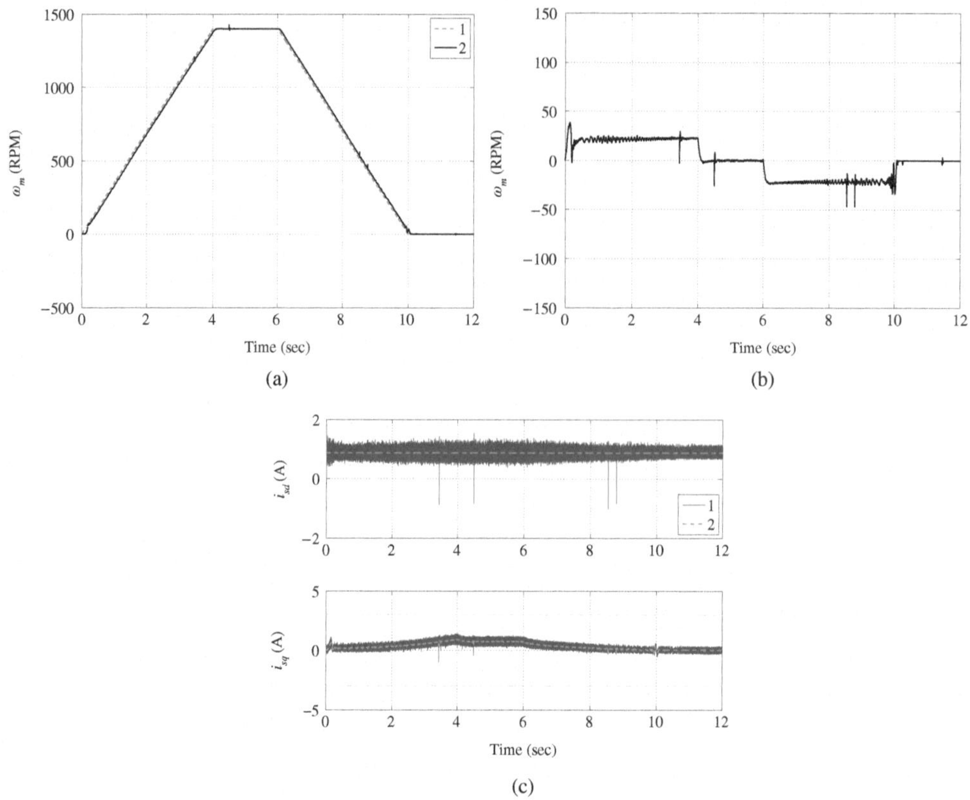

Figure 6.35 Experimental results for induction motor speed control using cascade I-FCS controller with ramp reference. Key: line (1): actual measurement; line (2): set-point signal. Sampling interval $\Delta t = 80 \times 10^{-6}$ sec, $k_d = k_q = 0.15$. (a) Velocity response, (b) Velocity error, and (c) i_{sd} and i_{sq} currents.

note that with this set of velocity reference signals, the constraints on the i_{sq} currents are not violated, that are not active. Because of the ramp velocity reference signals, the velocity changes slowly and it does not require a large current amplitude in the transient response.

6.13 Summary

This chapter proposed the I-FCS predictive current control system as well as the velocity, position and voltage control systems that use the I-FCS predictive controllers in a cascade feedback structure.

The traditional FCS predictive current controller uses an objective function with sum of absolute errors for $d - q$ currents. This kind of objective functions is conceptually simple, however, it does not readily yield an analytical optimal solution, hence hinders further analysis. Instead, this chapter has seen that with a sum of squared errors, the objective function for the traditional FCS predictive current controller has a simple analytical solution, leading to a closed-loop feedback control system that is equivalent to an optimal deadbeat control system. It is shown that without the limitations imposed on the control signals, this closed-loop feedback discrete-time control system has its eigenvalue at the origin of the complex plane.

With the limitations imposed on the control signals where at any given sampling instant t_i, there are only seven pairs of candidate control signal values available for the implementation. The proposed FCS predictive control scheme then selects the pair of control signal values that minimize the objective function measured by the distances from the optimal control signals and those candidate variables that are available for implementation. The error signals between the optimal control signals and the implemented control signals act as an input disturbance to the current control system, which results in variations of the current control signals as observed in the measured outputs.

The original FCS predictive controllers do not contain integrators in the design. There are drawbacks associated with the control systems without integral actions. One of the key problems is the steady-state performance of the current control systems that is not guaranteed error free in its mean value. Furthermore, the steady-state errors may vary with the physical parameters of the current system, leading to performance uncertainty. By considering the dynamics of the original FCS-MPC system as one sample delayed system, an integral control term is added to the FCS predictive controller via a cascade structure. Here, the choice of integral term is based on the selection of the closed-loop pole for the current control system in a simple fashion where the integral gain $k_d = 1 - p_{cl}$, and p_{cl} is the desired discrete-time closed-loop pole. For the implementation of the I-FCS predictive current controller, it is proposed to use incremental variables so as to avoid the use of steady-state values and to simplify the implementation.

When velocity, position and voltage control systems are needed in the applications, the cascaded control systems follow the identical procedures discussed in the previous Chapters 3–5 where the inner-loop current control is performed by the I-FCS predictive controller. However, because the I-FCS predictive controllers are designed and implemented in discrete-time, the inner-loop closed-loop time constant needs to be converted into its continuous-time equivalent form so as to quantify the unmodeled dynamics from the current control system. In addition, the sampling interval for the velocity, position and voltage control should be much larger in the cascade structure to reduce computational load.

The design parameters for the I-FCS predictive control systems are the sampling interval Δt and the location of the closed-loop pole of the current control system. The selections of these parameters are discussed below.

6.13.1 Selection of sampling interval Δt

The closed-loop performance of the current control system is dependent on the choice of sampling interval Δt. In general, the smaller Δt is, the smaller variations of the currents i_d and i_q will be. On the other hand, a smaller sampling interval Δt will lead to a larger computational load and more power losses due to the faster switching of the voltage inverters. A compromise should be reached between these factors. The size of Δt required to achieve satisfactory performance is relative to the value of inductance, as the ratios, $\frac{\Delta t}{L_d}$ and $\frac{\Delta t}{L_q}$, appear as the dominant factors in the proportional gains and as the weights in the objective functions. These quantities indicate that for a larger inductance, a larger Δt may be used. One way to determine the appropriate value of Δt to achieve satisfactory performance and to increase the economic benefit of the control system via reduced computational cost and power losses is to evaluate the variance of the current signals against the reduction of sampling interval Δt. One may also adapt the sampling interval Δt in real time to achieve the balance.

6.13.2 Selection of the Integral Gain

The selection of integral gain is straightforward. As the integral gain is a complement of the desired closed-loop pole ($k_d = 1 - p_d$), the larger the integral gain, the faster the closed-loop current response will be. Because the current control system is in a fast sampling environment, the integral gain is recommended to be between 0.05 to 0.2.

6.14 Further Reading

The model predictive torque control was proposed in Geyer *et al.* (2009) and implemented in Papafotiou *et al.* (2009). FCS-MPC method was used in controlling PMSM drive systems by Preindl and Bolognani (2013a), Preindl and Bolognani (2013b) and Chai and Wang (2012). Dead-beat control design technique was discussed in Kukrer (1996). Time-delay effects were examined in the application of FCS-MPC in Cortes *et al.* (2012). Steady-state performance of finite control set method was discussed in Lezana *et al.* (2009) and in Aguilera *et al.* (2013). Finite control set for active front end power converter was presented in Cortes *et al.* (2008b) and Kouro *et al.* (2009). Predictive current control of a voltage source inverter was proposed in Rodriguez *et al.* (2007) and extended to electric drives and power converters in Cortes *et al.* (2008a). Predictive torque control of induction machines was discussed in Nemec *et al.* (2007) and Correa *et al.* (2007). A comparative study was presented for current control of PMSM drives in Morel *et al.* (2009). A book was written based on FCS-MPC system (see Rodriguez and Cortes (2012)). In Goodwin *et al.* (2005), an approach to solve the problem caused by a limited number of candidate variables is proposed using model predictive control.

Dead-beat control techniques were discussed in Gokhale *et al.* (1987), Kawabata *et al.* (1990), Malesani *et al.* (1999), Mohamed and El-Saadany (2007) and Yang and Lee (2002). Geyer (2011) made comparative studies of control and modulation schemes for medium-voltage drives between predictive control concepts and PWM-based schemes.

References

Aguilera R, Lezana P and Quevedo D 2013 Finite-control-set model predictive control with improved steady-state performance. *IEEE Transactions on Industrial Informatics* **9**(2), 658–667.

Chai S and Wang L 2012 Finite control set model predictive control of 2LVSI-PMSM using interpolated switching states *IECON 2012 – 38th Annual Conference on IEEE Industrial Electronics Society*, pp. 1799–1804.

Correa P, Pacas M and Rodriguez J 2007 Predictive torque control for inverter-fed induction machines. *IEEE Transactions on Industrial Electronics* **54**, 1073–1079.

Cortes P, Kazmierkowski M, Kennel R, Quevedo D and Rodriguez J 2008a Predictive control in power electronics and drives. *IEEE Transactions on Industrial Electronics* **55**(12), 4312–4324.

Cortes P, Rodriguez J, Antoniewicz P and Kazmierkowski M 2008b Direct power control of an AFE using predictive control. *IEEE Transactions on Power Electronics* **23**(5), 2516–2523.

Cortes P, Rodriguez J, Silva C and Flores A 2012 Delay compensation in model predictive current control of a three-phase inverter. *IEEE Transactions on Industrial Electronics* **59**(2), 1323–1325.

Desoer CA 1969 Slowly varying system $\dot{x} = a(t)x$. *IEEE Transactions on Automatic Control* **14**, 780–781.

Geyer T 2011 A comparison of control and modulation schemes for medium-voltage drives: Emerging predictive control concepts versus PWM-based schemes. *IEEE Transactions on Industry Applications* **47**(3), 1380–1389.

Geyer T, Papafotiou G and Morari M 2009 Model predictive direct torque control part I: Concept, algorithm, and analysis. *IEEE Transactions on Industrial Electronics* **56**(6), 1894–1905.

Gokhale K, Kawamura A and Hoft R 1987 Dead beat microprocessor control of PWM inverter for sinusoidal output waveform synthesis. *IEEE Transactions on Industrial Applications* **IA-23**(5), 901–910.

Goodwin GC, Graebe SF and Salgado ME 2001 *Control System Design*. Prentice Hall, New Jersey.

Goodwin GC, Seron MM and De Dona JA 2005 *Constrained Control and Estimation: An Optimisation Approach*. Springer, London.

Kawabata T, Miyashita T and Yamamoto Y 1990 Dead beat control of three phase PWM inverter. *IEEE Transactions on Power Electronics* **5**(1), 21–28.

Kouro S, Cortes P, Vargas R, Ammann U and Rodriguez J 2009 Model predictive control–a simple and powerful method to control power converters. *IEEE Transactions on Industrial Electronics* **56**(6), 1826–1838.

Kukrer O 1996 Deadbeat control of a three-phase inverter with an output LC filter. *IEEE Transactions on Power Electronics* **11**(1), 16–23.

Lezana P, Aguilera R and Quevedo D 2009 Steady-state issues with finite control set model predictive control. *35th Annual Conference of IEEE Industrial Electronics* pp. 1776–1781.

Malesani L, Mattavelli P and Buso S 1999 Robust dead-beat current control for PWM rectifiers and active filters. *IEEE Transactions on Industry Applications* **35**(3), 613–620.

Mohamed YR and El-Saadany E 2007 An improved deadbeat current control scheme with a novel adaptive self-tuning load model for a three-phase PWM voltage-source inverter. *IEEE Transactions on Power Electronics* **54**(2), 747–759.

Morel F, Lin-Shi X, Retif JM, Allard B and Buttay C 2009 A comparative study of predictive current control schemes for a permanent-magnet synchronous machine drive. *IEEE Transactions on Industrial Electronics* **56**(7), 2715–2728.

Nemec M, Nedeljkovic D and Ambrozic V 2007 Predictive torque control of induction machines using immediate flux control. *IEEE Transactions on Industrial Electronics* **54**(4), 2009–2017.

Papafotiou G, Kley J, Papadopoulos K, Bohren P and Morari M 2009 Model predictive direct torque control part II: Implementation and experimental evaluation. *IEEE Transactions on Industrial Electronics* **56**(6), 1906–1915.

Preindl M and Bolognani S 2013a Model predictive direct speed control with finite control set of PMSM drive systems. *IEEE Transactions on Power Electronics* **28**(2), 1007–1015.

Preindl M and Bolognani S 2013b Model predictive direct torque control with finite control set for PMSM drive systems, part 2: Field weakening operation. *IEEE Transactions on Industrial Informatics* **9**(2), 648–657.

Rodriguez J and Cortes P 2012 *Predictive Control of Power Converters and Electrical Drives*. John Wiley & Sons, Inc., New York.

Rodriguez J, Pontt J, Silva C, Correa P, Lezana P, Cortes P and Ammann U 2007 Predictive current control of a voltage source inverter. *IEEE Transactions on Industrial Electronics* **54**(1), 495–503.

Yang S and Lee C 2002 A deadbeat current controller for field oriented induction motor drives. *IEEE Transactions on Power Electronics* **17**(5), 772–778.

7

FCS Predictive Control in $\alpha - \beta$ Reference Frame

In the previous chapter, the FCS current controllers in the $d - q$ reference frame have been designed for PMSM, induction motor and a two-level grid connected power converter. In the $d - q$ frame, the current reference signals are constant or piece-wise constant. Therefore, proportional plus integral controllers derived in the form of finite control set are appropriate to regulate the currents in the rotating $d - q$ reference frame.

In this chapter, we will investigate the finite control set (FCS) predictive control of current in the $\alpha - \beta$ reference frame (or the stationary frame). In the $\alpha - \beta$ reference frame, the currents $i_\alpha(t)$ and $i_\beta(t)$ are linear combinations of the three phase currents $i_a(t)$, $i_b(t)$ and $i_c(t)$. Thus, they are sinusoidal functions, and the voltage variables $v_\alpha(t)$ and $v_\beta(t)$ are also sinusoidal functions. The current reference signals to the FCS predictive control systems are sinusoidal signals, which differentiates the current control systems in the $\alpha - \beta$ reference frame from those in the $d - q$ reference frame. As we have discussed in Chapter 1, the mathematical models for AC motors and power converters in the $\alpha - \beta$ reference frame are de-coupled. Therefore, it will be shown in this chapter that the original FCS predictive controllers are single-input and single-output controllers with exceptionally simple forms. However, in order to track the sinusoidal current reference signals without steady-state errors, a controller with resonant characteristic is required in the $\alpha - \beta$ reference frame. In the case of electrical drive control, the frequency of the sinusoidal reference signals is dependent on the speed of the motor, which leads to time-varying frequency in the resonant controller. In the case of power converter control, the frequency of the sinusoidal reference signal is dependent on the grid frequency, which is assumed to be a constant, thus the frequency that is embedded in the resonant controller is also a constant.

This chapter consists of four sections. Using current control of PMSM as an example, in Section 7.1, the original FCS predictive control algorithm is presented in the $\alpha - \beta$ reference frame. In Section 7.2, a resonant FCS predictive control system is proposed to track the sinusoidal reference signals in the $\alpha - \beta$ reference frame. The resonant FCS predictive control algorithm is applied to the current control of an induction motor and a power converter in Section 7.3 and 7.4 respectively, both with experimental results.

7.1 FCS Predictive Current Control of PMSM

The original FCS predictive control of current in the $\alpha - \beta$ reference frame is designed identically to the controller in the $d - q$ reference frame. This means that the sum of squares errors between the current

PID and Predictive Control of Electrical Drives and Power Converters using MATLAB®/Simulink®, First Edition.
Liuping Wang, Shan Chai, Dae Yoo, Lu Gan and Ki Ng.

Companion Website: www.wiley.com/go/wang/pid

reference signals (i_α^*, i_β^*) and the one-step ahead prediction of the current signals (i_α, i_β) is minimized at the sampling time t_i to obtain the optimal voltage control signals (v_α, v_β). The receding horizon control principle is applied leading to feedback control.

7.1.1 Predictive Control Using One-step-ahead Prediction

From Chapter 1, the dynamic model of PMSM that relates the α-β voltages to the currents is described by the differential equations:

$$\frac{di_\alpha(t)}{dt} = -\frac{R_s}{L_s}i_\alpha(t) + \frac{1}{L_s}v_\alpha(t) + \frac{\phi_{mg}}{L_s}\omega_e(t)\sin\theta_e(t) \tag{7.1}$$

$$\frac{di_\beta(t)}{dt} = -\frac{R_s}{L_s}i_\beta(t) + \frac{1}{L_s}v_\beta(t) - \frac{\phi_{mg}}{L_s}\omega_e(t)\cos\theta_e(t). \tag{7.2}$$

It is seen from this model that in the $\alpha - \beta$ reference frame, there is no interaction between the currents i_α and i_β, which will effectively reduce the current controller in this reference frame to a single-input and single- output controller. Similar to the procedures used in Chapter 6, to calculate the control variables, at the sampling time t_i, the objective function is chosen as sum of the squared errors between the desired and the predicted signals:

$$\begin{aligned} J &= (i_\alpha^*(t_i) - i_\alpha(t_{i+1}))^2 + (i_\beta^*(t_i) - i_\beta(t_{i+1}))^2 \\ &= \left(\begin{bmatrix} i_\alpha^*(t_i) \\ i_\beta^*(t_i)\end{bmatrix} - \begin{bmatrix} i_\alpha(t_{i+1}) \\ i_\beta(t_{i+1})\end{bmatrix}\right)^T \left(\begin{bmatrix} i_\alpha^*(t_i) \\ i_\beta^*(t_i)\end{bmatrix} - \begin{bmatrix} i_\alpha(t_{i+1}) \\ i_\beta(t_{i+1})\end{bmatrix}\right), \end{aligned} \tag{7.3}$$

where $i_\alpha(t_{i+1})$ and $i_\beta(t_{i+1})$ are one-step-ahead predictions of $i_\alpha(t_i)$ and $i_\beta(t_i)$, respectively. The one-step-ahead predictions of the $i_\alpha(t_{i+1})$ and $i_\beta(t_{i+1})$ are expressed in matrix and vector forms:

$$\begin{aligned} \begin{bmatrix} i_\alpha(t_{i+1}) \\ i_\beta(t_{i+1})\end{bmatrix} &= (I + \Delta t A_m)\begin{bmatrix} i_\alpha(t_i) \\ i_\beta(t_i)\end{bmatrix} + \Delta t B_m \begin{bmatrix} v_\alpha(t_i) \\ v_\beta(t_i)\end{bmatrix} \\ &+ \begin{bmatrix} \frac{\phi_{mg}}{L_s}\omega_e(t_i)\sin\theta_e(t_i)\Delta t \\ -\frac{\phi_{mg}}{L_s}\omega_e(t_i)\cos\theta_e(t_i)\Delta t \end{bmatrix}, \end{aligned} \tag{7.4}$$

where I is the identity matrix with dimension 2×2 and the system matrices A_m and B_m are defined as

$$A_m = \begin{bmatrix} -\frac{R_s}{L_s} & 0 \\ 0 & -\frac{R_s}{L_s}\end{bmatrix}; B_m = \begin{bmatrix} \frac{1}{L_s} & 0 \\ 0 & \frac{1}{L_s}\end{bmatrix}.$$

By substituting the one-step-ahead prediction given by (7.4) into the objective function J (7.3), we obtain,

$$\begin{aligned} J &= \begin{bmatrix} f_\alpha(t_i) & f_\beta(t_i)\end{bmatrix}\begin{bmatrix} f_\alpha(t_i) \\ f_\beta(t_i)\end{bmatrix} - 2\begin{bmatrix} v_\alpha(t_i) & v_\beta(t_i)\end{bmatrix}\Delta t B_m^T \begin{bmatrix} f_\alpha(t_i) \\ f_\alpha(t_i)\end{bmatrix} \\ &+ \begin{bmatrix} v_\alpha(t_i) & v_\beta(t_i)\end{bmatrix}\Delta t^2 B_m^T B_m \begin{bmatrix} v_\alpha(t_i) \\ v_\beta(t_i)\end{bmatrix}, \end{aligned} \tag{7.5}$$

where the functions $f_\alpha(t_i)$ and $f_\beta(t_i)$ are defined as

$$\begin{bmatrix} f_\alpha(t_i) \\ f_\beta(t_i)\end{bmatrix} = \begin{bmatrix} i_\alpha^*(t_i) \\ i_\beta^*(t_i)\end{bmatrix} - (I + \Delta t A_m)\begin{bmatrix} i_\alpha(t_i) \\ i_\beta(t_i)\end{bmatrix} - \begin{bmatrix} \frac{\phi_{mg}}{L_s}\omega_e(t_i)\sin\theta_e(t_i)\Delta t \\ -\frac{\phi_{mg}}{L_s}\omega_e(t_i)\cos\theta_e(t_i)\Delta t \end{bmatrix}. \tag{7.6}$$

Following the similar derivation steps outlined in Chapter 6, we obtain the optimal control signals $v_\alpha(t_i)$ and $v_\beta(t_i)$ that minimizes the objective function J:

$$\begin{bmatrix} v_\alpha(t_i)^{opt} \\ v_\beta(t_i)^{opt} \end{bmatrix} = (\Delta t^2 B_m^T B_m)^{-1} \Delta t B_m^T \begin{bmatrix} f_\alpha(t_i) \\ f_\beta(t_i) \end{bmatrix}$$

$$= \frac{1}{\Delta t} \begin{bmatrix} L_s & 0 \\ 0 & L_s \end{bmatrix} \begin{bmatrix} f_\alpha(t_i) \\ f_\beta(t_i) \end{bmatrix}. \tag{7.7}$$

Note that in the $\alpha - \beta$ reference frame, the system matrix A_m is diagonal and there is no interaction between the i_α and i_β currents. The calculations of $v_\alpha(t_i)^{opt}$ and $v_\beta(t_i)^{opt}$ signals are scalar operations. More specifically, since the matrix $I + A_m \Delta t$ is a diagonal having the form:

$$I + A_m \Delta t = \begin{bmatrix} 1 - \frac{R_s \Delta t}{L_s} & 0 \\ 0 & 1 - \frac{R_s \Delta t}{L_s} \end{bmatrix},$$

from (7.7), we have

$$v_\alpha(t_i)^{opt} = \frac{L_s}{\Delta t} i_\alpha^*(t_i) - \frac{L_s}{\Delta t}\left(1 - \frac{R_s}{L_s}\Delta t\right) i_\alpha(t_i) - \phi_{mg}\omega_e(t_i)\sin\theta_e(t_i) \tag{7.8}$$

$$v_\beta(t_i)^{opt} = \frac{L_s}{\Delta t} i_\beta^*(t_i) - \frac{L_s}{\Delta t}\left(1 - \frac{R_s}{L_s}\Delta t\right) i_\beta(t_i) + \phi_{mg}\omega_e(t_i)\cos\theta_e(t_i) \tag{7.9}$$

where $i_\alpha^*(t_i)$ and $i_\beta^*(t_i)$ are current reference signals in the $\alpha - \beta$ reference frame, and $i_\alpha(t_i)$ and $i_\beta(t_i)$, $\omega_e(t_i)$, $\theta_e(t_i)$ are the electric velocity and angle of the PMSM. It is clearly seen that predictive controller uses a proportional feedback control with a feedforward compensation. Furthermore, the feedback control gain is dependent on the sampling interval of the current control system with the value

$$k_{fcs}^\alpha = k_{fcs}^\beta = \frac{L_s}{\Delta t}(1 - \frac{R_s}{L_s}\Delta t). \tag{7.10}$$

In order to ensure a negative feedback in the current control, the quantity $1 - \frac{R_s}{L_s}\Delta t > 0$, that is $\frac{R_s}{L_s}\Delta t < 1$.

7.1.2 FCS Current Control in $\alpha - \beta$ Reference Frame

A similar approach to the FCS in the $d - q$ reference frame is adopted to the FCS problem in the $\alpha - \beta$ reference frame. Following the same procedure as outlined in Chapter 6, it is easy to show that the objective function J (7.5) can also be expressed in terms of the optimal voltage signals in the $\alpha - \beta$ reference frame as

$$J = \left(\begin{bmatrix} v_\alpha(t_i) \\ v_\beta(t_i) \end{bmatrix} - \begin{bmatrix} v_\alpha(t_i)^{opt} \\ v_\beta(t_i)^{opt} \end{bmatrix}\right)^T (\Delta t^2 B_m^T B_m) \left(\begin{bmatrix} v_\alpha(t_i) \\ v_\beta(t_i) \end{bmatrix} - \begin{bmatrix} v_\alpha(t_i)^{opt} \\ v_\beta(t_i)^{opt} \end{bmatrix}\right)$$

$$= \frac{\Delta t^2}{L_s^2}(v_\alpha(t_i) - v_\alpha(t_i)^{opt})^2 + \frac{\Delta t^2}{L_s^2}(v_\beta(t_i) - v_\beta(t_i)^{opt})^2. \tag{7.11}$$

With both objective function and the optimal control signals defined, the next step in the FCS predictive control is to find the control signal $v_\alpha(t_i)$ and $v_\beta(t_i)$ that will minimize the objective function subject to the limited number of choices of voltage variables. In the $\alpha - \beta$ reference frame, there are seven pairs of candidate voltage values, which are also time-invariant. Their exact values are characterized by the values listed below:

$$\begin{bmatrix} 0 & 1 & \frac{1}{2} & -\frac{1}{2} & -1 & -\frac{1}{2} & \frac{1}{2} \\ 0 & 0 & \frac{\sqrt{3}}{2} & \frac{\sqrt{3}}{2} & 0 & -\frac{\sqrt{3}}{2} & -\frac{\sqrt{3}}{2} \end{bmatrix} \frac{2}{3} V_{dc}. \tag{7.12}$$

In FCS current control proposed in the $\alpha - \beta$ reference frame, the seven pairs of v_α and v_β values from (7.12) are used to evaluate the objective function (7.11). The pair of v_α and v_β that has yielded a minimum of the objective function J will be chosen as the FCS current control signals in the $\alpha - \beta$ reference frame.

As we may recall, in the $d - q$ reference frame, there are also seven candidate pairs that the $v_d(t_i)$ and $v_q(t_i)$ are permitted to take. However, the candidate variables are sinusoidal functions with respect to the electrical angle $\theta_e(t)$ (see (6.7)). Because the candidate variables in the $\alpha - \beta$ reference frame are constants, in addition to the optimal voltage variables being scalars (see (7.8), (7.9)), the real-time computational load for the control law is less than that in the $d - q$ reference frame.

It is worthwhile to emphasize that because the reference signals in the $\alpha - \beta$ frame are sinusoidal signals, the error signals $i_\alpha^*(t) - i_\alpha(t)$ and $i_\beta^*(t) - i_\beta(t)$ will not converge to zero as $t \to \infty$. However, as the sampling interval Δt reduces, then $|i_\alpha^*(t) - i_\alpha(t)|$ and $|i_\beta^*(t) - i_\beta(t)|$ will reduce as the feedback controller gain $\frac{L_s}{\Delta t}(1 - \frac{R_s}{L_s}\Delta t)$ increases.

7.1.3 Generating Current Reference Signals in $\alpha - \beta$ Frame

The reference signals to the FCS-MPC system in the $\alpha - \beta$ reference frame are sinusoidal signals in which their frequency is determined by the electrical velocity of the PMSM, $\omega_e(t)$. In the applications, the desired operational performance of a PMSM in a closed-loop current control is specified via the desired values of i_d and i_q currents. For instance, the desired value for i_d is chosen to be a constant and

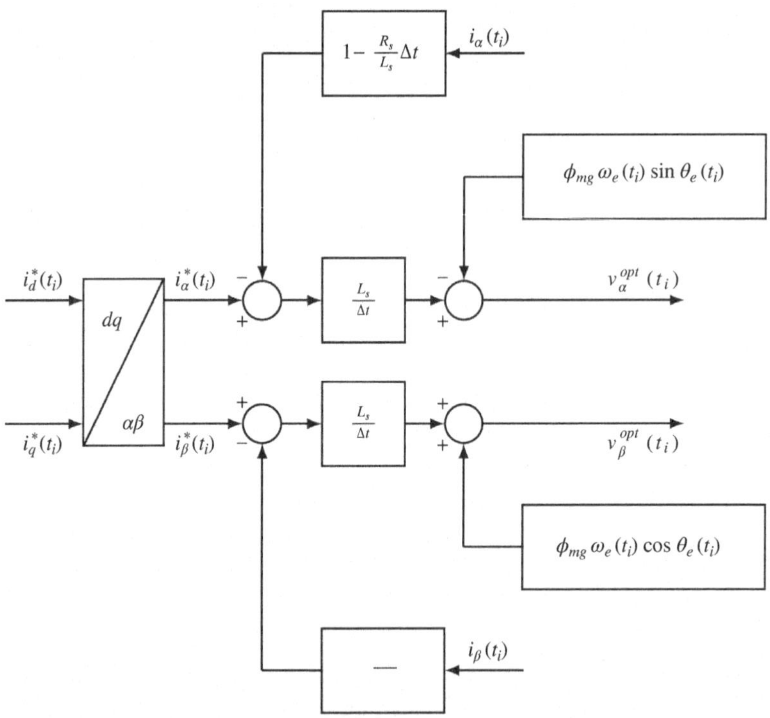

Figure 7.1 FCS-MPC in $\alpha - \beta$ frame.

the i_q current is related to the demanded electrical torque. The reference signals to the i_d and i_q currents are transparent to the applications and easy to choose. For these reasons, the reference signals to i_α and i_β currents are calculated using the inverse Park transformation:

$$\begin{bmatrix} i_\alpha^*(t) \\ i_\beta^*(t) \end{bmatrix} = \begin{bmatrix} \cos\theta_e(t) & -\sin\theta_e(t) \\ \sin\theta_e(t) & \cos\theta_e(t) \end{bmatrix} \begin{bmatrix} i_d^*(t) \\ i_q^*(t) \end{bmatrix}. \tag{7.13}$$

If the reference signal $i_d^*(t)$ is chosen to be zero, then $i_\alpha^*(t) = -\sin\theta_e(t) i_q^*(t)$, and $i_\beta^*(t) = \cos\theta_e(t) i_q^*(t)$.

Figure 7.1 shows the configuration of the feedback control system to generate the optimal control signals $v_\alpha(t)^{opt}$ and $v_\beta(t)^{opt}$.

7.2 Resonant FCS Predictive Current Control

The FCS predictive current controllers in the $\alpha - \beta$ reference frame can not completely eliminate the steady-state errors in the i_d and i_q current because it, in essence, is a proportional control. To reduce the steady-state errors in the i_d and i_q currents, the feedback errors $i_\alpha^*(t) - i_\alpha(t)$ and $i_\beta^*(t) - i_\beta(t)$ in the current control system need to be reduced.

7.2.1 *Control System Configuration*

It is known from the internal model control principle (Francis and Wonham (1975)) that in order for the feedback control system to track a periodic signal, the signal generator needs to be embedded in the controller. For the case of the current control in the $\alpha - \beta$ reference frame, because the reference current signals are sinusoidal signals, the generator of a sinusoidal signal should be embedded in the feedback control system so that the output current signals $i_\alpha(t)$ and $i_\beta(t)$ would track their reference signals without steady-state errors. In short, the controller should have a polynomial factor $1 - 2\cos\omega_d z^{-1} + z^{-2}$, which is $(1 - e^{j\omega_d} z^{-1})(1 - e^{-j\omega_d} z^{-1})$, contained in its denominator where ω_d is the discrete frequency of the sinusoidal reference signal. In other words, there is a pair of complex poles contained in the controller, where the locations of the poles are at $e^{\pm j\omega_d}$ on the complex plane.

The controller that has the capability to track a sinusoidal reference signal or to reject a sinusoidal disturbance signal is called a resonant controller in the power electronics community or a repetitive controller in a general control community. The resonant FCS current controller is proposed to have the feedback structure as illustrated in Figure 7.2. In the proposed control system structure, the feedback controllers k_{fcs}^α and k_{fcs}^α derived from the one-step-ahead prediction and optimization shown in (7.10) are used in the inner-loops for fast dynamic response, while two resonant controllers are used in the outer-loops to provide further compensations for the tracking errors between the reference and feedback current signals in the $\alpha - \beta$ reference frame.

The frequency ω_d is the discrete frequency, having the unit of radian. Assuming that a sinusoidal signal has a period of T, with a sampling interval Δt, the number of samples within this period T is $N_T = \frac{T}{\Delta t}$. The discrete frequency ω_d is calculated as

$$\omega_d = \frac{2\pi}{N_T} = \frac{2\pi\Delta t}{T}.$$

Suppose that the electrical velocity of the motor is 1000 RPM and the sampling interval is $\Delta t = 100 \times 10^{-6}$ sec. Then the frequency parameter ω_d is then

$$\omega_d = \frac{2\pi \times 1000 \times \Delta t}{60} = 0.0105.$$

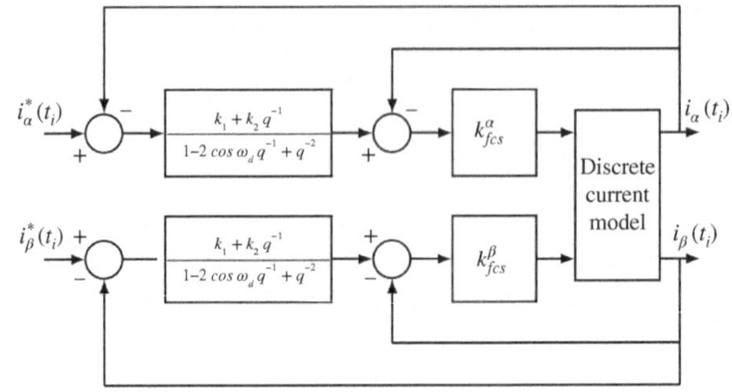

Figure 7.2 Resonant FCS-MPC in $\alpha - \beta$ frame.

The frequency parameter ω_d is time varying when the motor velocity changes. However, when the sampling interval Δt is small, this variation has a small effect on the locations of the complex poles in the controller. Let us say that the motor velocity varies from 100 to 10 000 RPM. When $\Delta t = 100 \times 10^{-6}$ sec, the corresponding ω_d for 100 RPM is approximately 0.001 rad and for 10 000 RPM is 0.1047 rad. The controller poles for the former case are approximately $z_{1,2} = 0.999995 \pm j0.001$ and the latter case approximately $z_{1,2} = 0.9945 \pm j0.1045$.

7.2.2 Outer-loop Controller Design

In Chapter 6 (see Section 6.4), the one-step-ahead predictive controller for the inner-loop system was shown to result in a closed-loop system with a transfer function z^{-1}. The same procedure is applied here to obtain the same result in the $\alpha - \beta$ reference frame. We leave the derivation of this result as an exercise.

With the inner-loop system modeled as a one sample delay (z^{-1}), the task of designing the resonant controller in the outer-loop becomes straightforward. Figure 7.3 illustrates the outer-loop system for controlling current i_α with the resonant controller and the inner-loop approximated by the transfer function z^{-1}.

It is clearly seen that the closed-loop system from the reference signal $I^*_\alpha(z)$ to $I_\alpha(z)$ is described by the z-transfer function

$$\frac{I_\alpha(z)}{I^*_\alpha(z)} = \frac{k_1z^{-1} + k_2z^{-2}}{1 - 2\cos\omega_d z^{-1} + z^{-2} + k_1z^{-1} + k_2z^{-2}}. \tag{7.14}$$

This is a second order discrete-time system with two closed-loop poles. Thus, the two coefficients from the resonant controller, k_1 and k_2, can be uniquely determined by using the technique of pole-assignment controller design.

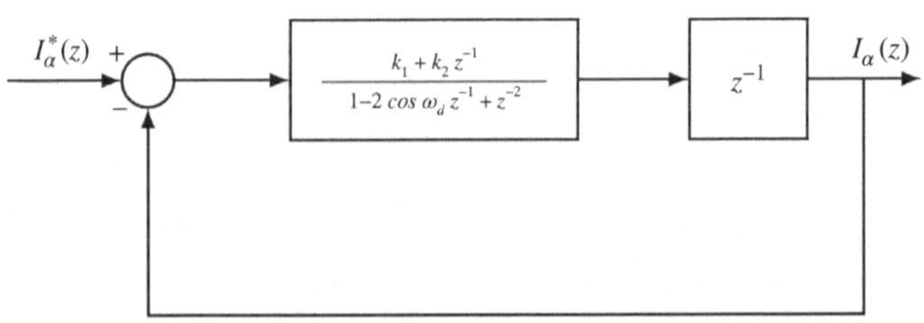

Figure 7.3 Outer-loop system with inner-loop approximated by a sample of time delay.

For simplicity, in the first approach, assuming that the desired closed-loops are identical real poles, denoted as $0 \leq \lambda < 1$, the desired closed-loop polynomial for the discrete system is given by

$$(1 - \lambda z^{-1})^2 = 1 - 2\lambda z^{-1} + \lambda^2 z^{-2}.$$

By comparing the desired closed-loop polynomial with the actual closed-loop polynomial given by the denominator of (7.14), we obtain the following equalities:

$$k_1 - 2\cos\omega_d = -2\lambda$$
$$k_2 + 1 = \lambda^2.$$

These equalities lead to the solutions for the gains of the resonant controller where

$$k_1 = 2\cos\omega_d - 2\lambda \tag{7.15}$$
$$k_2 = \lambda^2 - 1. \tag{7.16}$$

The performance tuning parameter for the resonant FCS controller is the location of the pair of desired discrete closed-loop poles $0 \leq \lambda < 1$. This parameter is selected in the design to reflect the closed-loop bandwidth of the control system, depending on the quality of the current model and current sensor noise level. A smaller λ corresponds to faster closed-loop response for the resonant FCS control system, which on the other hand, may cause noise amplification and the resulting closed-loop system to be less robust.

If one wishes to use the closed-loop performance specification in continuous-time that closely corresponds to the underlying physical system, then the desired closed-loop polynomial is chosen as $s^2 + 2\xi w_n s + w_n^2$. For $\xi = 0.707$ (or other damping coefficient less than one), the pair of continuous-time complex poles are $s_{1,2} = -\xi w_n \pm j w_n \sqrt{1 - \xi^2}$. With a sampling interval Δt, the pair of poles are converted from continuous-time to discrete-time via the following relationships:

$$z_1 = e^{-\xi w_n \Delta t + j w_n \sqrt{1-\xi^2}\Delta t}$$
$$z_2 = e^{-\xi w_n \Delta t - j w_n \sqrt{1-\xi^2}\Delta t}.$$

The desired closed-loop polynomial in discrete-time, but having a direct relation to the underlying continuous-time performance, becomes

$$\begin{aligned}&(1 - e^{-\xi w_n \Delta t + j w_n \sqrt{1-\xi^2}\Delta t} z^{-1})(1 - e^{-\xi w_n \Delta t - j w_n \sqrt{1-\xi^2}\Delta t} z^{-1})\\ &= 1 - 2e^{-\xi w_n \Delta t}\cos(w_n \sqrt{1-\xi^2}\Delta t) z^{-1} + e^{-2\xi w_n \Delta t} z^{-2}.\end{aligned} \tag{7.17}$$

When the desired closed-loop poles are selected this way, the coefficients of the resonant controller are found by comparing the desired closed-loop polynomial (7.17) with the actual closed-loop polynomial given by the denominator in (7.14):

$$k_1 = 2\cos\omega_d - 2e^{-\xi w_n \Delta t}\cos(w_n \sqrt{1-\xi^2}\Delta t)$$
$$k_2 = e^{-2\xi w_n \Delta t} - 1,$$

where w_n is the desired bandwidth for the closed-loop current control system specified in the continuous-time.

7.2.3 Resonant FCS Predictive Control System

The control system shown in Figure 7.2 is the resonant control system without the constraints of using the finite control set in the $\alpha - \beta$ frame. The resonant control algorithm for using the finite control set is an extension of the unconstrained control case. We will first summarize the algorithm, followed by its derivation.

Algorithm 5 *The resonant FCS control signals in the $\alpha - \beta$ reference frame at sampling time t_i, $v_\alpha(t_i)$ and $v_\beta(t_i)$, are found by finding the minimum of the objective function J with respect to the index k,*

$$J = \frac{\Delta t^2}{L_s^2}(v_\alpha(t_i)^k - v_\alpha(t_i)^{opt})^2 + \frac{\Delta t^2}{L_s^2}(v_\beta(t_i)^k - v_\beta(t_i)^{opt})^2, \tag{7.18}$$

where the values of $v_\alpha(t_i)^k$ and $v_\beta(t_i)^k$ ($k = 0, 1, 2, \ldots, 6$) are given by the finite control set,

$$\begin{bmatrix} 0 & 1 & \frac{1}{2} & -\frac{1}{2} & -1 & -\frac{1}{2} & \frac{1}{2} \\ 0 & 0 & \frac{\sqrt{3}}{2} & \frac{\sqrt{3}}{2} & 0 & -\frac{\sqrt{3}}{2} & -\frac{\sqrt{3}}{2} \end{bmatrix} \frac{2}{3} V_{dc}$$

and the signals $v_\alpha(t_i)^{opt}$ and $v_\beta(t_i)^{opt}$ are computed recursively using the following equations:

$$v_\alpha(t_i)^{opt} = 2\cos\omega_d v_\alpha(t_{i-1})^{opt} - v_\alpha(t_{i-2})^{opt} + v_\alpha(t_i)^{sopt} \tag{7.19}$$

$$v_\alpha(t_i)^{sopt} = k_{fcs}^{\alpha}[k_1(i_\alpha^*(t_i) - i_\alpha(t_i)) + k_2(i_\alpha^*(t_{i-1}) - i_\alpha(t_{i-1})) - (i_\alpha(t_i) - 2\cos\omega_d i_\alpha(t_{i-1}) + i_\alpha(t_{i-2}))] \tag{7.20}$$

$$v_\beta(t_i)^{opt} = 2\cos\omega_d v_\beta(t_{i-1})^{opt} - v_\beta(t_{i-2})^{opt} + v_\beta(t_i)^{sopt} \tag{7.21}$$

$$v_\beta(t_i)^{sopt} = k_{fcs}^{\beta}[k_1(i_\beta^*(t_i) - i_\beta(t_i)) + k_2(i_\beta^*(t_{i-1}) - i_\beta(t_{i-1})) - (i_\beta(t_i) - 2\cos\omega_d i_\beta(t_{i-1}) + i_\beta(t_{i-2}))]. \tag{7.22}$$

The feedback control gains used in the computation are defined as:

$$k_{fcs}^{\alpha} = k_{fcs}^{\beta} = \frac{L_s}{\Delta t}\left(1 - \frac{R_s}{L_s}\Delta t\right) \tag{7.23}$$

$$k_1 = 2\cos\omega_d - 2\lambda \tag{7.24}$$

$$k_2 = \lambda^2 - 1. \tag{7.25}$$

$0 \leq \lambda < 1$ *is the desired closed-loop pole location for the current control system.*

To derive how the control signals are chosen in the presence of constraints, we will resort to the solution via predictive control. Consider the difference equations that describe the currents i_α and i_β at the sampling time t_i:

$$i_\alpha(t_{i+1}) = \left(1 - \frac{R_s}{L_s}\Delta t\right) i_\alpha(t_i) + \frac{\Delta t}{L_s} v_\alpha(t_i) + \frac{\phi_{mg}}{L_s}\Delta t\omega_e(t_i)\sin(\theta_e(t_i)) \tag{7.26}$$

$$i_\beta(t_{i+1}) = \left(1 - \frac{R_s}{L_s}\Delta t\right) i_\beta(t_i) + \frac{\Delta t}{L_s} v_\beta(t_i) - \frac{\phi_{mg}}{L_s}\Delta t\omega_e(t_i)\cos\theta_e(t_i). \tag{7.27}$$

Because there is no interaction between the variables in the $\alpha - \beta$ reference frame, for simplicity, only the prediction for i_α is considered, and the results will extend naturally to the variable i_β.

Define the operator $D(q^{-1})$ as

$$D(q^{-1}) = 1 - 2\cos\omega_d q^{-1} + q^{-2},$$

with q^{-1} as the backward shift operator $q^{-1}f(t_i) = f(t_{i-1})$. When the operator $D(q^{-1})$ is applied to $\sin(\theta_e(t_i))$, we obtain the result that

$$D(q^{-1})\sin(\theta_e(t_i)) = 0. \tag{7.28}$$

To prove this relation is true, we note that

$$\sin\theta_e(t_i) = \sin\omega_e t_i = \sin\left(\frac{\omega_d}{\Delta t}k_i\Delta t\right) = \sin(\omega_d k_i)$$

also,

$$\sin(\omega_d k_i) = \frac{e^{j\omega_d k_i} - e^{-j\omega_d k_i}}{2j}$$

$$D(q^{-1}) = 1 - 2\cos\omega_d q^{-1} + q^{-2} = (1 - e^{j\omega_d}q^{-1})(1 - e^{-j\omega_d}q^{-1}),$$

since

$$(1 - e^{j\omega_d}q^{-1})e^{j\omega_d k_i} = e^{j\omega_d k_i} - e^{j\omega_d}e^{j\omega_d(k_i-1)} = 0$$

$$(1 - e^{-j\omega_d}q^{-1})e^{-j\omega_d k_i} = e^{-j\omega_d k_i} - e^{-j\omega_d}e^{-j\omega_d(k_i-1)} = 0.$$

Therefore,

$$D(q^{-1})\sin(\theta_e(t_i)) = (1 - e^{j\omega_d}q^{-1})(1 - e^{-j\omega_d}q^{-1})\frac{e^{j\omega_d k_i} - e^{-j\omega_d k_i}}{2j} = 0,$$

where we have used the backward shift relationship:

$$q^{-1}e^{j\omega_d k_i} = e^{j\omega_d(k_i-1)}; \quad q^{-1}e^{-j\omega_d k_i} = e^{-j\omega_d(k_i-1)}.$$

By assuming that the electrical velocity ω_e as a constant (say in steady-state operation), the last term of (7.26) vanishes when the operator $D(z^{-1})$ is applied to it. This conclusion is important because this last term is considered sinusoidal disturbance, which would be used as a feedforward compensation in the original FCS-MPC system. With the resonant FCS predictive controller, this feedforward compensation is no longer required.

Now, applying the operator $D(q^{-1})$ to both sides of (7.26) yields

$$i_\alpha(t_{i+1})^s = \left(1 - \frac{R_s}{L_s}\Delta t\right)i_\alpha(t_i)^s + \frac{\Delta t}{L_s}v_\alpha(t_i)^s, \tag{7.29}$$

where

$$i_\alpha(t_{i+1})^s = D(q^{-1})i_\alpha(t_{i+1}) \tag{7.30}$$

$$i_\alpha(t_i)^s = D(q^{-1})i_\alpha(t_i) \tag{7.31}$$

$$v_\alpha(t_i)^s = D(q^{-1})v_\alpha(t_i). \tag{7.32}$$

The variables $i_\alpha(t_i)^s$ and $v_\alpha(t_i)^s$ are the filtered current and voltage signals with the denominator of the resonant controller.

To include the resonant action into the controller, the weighted current errors

$$e_\alpha(t_i) = k_1(i_\alpha^*(t_i) - i_\alpha(t_i)) + k_2(i_\alpha^*(t_{i-1}) - i_\alpha(t_{i-1}))$$

is chosen as the steady-state of the $i_\alpha(t_i)^s$, where k_1 and k_2 are given by (7.24) and (7.25). The steady-state of $v_\alpha(t_i)^s$ is chosen to be zero. Subtracting the steady-state from the model (7.29) gives:

$$i_\alpha(t_{i+1})^s - e_\alpha(t_i) = \left(1 - \frac{R_s}{L_s}\Delta t\right)(i_\alpha(t_i)^s - e_\alpha(t_i)) + \frac{\Delta t}{L_s}v_\alpha(t_i)^s. \tag{7.33}$$

After applying the same procedure to the β-axis current, we obtain the formula for the i_β variable as

$$i_\beta(t_{i+1})^s - e_\beta(t_i) = \left(1 - \frac{R_s}{L_s}\Delta t\right)(i_\beta(t_i)^s - e_\beta(t_i)) + \frac{\Delta t}{L_s}v_\beta(t_i)^s, \tag{7.34}$$

where $e_\beta(t_i) = k_1(i_\beta^*(t_i) - i_\beta(t_i)) + k_2(i_\beta^*(t_{i-1}) - i_\beta(t_{i-1}))$.

The control objective is to minimize the error function J, where

$$J = \begin{bmatrix} i_\alpha(t_{i+1})^s - e_\alpha(t_i) \\ i_\beta(t_{i+1})^s - e_\beta(t_i) \end{bmatrix}^T \begin{bmatrix} i_\alpha(t_{i+1})^s - e_\alpha(t_i) \\ i_\beta(t_{i+1})^s - e_\beta(t_i) \end{bmatrix}, \tag{7.35}$$

which is to regulate the filtered current signals $i_\alpha(t_{i+1})^s$, $i_\beta(t_{i+1})^s$ to be as close as possible to $e_\alpha(t_i)$ and $e_\beta(t_i)$.

By substituting (7.33) and (7.34) into (7.35), and following the same derivation procedure as outlined in Section 6.6, it can be shown that the optimal solutions of $v_\alpha(t_i)^s$ and $v_\beta(t_i)^s$ that will minimize the objective function (7.35) are given by

$$v_\alpha(t_i)^{sopt} = k_{fcs}^\alpha(e_\alpha(t_i) - i_\alpha(t_i)^s) \tag{7.36}$$

$$v_\beta(t_i)^{sopt} = k_{fcs}^\beta(e_\beta(t_i) - i_\beta(t_i)^s), \tag{7.37}$$

where the feedback controller gains are defined as

$$k_{fcs}^\alpha = k_{fcs}^\beta = \frac{L_s}{\Delta t}\left(1 - \frac{R_s}{L_s}\Delta t\right). \tag{7.38}$$

Furthermore, the objective function J can be expressed via completing squares as

$$J = \frac{\Delta t^2}{L_s^2}(v_\alpha(t_i)^s - v_\alpha(t_i)^{sopt})^2 + \frac{\Delta t^2}{L_s^2}(v_\beta(t_i)^s - v_\beta(t_i)^{sopt})^2. \tag{7.39}$$

Now, note that $v_\alpha(t_i)^s$, $v_\alpha(t_i)^{sopt}$, $v_\beta(t_i)^s$, $v_\beta(t_i)^{sopt}$ are filtered voltage variables. Thus, by definition of the filtered control signals, the following relationships are true:

$$v_\alpha(t_i)^{opt} = \frac{v_\alpha(t_i)^{sopt}}{1 - 2\cos\omega_d q^{-1} + q^{-2}}$$

$$v_\beta(t_i)^{opt} = \frac{v_\beta(t_i)^{sopt}}{1 - 2\cos\omega_d q^{-1} + q^{-2}},$$

which leads to the expressions of $v_\alpha(t_i)^{sopt}$ and $v_\beta(t_i)^{sopt}$ in a recursive manner:

$$v_\alpha(t_i)^{sopt} = v_\alpha(t_i)^{opt} - 2\cos\omega_d v_\alpha(t_{i-1})^{opt} + v_\alpha(t_{i-2})^{opt} \tag{7.40}$$

$$v_\beta(t_i)^{sopt} = v_\beta(t_i)^{opt} - 2\cos\omega_d v_\beta(t_{i-1})^{opt} + v_\beta(t_{i-2})^{opt}. \tag{7.41}$$

By calculating the actual filtered control signals using the same past optimal control signal states, we obtain

$$v_\alpha(t_i)^s = v_\alpha(t_i) - 2\cos\omega_d v_\alpha(t_{i-1})^{opt} + v_\alpha(t_{i-2})^{opt} \tag{7.42}$$

$$v_\beta(t_i)^s = v_\beta(t_i) - 2\cos\omega_d v_\beta(t_{i-1})^{opt} + v_\beta(t_{i-2})^{opt}. \tag{7.43}$$

Substituting the filtered variables (7.40)–(7.43) into the objective function (7.39) yields

$$J = \frac{\Delta t^2}{L_s^2}(v_\alpha(t_i) - v_\alpha(t_i)^{opt})^2 + \frac{\Delta t^2}{L_s^2}(v_\beta(t_i) - v_\beta(t_i)^{opt})^2. \tag{7.44}$$

With the finite control set, the derived objective function here is identical to the one used in the Algorithm 5 (see (7.18)). This completes the derivation of the resonant FCS control algorithm.

It is emphasized that the resonant FCS control Algorithm 5 used the past optimal control signal states $(v_\alpha(t_{i-1})^{opt}, v_\alpha(t_{i-2})^{opt})$ together with the current $(v_\alpha(t_i))$ to predict the filtered $v_\alpha(t_i)^s$. If the past implemented control signal states $(v_\alpha(t_{i-1}), v_\alpha(t_{i-2}))$ were used in the prediction, then it could result in accumulated errors from the finite control set and lead to steady-state errors in the resonant FCS control system.

7.3 Resonant FCS Current Control of Induction Motor

In this section, we will discuss the FCS predictive control of induction motor in the $\alpha - \beta$ reference frame. The control algorithms are similar to those derived for the PMSM and they will be briefly summarized as the original FCS current control and the resonant current control. Experimental results will be used to demonstrate the algorithms.

7.3.1 The Original FCS Current Control of Induction Motor

The mathematical model of the induction motor in the $\alpha - \beta$ reference frame is governed by the following differential equations (see Section 1.5):

$$\frac{di_{s\alpha}(t)}{dt} = -\frac{1}{\tau'_\sigma} i_{s\alpha}(t) + \frac{k_r}{r_\sigma \tau'_\sigma \tau_r} \psi_{r\alpha}(t) + \frac{k_r}{r_\sigma \tau'_\sigma} \omega_e(t)\psi_{r\beta}(t) + \frac{1}{r_\sigma \tau'_\sigma} u_{s\alpha}(t) \tag{7.45}$$

$$\frac{di_{s\beta}(t)}{dt} = -\frac{1}{\tau'_\sigma} i_{s\beta}(t) - \frac{k_r}{r_\sigma \tau'_\sigma} \omega_e(t)\psi_{r\alpha}(t) + \frac{k_r}{r_\sigma \tau'_\sigma \tau_r} \psi_{r\beta}(t) + \frac{1}{r_\sigma \tau'_\sigma} u_{s\beta}(t). \tag{7.46}$$

In this model, the motor velocity $\omega_e(t)$ and the rotor fluxes in the α and β reference frame $(\psi_{r\alpha}(t), \psi_{r\beta}(t))$ are considered as disturbances to the current control system. In the $\alpha - \beta$ reference frame, the fluxes are sinusoidal functions.

The one-step predictions are expressed in matrix and vector form by using first order approximations in the model equations (7.45) and (7.46)

$$\begin{bmatrix} i_{s\alpha}(t_{i+1}) \\ i_{s\beta}(t_{i+1}) \end{bmatrix} = (I + \Delta t A_m) \begin{bmatrix} i_\alpha(t_i) \\ i_\beta(t_i) \end{bmatrix} + \Delta t B_m \begin{bmatrix} u_{s\alpha}(t_i) \\ u_{s\beta}(t_i) \end{bmatrix} + \Delta t \begin{bmatrix} \frac{k_r}{r_\sigma \tau'_\sigma \tau_r} \psi_{r\alpha}(t_i) + \frac{k_r}{r_\sigma \tau'_\sigma} \omega_e(t_i)\psi_{r\beta}(t_i) \\ -\frac{k_r}{r_\sigma \tau'_\sigma} \omega_e(t_i)\psi_{r\alpha}(t_i) + \frac{k_r}{r_\sigma \tau'_\sigma \tau_r} \psi_{r\beta}(t_i) \end{bmatrix}, \tag{7.47}$$

where I is a 2×2 identity matrix and the system matrices A_m and B_m are defined as

$$A_m = \begin{bmatrix} -\frac{1}{\tau'_\sigma} & 0 \\ 0 & -\frac{1}{\tau'_\sigma} \end{bmatrix}; \quad B_m = \begin{bmatrix} \frac{1}{r_\sigma \tau'_\sigma} & 0 \\ 0 & \frac{1}{r_\sigma \tau'_\sigma} \end{bmatrix}.$$

The final term in (7.47) is the disturbance that could be used in the feedforward compensation if the rotor fluxes were estimated.

The objective function selected for calculating the control variables at sampling time t_i, is given by

$$J = \begin{bmatrix} i^*_{s\alpha}(t_i) - i_{s\alpha}(t_{i+1}) \\ i^*_{s\beta}(t_i) - i_{s\beta}(t_{i+1}) \end{bmatrix}^T \begin{bmatrix} i^*_{s\alpha}(t_i) - i_{s\alpha}(t_{i+1}) \\ i^*_{s\beta}(t_i) - i_{s\beta}(t_{i+1}) \end{bmatrix}, \tag{7.48}$$

where $i_{s\alpha}(t_{i+1})$ and $i_{s\beta}(t_{i+1})$ are the one-step-ahead predictions of $i_{s\alpha}(t_i)$ and $i_{s\beta}(t_i)$, respectively. Substituting the one-step-ahead prediction (7.47) into the objective function J (7.48) leads to

$$J = \begin{bmatrix} f_\alpha(t_i) & f_\beta(t_i) \end{bmatrix} \begin{bmatrix} f_\alpha(t_i) \\ f_\beta(t_i) \end{bmatrix} - 2\begin{bmatrix} u_{s\alpha}(t_i) & u_{s\beta}(t_i) \end{bmatrix} \Delta t B_m^T \begin{bmatrix} f_\alpha(t_i) \\ f_\alpha(t_i) \end{bmatrix}$$
$$+ \begin{bmatrix} u_{s\alpha}(t_i) & u_{s\beta}(t_i) \end{bmatrix} \Delta t^2 B_m^T B_m \begin{bmatrix} u_{s\alpha}(t_i) \\ u_{s\beta}(t_i) \end{bmatrix}, \tag{7.49}$$

where the auxiliary functions $f_\alpha(t_i)$ and $f_\beta(t_i)$ are defined as:

$$\begin{bmatrix} f_\alpha(t_i) \\ f_\beta(t_i) \end{bmatrix} = \begin{bmatrix} i^*_{s\alpha}(t_i) \\ i^*_{s\beta}(t_i) \end{bmatrix} - (I + \Delta t A_m)\begin{bmatrix} i_{s\alpha}(t_i) \\ i_{s\beta}(t_i) \end{bmatrix}$$
$$- \Delta t \begin{bmatrix} \frac{k_r}{r_\sigma \tau'_\sigma \tau_r}\psi_{r\alpha}(t_i) + \frac{k_r}{r_\sigma \tau'_\sigma}\omega_e(t_i)\psi_{r\beta}(t_i) \\ -\frac{k_r}{r_\sigma \tau'_\sigma}\omega_e(t_i)\psi_{r\alpha}(t_i) + \frac{k_r}{r_\sigma \tau'_\sigma \tau_r}\psi_{r\beta}(t_i) \end{bmatrix}. \tag{7.50}$$

The optimal control signals $u_{s\alpha}(t_i)$ and $u_{s\beta}(t_i)$ that minimize the objective function J are computed as:

$$\begin{bmatrix} u_{s\alpha}(t_i)^{opt} \\ u_{s\beta}(t_i)^{opt} \end{bmatrix} = (\Delta t^2 B_m^T B_m)^{-1} \Delta t B_m^T \begin{bmatrix} f_\alpha(t_i) \\ f_\beta(t_i) \end{bmatrix}$$
$$= \frac{1}{\Delta t}\begin{bmatrix} r_\sigma \tau'_\sigma & 0 \\ 0 & r_\sigma \tau'_\sigma \end{bmatrix} \begin{bmatrix} f_\alpha(t_i) \\ f_\beta(t_i) \end{bmatrix}. \tag{7.51}$$

Note that in the $\alpha - \beta$ reference frame, the system matrix A_m is diagonal and there is no interaction between current $i_{s\alpha}$ and $i_{s\beta}$. Thus the calculations of optimal control signals $u_{s\alpha}^{opt}$ and $u_{s\beta}^{opt}$ are scalar operation. Since the matrix $I + A_m\Delta t$ is a diagonal having the form:

$$I + A_m \Delta t = \begin{bmatrix} 1 - \frac{\Delta t}{\tau'_\sigma} & 0 \\ 0 & 1 - \frac{\Delta t}{\tau'_\sigma} \end{bmatrix},$$

from (7.51), the optimal control signals are obtained as:

$$u_{s\alpha}(t_i)^{opt} = \frac{r_\sigma \tau'_\sigma}{\Delta t} i^*_{s\alpha}(t_i) - \frac{r_\sigma \tau'_\sigma}{\Delta t}\left(1 - \frac{\Delta t}{\tau'_\sigma}\right) i_{s\alpha}(t_i) - \frac{k_r}{r_\sigma \tau'_\sigma \tau_r}\psi_{r\alpha}(t_i)$$
$$- \frac{k_r}{r_\sigma \tau'_\sigma}\omega_e(t_i)\psi_{r\beta}(t_i) \tag{7.52}$$

$$u_{s\beta}(t_i)^{opt} = \frac{r_\sigma \tau'_\sigma}{\Delta t} i^*_{s\beta}(t_i) - \frac{r_\sigma \tau'_\sigma}{\Delta t}\left(1 - \frac{\Delta t}{\tau'_\sigma}\right) i_{s\beta}(t_i) + \frac{k_r}{r_\sigma \tau'_\sigma}\omega_e(t_i)\psi_{r\alpha}(t_i)$$
$$- \frac{k_r}{r_\sigma \tau'_\sigma \tau_r}\psi_{r\beta}(t_i), \tag{7.53}$$

where $i^*_{s\alpha}(t_i)$ and $i^*_{s\beta}(t_i)$ are current reference signals in the $\alpha - \beta$ reference frame, $\omega_e(t_i)$, $i_{s\alpha}(t_i)$ and $i_{s\beta}(t_i)$ are from the measurement at sampling time t_i, $\psi_{r\alpha}(t_i)$ and $\psi_{r\beta}(t_i)$ are estimated based on the model equations as

$$\psi_{r\alpha}(t_i) = \psi_{r\alpha}(t_{i-1}) + \Delta t\left(\frac{L_h}{\tau_r}\right) i_{s\alpha}(t_{i-1}) - \frac{1}{\tau_r}\psi_{r\alpha}(t_{i-1})$$
$$- \omega_e(t_{i-1})\psi_{r\beta}(t_{i-1})) \tag{7.54}$$

$$\psi_{r\beta}(t_i) = \psi_{r\beta}(t_{i-1}) + \Delta t \left(\frac{L_h}{\tau_r}\right) i_{s\beta}(t_{i-1}) + \omega_e(t_{i-1})\psi_{r\alpha}(t_{i-1})$$
$$- \frac{1}{\tau_r}\psi_{r\beta}(t_{i-1})). \tag{7.55}$$

It is clearly seen that the rotor fluxes in the $\alpha - \beta$ reference frame are sinusoidal functions because the stator current signals $i_{s\alpha}(t)$ and $i_{s\beta}(t)$ are sinusoidal functions.

The feedback control gains (see 7.52 and 7.53) are dependent on the sampling interval Δt of the current control system with the equal value

$$k^{\alpha}_{fcs} = k^{\beta}_{fcs} = \frac{r_\sigma \tau'_\sigma}{\Delta t}\left(1 - \frac{\Delta t}{\tau'_\sigma}\right). \tag{7.56}$$

In order to ensure a negative feedback in the current control, the quantity $1 - \frac{\Delta t}{\tau'_\sigma} > 0$, that is $\frac{\Delta t}{\tau'_\sigma} < 1$, the sampling time has to be smaller than the current model time constant τ'_σ.

Since the voltage control variables have limited values, the optimal control signals given in (7.52) and (7.53) are not exactly realized, thus, the objective function J needs to be further minimized. Following the same derivation procedure outlined in Section 6.4, the objective function (7.49) is expressed as

$$J = \left(\begin{bmatrix} u_{s\alpha}(t_i) \\ u_{s\beta}(t_i)\end{bmatrix} - \begin{bmatrix} u_{s\alpha}(t_i)^{opt} \\ u_{s\beta}(t_i)^{opt}\end{bmatrix}\right)^T (\Delta t^2 B_m^T B_m) \left(\begin{bmatrix} u_{s\alpha}(t_i) \\ u_{s\beta}(t_i)\end{bmatrix} - \begin{bmatrix} u_{s\alpha}(t_i)^{opt} \\ u_{s\beta}(t_i)^{opt}\end{bmatrix}\right)$$
$$= \frac{\Delta t^2}{(r_\sigma \tau'_\sigma)^2}(u_{s\alpha}(t_i) - u_{s\alpha}(t_i)^{opt})^2 + \frac{\Delta t^2}{(r_\sigma \tau'_\sigma)^2}(u_{s\beta}(t_i) - u_{s\beta}(t_i)^{opt})^2, \tag{7.57}$$

where $u_{s\alpha}(t_i)$ and $u_{s\beta}(t_i)$ are the pair of control signals to be selected among the seven pair of candidate variables.

For induction motor control, the candidate variables are in identical forms to those used in the PMSM control, whose values are restricted to the following seven pairs:

$$\begin{bmatrix} 0 & 1 & \frac{1}{2} & -\frac{1}{2} & -1 & -\frac{1}{2} & \frac{1}{2} \\ 0 & 0 & \frac{\sqrt{3}}{2} & \frac{\sqrt{3}}{2} & 0 & -\frac{\sqrt{3}}{2} & -\frac{\sqrt{3}}{2} \end{bmatrix} \frac{2}{3} V_{dc}. \tag{7.58}$$

It is seen that once the voltage from the power supply V_{dc} is determined, the candidate variables are time-invariant. Because the candidate variables in the $\alpha - \beta$ reference frame are constants, in addition to the optimal voltage variables being scalars, the real-time computational load for the control law is less than that in the $d - q$ reference frame.

Figure 7.4 shows the configuration of the feedback control system to generate optimal control signals $u_{s\alpha}(t)^{opt}$ and $u_{s\beta}(t)^{opt}$ when using the proportional feedback and feedforward compensations.

From this figure it is seen that the original FCS predictive controller in the $\alpha - \beta$ reference frame is a proportional controller with feedforward compensation. Because the reference signals in the $\alpha - \beta$ frame are sinusoidal signals, the proportional controller will not completely eliminate the steady-state errors in the closed-loop control systems. Furthermore, the steady-state errors reduce as the feedback controller gain $\frac{r_\sigma \tau'_\sigma}{\Delta t}(1 - \frac{\Delta t}{\tau'_\sigma})$ increases which can be achieved by reducing the sampling interval Δt.

The reference signals to the FCS current control in the $\alpha - \beta$ reference signals are sinusoidal signals in which their frequency is determined by the synchronous velocity of the induction motor $\omega_s(t)$. As explained in Section 1.5, the stator current signals in the $d - q$ reference frame are DC signals, therefore, their reference signals are constant or piece-wise constant, which are easier to specify. Therefore, similar to the PMSM applications, the desired operational performance in a closed-loop current control is specified via the desired reference values of stator currents i^*_{sd} and i^*_{sq} in the $d - q$ reference frame. For instance, the desired value for i^*_{sd} is chosen related to the rotor flux as $\frac{\psi^*_{rd}}{L_h}$ and the i^*_{sq} current is related to

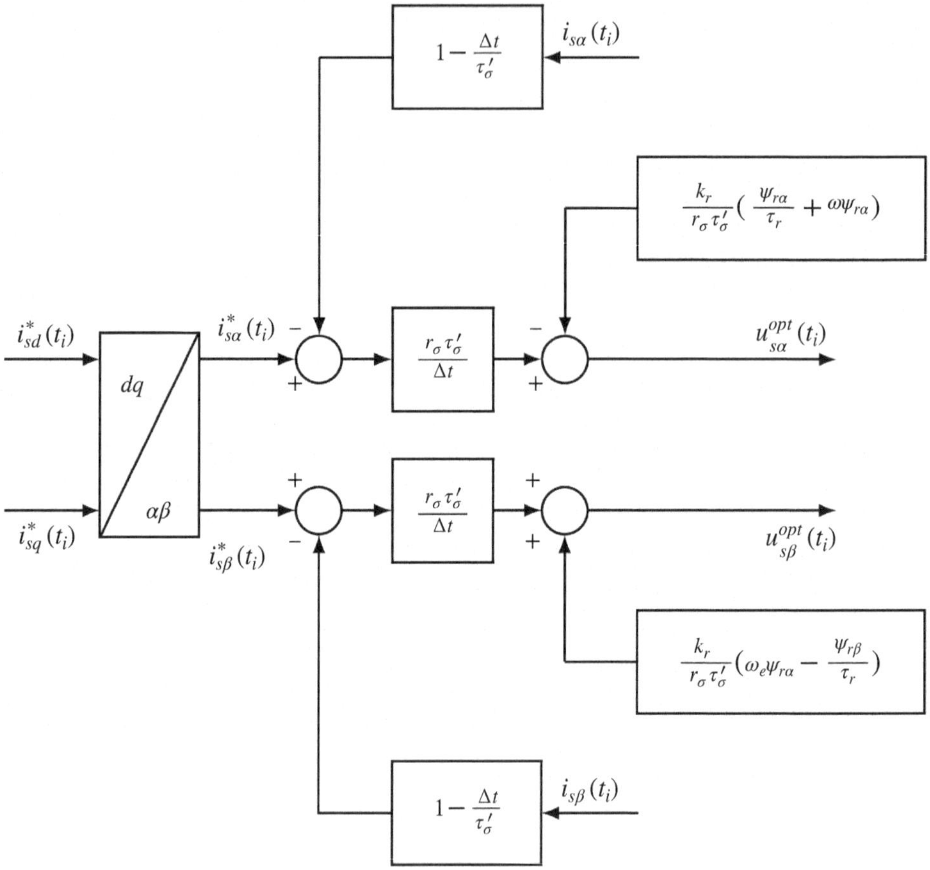

Figure 7.4 FCS-MPC in $\alpha - \beta$ frame.

required electromagnetic torque. With stator current reference signals specified, the reference signals to $i^*_{s\alpha}$ and $i^*_{s\beta}$ currents are calculated using the inverse Park transformation:

$$\begin{bmatrix} i^*_{s\alpha}(t) \\ i^*_{s\beta}(t) \end{bmatrix} = \begin{bmatrix} \cos\theta_s(t) & -\sin\theta_s(t) \\ \sin\theta_s(t) & \cos\theta_s(t) \end{bmatrix} \begin{bmatrix} i^*_{sd}(t) \\ i^*_{sq}(t) \end{bmatrix}. \tag{7.59}$$

7.3.2 Resonant FCS Predictive Current Control of Induction Motor

The resonant controller for current control of an induction motor has a polynomial factor of $1 - 2\cos(\omega_d(t))z^{-1} + z^{-2}$, which is $(1 - e^{j\omega_d(t)}z^{-1})(1 - e^{-j\omega_d(t)}z^{-1})$, where $\omega_d(t)$ is the discrete frequency of the sinusoidal reference signal that is dependent on the synchronous system's $\omega_s(t)$. The resonant FCS current controller is proposed to have the feedback structure as illustrated in Figure 7.5. In the proposed control system structure, the feedback controllers k^α_{fcs} and k^α_{fcs} derived from the one-step-ahead prediction are calculated using

$$k^\alpha_{fcs} = k^\beta_{fcs} = \frac{r_\sigma \tau'_\sigma}{\Delta t}\left(1 - \frac{\Delta t}{\tau'_\sigma}\right).$$

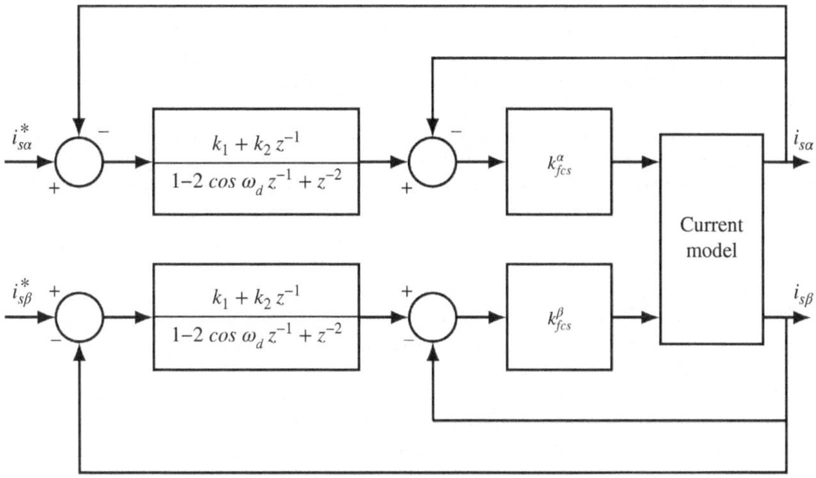

Figure 7.5 Resonant FCS-MPC in $\alpha - \beta$ frame.

These two proportional controllers are located in the inner-loops for fast dynamic response with a fast sampling rate, while two resonant controllers are used in the outer-loops to provide further compensations for the tracking errors between the reference and feedback current signals in the $\alpha - \beta$ reference frame. Note that with the resonant controllers, the feedforward compensation terms in the original FCS current controller are no longer required as the controllers have the capability to completely reject the sinusoidal disturbances caused by the rotor fluxes. Thus the estimation of the rotor fluxes in the $\alpha - \beta$ reference frame is avoided and so too are the associated estimation errors.

The frequency $\omega_d(t)$ is the discrete frequency, which has the unit of radian per sample interval. Unlike the PMSM, the frequency ω_d is the discrete frequency of the synchronous frequency ω_s. For example, if the synchronous velocity of the induction motor is $\omega_s = 200$ rad/s, and the sampling interval is $\Delta t = 20\ \mu s$, then the discrete frequency is

$$\omega_d = \omega_s \times \Delta t = 0.004\ \text{rad}.$$

The frequency parameter ω_d is time-varying when the synchronous velocity changes. However, when the sampling interval Δt is small, this variation has a small effect on the locations of the complex poles in the controller.

The design of the resonant controller is identical to that used in the application of PMSM control (see Section 7.2), where the coefficients of the resonant controller are calculated as

$$k_1 = 2\cos\omega_d - 2\lambda$$
$$k_2 = \lambda^2 - 1,$$

where $0 \leq \lambda < 1$ is the desired discrete closed-loop pole location for the current control system, or

$$k_1 = 2\cos\omega_d - 2e^{-\xi w_n \Delta t}\cos(w_n\sqrt{1-\xi^2}\Delta t)$$
$$k_2 = e^{-2\xi w_n \Delta t} - 1,$$

where $\xi = 0.707$ and w_n is the desired bandwidth for the closed-loop current control system specified in the continuous-time.

Figure 7.5 shows the configuration of the resonant FCS current control system in the $\alpha - \beta$ reference frame. The implementation of the resonant FCS predictive controller is summarized below. The derivation of this algorithm follows a similar procedure presented in the resonant FCS predictive control of PMSM (see Section 7.2.3).

Algorithm 6 *The resonant FCS control signals in the $\alpha - \beta$ reference frame at sampling time t_i, $u_{s\alpha}(t_i)$ and $u_{s\beta}(t_i)$, are found by finding the minimum of the objective function J with respect to the index k,*

$$J = \frac{\Delta t^2}{(r_\sigma \tau'_\sigma)^2}(u_{s\alpha}(t_i)^k - u_{s\alpha}(t_i)^{opt})^2 + \frac{\Delta t^2}{(r_\sigma \tau'_\sigma)^2}(u_{s\beta}(t_i)^k - u_{s\beta}(t_i)^{opt})^2,$$

where the values of $u_{s\alpha}(t_i)^k$ and $u_{s\beta}(t_i)^k$ ($k = 0, 1, 2, \ldots, 6$) are given by the finite control set,

$$\begin{bmatrix} 0 & 1 & \frac{1}{2} & -\frac{1}{2} & -1 & -\frac{1}{2} & \frac{1}{2} \\ 0 & 0 & \frac{\sqrt{3}}{2} & \frac{\sqrt{3}}{2} & 0 & -\frac{\sqrt{3}}{2} & -\frac{\sqrt{3}}{2} \end{bmatrix} \frac{2}{3} V_{dc}$$

and the signals $u_{s\alpha}(t_i)^{opt}$ and $u_{s\beta}(t_i)^{opt}$ are computed recursively using the following equations:

$$u_{s\alpha}(t_i)^{opt} = 2\cos\omega_d u_{s\alpha}(t_{i-1})^{opt} - u_{s\alpha}(t_{i-2})^{opt} + u_{s\alpha}(t_i)^{sopt} \tag{7.60}$$

$$\begin{aligned} u_{s\alpha}(t_i)^{sopt} = {} & k^{\alpha}_{fcs}[k_1(i^*_{s\alpha}(t_i) - i_{s\alpha}(t_i)) + k_2(i^*_{s\alpha}(t_{i-1}) - i_{s\alpha}(t_{i-1})) \\ & - (i_{s\alpha}(t_i) - 2\cos\omega_d i_{s\alpha}(t_{i-1}) + i_{s\alpha}(t_{i-2}))] \end{aligned} \tag{7.61}$$

$$u_{s\beta}(t_i)^{opt} = 2\cos\omega_d u_{s\beta}(t_{i-1})^{opt} - u_{s\beta}(t_{i-2})^{opt} + u_{s\beta}(t_i)^{sopt} \tag{7.62}$$

$$\begin{aligned} u_{s\beta}(t_i)^{sopt} = {} & k^{\beta}_{fcs}[k_1(i^*_{s\beta}(t_i) - i_{s\beta}(t_i)) + k_2(i^*_{s\beta}(t_{i-1}) - i_{s\beta}(t_{i-1})) \\ & - (i_{s\beta}(t_i) - 2\cos\omega_d i_{s\beta}(t_{i-1}) + i_{s\beta}(t_{i-2}))]. \end{aligned} \tag{7.63}$$

The feedback control gains used in the computation are defined as:

$$k^{\alpha}_{fcs} = k^{\beta}_{fcs} = \frac{r_\sigma \tau'_\sigma}{\Delta t}(1 - \frac{\Delta t}{\tau'_\sigma}) \tag{7.64}$$

$$k_1 = 2\cos\omega_d - 2\lambda \tag{7.65}$$

$$k_2 = \lambda^2 - 1. \tag{7.66}$$

$0 \leq \lambda < 1$ *is the desired closed-loop pole location for the current control system.*

7.3.3 Experimental Evaluations of Resonant FCS Predictive Control

There are two sets of experimental evaluation results presented in this section. One is the original FCS predictive control of induction motor while the other is the resonant FCS predictive control. Both evaluations use the same motor parameters and the same sampling interval where $\Delta t = 80 \times 10^{-6}$ sec. With the motor parameters and the sampling interval, the proportional controller gains are calculated as

$$k^{\alpha}_{fcs} = k^{\beta}_{fcs} = 1310.3.$$

In the experiments, the reference signals $i^*_{s\alpha}$ and $i^*_{s\beta}$ are calculated using the inverse Park Transform

$$\begin{bmatrix} i^*_{s\alpha}(t) \\ i^*_{s\beta}(t) \end{bmatrix} = \begin{bmatrix} \cos\theta_s(t) & -\sin\theta_s(t) \\ \sin\theta_s(t) & \cos\theta_s(t) \end{bmatrix} \begin{bmatrix} i^*_{sd}(t) \\ i^*_{sq}(t) \end{bmatrix}, \tag{7.67}$$

where the reference signal to the d-axis stator current i^*_{sd} is 0.8772 A, and the reference signal to the q-axis stator current i^*_{sq} is 0 A for an initial period and a step change to 1.5 A. The induction motor has an unknown load disturbance.

7.3.3.1 Results from the Original FCS Predictive Current Control

The experimental results of the original FCS predictive control in the $\alpha - \beta$ reference frame are shown in Figure 7.6 (a)–(c). In Figure 7.6(a), the stator currents, $i_{s\alpha}$ and $i_{s\beta}$, are shown to track the sinusoidal reference signals and these signals are converted to their corresponding current signals in the $d - q$ reference frame as seen in Figure 7.6(b). The frequency analysis of the A-phase current is performed and its magnitude is presented in Figure 7.6(c). With respect to this set of physical parameters and the selection of sampling interval Δt, the tracking errors in both $\alpha - \beta$ reference frame and the $d - q$ reference frame are very small, not obvious from reading of the graphs. Indeed, the mean error between $i_{s\alpha}^*$ and $i_{s\alpha}$ is -1.5×10^{-3} A and the one between $i_{s\beta}^*$ and $i_{s\beta}$ is -1.3×10^{-3} A, which are quite small. It may be concluded from this test environment that the original FCS predictive current control system produces a relatively small steady-state error for this choice of sampling interval. However, further tests need to be conducted for performance robustness. It is envisaged that the closed-loop performance will degrade when the physical parameters are varying and a larger load is added.

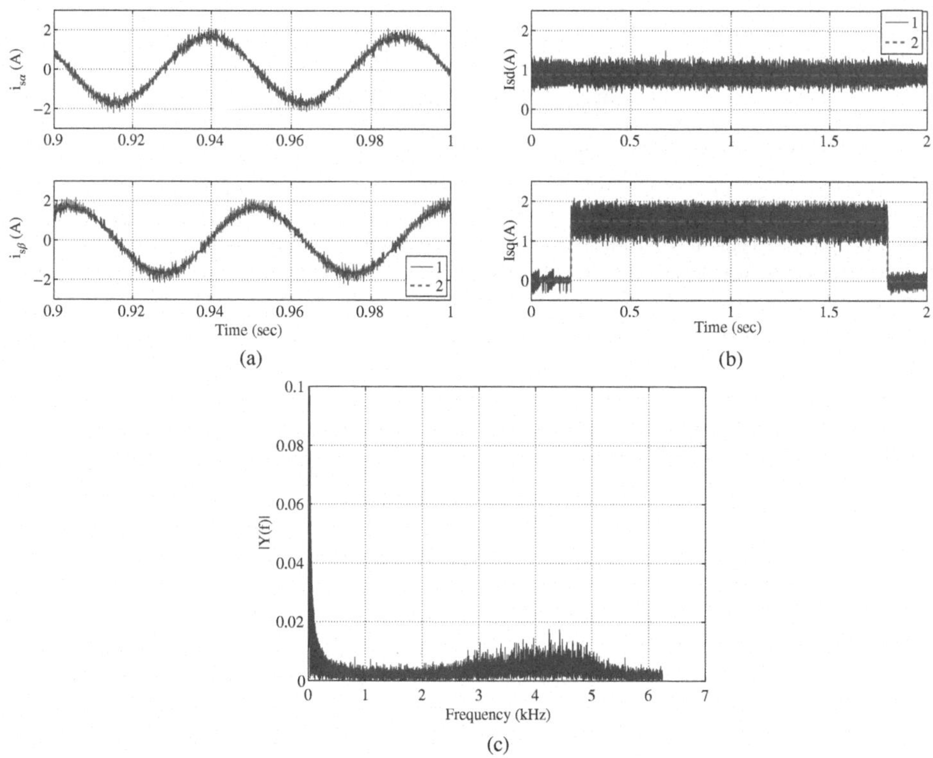

Figure 7.6 Experimental results of current control using the original FCS-MPC scheme in $\alpha - \beta$ reference frame. $\Delta t = 80\,\mu$s. Key: line (1) Actual feedback; line (2) set-point signal. (a) Stator current response in $\alpha - \beta$ reference frame, (b) Stator current response in $d - q$ reference frame, and (c) Amplitude of the frequency response of A phase current.

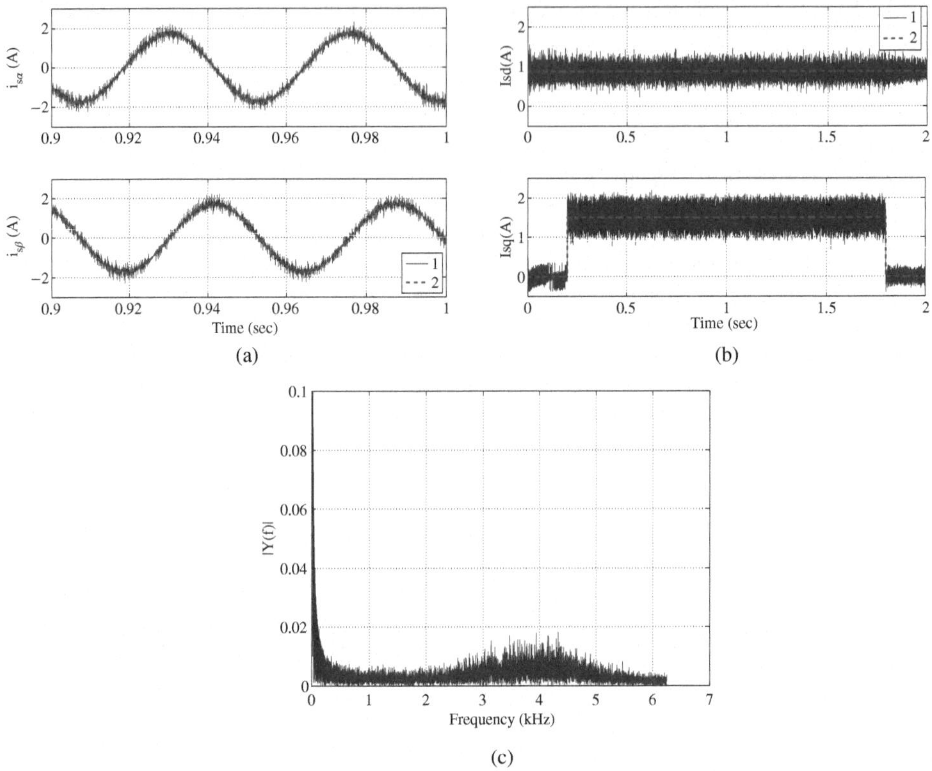

Figure 7.7 Experimental results of current control using the resonant FCS-MPC scheme in $\alpha - \beta$ reference frame. $\Delta t = 80\,\mu$s. The closed-loop poles for the resonant controller are located at 0.92. Key: line (1) Actual feedback; line (2) set-point signal. (a) Stator current response in $\alpha - \beta$ reference frame, (b) Stator current response in $d - q$ reference frame, and (c) Amplitude of the frequency response of A phase current.

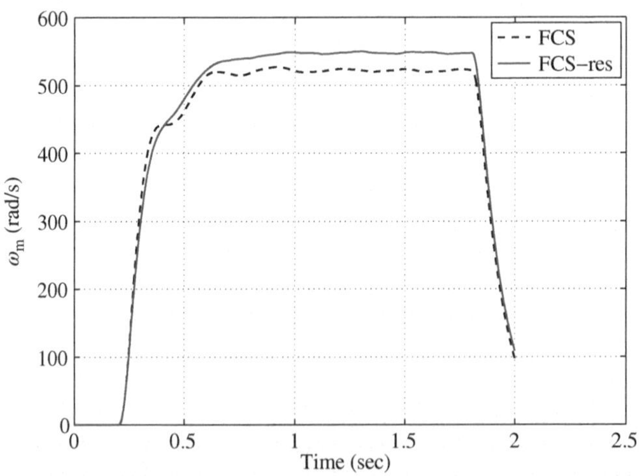

Figure 7.8 Comparison of open-loop response of the motor speed.

7.3.3.2 Results from the Resonant FCS Predictive Current Control

The experimental results of the resonant FCS predictive control system are shown in Figures 7.7(a)–(c). In Figure 7.7(a), it is seen that the stator currents in the $\alpha - \beta$ reference frame track their sinusoidal current reference signals, and in Figure 7.7(b), the corresponding current signals are shown in the $d - q$ reference frame. Figure 7.7(c) shows the magnitude of the frequency response of A phase current. All three figures confirm that the resonant FCS current control system provides a satisfactory closed-loop performance. The mean error between $i^*_{s\alpha}$ and $i_{s\alpha}$ is -1.2358×10^{-4} A and the one between $i^*_{s\beta}$ and $i_{s\beta}$ is -2.2802×10^{-4} A, which are about 10 percent of the mean errors produced by the original FCS predictive controller. Furthermore, by comparing the open-loop velocity responses as shown in Figure 7.8, it is seen that the two FCS current control schemes result in different motor velocity responses, particularly in the steady-state. For the original FCS current controller, the steady-state speed is about 522 RPM and for the resonant FCS current controller, the steady-state speed is about 548 RPM, which is a higher speed given all other physical conditions being identical. This could mean that by regulating the steady-state current responses closer to the desired reference signals, the induction motor has an improved efficiency in operation.

7.4 Resonant FCS Predictive Power Converter Control

In this section, we will firstly discuss the original FCS predictive current control of a power converter in the $\alpha - \beta$ reference frame, secondly propose the resonant FCS predictive control of power converter, and thirdly present the experimental evaluation results of both control schemes.

7.4.1 FCS Predictive Current Control of Power Converter

The mathematical model of a power converter in the stationary frame is described by

$$\frac{di_\alpha(t)}{dt} = -\frac{R_s}{L_s} i_\alpha(t) - \frac{1}{L_s} v_\alpha(t) + \frac{1}{L_s} E_\alpha(t) \tag{7.68}$$

$$\frac{di_\beta(t)}{dt} = -\frac{R_s}{L_s} i_\beta(t) - \frac{1}{L_s} v_\beta(t) - \frac{1}{L_s} E_\beta(t) \tag{7.69}$$

$$\frac{dv_{dc}(t)}{dt} = \frac{3}{4C}(S_\alpha(t) i_\alpha(t) + S_\beta(t) i_\beta(t)) - i_L, \tag{7.70}$$

where E_α and E_β are the stationary frame grid voltages, and $i_\alpha(t)$ and $i_\beta(t)$ are the stationary frame grid currents. v_{dc} is the *DC* bus voltage as illustrated previously in Section 1.6 of Chapter 1.

The manipulated variables for the current control are the voltages v_α, v_β in the $\alpha - \beta$ reference frame, and they are related S_α, S_β and $v_{dc}(t)$ via the following relationships:

$$v_\alpha(t) = \frac{S_\alpha(t) v_{dc}(t)}{2} \tag{7.71}$$

$$v_\beta(t) = \frac{S_\beta(t) v_{dc}(t)}{2}. \tag{7.72}$$

In the $\alpha - \beta$ reference frame, there are seven pairs of candidate voltage values v_α and v_β, which are dependent on the *DC* bus voltage $v_{dc}(t)$. Their exact values are characterized by the values listed below:

$$\begin{bmatrix} 0 & 1 & \frac{1}{2} & -\frac{1}{2} & -1 & -\frac{1}{2} & \frac{1}{2} \\ 0 & 0 & \frac{\sqrt{3}}{2} & \frac{\sqrt{3}}{2} & 0 & -\frac{\sqrt{3}}{2} & -\frac{\sqrt{3}}{2} \end{bmatrix} \frac{2}{3} v_{dc}(t). \tag{7.73}$$

The variations of the candidate voltage values are caused by the changes of DC bus voltage v_{dc}. This is the key difference between the control of power converter and that of the AC drives as discussed in the previous sections of this chapter.

It is seen from this model that in the $\alpha - \beta$ reference frame, there is no interaction between the currents i_α and i_β, which will effectively reduce the current controller in this reference frame to two single-input and single-output controllers. To calculate the control variables, at the sampling time t_i, the objective function is chosen as sum of the squared errors between the desired and predicted signals:

$$J = (i^*_\alpha(t_i) - i_\alpha(t_{i+1}))^2 + (i^*_\beta(t_i) - i_\beta(t_{i+1}))^2$$
$$= \left(\begin{bmatrix} i^*_\alpha(t_i) \\ i^*_\beta(t_i) \end{bmatrix} - \begin{bmatrix} i_\alpha(t_{i+1}) \\ i_\beta(t_{i+1}) \end{bmatrix}\right)^T \left(\begin{bmatrix} i^*_\alpha(t_i) \\ i^*_\beta(t_i) \end{bmatrix} - \begin{bmatrix} i_\alpha(t_{i+1}) \\ i_\beta(t_{i+1}) \end{bmatrix}\right), \tag{7.74}$$

where $i_\alpha(t_{i+1})$ and $i_\beta(t_{i+1})$ are one-step-ahead predictions of $i_\alpha(t_i)$ and $i_\beta(t_i)$, respectively. The one-step-ahead predictions of the $i_\alpha(t_{i+1})$ and $i_\beta(t_{i+1})$ are expressed in matrix and vector forms:

$$\begin{bmatrix} i_\alpha(t_{i+1}) \\ i_\beta(t_{i+1}) \end{bmatrix} = (I + \Delta t A_m) \begin{bmatrix} i_\alpha(t_i) \\ i_\beta(t_i) \end{bmatrix} + \Delta t B_m \begin{bmatrix} v_\alpha(t_i) \\ v_\beta(t_i) \end{bmatrix} + \begin{bmatrix} \frac{1}{L_s} E_\alpha \Delta t \\ \frac{1}{L_s} E_\beta \Delta t \end{bmatrix}, \tag{7.75}$$

where I is the identity matrix with dimension 2×2 and the system matrices A_m and B_m are defined as

$$A_m = \begin{bmatrix} -\frac{R_s}{L_s} & 0 \\ 0 & -\frac{R_s}{L_s} \end{bmatrix}; \; B_m = \begin{bmatrix} -\frac{1}{L_s} & 0 \\ 0 & -\frac{1}{L_s} \end{bmatrix}.$$

Substituting the one-step-ahead prediction given by (7.75) into the objective function J (7.74) leads to

$$J = \begin{bmatrix} f_\alpha(t_i) & f_\beta(t_i) \end{bmatrix} \begin{bmatrix} f_\alpha(t_i) \\ f_\beta(t_i) \end{bmatrix} - 2 \begin{bmatrix} v_\alpha(t_i) & v_\beta(t_i) \end{bmatrix} \Delta t B^T_m \begin{bmatrix} f_\alpha(t_i) \\ f_\alpha(t_i) \end{bmatrix}$$
$$+ \begin{bmatrix} v_\alpha(t_i) & v_\beta(t_i) \end{bmatrix} \Delta t^2 B^T_m B_m \begin{bmatrix} v_\alpha(t_i) \\ v_\beta(t_i) \end{bmatrix}, \tag{7.76}$$

where the functions $f_\alpha(t_i)$ and $f_\beta(t_i)$ are defined as

$$\begin{bmatrix} f_\alpha(t_i) \\ f_\beta(t_i) \end{bmatrix} = \begin{bmatrix} i^*_\alpha(t_i) \\ i^*_\beta(t_i) \end{bmatrix} - (I + \Delta t A_m) \begin{bmatrix} i_\alpha(t_i) \\ i_\beta(t_i) \end{bmatrix} - \begin{bmatrix} \frac{1}{L_s} E_\alpha(t_i) \Delta t \\ -\frac{1}{L_s} E_\beta(t_i) \Delta t \end{bmatrix}. \tag{7.77}$$

The optimal control signals $v_\alpha(t_i)$ and $v_\beta(t_i)$ that minimize the objective function J are given by the following expression:

$$\begin{bmatrix} v_\alpha(t_i)^{opt} \\ v_\beta(t_i)^{opt} \end{bmatrix} = (\Delta t^2 B^T_m B_m)^{-1} \Delta t B^T_m \begin{bmatrix} f_\alpha(t_i) \\ f_\beta(t_i) \end{bmatrix}$$
$$= -\frac{1}{\Delta t} \begin{bmatrix} L_s & 0 \\ 0 & L_s \end{bmatrix} \begin{bmatrix} f_\alpha(t_i) \\ f_\beta(t_i) \end{bmatrix}. \tag{7.78}$$

Note that in the $\alpha - \beta$ reference frame, the system matrix A_m is diagonal and there is no interaction between the i_α and i_β currents. The calculations of $v_\alpha(t_i)^{opt}$ and $v_\beta(t_i)^{opt}$ signals are scalar operations, which are given by

$$v_\alpha(t_i)^{opt} = -\frac{L_s}{\Delta t} i^*_\alpha(t_i) + \frac{L_s}{\Delta t}\left(1 - \frac{R_s}{L_s}\Delta t\right) i_\alpha(t_i) + \frac{1}{\Delta t} E_\alpha(t_i) \tag{7.79}$$

$$v_\beta(t_i)^{opt} = -\frac{L_s}{\Delta t} i^*_\beta(t_i) + \frac{L_s}{\Delta t}(1 - \frac{R_s}{L_s}\Delta t) i_\beta(t_i) - \frac{1}{\Delta t} E_\beta(t_i), \tag{7.80}$$

where $i_\alpha^*(t_i)$ and $i_\beta^*(t_i)$ are current reference signals in the $\alpha - \beta$ reference frame, and $i_\alpha(t_i)$ and $i_\beta(t_i)$ are the measured current of the power converter. It is clearly seen that predictive controller uses a proportional feedback control with a feedforward compensation. Furthermore, the feedback control gain is dependent on the sampling interval of the current control system, the gains being given by

$$k_{fcs}^\alpha = k_{fcs}^\beta = -\frac{L_s}{\Delta t}(1 - \frac{R_s}{L_s}\Delta t). \tag{7.81}$$

In order to ensure a negative feedback in the current control, the quantity $1 - \frac{R_s}{L_s}\Delta t > 0$, that is $\frac{R_s}{L_s}\Delta t < 1$. Here the feedback controllers have a negative gain because of the negative diagonal elements in B_m matrix.

Taking a similar approach to that given in the derivation of $d - q$ reference frame, it is easy to show that the objective function J (7.76) can also be expressed in terms of the optimal voltage signals in the $\alpha - \beta$ reference frame as

$$\begin{aligned} J &= \left(\begin{bmatrix} v_\alpha(t_i) \\ v_\beta(t_i) \end{bmatrix} - \begin{bmatrix} v_\alpha(t_i)^{opt} \\ v_\beta(t_i)^{opt} \end{bmatrix}\right)^T (\Delta t^2 B_m^T B_m) \left(\begin{bmatrix} v_\alpha(t_i) \\ v_\beta(t_i) \end{bmatrix} - \begin{bmatrix} v_\alpha(t_i)^{opt} \\ v_\beta(t_i)^{opt} \end{bmatrix}\right) \\ &= \frac{\Delta t^2}{L_s^2}(v_\alpha(t_i) - v_\alpha(t_i)^{opt})^2 + \frac{\Delta t^2}{L_s^2}(v_\beta(t_i) - v_\beta(t_i)^{opt})^2. \end{aligned} \tag{7.82}$$

With both objective function and the optimal control signals defined, the next step in the FCS predictive control is to find the control signal $v_\alpha(t_i)$ and $v_\beta(t_i)$ that will minimize the objective function subject to the limited number of choices of voltage variables as mentioned in (7.73). The pair of v_α and v_β that yields a minimum of the objective function J will be chosen as the FCS-MPC current control signals in the $\alpha - \beta$ reference frame.

The reference signals to the FCS current control in the $\alpha - \beta$ reference frame are sinusoidal signals in which their frequency is determined by the grid frequency ω_g. In the applications, the desired operational performance of a power converter in a closed-loop current control is often specified via the i_d and i_q currents in the $d - q$ reference frame for the reason that these reference signals are transparent to the applications and easy to choose. For instance, the desired value for the i_d current is related to real power in demand and i_q is chosen to be 0 for unity power factor. With the reference signals $i_d^*(t)$ and $i_q^*(t)$ determined, the reference signals to i_α and i_β currents are calculated using the inverse Park transformation:

$$\begin{bmatrix} i_\alpha^*(t) \\ i_\beta^*(t) \end{bmatrix} = \begin{bmatrix} \cos\omega_g t & -\sin\omega_g t \\ \sin\omega_g t & \cos\omega_g t \end{bmatrix} \begin{bmatrix} i_d^*(t) \\ i_q^*(t) \end{bmatrix}. \tag{7.83}$$

In the majority of applications where $i_q^*(t) = 0$, simply, $i_\alpha^*(t) = i_d^*(t)\cos\omega_g t$, and $i_\beta^*(t) = i_d^*(t)\sin\omega_g t$.

Figure 7.9 shows the configuration of the feedback controller to generate optimal control signals $v_\alpha(t)^{opt}$ and $v_\beta(t)^{opt}$. The resonant FCS current controller for the power converter is proposed to have the feedback structure as illustrated in Figure 7.10. In the proposed control system structure, the feedback controllers k_{fcs}^α and k_{fcs}^β derived from the one-step-ahead prediction and optimization are used in the inner-loops for fast dynamic response, while two resonant controllers are used in the outer-loops to provide further compensations for the tracking errors between the reference and feedback current signals in the $\alpha - \beta$ reference frame.

The frequency ω_d is the discrete frequency of the grid, having the unit of radian. Suppose that the frequency of the grid is 50 Hz and the sampling interval is $\Delta t = 80 \times 10^{-6}$ second. Then the frequency parameter ω_d is

$$\omega_d = \frac{2\pi\Delta t}{T} = \frac{2\pi\Delta t}{0.02} = 0.0251.$$

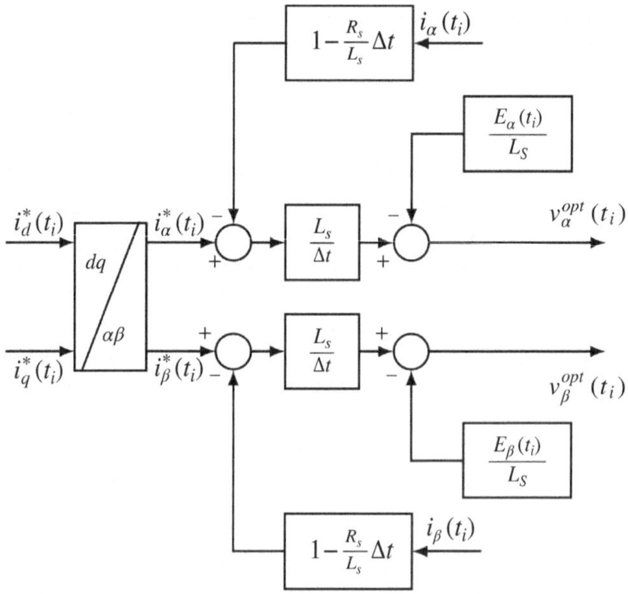

Figure 7.9 FCS predictive current control in $\alpha - \beta$ frame.

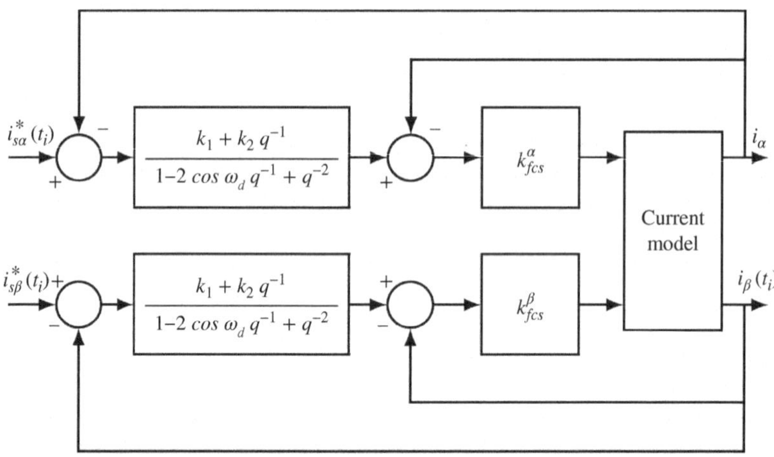

Figure 7.10 Resonant FCS predictive control of power converter in $\alpha - \beta$ reference frame.

The frequency parameter ω_d is time invariant in the design because the grid frequency is generally assumed unchanged. In reality, it may change with respect to time. However, when the sampling interval Δt is small, this variation has a small effect on the locations of the complex poles in the controller. Let us say that the grid frequency varies from 50 to 50.5 Hz. When $\Delta t = 80 \times 10^{-6}$ sec, the corresponding ω_d to 50 Hz is approximately 0.0251 rad and 50.5 Hz is 0.0254 rad. The controller poles for the former case are approximately $0.9997 \pm j0.0251$ and the latter case approximately 0.9997 ± 0.0254.

Therefore, there is no need to track the change of the grid frequency and incorporate it in the resonant controller.

Assuming that the inner-loop system is modeled as one sample of delay, z^{-1}, the task of designing the resonant controller in the outer-loop becomes straightforward, which is identical to the design used in the application of PMSM control (see Section 7.2). More specifically, the coefficients of the resonant controller in Figure 7.10 are calculated as

$$k_1 = 2\cos\omega_d - 2\lambda$$
$$k_2 = \lambda^2 - 1,$$

where $0 \leq \lambda < 1$ is the desired discrete closed-loop pole location for the current control system.

The performance tuning parameter for the resonant FCS controller is the location of the pair of desired discrete closed-loop poles $0 \leq \lambda < 1$. The performance parameter is selected in the design to reflect the closed-loop bandwidth of the control system, depending on the quality of the current model and current sensor noise level. A smaller λ corresponds to faster closed-loop response for the resonant FCS predictive control system, which on the other hand, it may cause noise amplification and the resulting closed-loop system being less robust.

The algorithm for the resonant FCS predictive current controller is summarized as follows. The derivation of the algorithm follows a similar procedure as presented in the case of resonant FCS predictive control of PMSM (see Section 7.2.3) and is not given here.

Algorithm 7 *The resonant FCS control signals in the $\alpha - \beta$ reference frame at sampling time t_i, $v_\alpha(t_i)$ and $v_\beta(t_i)$, are found by finding the minimum of the objective function J with respect to the index k,*

$$J = \frac{\Delta t^2}{L_s^2}(v_\alpha(t_i)^k - v_\alpha(t_i)^{opt})^2 + \frac{\Delta t^2}{L_s^2}(v_\beta(t_i)^k - v_\beta(t_i)^{opt})^2, \tag{7.84}$$

where the values of $v_\alpha(t_i)^k$ and $v_\beta(t_i)^k$ ($k = 0, 1, 2, \ldots, 6$) are given by the finite control set,

$$\begin{bmatrix} 0 & 1 & \frac{1}{2} & -\frac{1}{2} & -1 & -\frac{1}{2} & \frac{1}{2} \\ 0 & 0 & \frac{\sqrt{3}}{2} & \frac{\sqrt{3}}{2} & 0 & -\frac{\sqrt{3}}{2} & -\frac{\sqrt{3}}{2} \end{bmatrix} \frac{2}{3} V_{dc}$$

and the signals $v_\alpha(t_i)^{opt}$ and $v_\beta(t_i)^{opt}$ are computed recursively using the following equations:

$$\begin{aligned} v_\alpha(t_i)^{opt} &= 2\cos\omega_d v_\alpha(t_{i-1})^{opt} - v_\alpha(t_{i-2})^{opt} + v_\alpha(t_i)^{sopt} \\ v_\alpha(t_i)^{sopt} &= k_{fcs}^\alpha[k_1(i_\alpha^*(t_i) - i_\alpha(t_i)) + k_2(i_\alpha^*(t_{i-1}) - i_\alpha(t_{i-1})) \\ &\quad - (i_\alpha(t_i) - 2\cos\omega_d i_\alpha(t_{i-1}) + i_\alpha(t_{i-2}))] \\ v_\beta(t_i)^{opt} &= 2\cos\omega_d v_\beta(t_{i-1})^{opt} - v_\beta(t_{i-2})^{opt} + v_\beta(t_i)^{sopt} \\ v_\beta(t_i)^{sopt} &= k_{fcs}^\beta[k_1(i_\beta^*(t_i) - i_\beta(t_i)) + k_2(i_\beta^*(t_{i-1}) - i_\beta(t_{i-1})) \\ &\quad - (i_\beta(t_i) - 2\cos\omega_d i_\beta(t_{i-1}) + i_\beta(t_{i-2}))]. \end{aligned}$$

The feedback control gains used in the computation are defined as:

$$k_{fcs}^\alpha = k_{fcs}^\beta = \frac{L_s}{\Delta t}(1 - \frac{R_s}{L_s}\Delta t) \tag{7.85}$$

$$k_1 = 2\cos\omega_d - 2\lambda \tag{7.86}$$

$$k_2 = \lambda^2 - 1. \tag{7.87}$$

$0 \leq \lambda < 1$ *is the desired closed-loop pole location for the current control system.*

7.4.2 Experimental Results of Resonant FCS Predictive Control

There are two cases to be investigated in the experimental evaluations. One is the original FCS predictive control and the other is the resonant FCS predictive control for the power converter. The model parameters used in the computation of controller gains $k^{\alpha}_{fcs} = k^{\beta}_{fcs}$ are converter side resistance $R = 0.2\ \Omega$, inductance $L_s = 6.3 \times 10^{-3}$ H. The grid frequency is assumed to 50 Hz that gives the parameter $\omega_g = 100\pi$ rad/ sec. The sampling interval used in the experiments is $\Delta t = 80 \times 10^{-6}$ sec. With these choices of physical parameters, the discrete frequency $\omega_d = 100\pi\Delta t = 0.0251$ rad, and the feedback control gain

$$k^{\alpha}_{fcs} = k^{\beta}_{fcs} = -\frac{L_s}{\Delta t}\left(1 - \frac{R_s}{L_s}\Delta t\right) = -157.3.$$

The reference signals to the $i_\alpha(t)$ and $i_\beta(t)$ current signals at sampling time t_i are calculated using the inverse Park transformation:

$$\begin{bmatrix} i^*_\alpha(t) \\ i^*_\beta(t) \end{bmatrix} = \begin{bmatrix} \cos\omega_g t & -\sin\omega_g t \\ \sin\omega_g t & \cos\omega_g t \end{bmatrix} \begin{bmatrix} i^*_d(t) \\ i^*_q(t) \end{bmatrix}, \tag{7.88}$$

where the reference signal $i^*_d(t)$ is a step signal that is changed from 3 A to 5 A and the reference signal $i^*_q(t)$ is zero.

7.4.2.1 The Original FCS Predictive Control Results

Figure 7.11 shows the closed-loop current responses in the $\alpha - \beta$ reference frame and their errors in the steady-state operation at $i^*_d = 3$ A and $i^*_d = 5$ A and for a step change in the $i^*_d(t)$. The error signals in the figure clearly show that the responses contain periodic components, and therefore, they do not entirely track the periodic current reference signals.

7.4.2.2 The Resonant FCS Predictive Control Results

In the design of the resonant FCS predictive controller, the closed-loop poles for the resonant control system are chosen to be $\lambda = 0.95$, which leads to the controller gains:

$$k_1 = 2\cos\omega_d - 2\lambda = 0.0994 \tag{7.89}$$

$$k_2 = \lambda^2 - 1 = -0.0975. \tag{7.90}$$

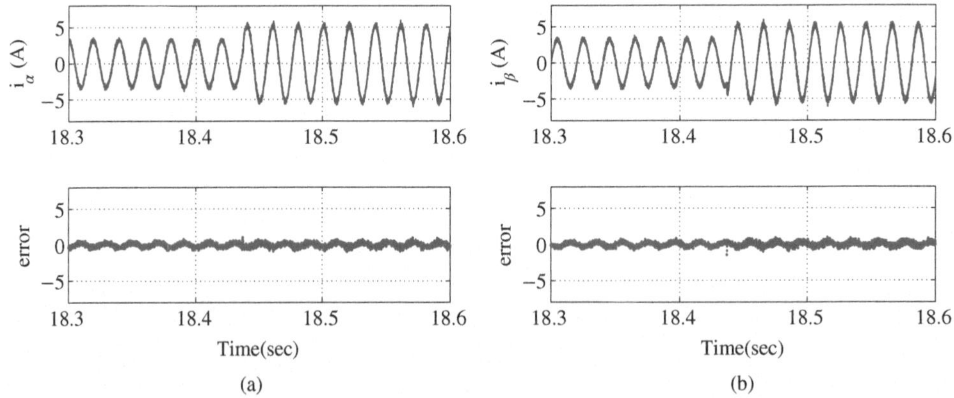

Figure 7.11 Experimental results of current control of power converter using the original FCS scheme in $\alpha - \beta$ reference frame. $\Delta t = 80\ \mu$s. Key: line (1) Actual feedback; line (2) set-point signal. (a) i_α and i_α current error and (b) i_β and i_β current error.

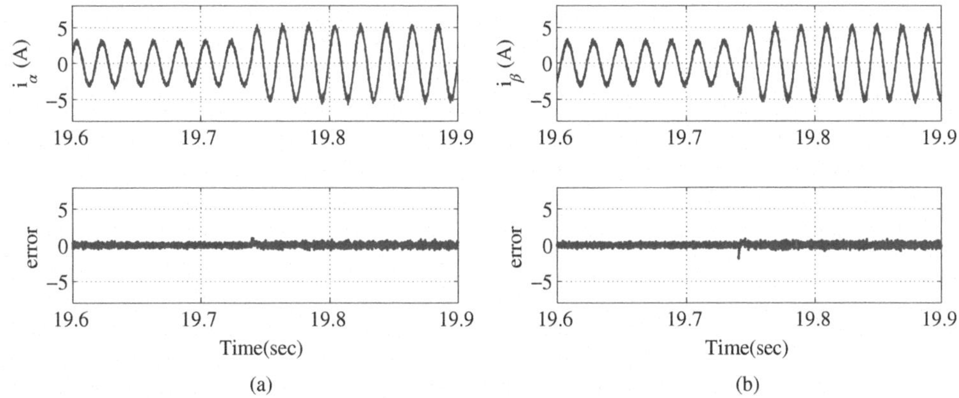

Figure 7.12 Experimental results of current control of power converter using the resonant FCS scheme in $\alpha - \beta$ reference frame. $\Delta t = 80$ μs. Key: line (1) Actual feedback; line (2) set-point signal. (a) i_α and i_α current error and (b) i_β and i_β current error.

Figure 7.12 shows the closed-loop current responses in the $\alpha - \beta$ reference frame and their errors for steady-state operation at $i_d^* = 3$ A and $i_d^* = 5$ A as well as the transient responses at the change from 3 A to 5 A. From the error signals, it is clearly seen that the periodic components exhibited in the original FCS predictive control system are removed. The removal of the periodic components in the error signals is translated into a better tracking of the i_d^* and i_q^* reference signals. Here, the original FCS controller had a mean value of 3.3129 A and 5.2666 A for the set point 3 A and 5 A respectively, which is about 10.43% and 5.32% errors, respectively. In comparison, the resonant FCS controller had a mean value of 3.0008 A and 4.9611 A for the same reference signal, which is about 0.0267% and 0.778% errors, respectively.

7.5 Summary

This chapter has presented the design and implementation of resonant FCS predictive current control in the $\alpha - \beta$ reference frame.

It is shown that the traditional FCS predictive current controller that deploys a cost function using sum of squared errors is essentially a high gain proportional controller in the absence of constraints. Thus, in the steady-state operation, the current outputs cannot entirely follow the sinusoidal current reference signals.

Taking a similar approach as proposed in Chapter 6, this chapter has analyzed the characteristics of the closed-loop current control system that uses the original FCS current controller in the $\alpha - \beta$ reference frame, which is a deadbeat control system with its closed-loop pole located at the origin of the complex plane. In a cascade control system structure, a resonant controller is designed to track the sinusoidal reference signal without steady-state errors. In the design of the resonate controller, the original FCS predictive control system is modeled by one sample of delay. Since in the $\alpha - \beta$ reference frame, there are no interactions between the $\alpha - \beta$ variables, the FCS proportional controller is decoupled into two single input and single output controllers, leading to a reduced computational load and a simpler controller structure in comparison with the its counterpart in the $d - q$ reference frame.

With the limitations of the IGBT voltage inverter, the candidate variables in the $\alpha - \beta$ reference frame are a set of constant variables for the electrical drives once the voltage of the power supply is given. For the power converters, the candidate variables are functions of the voltage converter. In any case, the candidate variables are not sinusoidal functions, which again leads to potential savings of computational cost. Furthermore, the inverse Park transformation that converters the variables in the $d - q$ reference frame to the $\alpha - \beta$ reference frame is performed to generate the sinusoidal reference signals; however, the nonlinear transformation occurs outside the current feedback-loop.

Similar to the FCS predictive control in the $d - q$ reference frame, the closed-loop performance of the resonant FCS predictive control system is dependent on the choice of sampling rate. In general, a smaller Δt will lead to smaller current variations. The selection of the desired closed-loop poles for the resonant controller follows the specification on the desired closed-loop current response. In general, a faster current response will demand a pair of desired closed-loop poles with a smaller magnitude.

7.6 Further Reading

The model predictive torque control was proposed in Geyer *et al.* (2009) and implemented in Papafotiou *et al.* (2009). Finite control set predictive control was used in controlling PMSM drive systems (Preindl and Bolognani (2013a), Preindl and Bolognani (2013b)). The dead-beat control design technique was discussed in Kukrer (1996). Time-delay was examined in the application of finite control set in Cortes *et al.* (2012). Finite control set predictive control for active front end power converter was presented in Cortes *et al.* (2008b) and Kouro *et al.* (2009). Predictive current control of a voltage source inverter was proposed in Rodriguez *et al.* (2007) with extension to electric drives and power converters in Cortes *et al.* (2008a). A comparative study was presented for current control of PMSM drives by Morel *et al.* (2009). Stationary frame three phase *AC* current regulators were designed in Holmes *et al.* (2009). Predictive torque control of induction machines was discussed in Nemec *et al.* (2007), Correa *et al.* (2007). A book was written based on finite control set predictive control (see Rodriguez and Cortes (2012)).

Proportional-resonant controller and filters for grid-connected voltage-source converters were discussed in Teodorescu *et al.* (2006), PI-resonant controller in Liserre *et al.* (2006), adaptive resonant controller in Timbus *et al.* (2006). Dead-beat control techniques were discussed in Gokhale *et al.* (1987), in Kawabata *et al.* (1990), in Malesani *et al.* (1999), in Mohamed and El-Saadany (2007) and in Yang and Lee (2002). Geyer (2011) made comparative studies of control and modulation schemes for medium-voltage drives between predictive control concepts and PWM-Based Schemes.

Internal model control principle was applied to minimize torque ripples Gan and Qiu (2004). Repetitive predictive control algorithms were discussed in Cao and Low (2009), Wang *et al.* (2012), in Wang and Rossiter (2008), in Wang *et al.* (2013), application to inverters in Zhou *et al.* (2009), in Tzou *et al.* (1997), in Zhang *et al.* (2003), the continuous-time counter part was proposed in Wang *et al.* (2010).

References

Cao R and Low KS 2009 A repetitive model predictive control approach for precision tracking of a linear motion system. *IEEE Transactions on Industrial Electronics* **56**(6), 1955–1962.

Correa P, Pacas M and Rodriguez J 2007 Predictive torque control for inverter-fed induction machines. *IEEE Transactions on Industrial Electronics* **54**, 1073–1079.

Cortes P, Kazmierkowski M, Kennel R, Quevedo D and Rodriguez J 2008a Predictive control in power electronics and drives. *IEEE Transactions on Industrial Electronics* **55**(12), 4312–4324.

Cortes P, Rodriguez J, Antoniewicz P and Kazmierkowski M 2008b Direct power control of an AFE using predictive control. *IEEE Transactions on Power Electronics* **23**(5), 2516–2523.

Cortes P, Rodriguez J, Silva C and Flores A 2012 Delay compensation in model predictive current control of a three-phase inverter. *IEEE Transactions on Industrial Electronics* **59**(2), 1323–1325.

Francis BA and Wonham WM 1975 The internal model principle for linear multivariable regulators. *Applied Mathematics and Optimization* **2**(2), 170–194.

Gan WC and Qiu L 2004 Torque and velocity ripple elimination of AC permanent magnet motor control systems using the internal model principle. *IEEE/ASME Transactions on Mechatronics* **9**(2), 436–447.

Geyer T 2011 A comparison of control and modulation schemes for medium-voltage drives: Emerging predictive control concepts versus PWM-based schemes. *IEEE Transactions on Industry Applications* **47**(3), 1380–1389.

Geyer T, Papafotiou G and Morari M 2009 Model predictive direct torque control part I: Concept, algorithm, and analysis. *IEEE Transactions on Industrial Electronics* **56**(6), 1894–1905.

Gokhale K, Kawamura A and Hoft R 1987 Dead beat microprocessor control of PWM inverter for sinusoidal output waveform synthesis. *IEEE Transactions on Industrial Applications* **IA-23**(5), 901–910.

Holmes D, Lipo T, McGrath B and Kong W 2009 Optimized design of stationary frame three phase AC current regulators. *IEEE Transactions on Power Electronics* **24**(11), 2417–2426.

Kawabata T, Miyashita T and Yamamoto Y 1990 Dead beat control of three phase PWM inverter. *IEEE Transactions on Power Electronics* **5**(1), 21–28.

Kouro S, Cortes P, Vargas R, Ammann U and Rodriguez J 2009 Model predictive control – a simple and powerful method to control power converters. *IEEE Transactions on Industrial Electronics* **56**(6), 1826–1838.

Kukrer O 1996 Deadbeat control of a three-phase inverter with an output LC filter. *IEEE Transactions on Power Electronics* **11**(1), 16–23.

Liserre M, Teodorescu R and Blaabjerg F 2006 Multiple harmonics control for three-phase grid converter systems with the use of PI-RES current controller in a rotating frame. *IEEE Transactions on Power Electronics* **21**(3), 836–841.

Malesani L, Mattavelli P and Buso S 1999 Robust dead-beat current control for PWM rectifiers and active filters. *IEEE Transactions on Industry Applications* **35**(3), 613–620.

Mohamed YR and El-Saadany E 2007 An improved deadbeat current control scheme with a novel adaptive self-tuning load model for a three-phase PWM voltage-source inverter. *IEEE Transactions on Power Electronics* **54**(2), 747–759.

Morel F, Lin-Shi X, Retif JM, Allard B and Buttay C 2009 A comparative study of predictive current control schemes for a permanent-magnet synchronous machine drive. *IEEE Transactions on Industrial Electronics* **56**(7), 2715–2728.

Nemec M, Nedeljkovic D and Ambrozic V 2007 Predictive torque control of induction machines using immediate flux control. *IEEE Transactions on Industrial Electronics* **54**(4), 2009–2017.

Papafotiou G, Kley J, Papadopoulos K, Bohren P and Morari M 2009 Model predictive direct torque control part II: Implementation and experimental evaluation. *IEEE Transactions on Industrial Electronics* **56**(6), 1906–1915.

Preindl M and Bolognani S 2013a Model predictive direct speed control with finite control set of PMSM drive systems. *IEEE Transactions on Power Electronics* **28**(2), 1007–1015.

Preindl M and Bolognani S 2013b Model predictive direct torque control with finite control set for PMSM drive systems, part 2: Field weakening operation. *IEEE Transactions on Industrial Informatics* **9**(2), 648–657.

Rodriguez J and Cortes P 2012 *Predictive Control of Power Converters and Electrical Drives*. John Wiley & Sons, Inc., New York.

Rodriguez J, Pontt J, Silva C, Correa P, Lezana P, Cortes P and Ammann U 2007 Predictive current control of a voltage source inverter. *IEEE Transactions on Industrial Electronics* **54**(1), 495–503.

Teodorescu R, Blaabjerg F, Liserre M and Loh P 2006 Proportional-resonant controllers and filters for grid-connected voltage-source converters. *IEE Proceedings Electric Power Applications* **153**: 750–762.

Timbus A, Ciobotaru M, Teodorescu R and Blaabjerg F 2006 Adaptive resonant controller for grid-connected converters in distributed power generation systems *Applied Power Electronics Conference and Exposition, 2006. APEC '06. Twenty-First Annual IEEE.*

Tzou Y, Ou R, Jung SL and Chang M 1997 High-performance programmable AC power source with low harmonic distortion using DSP-based repetitive control technique. *IEEE Transactions on Power Electronics* **12**(4), 715–725.

Wang L and Rossiter J 2008 Disturbance rejection and set-point tracking of sinusoidal signals using generalized predictive control *47th IEEE Conference on Decision and Control, 2008. CDC 2008*, pp. 4079–4084.

Wang L, Chai S, Rogers E and Freeman CT 2012 Multivariable repetitive-predictive controllers using frequency decomposition. *IEEE Transactions on Control Systems Technology* **20**(6), 1597–1604.

Wang L, Freeman CT, Chai S and Rogers E 2013 Predictive-repetitive control with constraints: from design to implementation. *Journal of Process Control*.

Wang L, Gawthrop P, Owens DH and Rogers E 2010 Switched linear predictive controllers for periodic exogenous signals. *International Journal of Control* **83**(4), 848–861.

Yang S and Lee C 2002 A deadbeat current controller for field oriented induction motor drives. *IEEE Transactions on Power Electronics* **17**(5), 772–778.

Zhang K, Kang Y, Xiong J and Chen J 2003 Direct repetitive control of SPWM inverter for UPS purpose. *IEEE Transactions on Power Electronics* **18**(3), 784–792.

Zhou K, Wang D, Zhang B and Wang Y 2009 Plug-in dual-mode-structure repetitive controller for CVCF PWM inverters. *IEEE Transactions on Industrial Electronics* **56**(3), 784–791.

8

Discrete-time Model Predictive Control (DMPC) of Electrical Drives and Power Converter

The previous two chapters (Chapters 6 and 7) have seen the development of FCS predictive control algorithms with experimental demonstrations. The FCS predictive control algorithms are limited to the applications of current control because they are designed using a high gain feedback control approach, and with an inversion of the dynamic model to reach a deadbeat control status. FCS predictive control algorithms work well for current control systems since these systems are first order systems that are invertible and also benefit from high gain feedback control. In the next two chapters, the traditional discrete-time and continuous-time model predictive controller will be used to control the AC drives and power converters. These predictive control algorithms were derived for general applications without those restrictions imposed on system dynamics. The MATLAB programs used in the applications were given in Wang (2009). Although the traditional predictive control algorithms could be applied to current control, their advantages are perhaps lost to the simpler and more effective FCS predictive control approaches, also to the simpler PI controllers. Therefore, in Chapters 8 and 9, velocity control in AC drives and voltage control in power converters are considered for in these cases where the traditional model predictive controllers (MPC) offer the advantages of designing the control systems using multi-input and multi-output approaches in the presence of constraints.

The traditional MPC design is generally based on a linear model, therefore the nonlinearities existing in AC motors and power converters will be linearized at their steady-state operating conditions. Taking PMSM as an example, Section 8.1 shows how to obtain the discrete-time linear model with respect to operating conditions. In Section 8.2, discrete-time MPC design with constraints is introduced based on the PMSM model. In particular, the linearized model from Section 8.1 is used to formulate an augmented model that directly uses the incremental voltage variables as the control inputs. This formulation leads to the embedding of integrators into the predictive controller and simplifies the control implementation process because it does not require the steady-state information and uses the actual physical variables in the computation. In Section 8.3, experimental evaluations of the discrete-time model predictive control system is presented using the PMSM test-bed. In the evaluations, the tuning of the discrete-time MPC controller is discussed. In Section 8.4, discrete-time predictive control system is designed for the power converter and experimental results are presented for demonstration.

PID and Predictive Control of Electrical Drives and Power Converters using MATLAB®/Simulink®, First Edition.
Liuping Wang, Shan Chai, Dae Yoo, Lu Gan and Ki Ng.

Companion Website: www.wiley.com/go/wang/pid

8.1 Linear Discrete-time Model for PMSM

The first step in the design of a continuous-time model predictive controller is to determine a linear model based on the physical models of the electrical drives and power converters. The linearization step was not required when the PID controllers were designed earlier. This material has been extensively covered in textbooks (Goodwin *et al.* (2001), Bay (1999)). Perhaps, an interesting point is that the steady-state values of the physical variables are not removed from the linearized physical model so that all the variables remain to be the actual variables not the deviation variables as in the textbooks.

8.1.1 Linear Model for PMSM

A PMSM is described by the differential equations in the d-q rotating reference frame

$$\frac{di_d(t)}{dt} = \frac{1}{L_d}(v_d(t) - R_s i_d(t) + \omega_e(t) L_q i_q(t)) \tag{8.1}$$

$$\frac{di_q(t)}{dt} = \frac{1}{L_q}(v_q(t) - R_s i_q(t) - \omega_e(t) L_d i_d(t) - \omega_e(t)\phi_{mg}) \tag{8.2}$$

$$\frac{d\omega_e(t)}{dt} = \frac{Z_p}{J_m}(T_e - \frac{B_v}{Z_p}\omega_e(t) - T_L) \tag{8.3}$$

$$T_e = \frac{3}{2} Z_p[\phi_{mg} i_q + (L_d - L_q) i_d(t) i_q(t)]. \tag{8.4}$$

Examining the physical model equations, it is seen that there are three terms having nonlinear expressions: $\omega_e(t)i_q(t)$ (8.1), $\omega_e(t)i_d(t)$ (8.2) and $i_d(t)i_q(t)$ (8.4). These are bilinear functions. In order to obtain the linear approximations for these three terms, three operating points for these variables are selected as i_d^0, i_q^0 and ω_e^0, which are constant in the linearized model and form part of the system parameters.

The bilinear terms are approximated at the operating points using a first order Taylor series expansion as

$$\omega_e(t)i_q(t) \approx \omega_e^0 i_q^0 + i_q^0(\omega_e(t) - \omega_e^0) + \omega_e^0(i_q(t) - i_q^0) \tag{8.5}$$

$$\omega_e(t)i_d(t) \approx \omega_e^0 i_d^0 + i_d^0(\omega_e(t) - \omega_e^0) + \omega_e^0(i_d(t) - i_d^0). \tag{8.6}$$

By substituting these approximations into (8.1) and (8.2), it can be verified that the linearized model has the following form,

$$\frac{dx_m(t)}{dt} = A_m x_m(t) + B_m u(t) + \mu^0 \tag{8.7}$$

$$y(t) = C_m x_m(t), \tag{8.8}$$

where $x_m(t) = [i_d(t) \quad i_q(t) \quad \omega_e(t)]^T$, $u(t) = [v_d(t) \quad v_q(t)]^T$, $y(t) = [i_d(t) \quad \omega_e(t)]^T$ and the matrices A_m, B_m and C_m are defined as

$$A_m = \begin{bmatrix} -\frac{R_s}{L_d} & \frac{L_q}{L_d}\omega_e^0 & \frac{L_q}{L_d}i_q^0 \\ -\frac{L_d}{L_q}\omega_e^0 & -\frac{R_s}{L_q} & -(\frac{L_d}{L_q}i_d^0 + \frac{\phi_{mg}}{L_q}) \\ 0 & \frac{3Z_p^2\phi_{mg}}{2J_m} & -\frac{B_v}{J_m} \end{bmatrix}; \quad B_m = \begin{bmatrix} \frac{1}{L_d} & 0 \\ 0 & \frac{1}{L_q} \\ 0 & 0 \end{bmatrix}$$

$$C_m = \begin{bmatrix} 1 & 0 & 0 \\ 0 & 0 & 1 \end{bmatrix}; \quad \mu^0 = \begin{bmatrix} -\frac{L_q}{L_d}\omega_e^0 i_q^0 \\ \frac{L_d}{L_q}\omega_e^0 i_d^0 \\ -\frac{Z_p T_L}{J_m} \end{bmatrix}.$$

μ^0 is a constant vector that contains the steady-state parameters and the load torque. This term is regarded as an input disturbance to the system and it is overcome when integrators are used in the predictive control as shown in the sequel.

The control objective is to achieve the desired speed set-point whilst maintaining the i_d component at zero and the d-axis and q-axis voltages within operational limits. Therefore, in the state-space formulation, as seen above, the outputs are the d-axis current i_d and the electric velocity ω_e, and the i_q current is one of the state variables that will be measured in the implementation.

8.1.2 Discretization of the Continuous-time Model

The continuous-time state-space model for the PMSM given by (8.7) and (8.8) will be discretized for the purpose of designing a discrete model predictive controller. In Chapters 6 and 7 the discretization of a continuous-time model is performed using an approximation to the differential equation (see for example, (6.1) and (6.12)), and the approximation error is dependent on the sampling interval Δt used in the calculation. The smaller Δt is, the smaller the approximation error. Furthermore, as $\Delta t \to 0$, the coefficients of the discrete-time model converge to their continuous-time counter-part (Middleton and Goodwin (1990)).

Here, instead of using the approximation of differential equations, the discretization of continuous-time model is performed together with the mechanism of zero-order hold, which is a commonly used approach in the computer controlled systems (Astrom and Wittenmark (1997)). This discretization method is reviewed in the following discussions.

We begin with the linearized differential equations (8.7) and (8.8) that describe the dynamics of the PMSM at a set of operating points specified by the user. Assuming a constant sampling interval Δt and sampling index i, the sampling time is expressed as $t_0 = 0$, $t_1 = \Delta t$, $t_2 = 2\Delta t$, $\ldots$, $t_i = i\Delta t$. Now, given an initial condition vector $x_m(0)$, the analytical solution of the differential equation (8.7) at time $t_1 = \Delta t$ is obtained via the expression

$$x_m(t_1) = e^{A_m\Delta t}x_m(0) + \int_0^{\Delta t} e^{A_m(\Delta t-\tau)}B_m u(\tau)d\tau + \int_0^{\Delta t} e^{A_m(\Delta t-\tau)}\mu^0 d\tau, \tag{8.9}$$

where the first term on the left-hand side is the solution generated by the initial condition vector, the second term is the one from the input signal and the third from the constant term. We emphasize that the description of the response $x_m(t_1)$ is exact. As the time progresses to the next sampling time $t_2 = 2\Delta t$, with the initial condition vector $x_m(t_1)$, the analytical solution of the differential equation (8.7) at time t_2 is obtained via the expression

$$x_m(t_2) = e^{A_m\Delta t}x_m(t_1) + \int_{\Delta t}^{2\Delta t} e^{A_m(2\Delta t-\tau)}B_m u(\tau)d\tau + \int_{\Delta t}^{2\Delta t} e^{A_m(2\Delta t-\tau)}\mu^0 d\tau. \tag{8.10}$$

For an arbitrary time $t_i = i\Delta t$, with the initial condition vector $x_m(t_{i-1})$ the analytical solution of $x_m(t_i)$ is given by

$$x_m(t_i) = e^{A_m\Delta t}x_m(t_{i-1}) + \int_{(i-1)\Delta t}^{i\Delta t} e^{A_m(i\Delta t-\tau)}B_m u(\tau)d\tau + \int_{(i-1)\Delta t}^{i\Delta t} e^{A_m(i\Delta t-\tau)}\mu^0 d\tau. \tag{8.11}$$

Note that the first term of the analytical solution is dependent on the sampling interval Δt and the previous sampled state vector $x_m(t_{i-1})$. However, for the convolutional integrals, some manipulations are required to obtain the discrete model. Firstly, assuming that a zero order hold is used in the control signal implementation so that for $(i-1)\Delta t \le \tau < i\Delta t$, the control signal $u(\tau)$ is a constant vector with its value equal to the beginning of the sample period, $u(\tau) = u(t_{i-1})$. With this assumption, the convolutional integral is simplified as the control signal $u(\tau)$ is taken out of the integration. Next, it can be shown that the following integral equality holds:

$$\int_{(i-1)\Delta t}^{i\Delta t} e^{A_m(i\Delta t-\tau')}d\tau' = \int_0^{\Delta t} e^{A_m\tau}d\tau, \tag{8.12}$$

where substitution of variable $i\Delta t - \tau' = \tau$ is used in the derivation. With both the zero-order hold and the integral equality (8.12), the discrete-time model is obtained:

$$x_m(t_i) = e^{A_m\Delta t}x_m(t_{i-1}) + \int_0^{\Delta t} e^{A_m\tau}d\tau B_m u(t_{i-1}) + \int_0^{\Delta t} e^{A_m\tau}d\tau \mu^0. \quad (8.13)$$

Let the discrete system matrices be denoted by

$$A_d = e^{A_m\Delta t}; \quad B_d = \int_0^{\Delta t} e^{A_m\tau}d\tau B_m$$

and the discrete constant vector by $\mu^d = \int_0^{\Delta t} e^{A_m\tau}d\tau\mu^0$. Then, the discrete time model (8.13) is written in a compact form

$$x_m(t_i) = A_d x_m(t_{i-1}) + B_d u(t_{i-1}) + \mu^d. \quad (8.14)$$

The output $y(t_i) = C_d x_m(t_i)$ where $C_d = C_m$. Equation (8.14) can also be expressed in terms of a one-step ahead prediction as

$$x_m(t_{i+1}) = A_d x_m(t_i) + B_d u(t_i) + \mu^d. \quad (8.15)$$

Note that the constant vector μ^0 that contains the steady-state parameters and the torque disturbance remains a constant vector, which is the constant input disturbance vector in the discrete-time model.

The computation of the quantity $\int_0^{\Delta t} e^{A_m\tau}d\tau$ uses the Taylor series expansion of the exponential matrix $e^{A_m\tau}$ followed by integration, yielding to

$$\int_0^{\Delta t} e^{A_m\tau}d\tau = I\Delta t + \frac{A_m\Delta t^2}{2!} + \frac{A_m^2\Delta t^3}{3!} + \ldots + \frac{A_m^j\Delta t^{j+1}}{(j+1)!} + \ldots$$

With this quantity computed, the matrices A_d and B_d are obtained via the following relations:

$$A_d = I + A_m\int_0^{\Delta t} e^{A_m\tau}d\tau; \quad B_d = \int_0^{\Delta t} e^{A_m\tau}d\tau B_m.$$

It is seen that the matrices A_d and B_d can be approximated using the relationships,

$$A_d \approx I + A_m\Delta t; \quad B_d \approx B_m\Delta t,$$

when sampling interval Δt is small. These approximations are equivalent to the previous discrete-time models used in Chapters 6 and 7 (see for example, (6.11) and (6.12)).

MATLAB functions such as "c2d" are commonly used in the computation of the A_d and B_d matrices.

Since there is only one predictive controller to control the i_d, i_q currents and the velocity ω_e, the choice of sampling interval Δt is critical for the implementation of the control system. The time constants for the electrical system are very small in comparison with the mechanical system, as we discussed in Chapter 3, thus, in the predictive control system design, a balance needs to be reached to consider both cases. A small Δt is needed in order to capture the dynamic responses of the i_d and i_q currents, which consequently will increase computational load.

8.2 Discrete-time MPC Design with Constraints

The MPC design used in this chapter is from Wang (2009) and this section summarizes the main steps involved for an m-input, q-output and n-state model where in the experimental results section these are specialized to the case of speed regulation for a PMSM when $m = 2$, $q = 2$, and $n = 3$. The MATLAB programs used in this section can be found in Wang (2009).

8.2.1 Augmented Model

In common with the PI controller, an integrator is usually embedded into the MPC design. Here the benefit of the embedded integrator is exploited to eliminate the vector μ^d in (8.14) or μ^0 in (8.7), which contains motor parameters that have a certain degree of uncertainty associated with them in applications, and the load disturbance torque which is assumed to be an unknown constant. Once the integrators have been added to the model, the optimization analysis is applied to the model expressed in terms of incremental, rather than the actual variables. One of the advantages by using the incremental model is that the steady-state values of the state variables as well as those of the control variables are not required in the implementation. In addition, the differenced constant terms in linearized model, which are denoted by the constant vector μ^d, are zero, when the state variables are expressed in terms of incremental variables.

Taking the difference between the current and next sample instances of the state-space model, that is the difference between Equations (8.15) and (8.14), gives incremental state dynamics as

$$\Delta x_m(t_{i+1}) = A_d \Delta x_m(t_i) + B_d \Delta u(t_i), \tag{8.16}$$

where $\Delta x_m(t_i) = x_m(t_i) - x_m(t_{i-1})$, $\Delta u(t_i) = u(t_i) - u(t_{i-1})$, $\mu^d - \mu^d = 0$. In a similar manner the output incremental dynamics are given by

$$\begin{aligned} y(t_{i+1}) - y(t_i) &= C_d(x_m(t_{i+1}) - x_m(t_i)) = C_d \Delta x_m(t_{i+1}) \\ &= C_d A_d \Delta x_m(t_i) + C_d B_d \Delta u(t_i) \end{aligned} \tag{8.17}$$

or

$$y(t_{i+1}) = y(t_i) + C_d A_d \Delta x_m(t_i) + C_d B_d \Delta u(t_i). \tag{8.18}$$

By choosing a new state vector $x(t_i) = \left[\Delta x_m(t_i) \ y(t_i)\right]^T$ the augmented state-space model is obtained by combining (8.16) with (8.17)

$$x(t_{i+1}) = Ax(t_i) + B\Delta u(t_i) \tag{8.19}$$

$$y(t_i) = Cx(t_i), \tag{8.20}$$

where

$$A = \begin{bmatrix} A_d & 0_1 \\ C_d A_d & I \end{bmatrix}; \quad B = \begin{bmatrix} B_d \\ C_d B_d \end{bmatrix}; \quad C = \begin{bmatrix} 0_2 & I \end{bmatrix}$$

and the 0_1, 0_2 and I, respectively, denote the zero and identity matrices of compatible dimensions. Since the PMSM has two outputs and three state variables in the original model, the 0_1 matrix has the dimensions of 3×2, 0_2 has the dimensions of 2×3 and I has the dimensions of 2×2.

It is emphasized that the constant vector term μ^d resulting from the linearization of the nonlinear model and the existence of the unknown torque does not appear in the augmented model (8.19). As for the output signal expressed by (8.20), $y(t_i)$ is the actual measured output signal as the steady-state values of its individual components (i_d, i_q currents and velocity ω_e) have not been removed in the process of derivation. If the control objective is to maintain plant steady-state operation and reject the unknown disturbances, in the design of the regulatory type of control systems, the steady-state values of the augmented state variable $x(t_i)$ are required to be removed in order to obtain a valid linear model at its operating conditions. Because the steady-state value of the vector $\Delta x_m(t_i)$ is a zero vector where the difference of a constant vector is considered, the remaining steady-state values in $x(t_i)$ to be considered are the steady-states of the output signals. A simple approach is to use the reference signals as the steady-state values of the output signals. This means that to consider designing the predictive control system for maintaining steady-state operation and rejecting the unknown disturbances requires the output equation (8.18) be modified as

$$y(t_{i+1}) - y^{ss} = y(t_i) - y^{ss} + C_d A_d \Delta x_m(t_i) + C_d B_d \Delta u(t_i). \tag{8.21}$$

For PMSM control, the steady-state vector y^{ss} can be selected as $y^{ss} = [i_d^* \quad \omega_e^*]^T$ where i_d^* and ω_e^* are the reference signals for i_d current and the motor velocity.

8.2.2 Design without Constraints

Given the augmented state-space model, the next task is to optimize the incremental control signal $\Delta u(t_i)$ using the framework of model predictive control. Model predictive control is a well established control technology (see for example, Maciejowski (2000), Rawlings (2000), and Wang (2009)). Assuming that at the sampling instant t_i, $t_i > 0$, the state vector $x(t_i)$ is available through measurement where the state vector $x(t_i)$ provides the current plant information. The future control trajectory is denoted by the vector

$$\Delta U = \left[\Delta u(t_i)\ \Delta u(t_{i+1})\ \ldots\ \Delta u(t_{i+N_c-1})\right]^T,$$

where N_c is the control horizon dictating the number of parameters used to capture the future control trajectory. With this given information $x(t_i)$, the future state vectors are predicted for N_p samples, where N_p is termed the prediction horizon ($N_c \leq N_p$). It is assumed that after N_c samples, the incremental control signal $\Delta u(t_{i+k})$ becomes zero for all future samples ($k \geq N_c$). The state vectors so obtained are written as denoting the future state variables,

$$X = \left[x(t_{i+1}|t_i)^T\ \ldots\ x(t_{i+N_p}|t_i)^T\right]^T.$$

Using the augmented state-space model (8.19), the future state variables are calculated sequentially using $\Delta u(t_i)$, $\Delta u(t_{i+1})$, … , $\Delta u(t_{i+N_c-1})$, leading to the following equations:

$$\begin{aligned} x(t_{i+1}|t_i) &= Ax(t_i) + B\Delta u(t_i) \\ x(t_{i+2}|t_i) &= A^2x(t_i) + AB\Delta u(t_i) + B\Delta u(t_{i+1}) \\ &\vdots \\ x(t_{i+N_p}|t_i) &= A^{N_p}x(t_i) + A^{N_p-1}B\Delta u(t_i) \\ &\quad + \cdots + A^{N_p-N_c}B\Delta u(t_{i+N_c-1}), \end{aligned}$$

which are written in a compact matrix and vector form, as

$$X = F_x x(t_i) + \Phi\Delta U, \tag{8.22}$$

where

$$F_x = \begin{bmatrix} A \\ A^2 \\ \vdots \\ A^{N_p} \end{bmatrix};\quad \Phi = \begin{bmatrix} B & 0 & .. & 0 \\ AB & B & .. & 0 \\ A^2B & AB & .. & 0 \\ \vdots & \vdots & .. & \vdots \\ A^{N_p-1}B & A^{N_p-2}B & .. & A^{N_p-N_c}B \end{bmatrix}$$

The design criterion for the predictive controller is to find the control vector ΔU such that the following cost function is minimized

$$J = X^T\bar{Q}X + \Delta U^T\bar{R}\Delta U,$$

where $\bar{Q} = diag[Q, \ldots, Q]$ and $\bar{R} = diag[R, \ldots, R]$ are block diagonal matrices that have identical component matrices Q and R respectively. Here Q is a positive semi-definite matrix (with (Q, A) detectable) and R is a positive definite matrix. The dimensions of the Q and R matrices are 5×5 and 2×2 for the application of PMSM control. This expression for the objective function is used for simplicity in formulation of the MPC problem in which, additionally, the application lies in disturbance rejection.

Substituting (8.22) into the cost function gives

$$\begin{aligned} J =& \Delta U^T(\Phi^T\bar{Q}\Phi + \bar{R})\Delta U + 2\Delta U^T\Phi^T\bar{Q}F_x x(t_i) \\ &+ x(t_i)^T F_x^T\bar{Q}F_x x(t_i), \end{aligned} \tag{8.23}$$

or simply

$$J = \Delta U^T \Omega_{mpc} \Delta U + 2\Delta U^T \Psi_{mpc} x(t_i) + constant \tag{8.24}$$

where the data matrices are defined as

$$\Omega_{mpc} = \Phi^T \bar{Q} \Phi + \bar{R}; \quad \Psi_{mpc} = \Phi^T \bar{Q} F_x.$$

From the first derivative of the objective function J:

$$\frac{\partial J}{\partial \Delta U} = 2\Omega_{mpc} \Delta U + 2\Psi_{mpc} x(t_i), \tag{8.25}$$

the necessary condition of an extremum J is obtained as

$$\frac{\partial J}{\partial \Delta U} = 0. \tag{8.26}$$

Since the second derivative $\frac{\partial^2 J}{\partial^2 \Delta U} = 2\Omega_{mpc}$ which is positive definite based on the assumptions for $\bar{Q}$ and $\bar{R}$, without constraints, the minimum of the cost function J is achieved with the optimal control vector

$$\Delta U = -\Omega_{mpc}^{-1} \Psi_{mpc} x(t_i). \tag{8.27}$$

Using receding horizon control, only the components (first two rows) in ΔU corresponding to $\Delta u(t_i)$ are used, so that

$$\Delta u(t_i) = -\begin{bmatrix} I & 0 & \ldots & 0 \end{bmatrix} \Omega_{mpc}^{-1} \Psi_{mpc} x(t_i) \tag{8.28}$$

$$= -K_{mpc} x(t_i), \tag{8.29}$$

where the state feedback control gain matrix is defined as

$$K_{mpc} = \begin{bmatrix} I & 0 & \ldots & 0 \end{bmatrix} \Omega_{mpc}^{-1} \Psi_{mpc},$$

which is the first two rows of matrix $\Omega_{mpc}^{-1} \Psi_{mpc}$.

The actual control signal applied to the plant is constructed using

$$u(t_i) = \Delta u(t_i) + u(t_{i-1}), \tag{8.30}$$

where both the current optimal control $\Delta u(t_i)$ and the past value $u(t_{i-1})$ are known.

When the predictive controller is used for disturbance rejection, the control objective is to maintain the plant in steady-state operation, and the incremental state vector $\Delta x_m(t_i)$ has zero steady-state whilst the steady-state of the plant output is a constant vector in the MIMO case (see 8.21). This control objective is reflected in the selection of the cost function (8.23). When the predictive controller is used for tracking reference signals, the references signals will enter the computation through the augmented output variables. This can be readily achieved by generalizing (8.23) where the steady-state vector y^{ss} is replaced by the actual reference signal vector $r(t_i)$ at the sampling instant t_i to form the feedback error $y(t_i) - r(t_i)$. Thus, the state vector in (8.23) or (8.27) becomes

$$x(t_i) = [\Delta x_m(t_i)^T \quad (y(t_i) - r(t_i))^T]^T.$$

It is worthwhile to emphasize that both $y(t_i)$ and $r(t_i)$ are the actual physical quantities without removing their steady-state values, avoiding computation of the steady-state values of the outputs.

The data matrices Ω_{mpc} and Ψ_{mpc} are calculated off-line to reduce computational demand in the the implementation of the predictive controller.

A key strength of predictive control lies in its ability to systematically impose constraints on plant input and output variables. The next two sections will detail how the constrained control problem is systematically formulated and solved in a predictive control system.

8.2.3 *Formulation of the Constraints*

As a first step in the introduction of constraints we consider control amplitude constraints imposed at the sampling instant k by writing them in the form of a set of linear inequalities

$$u^{min} \leq u(t_i) \leq u^{max}, \tag{8.31}$$

where u^{min} and u^{max} are 2×1 data vectors containing the required lower and upper limits of the control amplitude for each input signal, respectively. Also, the incremental changes in the control signal ($\Delta u(t_i) = u(t_i) - u(t_{i-1})$) are limited in the applications, where the constraints are written in the form,

$$\Delta u^{min} \leq \Delta u(t_i) \leq \Delta u^{max}. \tag{8.32}$$

To impose the constraints at the sampling instant t_i using the moving horizon window, $\Delta u(t_i)$ is related to the parameter vector ΔU as given in (8.23). Firstly, note that the amplitude constraints (8.31) can be written as

$$\Delta u(t_i) \leq u^{max} + u(t_{i-1}) \tag{8.33}$$

$$-\Delta u(t_i) \leq -(u^{min} + u(t_{i-1})) \tag{8.34}$$

and the constraints on $\Delta u(t_i)$ as

$$\Delta u(t_i) \leq \Delta u^{max} \tag{8.35}$$

$$-\Delta u(t_i) \leq -\Delta u^{min}. \tag{8.36}$$

Secondly, the incremental control signal $\Delta u(t_i)$ is of the first 2 elements of the parameter vector ΔU. By combining the constraints together, the following linear inequalities are obtained:

$$M \Delta U \leq \Gamma \tag{8.37}$$

where

$$M = \begin{bmatrix} I & 0 & \dots & 0 \\ -I & 0 & \dots & 0 \\ I & 0 & \dots & 0 \\ -I & 0 & \dots & 0 \end{bmatrix}; \quad \Gamma = \begin{bmatrix} u^{max} + u(t_{i-1}) \\ -(u^{min} + u(t_{i-1})) \\ \Delta u^{max} \\ -\Delta u^{min} \end{bmatrix}.$$

8.2.4 *On-line Solution for Constrained MPC*

Iterative computation in the solution of the quadratic programming problem arises from the inequalities (8.37) that represent the constraints. This formulation also means that at sampling instant t_i (8.28) and (8.30) must satisfy all of the constraints. If this is the case, the constrained and unconstrained optimal solutions are identical but if (8.28) produces a control signal that partly violates the constraints then iterative computation is needed to find the active constraints and hence the solution that minimizes the cost function with these in place. Because the optimization problem is solved in real-time in conjunction with the control signal calculations, the key to solving the constrained optimization problem lies in finding an algorithm that can be coded by the user for convenience of adding safety protection and other computational requirements and has reliability in the presence of conflict constraints. Additionally, its computational demand must be relatively low.

One established approach in the literature is to directly search for the active constraints using optimization of Lagrange multipliers, see, for example, Luenberger (1969, 2003) and this algorithm is summarized next.

Using the Kuhn-Tucker conditions (Luenberger (2003), Boyd and Vandenberghe (2004)), the optimization problem is to find the parameter vector ΔU by solving the following problem:

$$\max_{\lambda \geq 0} \min_{\Delta U} [\Delta U^T \Omega_{mpc} \Delta U + 2\Delta U^T \Psi_{mpc} x(t_i) + 2\lambda^T (M\Delta U - \Gamma)], \tag{8.38}$$

where λ is the vector of Lagrange multipliers, Ω and Ψ are the data matrices from the predictive control's cost function (8.24) and M and Γ are given by (8.37). Also the cost function is minimized with ΔU unconstrained when

$$\Delta U = -\Omega_{mpc}^{-1}(\Psi_{mpc} x(t_i) + M^T \lambda) \tag{8.39}$$

and, on substituting (8.39) into (8.38), the Lagrange multiplier vector λ is obtained by solving

$$J = \min_{\lambda \geq 0} [\lambda^T M \Omega_{mpc}^{-1} M^T \lambda + 2\lambda^T (M \Omega_{mpc}^{-1} \Psi_{mpc} x(t_i) + \Gamma) + x(t_i)^T \Psi_{mpc}^T \Omega_{mpc}^{-1} \Psi_{mpc} x(t_i)], \tag{8.40}$$

where the maximization problem (8.38) has been converted to this minimization problem by multiplying the objective function by -1. For notational simplicity, (8.40) is written as

$$\hat{J} = \min_{\lambda \geq 0} [\lambda^T H \lambda + 2\lambda^T K], \tag{8.41}$$

from this point onwards where $H = M\Omega_{mpc}^{-1} M^T$ and $K = M\Omega_{mpc}^{-1} \Psi_{mpc} x(t_i) + \Gamma$ and the final term is independent of λ and hence can be neglected. Although this is still a quadratic programming problem that requires an iterative solution, the constraints are much simpler ($\lambda \geq 0$) and iterative solutions much easier to find.

Hildreth's algorithm (Hildreth (1957), Wismer and Chattergy (1978)) is one of the simple algorithms for solving the quadratic optimization problem given by (8.41). This algorithm is suitable for real-time computation and can be easily coded by the user for applications. The algorithm searches the active constraints by finding the positive Lagrange multipliers element by element. Therefore, it entirely avoids matrix operations in the process of finding the positive Lagrange multipliers. Taking matrix inversion in real-time where there is no guarantee that the inversion exists leads to problems, which is the case when the active constraints become conflict. The algorithm can be summarized using the following equations. On iteration $m+1$ this algorithm solves for the *ith* entry, denoted by λ^i, in the Lagrange multiplier vector λ using

$$\lambda_i^{m+1} = \max(0, \overset{m+1}{\underset{i}{\mathrm{w}}}), \tag{8.42}$$

with

$$w_i^{m+1} = -\frac{1}{h_{ii}} \left[k_i + \sum_{j=1}^{i-1} h_{ij} \lambda_j^{m+1} + \sum_{j=i+1}^{n} h_{ij} \lambda_j^m \right], \tag{8.43}$$

where n is the dimension of the Lagrange multiplier vector, h_{ij} is the element in row i and column j of H and k_i is the ith element in K. MATLAB program for this algorithm was given in Wang (2009).

Once convergence is achieved, the resulting λ is used to compute the parameter vector ΔU and hence the predictive controller using (8.39) where the constraints corresponding to the nonzero Lagrange multipliers are active and the rest inactive. Because the Lagrange multiplier vector λ contains the zero elements for inactive constraints and positive elements for active constraints upon convergence, the constrained control solution is simply computed using

$$\Delta U = -\Omega_{mpc}^{-1}(\Psi_{mpc} x(t_i) + M^T \lambda), \tag{8.44}$$

where λ is the computed Lagrange multiplier vector.

This algorithm has been used in this book to solve all the constrained optimization problems in real-time for both discrete-time MPC and continuous-time MPC systems.

8.3 Experimental Evaluation of DMPC of PMSM

In this section, the predictive control of PMSM is evaluated experimentally in which the experimental results for three different cases are obtained and discussed. In particular, the performance of the MPC controller is analyzed in terms of the constraints, tuning of controller feedback gain and load disturbance rejection.

8.3.1 The MPC Parameters

The sampling interval Δt is selected as 200×10^{-6} second in order to capture the dynamics of the current system. This sampling rate is relatively fast for the mechanical system, but slow for the current system. Not being flexible in selecting the sampling rates for the electrical and mechanical systems is the first drawback of the MPC approach.

To determine the steady-state operating conditions, the reference signal to d-axis current (i_d^*) is used as $i_d^0 = i_d^* = 0$ and the reference signal ω_e^* is used as $\omega_e^0 = \omega_e^* = 800$ RPM. However, there is some uncertainty associated with determining the steady-state value i_q^0 because this value depends on the load torque T_L that is unknown (see 8.3). Thus, the steady-state value i_q^0 is estimated as $i_q^0 = 1\ A$. As the parameters in the state-space model of a PMSM are dependent on steady-state operating conditions, they all have more or less uncertainties associated with them. The DC power supply has voltage $V_{dc} = 100\ V$.

Now, with the steady-state parameters determined, the continuous-time linear model (8.7) for the PMSM has the following system matrices

$$A_m = \begin{bmatrix} -\frac{R_s}{L_d} & \frac{L_q}{L_d}\omega_e^0 & \frac{L_q}{L_d} i_q^0 \\ -\frac{L_d}{L_q}\omega_e^0 & -\frac{R_s}{L_q} & -(\frac{L_d}{L_q} i_d^0 + \frac{\phi_{mg}}{L_q}) \\ 0 & \frac{3Z_p^2\phi_{mg}}{2J_m} & -\frac{B_v}{J_m} \end{bmatrix} = \begin{bmatrix} -425.7 & 83.8 & 1 \\ -83.8 & --425.7 & -17.9 \\ 0 & 15957.4 & -2.3 \end{bmatrix}$$

$$B_m = \begin{bmatrix} \frac{1}{L_d} & 0 \\ 0 & \frac{1}{L_q} \\ 0 & 0 \end{bmatrix} = \begin{bmatrix} 142.9 & 0 \\ 0 & 142.9 \\ 0 & 0 \end{bmatrix}$$

$$C_m = \begin{bmatrix} 1 & 0 & 0 \\ 0 & 0 & 1 \end{bmatrix}.$$

There is no need to calculate the μ^0 term as it is not required in the predictive controller design.

With sampling interval $\Delta t = 200 \times 10^{-6}$ second, the discrete-time model

$$x_m(t_{i+1}) = A_d x_m(t_i) + B_d u(t_i)$$

has the system matrices calculated as

$$A_d = e^{A_m \Delta t} = \begin{bmatrix} 0.918 & 0.016 & 0.00016 \\ -0.025 & 0.913 & -0.0034 \\ 0 & 3.05 & 0.994 \end{bmatrix}$$

$$B_d = \int_0^{\Delta t} e^{A_m \tau} d\tau B_m = \begin{bmatrix} 0.027 & 0.00023 \\ -0.00023 & 0.027 \\ -0.00024 & 0.044 \end{bmatrix}.$$

The matrices A_d and B_d here were computed using MATLAB function "c2d".

In the predictive controller design, the control horizon N_c is selected as $N_c = 5$ for both inputs and the prediction horizon N_p is selected as $N_p = 20$ for both outputs. Because of the large differences between the time constants in the electrical system and the mechanical system, parameters in the weight matrix Q play an important role in determining the closed-loop performance. In order to ensure faster dynamic responses in the electrical system, larger weights need to be placed on the coefficients corresponding to the i_d and i_q components of the Q. However, this task is time consuming, and is perhaps another drawback of the approach.

To simplify the task, the weight matrices in the cost function (8.23) are

$$Q = \begin{bmatrix} 0 & 0 & 0 & 0 & 0 \\ 0 & 0 & 0 & 0 & 0 \\ 0 & 0 & 0 & 0 & 0 \\ 0 & 0 & 0 & q_{id} & 0 \\ 0 & 0 & 0 & 0 & q_w \end{bmatrix}; \quad R = r_w \begin{bmatrix} 1 & 0 \\ 0 & 1 \end{bmatrix}, \tag{8.45}$$

where q_{id} and q_w will be used as performance tuning parameters. The components in R matrix also affect the closed-loop performance. The smaller the parameter r_w is, the faster the closed-loop response will be. However, it seems that for controlling the electrical drives and power converters using the discrete-time MPC, the tuning parameter r_w is less effective. In this application, r_w is chosen to be 100.

8.3.2 Constraints

The maximum voltages in the servo drive are determined by the DC bus voltage (V_{dc}) and the pulse width modulation techniques are adopted in the implementation in order to prevent the over-modulation from occurring as discussed in Chapter 2. The linear modulation range limits amplitudes of maximum achievable voltage so that

$$\sqrt{v_d^2 + v_q^2} \leq \frac{V_{dc}}{\sqrt{3}} \tag{8.46}$$

and this nonlinear limitation can be imposed using a rectangular area approximation to the circular area (see Section 4.1.1 for discussions), as

$$|v_q| \leq \epsilon \frac{V_{dc}}{\sqrt{3}}$$
$$|v_d| \leq \sqrt{(1-\epsilon^2)} \frac{V_{dc}}{\sqrt{3}},$$

where $0 \leq \epsilon \leq 1$. In the experimental evaluation, ϵ is selected as 0.5 and the DC supply voltage is $V_{dc} =$ 100 V. With these specifications, the constraints on v_d and v_q become

$$-28.87 \leq v_q \leq 28.87$$
$$-50 \leq v_d \leq 50. \tag{8.47}$$

The constraints on the incremental changes of the v_d and v_q are imposed as

$$-10 \leq \Delta v_q \leq 10 \qquad -10 \leq \Delta v_d \leq 10.$$

8.3.3 Response to Load Disturbances

The load disturbance is represented as T_L in (8.3), which can be modeled as a slowly varying signal, for example, the external friction forces in a grinding process. Under the assumption that the load disturbance mainly contains low frequency components, the design of MPC with the embedded integrator can reject the impact of external load disturbances. In order to achieve a fast response for the load disturbance rejection, the weight coefficients are chosen to be $q_{id} = 1$ and $q_\omega = 1$.

In this case, the constraint on the q-axis voltage is specified by choosing $\epsilon = 0.9$, leading to

$$-51.96 \leq v_q \leq 51.96, \quad -25.17 \leq v_d \leq 25.17.$$

Because of the load torque disturbance, the limits on the q-axis voltage could prevent the velocity from returning to the set-point ω_e^* if they were too restricted. By choosing a larger ϵ so that a larger amplitude is permitted for the q-axis voltage, this problem is avoided.

The experiment is carried out by suddenly switching on the resistor load when the coupled motor is running at the constant speed (see Section 10.5 for description of the experimental setup). Using the setup

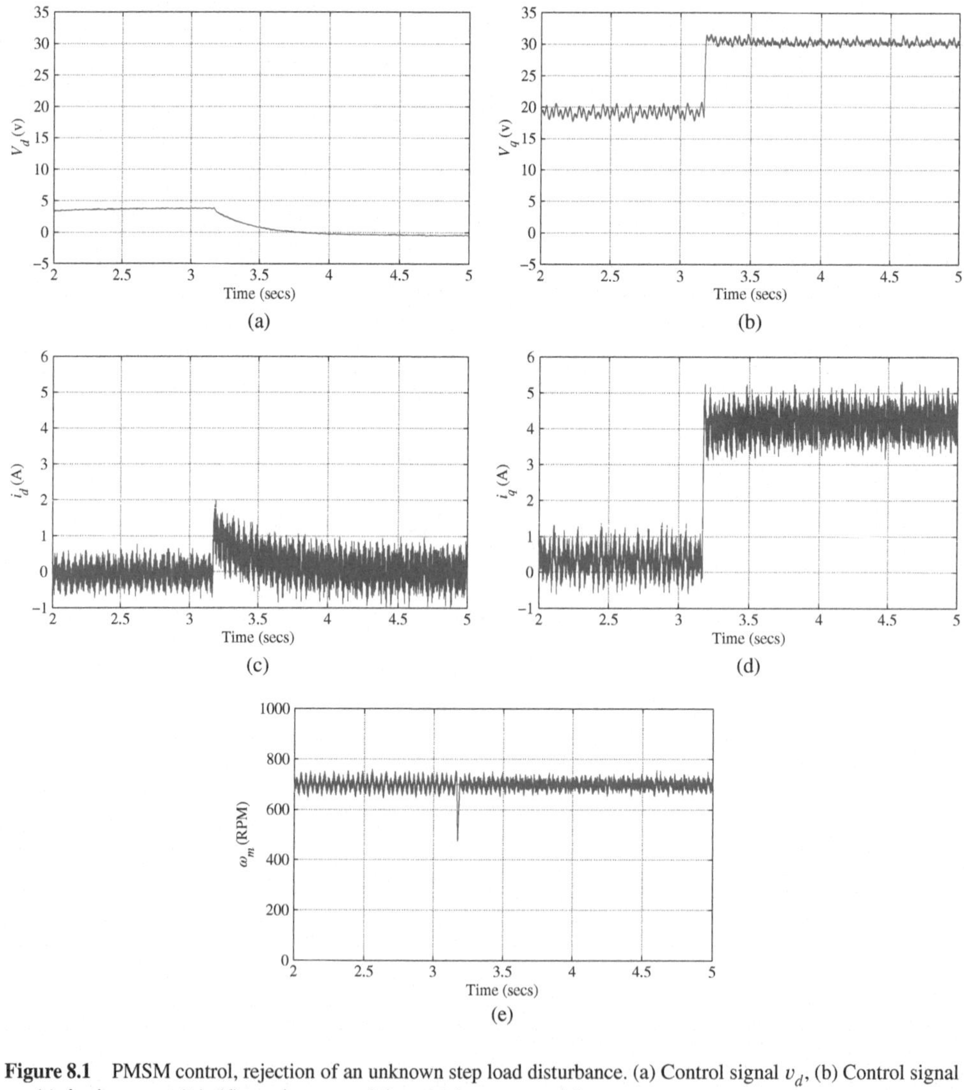

Figure 8.1 PMSM control, rejection of an unknown step load disturbance. (a) Control signal v_d, (b) Control signal v_q, (c) d-axis current (i_d), (d) q-axis current (i_q), and (e) motor speed (ω_e.

shown in Figure 10.6, a positive step load disturbance is added to the motor shaft at the sampling time $t =$ 3.2 second (see Figure 8.1(e)). This equivalent step load disturbance T_L has unknown amplitude, which causes the i_d and i_q currents to suddenly increase (see Figures 8.1(c)–(d)). To reject this disturbance, the control signal v_q increases from about 19 V to 30 V, and the control signal v_d decreases from about 4 V to close to 0 (see Figures 8.1(a)–(b)). After the transient response, the i_d current returns to its set-point value at 0 and the i_q current finds a new state-steady value that will compensate for the effect of the load disturbance. It is seen that the response time for the motor speed ω_m is very fast (see Figure 8.1(e)); however, it takes a longer time for the i_d current to return to its desired value (see Figure 8.1(c)). For this experiment of disturbance rejection, all the constraints on the control signals were naturally satisfied, namely, they were not active.

8.3.4 Response to a Staircase Reference

Figure 8.2 shows the performance of the MPC controller in response to a staircase reference speed. The control objective is to regulate the motor to a different reference speed while maintaining i_d at zero. The controller gain for this case was selected to obtain a fast response while minimizing the noise level, where the weight coefficients were chosen to be $q_{id} = 1$ and $q_\omega = 1$. Figure 8.2(c) shows that the steady-state d-axis current is zero but corrupted by noise and high frequency ripples. The high frequency ripples appear as pulsing torque (Jahns and Soong (1996)) in the closed-loop system due to various reasons, such as current measurement errors (Chung and Sul (1998)), flux harmonics and cogging torques (Xu *et al.* (2004)). It could also be caused by the modeling error when the chosen operating conditions for linearization differ from the motor's operating condition at the low speed. It is also seen from Figures 8.2(c)–(d) that the frequency of oscillation increases as the synchronous speed increases.

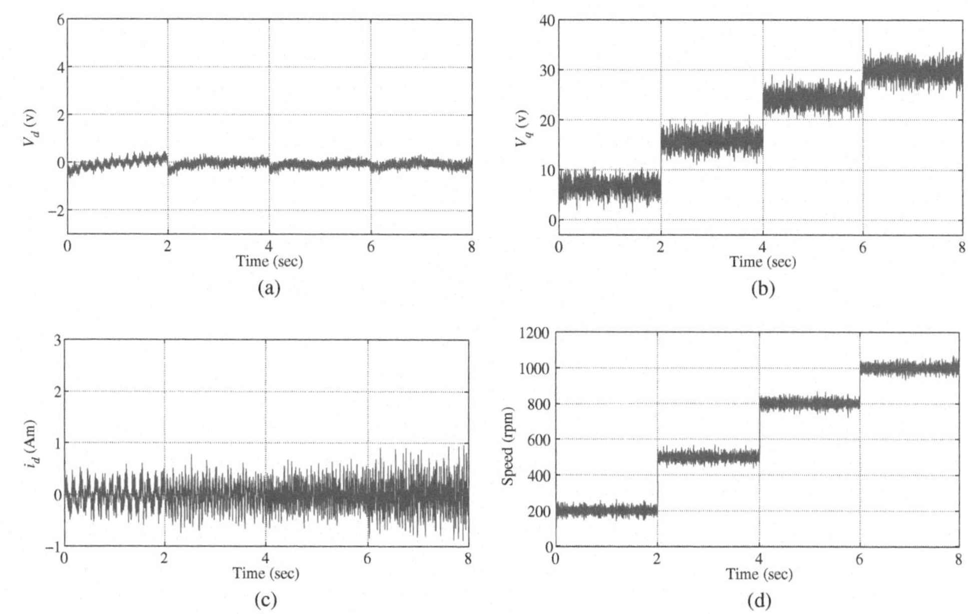

Figure 8.2 PMSM control: closed-loop response to a staircase reference speed. (a) Control signal v_d, (b) Control signal v_q, (c) d-axis current (i_d), and (d) motor speed (ω_e.

8.3.5 *Tuning of the MPC controller*

The desired closed-loop performance in the MPC design is dependent on the weight matrix Q and R matrices in the cost function (8.23). With the special diagonal form, there are two parameters to be selected in the performance specification, where q_{id} is the weighting parameter on the d-axis current and q_ω is the weighting parameter on the velocity ω_e. Their relatively large values produce high gains in the closed-loop control that, in turn, amplify the noise and high frequency ripples in the drive. Hence the selection of these weights is a trade-off between response and the attenuation of noise and high frequency ripples.

There are two cases to be investigated here with respect to the weight coefficients q_{id} and q_ω. For both cases, the value of q_{id} is chosen to be 1, but q_ω varies. It is found that the ratio between the weight coefficients q_ω and q_{id} significantly affects the closed-loop responses.

8.3.5.1 Case A

$q_\omega = q_{id} = 1$. When $q_\omega = q_{id}$, which means that the reduction of the errors for the current control and the velocity control has an equal importance, the closed-loop system has higher gains. Here, with these weight matrices, the state feedback control gain matrix is

$$K_{mpc} = \begin{bmatrix} 0.3146 & -0.6139 & -0.0181 & 0.0350 & -0.0009 \\ -0.6700 & 10.4550 & 0.6348 & 0.0022 & 0.0791 \end{bmatrix}$$

and the closed-loop eigenvalues are

$$Eig = \begin{bmatrix} 0.8942 + 0.1534i \\ 0.8942 - 0.1534i \\ 0.8008 \\ 0.9211 \\ 0.9879 \end{bmatrix}.$$

8.3.5.2 Case B

$q_\omega = 0.01$ and $q_{id} = 1$. When q_ω is relatively smaller than q_{id}, the emphasis is put on the reduction of the error in the current i_d. The MPC controller has lower gains with this set of performance parameters. Here, with this selection of the weight coefficients, the state feedback gain matrix is

$$K_{mpc} = \begin{bmatrix} 0.2536 & -0.0723 & -0.0046 & 0.0350 & -0.0004 \\ -0.0771 & 1.2510 & 0.0614 & 0.0029 & 0.0064 \end{bmatrix}$$

and the eigenvalues of the closed-loop predictive control system are

$$Eig = \begin{bmatrix} 0.9548 + 0.1081i \\ 0.9548 - 0.1081i \\ 0.9250 \\ 0.9577 \\ 0.9877 \end{bmatrix}.$$

The experimental results can be compared from Figures 8.3–8.6. With the selection of performance parameters in Case A, the impact of high gain feedback control is evident from the responses of the current control signals and their differences (see Figure 8.3). In particular, the second row of the feedback controller gain matrix K_{mpc} has components almost 10 times larger than those in Case B (except the second last component). This results in much larger i_q current response. It is seen that the constraints on v_q and Δv_q become active in the start-up phase. In comparison, when q_ω is reduced to 0.01, the magnitudes of both v_d and v_q have been significantly reduced (see Figure 8.4), and the control signals are far away

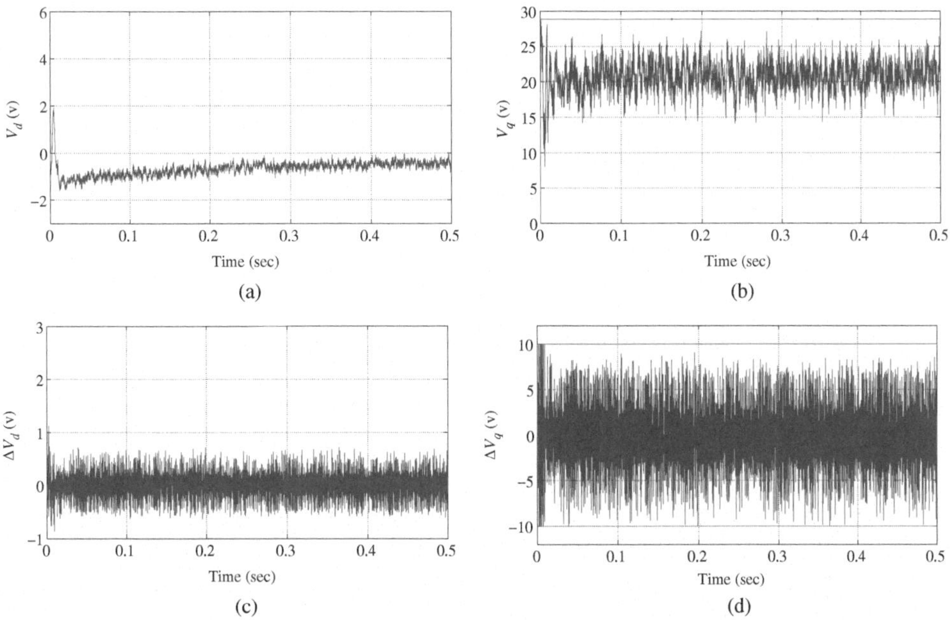

Figure 8.3 Control signal when $q_\omega = 1$. (a) Control signal v_d, (b) Control signal v_q, (c) Difference of the control signal Δv_d, and (d) Difference of the control signal Δv_q.

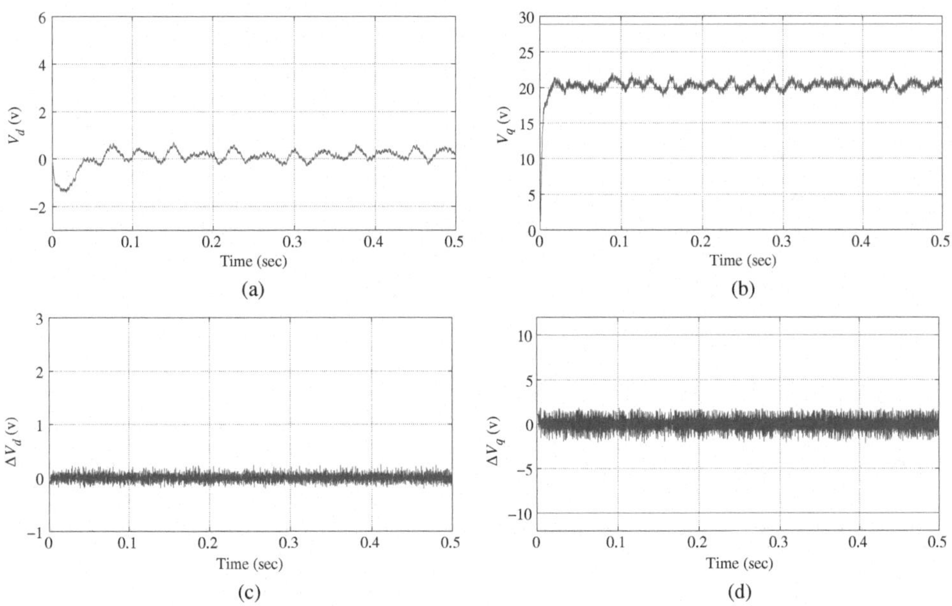

Figure 8.4 Control signal when $q_\omega = 0.01$. (a) Control signal v_d, (b) Control signal v_q, (c) Difference of the control signal Δv_d, and (d) Difference of the control signal Δv_q.

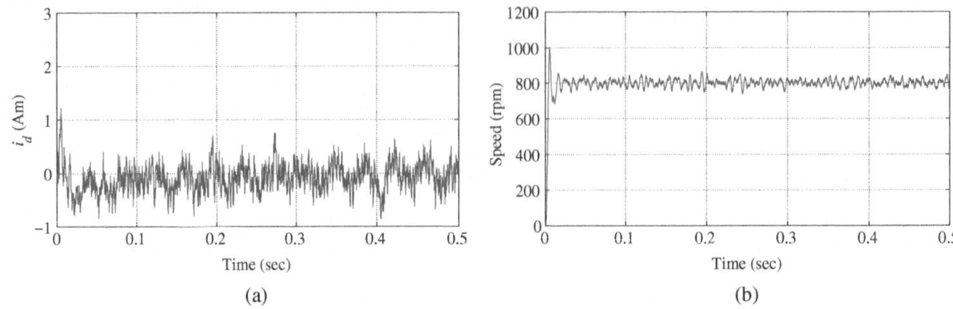

Figure 8.5 Output response when $q_\omega = 1$. (a) d-axis current (i_d), and (b) Motor speed (ω_e).

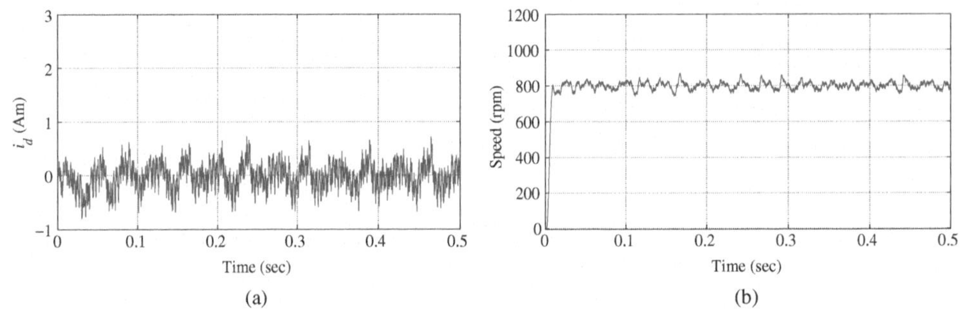

Figure 8.6 Output response when $q_\omega = 0.01$. (a) d-axis current (i_d), and (b) Motor speed (ω_e).

from their constraints. For the two choices of q_ω parameter, the closed-loop output responses, i_d current and ω_m, are compared in Figures 8.5 and 8.6. It is seen from these two figures that the closed-loop output responses from the smaller q_ω parameter case are more aggressive than those using a larger q_ω.

This case study indicates that because of the much faster dynamics in the electrical system, the error reduction on the i_d current response should have a much faster rate than the reduction on the velocity response, which is translated into a choice of larger q_ω relative to the value of q_{id}.

8.4 Power Converter Control Using DMPC with Experimental Validation

The DMPC algorithms have also been successfully used in controlling the power converter with constraints.

For the power converter control, the linearized continuous-time state-space model is described by the differential equation:

$$\frac{dx_m(t)}{dt} = A_m x_m(t) + B_m u(t) \tag{8.48}$$

$$y(t) = C_m x_m(t), \tag{8.49}$$

where the state vector is $x_m(t) = [i_d(t) \quad i_q(t) \quad v_{dc}(t)]^T$, the control signal vector is $u(t) = [S_d(t) \quad S_q(t)]^T$, and the output vector is $y(t) = [i_q(t) \quad v_{dc}]^T$.

$$A_m = \omega_b \begin{bmatrix} -\frac{\bar{R}_s}{\bar{L}_s} & \omega_g & -\frac{S_{do}}{\bar{L}_s} \\ -\omega_g & -\frac{\bar{R}_s}{\bar{L}_s} & -\frac{S_{qo}}{\bar{L}_s} \\ \frac{S_{do}}{\bar{C}_{dc}} & \frac{S_{qo}}{\bar{C}_{dc}} & 0 \end{bmatrix}, B_m = \omega_b \begin{bmatrix} -\frac{v_{dco}}{\bar{L}_s} & 0 \\ 0 & -\frac{v_{dco}}{\bar{L}_s} \end{bmatrix}, \tag{8.50}$$

$$C_m = \begin{bmatrix} 0 & 1 & 0 \\ 0 & 0 & 1 \end{bmatrix},$$

where ω_b is the base value used in the per unit model ($\omega_b = 2\pi \times 50$), $S_{do} = 0.6532$, $S_{qo} = -0.3232$ are the steady state control signals; $v_{dco} = 1.2247$ is the steady state voltage in per unit value. By choosing the steady-state control signal values as $S_{do} = 0.6532$, $S_{qo} = -0.3232$ and the steady-state voltage value as $v_{dco} = 60$ V, the A_m and B_m matrices are calculated as

$$A_m = \begin{bmatrix} -31.7460 & 314.1593 & -622.0926 \\ -314.1593 & -31.7460 & 256.5100 \\ 198.6073 & -81.8925 & 0 \end{bmatrix}; \quad B_m = \begin{bmatrix} -972.0197 & 0 \\ 0 & -972.0197 \\ 253.3784 & 0 \end{bmatrix},$$

where the parameters in the physical per unit model are $\bar{R}_s = 0.04$, $\bar{L}_s = 0.3958$ and $\bar{C}_{dc} = 1.2399$. Here, with the per unit model, the parameters in the system matrices are scaled to a better numerically conditioned system model. The continuous-time state space system matrices A_m, B_m, C_m are further discretized into A_d, B_d, C_d with the current loop sampling interval $\Delta t = 100 \times 10^{-6}$ sec, leading to the system matrices,

$$A_d = \begin{bmatrix} 0.9957 & 0.0316 & -0.0617 \\ -0.0310 & 0.9962 & 0.0266 \\ 0.0199 & -0.0079 & 0.9993 \end{bmatrix}; \quad B_d = \begin{bmatrix} -0.0978 & -0.0015 \\ 0.0018 & -0.0970 \\ 0.0244 & 0.0004 \end{bmatrix}.$$

The reference signal to current i_q is chosen to be 0 and the reference signal to the voltage v_{dc} is called v_{dc}^* specified in the applications. Constraints are imposed on the control signals S_d and S_q to limit them in the range of −1 to 1. In the discrete-time MPC system, the prediction horizon is selected as $N_p = 100$ and the control horizon $N_c = 10$. The weight matrices Q and R are selected to be in identical forms as those in (8.45), however with the different value $r_w = 3500$, and the last two parameters in the diagonal of Q matrix as $q_{id} = 1$ and $q_\omega = 0.1$ respectively. As the current and voltage loops have different time constants, the element in Q matrix corresponding to the current is chosen to be much larger than the element corresponding to the voltage variable in order to achieve satisfactory performance.

In the experimental evaluation, the reference to the *DC* bus voltage, v_{dc}^*, is set to 55 V, then changed to 60 V as a step signal. The load disturbance is introduced by changing the *DC* load from 20 Ω to 30 Ω. Figure 8.7 shows the closed-loop responses of the *DC* bus voltage, i_d and i_q currents and control signal responses. From Figure 8.7(a) it is seen that the predictive control system is successful in tracking the reference signal v_{dc}^* without steady-state errors, also the load disturbance is rejected without steady-state error. Figure 8.7(b) shows that the q-axis current follows a zero reference signal, however the d-axis current varies in responses to the voltage reference signal and to the load change.

8.5 Summary

This chapter has presented the discrete-time model predictive control of electrical drives and power converter. In this design, the predictive control system is centralized and integrators are embedded into the predictive controller to overcome steady-state errors in the output responses. Experimental results are obtained to demonstrate the control system design and implementations.

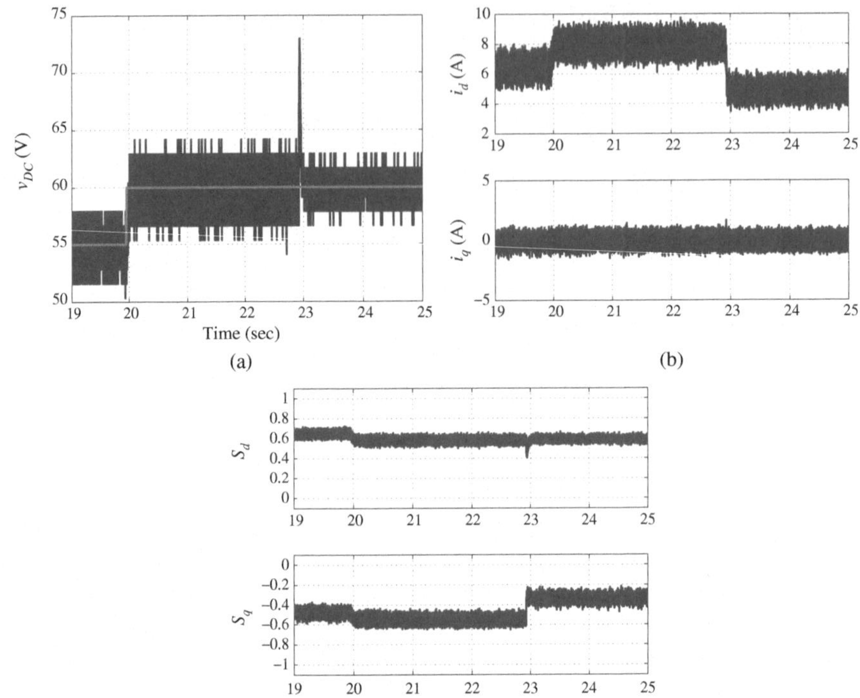

Figure 8.7 Experimental results for discrete-time model predictive control of power converter. (a) *DC* bus voltage response to step reference change and load change, (b) i_d and i_q current response to step reference voltage change and load change, and (c) Control signal responses to step reference voltage change and load change.

Although one might say that the design is straightforward with just one model, there are several drawbacks with this particular predictive control approach. Because it uses only one model and one controller in the design and implementation, it only permits one sampling rate for both currents, velocity or voltage, which means that a suitable sampling rate is hard to find. For instance, a sampling interval is too large for the current control system, yet is too small for the mechanical or the voltage system. This causes numerical problems in the discretized model, although a per unit model helps to reduce the impact of the numerical problems to some degree. The second drawback with this type of predictive control system is that it requires a large "trial and error" effort in selecting the suitable weight matrices Q and R, which could be caused again by the large discrepancies between the electrical time constants and the time constants in the mechanical and voltage systems. The third drawback is that because the current i_q in the AC drives is a state variable, it is difficult to impose constraints on its value. The fourth drawback is that this approach requires a larger computational load than the FCS-MPC or the PID control systems.

8.6 Further Reading

General literature in predictive control includes Rawlings (2000), Maciejowski (2000), Qin and Badgwell (2003), Wang (2009), Rossiter (2003). A book was published on predictive control of electric drives by Linder *et al.* (2010). Discrete-time model predictive control of PMSM drives was discussed in Linder and Kennel (2005), Moon *et al.* (2003), in Liu and Li (2012), in Chai *et al.* (2013), for induction machine drives in de Santana *et al.* (2008), Maaziz *et al.* (2000) in Zhang *et al.* (1997), Kennel *et al.* (2001) for induction motor control using Generalized Predictive Control. Design and implementation was discussed in Bolognani *et al.* (2009), with field weakening features in Bolognani *et al.* (2008), with torque control over modulation in Ishida *et al.* (2010), speed and position control in Bolognani *et al.* (2011). Applications for induction motor control can be found in de Santana *et al.* (2008) and *DC-DC* power supplies in Beccuti *et al.* (2009). Cascade control of power converter using DMPC for both inner-loop and outer-loop systems was presented in Ng and Wang (2013) with comparison between the cascade predictive control system and the centralized predictive control system that was discussed in this chapter.

Model predictive controllers were compared with PID controllers in Thomsen *et al.* (2011). Direct current control was discussed in Ambrozic *et al.* (2003). Hildreth's quadratic programming procedure was first proposed in Hildreth (1957). Another approach to solving (8.41) is to use Parallel Quadratic Programming (PQP) (Brand *et al.* (2011)). In Wang and Boyd (2010), online optimization algorithms were investigated for model predictive control. The characteristics of a discrete-time model were discussed in Astrom *et al.* (1984).

References

Ambrozic V, Fiser R and Nedeljkovic D 2003 Direct current control–a new current regulation principle. *IEEE Transactions on Power Electronics* **18**(1), 495–503.

Astrom KJ and Wittenmark B 1997 *Computer Controlled Systems: Theory and Design*. Prentice-Hall, Englewood Cliffs, N. J., 3rd edition.

Astrom KJ, Hagander P and Sternby J 1984 Zeros of sampled systems. *Automatica* Vol.**20**, 31–38.

Bay JS 1999 *Fundamentals of Linear State Space Systems*. McGraw-Hill.

Beccuti A, Mariethoz S, Cliquennois S, Wang S and Morari M 2009 Explicit model predictive control of DC-DC switched-mode power supplies with extended kalman filtering. *IEEE Transactions on Industrial Electronics* **56**(6), 1864–1874.

Bolognani S, Kennel R, Kuehl S and Paccagnella G 2011 Speed and current model predictive control of an IPM synchronous motor drive *Electric Machines Drives Conference (IEMDC), 2011 IEEE International*, pp. 1597–1602.

Bolognani S, Peretti L and Zigliotto M 2008 Combined speed and current model predictive control with inherent field-weakening features for PMSM drives *Electrotechnical Conference, 2008. MELECON 2008. The 14th IEEE Mediterranean*, pp. 472–478.

Bolognani S, Peretti L and Zigliotto M 2009 Design and implementation of model predictive control for electrical motor drives. *IEEE Transactions on Industrial Electronics* **56**(6), 1925–1936.

Boyd S and Vandenberghe L 2004 *Convex Optimization*. Cambridge University Press.

Brand M, Shilpiekandula V, Yao C and Bortoff S 2011 A parallel quadratic programming algorithm for model predictive control *18th IFAC World Congress, Milan, Italy*, vol. 18, pp. 1031–1039.

Chai S, Wang L and Rogers E 2013 Model predictive control of a permanent magnet synchronous motor with experimental validation. *Control Engineering Practice* **21**(11), 1584–1593.

Chung DW and Sul SK 1998 Analysis and compensation of current measurement error in vector-controlled AC motor drives. *IEEE Transactions on Industry Applications* **34**(2), 340–345.

de Santana E, Bim E and do Amaral W 2008 A predictive algorithm for controlling speed and rotor flux of induction motor. *IEEE Transactions on Industrial Electronics* **55**(12), 4398–4407.

Goodwin GC, Graebe SF and Salgado ME 2001 *Control System Design*. Prentice Hall, New Jersey.

Hildreth C 1957 A quadratic programming procedure. *Naval Research Logistics* **4**(1), 79–85.

Ishida J, Doki S and Okuma S 2010 Fast torque control system of PMSM based on model predictive control considering overmodulation region *2010 International Power Electronics Conference (IPEC)*, pp. 1403–1406.

Jahns T and Soong W 1996 Pulsating torque minimization techniques for permanent magnet AC motor drives-a review. *IEEE Transactions on Industrial Electronics* **43**(2), 321–330.

Kennel R, Linder A and Linke M 2001 Generalized predictive control (GPC)-ready for use in drive applications *2001. PESC. 2001 IEEE 32nd Annual Power Electronics Specialists Conference*, vol. 4, pp. 1839–1844.

Linder A and Kennel R 2005 Model predictive control for electrical drives *Power Electronics Specialists Conference, 2005. PESC '05. IEEE 36th*, pp. 1793–1799.

Linder A, Kanchan R, Kennel R and Stolze P 2010 *Model-Based Predictive Control of Electric Drives*. Cuvillier.

Liu H and Li S 2012 Speed control for PMSM servo system using predictive functional control and extended state observer. *IEEE Transactions on Industrial Electronics* **59**(2), 1171–1183.

Luenberger D 2003 *Linear and Nonlinear Programming* 2nd edn. Springer.

Luenberger DG 1969 *Optimization by Vector Space Methods*. John Wiley & Sons, Inc., New York.

Maaziz M, Boucher P and Dumur D 2000 A new control strategy for induction motor based on non-linear predictive control and feedback linearization. *International Journal of Adaptive Control and Signal Processing* **14**(2-3), 313–329.

Maciejowski JM 2000 *Predictive Control with Constraints* 1st edn. Prentice Hall.

Middleton RH and Goodwin GC 1990 *Digital Control and Estimation A Unified Approach*. Prentice-Hall, Englewood Cliffs, NJ.

Moon H, Kim H and Youn M 2003 A discrete-time predictive current control for PMSM. *IEEE Transactions on Power Electronics* **18**(1), 464–472.

Ng K and Wang L 2013 Cascaded and centralized model predictive control of power converter *Industrial Electronics Society, IECON 2013 - 39th Annual Conference of the IEEE*, pp. 5782–5787.

Qin SJ and Badgwell TA 2003 A survey of industrial model predictive control technology. *Control Engineering Practice* **11**(7), 733–764.

Rawlings J 2000 Tutorial overview of model predictive control. *IEEE Control Systems* **20**(3), 38–52.

Rossiter J 2003 *Model-based Predictive Control: a Practical Approach*. CRC Press.

Thomsen S, Hoffmann N and Fuchs F 2011 PI control, PI-based state space control, and model-based predictive control for drive systems with elastically coupled loads – a comparative study. *IEEE Transactions on Industrial Electronics* **58**(8), 3647–3657.

Wang L 2009 *Model Predictive Control System Design and Implementation Using MATLAB* 1st edn. Springer London.

Wang Y and Boyd S 2010 Fast model predictive control using online optimization. *IEEE Transactions on Control System Technology* **18**(2), 267–278.

Wismer D and Chattergy R 1978 *Introduction to Nonlinear Optimization, a Problem Solving Approach*. North-Holland, New York.

Xu J, Panda S, Pan Y, Lee T and Lam B 2004 A modular control scheme for pmsm speed control with pulsating torque minimization. *IEEE Transactions on Industrial Electronics* **51**(3), 526–536.

Zhang L, Norman R and Shepherd W 1997 Long-range predictive control of current regulated PWM for induction motor drives using the synchronous reference frame. *IEEE Transactions on Control Systems Technology* **5**(1), 119–126.

9

Continuous-time Model Predictive Control (CMPC) of Electrical Drives and Power Converter

Traditional model predictive control systems are designed based on a discrete-time model as shown in Chapter 8. More recently, it has been established that there is no explicit obstacle to designing predictive control systems using the models described in the continuous-time while the discretization comes later at the implementation stage (see Wang (2009)). This may be viewed as an advantage in the control design of electrical drives and power converters because the choice of sampling interval is particularly important for the reasons of computational cost and the closed-loop performance, which then could be tested without changing the original design in the continuous-time. The drawback of a continuous-time design and a discrete-time implementation is the existence of approximation error in the implementation stage, which needs to be considered carefully in the choice of sampling interval.

Control systems using model predictive control have the capability of handling constraints and optimizing the closed-loop dynamic response in the presence of constraints. Because of these strengths, the potential advantages of model predictive control are the improvement of power density for motors, namely a smaller motor for a larger load. In these applications, when a larger load occurs, the constraint on the torque current becomes active in the dynamic response phase, and when the control signal tends to its steady-state value, the constraints are satisfied. With the characteristics of model predictive control, the specification of motors could be smaller, where the maximum current is related to the steady-state requirement of the torque current.

There are two major topics to be discussed in this chapter. Firstly, the basic approach to the design of a continuous model predictive controller is introduced in which a single linearized state-space model is used in the design of the control system for the AC motors and power converters. This is the simplest approach with full consideration of the interactions between the variables. The main shortcoming of this design is that it neglects the large difference between the time scales of the electrical and mechanical systems, and it is not computationally effective. The second topic is to discuss how the nonlinearity may affect the predictive control systems as the electrical drive models and power converter models are linearized, and if it is necessary, how a gain scheduled MPC is designed and implemented so that the predictive control system covers a large range of operating conditions.

In this chapter, an induction motor is used as the basis of the continuous-time predictive control system design and implementation, followed by the application to power converters. Using induction motor as

PID and Predictive Control of Electrical Drives and Power Converters using MATLAB®/Simulink®, First Edition.
Liuping Wang, Shan Chai, Dae Yoo, Lu Gan and Ki Ng.

Companion Website: www.wiley.com/go/wang/pid

an example, Section 9.1 introduces the continuous-time MPC design based on one predefined operating condition. In Section 9.3, simulation and experimental results are presented for the continuous-time predictive control of induction motor that demonstrates the selection of the performance tuning parameters. The same continuous-time MPC algorithm is applied to control a power converter in Section 9.4 and experimental validation results are also presented in the same section. Because of the existing non-linearities in the electrical drives and power converter, the gain scheduled model predictive controller is introduced in Section 9.5 based on the linear models derived from several operating conditions. This gain scheduled continuous-time predictive controller is then implemented on the induction motor test-bed and the experimental results presented in Section 9.6 show the closed-loop control performance over a large operating region and significant improvement over the performance produced by the predictive controller designed based on a single operating condition.

The MATLAB programs for the continuous-time predictive control algorithms can be found in Wang (2009).

9.1 Continuous-time MPC Design

In this section, the Model Predictive Controller is designed based on a continuous-time model. The advantages of continuous-time MPC include that the physical model is directly used for control design. In order to achieve zero steady-state error and disturbance rejection, integrators are embedded in the predictive controller.

9.1.1 Augmented Model

The linearized model of an induction motor for speed control purpose is described by the following differential equations:

$$\begin{aligned}\frac{dx_m(t)}{dt} &= A_m x_m(t) + B_m u(t) + \mu^0 \\ y(t) &= C_m x_m(t),\end{aligned} \tag{9.1}$$

where $x_m(t) = [i_{sd}(t) \quad i_{sq}(t) \quad \psi_{rd}(t) \quad \omega_m(t)]^T$, $u(t) = [u_{sd}(t) \quad u_{sq}(t)]^T$, and with the coefficient $\kappa_t = \frac{3Z_pL_h}{2L_rJ_m}$, the matrices A_m and B_m are defined as

$$A_m = \begin{bmatrix} -\frac{1}{\tau'_\sigma} & \omega_m^0 + \frac{2L_h}{\tau_r}\frac{i_{sq}^0}{\psi_{rd}^0} & \frac{k_r}{r_\sigma \tau_r \tau'_\sigma} - \frac{L_h}{\tau_r}\frac{(i_{sq}^0)^2}{(\psi_{rd}^0)^2} & i_{sq}^0 \\ -\omega_m^0 - \frac{L_h}{\tau_r}\frac{i_{sq}^0}{\psi_{rd}^0} & -\frac{1}{\tau'_\sigma} - \frac{L_h}{\tau_r}\frac{i_{sd}^0}{\psi_{rd}^0} & -\frac{k_r}{r_\sigma \tau'_\sigma}\omega_m^0 + \frac{L_h}{\tau_r}\frac{i_{sq}^0 i_{sd}^0}{(\psi_{rd}^0)^2} & -\frac{k_r}{r_\sigma \tau'_\sigma}\psi_{rd}^0 - i_{sd}^0 \\ \frac{L_h}{\tau_r} & 0 & -\frac{1}{\tau_r} & 0 \\ 0 & \kappa_t \psi_{rd}^0 & \kappa_t i_{sq}^0 & -\frac{f_d}{J_m} \end{bmatrix}$$

$$B_m = \begin{bmatrix} \frac{1}{r_\sigma \tau'_\sigma} & 0 \\ 0 & \frac{1}{r_\sigma \tau'_\sigma} \\ 0 & 0 \\ 0 & 0 \end{bmatrix}; C_m = \begin{bmatrix} 0 & 0 & 1 & 0 \\ 0 & 0 & 0 & 1 \end{bmatrix}.$$

The constant vector μ^0, consisting of the steady-state parameters, is given by

$$\mu^0 = \begin{bmatrix} -\omega_m^0 i_{sq}^0 + \frac{L_h}{\tau_r} i_{sq}^0 \psi_{rd}^0 - \frac{L_h}{\tau_r}\frac{(i_{sq})^0}{\psi_{rd}^0} & \omega_m^0 i_{sd}^0 + \frac{k_r}{r_\sigma \tau'_\sigma}\psi_{rd}^0 & 0 & \kappa_t i_{sq}^0 \psi_{rd}^0 - \frac{T_L}{J_m} \end{bmatrix}^T.$$

Based on the above model, the operating condition of parameters ω_m^0, ψ_{rd}^0, i_{sd}^0 and i_{sq}^0 need to be pre-defined, since ψ_{rd}^0 is equal to the set-point value, i_{sd}^0 is derived based on the steady-state relationship:

$$\frac{d\psi_{rd}(t)}{dt} = 0,$$

which leads to

$$i_{sd}^0 = \frac{\psi_{rd}^0}{L_h}.$$

The i_{sq}^0 value is calculated using the steady-state velocity ω_m^0 and the estimated steady-state torque value T_L:

$$i_{sq}^0 = \frac{(f_d/J_m)\omega_m^0 + T_L}{\kappa_t \psi_{rd}^0},$$

where T_L is the load torque that often changes with respect to time, so there is some uncertainty associated with the parameter i_{sq}^0.

In order to embed integrators into the continuous-time model predictive controller, an augmented model is formulated. Taking a derivative operation on the state-space equation of the linearised continuous-time state space model (9.1) leads to

$$\ddot{x}_m(t) = A_m \dot{x}_m(t) + B_m \dot{u}(t), \tag{9.2}$$

where the derivative of the constant vector μ^0 is zero. To obtain the augmented model, firstly the auxiliary variable vector $z(t)$ is selected as $z(t) = \dot{x}_m(t)$, then, the new state variable vector is chosen as $x(t) = [z(t)^T \quad y(t)^T]^T$. With the new state variable vector $x(t)$,the augmented state-space model of the induction motor becomes

$$\dot{x}(t) = Ax(t) + B\dot{u}(t) \tag{9.3}$$

$$y(t) = Cx(t), \tag{9.4}$$

where the augmented system matrices are given by

$$A = \begin{bmatrix} A_m & O_{4\times2} \\ C_m & O_{2\times2} \end{bmatrix}; \quad B = \begin{bmatrix} B_m \\ O_{2\times2} \end{bmatrix}; \quad C = \begin{bmatrix} O_{2\times4} & I_{2\times2} \end{bmatrix}$$

and $I_{*\times*}$ and $O_{*\times*}$ are the identity and zero matrices with dimensions denoted by the sub-indices, respectively. Note that in the augmented state-space model, the constant vector μ^0 is eliminated.

Several comments are in order about the augmented state-space model. Firstly, in the augmented model, the first part of the state variables consists of $\dot{x}_m(t)$. Because the predictive controller is designed to follow step reference signals, the steady-state vector of $x_m(t)$ is a constant vector, and as a result, the steady-state vector of $\dot{x}_m(t)$, is ensured to be a zero vector for all operating conditions. This information provides the convenience in the implementation of the predictive controller as it could be a nontrivial task to find the steady-state values of the original state vector. The second part of the state variables consists of the plant outputs whose steady-state values are the desired reference signals to the control system. At this point, we could write the corresponding small signal model for the augmented model (9.4) by subtracting the steady-state values of the outputs from the second part of state variables ($\tilde{y}_1(t) = \psi_{rd}(t) - \psi_{rd}^0$, $\tilde{y}_2(t) = \omega_m(t) - \omega_m^0$). Furthermore, the steady-state values ψ_{rd}^0 and ω_m^0 are taken as the reference signals to the control system and they can be included inside the objective function in the sequel.

9.1.2 Description of the Control Trajectories Using Laguerre Functions

When designing a continuous-time model predictive control system, the trajectories of the future control signals need to be described using orthonormal basis functions so that the optimization problem is converted to finding a set of coefficients that will minimize an objective function subject to constraints (see Wang (2009)). Laguerre functions form a set of orthonormal basis functions that will be used to describe the control trajectories.

The set of Laguerre functions is defined as (see Lee (1960)), for any $p > 0$,

$$
\begin{aligned}
l_1(t) &= \sqrt{2p} \times e^{-pt} \\
l_2(t) &= \sqrt{2p}(-2pt + 1) \quad e^{-pt} \\
\vdots &= \vdots \\
l_i(t) &= \sqrt{2p}\frac{e^{pt}}{(i-1)!}\frac{d^{i-1}}{dt^{i-1}}[t^{i-1}e^{-2pt}].
\end{aligned} \tag{9.5}
$$

In the literature, parameter p here is called the time scaling factor for the Laguerre functions. This time scaling factor plays an important role in the application of Laguerre functions since it determines their exponential decay rate. It is used as a design parameter that the user will specify as part of the design requirement.

Laguerre functions are one set of the orthonormal functions that satisfy the following orthonormal properties:

$$\int_0^\infty l_i^2(t)dt = 1 \tag{9.6}$$

and

$$\int_0^\infty l_i(t)l_j(t)dt = 0 \quad i \neq j. \tag{9.7}$$

Assuming initial conditions of the state vector as $L(0) = \sqrt{2p}\begin{bmatrix}1 & 1 & \dots & 1\end{bmatrix}^T$, then the Laguerre functions satisfy the state-space equation:

$$
\begin{bmatrix} \dot{l}_1(t) \\ \dot{l}_2(t) \\ \vdots \\ \dot{l}_N(t) \end{bmatrix} = \begin{bmatrix} -p & 0 & \dots & 0 \\ -2p & -p & \dots & 0 \\ \vdots & \dots & \dots & \vdots \\ -2p & \dots & -2p & -p \end{bmatrix} \begin{bmatrix} l_1(t) \\ l_2(t) \\ \vdots \\ l_N(t) \end{bmatrix}. \tag{9.8}
$$

Figure 9.1 shows the first three Laguerre functions for the scaling factor $p = 1$.

Because there are two inputs in the control system of the induction motor, each input will be described using the expansion of Laguerre functions. For instance, within one optimization window of length T_p, the derivative of the first control signal $\dot{u}_1(\tau)$, $0 \le \tau \le T_p$ is described by a set of Laguerre functions

$$\dot{u}_1(\tau) = L_1(\tau)^T \eta_1,$$

where $L_1(\tau)^T = \begin{bmatrix} l_1(\tau) & l_2(\tau) & \dots & l_N(\tau) \end{bmatrix}$ and $\eta_1 = \begin{bmatrix} \xi_1 & \xi_2 & \dots & \xi_N \end{bmatrix}^T$.

Similarly, the derivative of the second control signal is described by

$$\dot{u}_2(\tau) = L_2(\tau)^T \eta_2,$$

where $L_2(\tau)$ may differ from $L_1(\tau)$ in terms of the scaling factor p and the number of terms used in the expansion. To distinguish these Laguerre function sets, N_1 and p_1 denote the number of terms and scaling factor for L_1, likewise, N_2 and p_2 for L_2.

Because of the orthonormal properties and completeness of the Laguerre functions, their applications in continuous-time model predictive control will ensure that the predictive control trajectories will converge to the underlying optimal control trajectories as the numbers (N_1, N_2) of Laguerre functions increase (Wang (2009)).

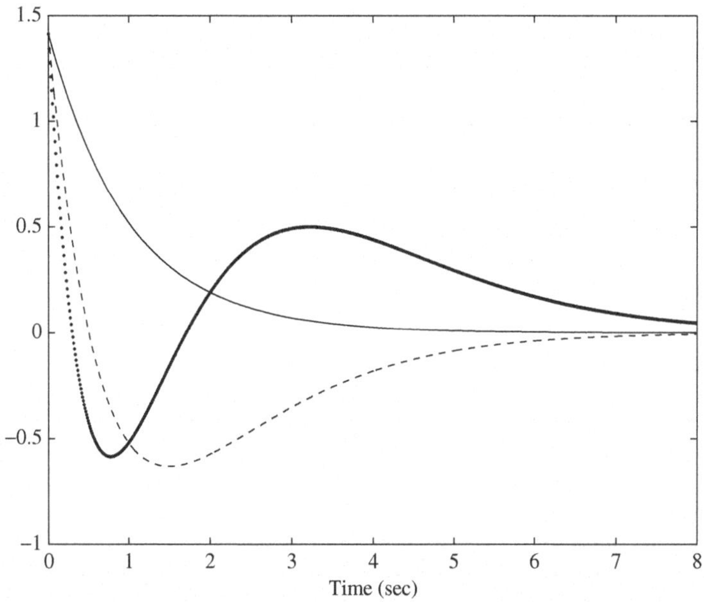

Figure 9.1 Laguerre functions ($p = 1$). Solid line: $l_1(t)$; dashed line: $l_2(t)$; dotted line: $l_3(t)$.

9.1.3 Continuous-time Predictive Control without Constraints

In order to compute the optimal control solution, several steps are discussed in this section. Firstly, at the current time t_i, based on the plant model, the predicted response trajectory of the error state variable vector $\tilde{x}(t) = x(t) - x^0$, where x^0 is the steady-state vector of $x(t)$. At future time τ, $\tau > 0$, $\tilde{x}(t_i + \tau | t_i)$ is described as follows:

$$\tilde{x}(t_i + \tau | t_i) = e^{A\tau}\tilde{x}(t_i) + \int_0^{\tau} e^{A(\tau-\gamma)}B\dot{u}(\gamma)d\gamma, \tag{9.9}$$

where γ presents the time variable within the prediction window to distinguish it from τ.

Since the induction motor model has 2 inputs, the control signal vector and the input matrix B can be written as

$$\dot{u}(\tau) = [\dot{u}_1(\tau)^T \quad \dot{u}_1(\tau)^T]^T$$
$$B = [B_1^T \quad B_2^T]^T.$$

The approximation by using Laguerre functions to both control signals leads to

$$\dot{u}_1(\tau) = L_1(\tau)^T \eta_1$$
$$\dot{u}_2(\tau) = L_2(\tau)^T \eta_2.$$

Therefore, the prediction of future state at time τ (see (9.9)) is rewritten with the orthonormal expansion, as

$$\tilde{x}(t_i + \tau | t_i) = e^{A\tau}\tilde{x}(t_i) + \phi(\tau)^T \eta, \tag{9.10}$$

where $\phi(\tau)^T$ is the convolution integral

$$\phi(\tau)^T = \int_0^{\tau} e^{A(\tau-\gamma)}[B_1 L_1(\gamma)^T \quad B_2 L_2(\gamma)^T]d\gamma$$

and the coefficient vector η contains the two sub-coefficient vectors η_1 and η_2:

$$\eta = [\eta_1^T \quad \eta_2^T]^T;$$

therefore, the matrix $\phi(\tau)^T$ has dimension of $n \times (N_1 + N_2)$ and the optimal control coefficient vector has dimension of $N_1 + N_2$.

By substituting the predicted state trajectory into the state-space model, the prediction of future output at time τ is written as

$$y(t_i + \tau | t_i) = Ce^{A\tau}\tilde{x}(t_i) + C\phi(\tau)^T\eta. \tag{9.11}$$

The cost function, which is similar to the classical Linear Quadratic Regulator (LQR)'s cost function, is chosen as

$$J = \int_0^{T_p} (\tilde{x}(t_i + \tau | t_i)^T Q\tilde{x}(t_i + \tau | t_i) + \dot{u}(\tau)^T R\dot{u}(\tau) d\tau, \tag{9.12}$$

where $\tilde{x}(t_i + \tau | t_i) = x(t_i + \tau | t_i) - x^0$, Q and R are semi-positive and positive definite matrices. Note as stated before that the first part of state vector $x(t)$ is $\dot{x}_m(t)$ having a zero steady-state vector and the second part of $x(t)$ has its steady-state vector corresponding to the reference signals because of the augmented model used in the design (see (9.4)). More specifically, for the induction motor control application, the steady-state vector is defined as

$$x^0 = \begin{bmatrix} 0 & 0 & 0 & 0 & \psi_{rd}^0 & \omega_m^0 \end{bmatrix}^T. \tag{9.13}$$

From the property of the orthonormal functions, $\int_0^\infty L_1(\tau)L_1(\tau)^T d\tau$ is equal to the identity matrix, which is also true for $\int_0^\infty L_2(\tau)L_2(\tau)^T d\tau$. As a result, the second term of the cost function (9.12) is expressed, by assuming R to be a diagonal matrix and sufficiently large prediction horizon T_p, as

$$\int_0^{T_p} \dot{u}(\tau)^T R\dot{u}(\tau) d\tau = \eta^T R_L \eta, \tag{9.14}$$

where R_L is a block diagonal matrix with two blocks corresponding to the weights on the control signals.

By substituting the prediction equation (9.10) into the cost function (9.12), it becomes:

$$J = \int_0^{T_p} (e^{A\tau}\tilde{x}(t_i) + \phi(\tau)^T\eta)^T Q(e^{A\tau}\tilde{x}(t_i) + \phi(\tau)^T\eta) d\tau + \eta^T R_L \eta, \tag{9.15}$$

which is a quadratic function with respect to η:

$$J = \eta^T [\int_0^{T_p} \phi(\tau)Q\phi(\tau)^T d\tau + R_L]\eta + 2\eta^T \int_0^{T_p} \phi(\tau)Qe^{A\tau} d\tau \tilde{x}(t_i)$$
$$+\tilde{x}(t_i)^T \int_0^{T_p} e^{A^T\tau}Qe^{A\tau} d\tau \tilde{x}(t_i). \tag{9.16}$$

By defining two matrices as follows:

$$\Omega_{mpc} = \int_0^{T_p} \phi(\tau)Q\phi(\tau)^T d\tau + R_L \tag{9.17}$$

$$\Psi_{mpc} = \int_0^{T_p} \phi(\tau)Qe^{A\tau} d\tau, \tag{9.18}$$

the objection function (9.16) has the compact expression:

$$J = \eta^T\Omega_{mpc}\eta + 2\eta^T\Psi_{mpc}\tilde{x}(t_i) + \tilde{x}(t_i)^T \int_0^{T_p} e^{A^T\tau}Qe^{A\tau} d\tau \tilde{x}(t_i). \tag{9.19}$$

Completing the squares in the cost function leads to

$$J = [\eta + \Omega_{mpc}^{-1} \Psi_{mpc} \tilde{x}(t_i)]^T \Omega_{mpc} [\eta + \Omega_{mpc}^{-1} \Psi_{mpc} \tilde{x}(t_i)]$$
$$+ \tilde{x}(t_i)^T \int_0^{T_p} e^{A^T \tau} Q e^{A\tau} d\tau \tilde{x}(t_i) - \tilde{x}(t_i)^T \Omega_{mpc}^T \Psi_{mpc}^{-1} \Omega_{mpc} \tilde{x}(t_i). \tag{9.20}$$

Since the last two terms are independent of η and Ω_{mpc} is a positive definite matrix, the minimum of the cost function with respect to η is achieved if the first term is set to zero, that is,

$$\eta = -\Omega_{mpc}^{-1} \Psi_{mpc} \tilde{x}(t_i). \tag{9.21}$$

After the computation of optimal control coefficient vector η, the control trajectory $\dot{u}(\tau)$ is reconstructed with the Laguerre functions

$$\dot{u}(\tau) = \begin{bmatrix} L_1(\tau)^T & O_2 \\ O_1 & L_2(\tau)^T \end{bmatrix} \eta,$$

where O_1 and O_2 are zeros vectors with their dimensions equal to those of $L_1(\tau)^T$ and $L_2(\tau)^T$.

The principle of receding horizon control strategy is to use the information from the first sample of the control trajectory. Hence, at the sampling time t_i, the optimal control $\dot{u}(t_i)$ for the unconstrained problem is

$$\dot{u}(t_i) = \begin{bmatrix} L_1(0)^T & O_2 \\ O_1 & L_2(0)^T \end{bmatrix} \eta. \tag{9.22}$$

The actual control signal is computed using

$$u(t_i) = u(t_{i-1}) + \dot{u}(t_i)\Delta t, \tag{9.23}$$

where Δt is the sampling interval used in the implementation of the continuous-time predictive control system.

Without constraints, the optimal control solution can also be expressed as the state feedback control

$$\dot{u}(t) = -K_{mpc} \tilde{x}(t), \tag{9.24}$$

where the feedback control gain matrix is

$$K_{mpc} = \begin{bmatrix} L_1(0)^T & O_2 \\ O_1 & L_2(0)^T \end{bmatrix} \Omega_{mpc}^{-1} \Psi_{mpc}. \tag{9.25}$$

The data matrices Ω_{mpc} and Ψ_{mpc} are computed off-line as the process of continuous-time MPC design. Therefore, in the unconstrained case, while the feedback gain matrix K_{mpc} is computed off-line as shown in (9.25), the on-line computation only involves Equation (9.24). Furthermore, the location of closed-loop poles is evaluated by calculating the eigenvalues of $(A - BK_{mpc})$. When the numbers of terms are large, with a long prediction horizon T_p, the derivative of the control trajectory $\dot{u}(.)$ closely matches the underlying optimal control trajectory defined by the linear quadratic regulator (LQR) (Wang (2009)).

In the application of the induction motor, for a given operating condition, the time varying components of the dynamic model are calculable, so that a linear time invariant model will be obtained for the MPC design. For example, the velocity and flux references are set as, $\omega_m^* = 1400$ RPM and $\psi_{rd} = 0.6$ Wb. The CMPC design parameters include: the Laguerre function parameters $N_1 = N_2 = 6$ and $p_1 = p_2 = 20$, prediction horizon $T_p = 0.5$, the weight matrices $R_L = r_k I$ ($r_k = 1$) and $Q = C^T C$, then the closed-loop eigenvalues are calculated as

$$[-139.72 \pm j288.84 \quad -68.46 \quad -1.43 \pm j2.16 \quad -5.39 \times 10^{-6}],$$

where one pole is almost located at the origin. Thus, the feedback control is considered marginally stable, the change of tuning parameter r_k could affect the closed-loop eigenvalues, Table 9.1 shows that smaller weighting parameter r_k does push the dominant eigenvalue away from the origin but not significantly. Therefore, the method called exponential data weighting is used for such a situation.

Table 9.1 Closed-loop eigenvalues with different weight coefficient in R_L matrix

$r_k = 0.1$	$-139.72 \pm j288.84$	-68.46	$-3.74 \pm j3.74$	-5.38×10^{-5}
$r_k = 0.01$	$-139.72 \pm j288.84$	-68.46	$-6.57 \pm j6.53$	-5.36×10^{-4}
$r_k = 0.001$	$-139.72 \pm j288.84$	-68.45	$-11.92 \pm j11.88$	-0.0053
$r_k = 0.0001$	$-139.72 \pm j288.84$	-68.35	$-21.07 \pm j20.88$	-0.0529

9.1.4 Tuning of CMPC Control System Using Exponential Data Weighting and Prescribed Degree of Stability

From the previous section, the closed-loop eigenvalues are heavily dependent on weight matrices Q and R_L. For example, in this induction motor case study, there are 36 elements in the Q matrix, and it is a complicated matter to find the individual elements and the combinations of them to achieve desired closed-loop performance. As demonstrated in Table 9.1, the variation of R_L does not sufficiently change the closed-loop performance as desired.

Exponential weighting is to use a time dependent weighting $e^{-\alpha t}$ ($\alpha > 0$) in the cost function of the predictive control system design, in order to produce a numerically well-conditioned Ω_{mpc} matrix. On the basis of exponential data weighting, a prescribed degree of stability is to ensure that the eigenvalues of the closed-loop predictive control system are on the left-hand side of the $-\beta$ line ($\beta > 0$) in the complex plane.The locations of the desired closed-loop eigenvalues are illustrated in Figure 9.2.

The detailed information about the exponential data weighting and prescribed degree of stability can be found in Wang (2009).

9.1.4.1 Selection of α

The idea behind the selection of α is to make sure that the design model with $(A - \alpha I)$ is stable with all eigenvalues on the left-half of the complex plane. The computation of the prediction when using $A - \alpha I$ is numerically sound.

From a given augmented state-space model (A, B), the eigenvalues of A are determined. Because the induction motor is a stable system, the unstable eigenvalues of A come from the integrators that have been embedded in the model. In this case, any $\alpha > 0$ will serve the purpose of exponential data weighting.

Once the exponential weight factor α is selected, the eigenvalues of the matrix $A - \alpha I$ are fixed. Since this matrix is stable with an appropriate choice of α, the prediction of the state variables is numerically sound. In general, if the eigenvalues of $A - \alpha I$ were further away from the imaginary axis on the complex plane, then a smaller prediction horizon T_p would be required.

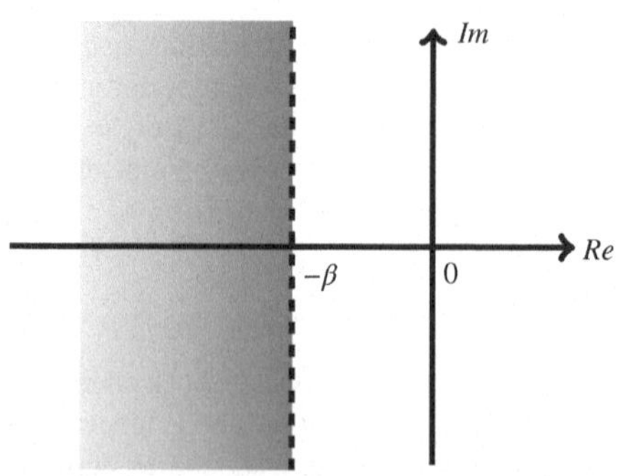

Figure 9.2 Shaded area gives the prescribed degree of stability of $-\beta$.

9.1.4.2 Selection of degree of stability β

So far the closed-loop performance of a predictive control system is determined by the choice of Q and R matrices. Tuning via Q and R could be very time consuming as it often requires finding the off-diagonal elements in Q and R to achieve satisfactory performance. This is often performed in a trial-and-error manner as demonstrated in Chapter 8. Now, with the additional parameter β that dictates the degree of stability, the closed-loop eigenvalues of the predictive control system are effectively positioned to the left-hand side of the $-\beta$ line on the complex plane. The parameter β will be used to shift the closed-loop eigenvalues of the predictive control system. In the design, the user will specify the parameter β for the location of the closed-loop eigenvalues, leading to the specification of the closed-loop response speed. For instance, a larger β will result in a faster closed-loop response speed.

Once a suitable value for β is chosen, which is the degree of stability, the following Riccati equation is solved for the P matrix:

$$P(A + \beta I) + (A + \beta I)^T P - PBR^{-1}B^T P + Q = 0. \tag{9.26}$$

A MATLAB script can be used for this solution:

```
[K,P,E]= lqr(A+beta*eye(n,n), B, Q, R);
```

The matrix Q_α is then determined, with the values of α, β and P, using

$$Q_\alpha = Q + 2(\alpha + \beta)P.$$

The augmented state-space model (A, B) is modified for use in the design. The matrix B is unchanged; however, the matrix A is modified to become $A - \alpha I$. With this set of performance parameters (Q_α, R) and the design model $(A - \alpha I, B)$, the predictive control problem is converted back to the original problem stated in Sections 9.1 to 9.1.3.

9.1.4.3 The Parameters in Laguerre Functions

When N increases, the predictive control trajectory converges to the underlying optimal control trajectory of the linear quadratic regulator. However, with a small N, the scaling factor in the Laguerre functions p will affect the closed-loop response. The parameter p should be chosen close to the smallest magnitude of the eigenvalue from the LQR design, then increasing the parameter N until the closed-loop eigenvalues from predictive control system become close to those produced by the LQR system.

To demonstrate the effectiveness of the tuning procedure, we redo the example presented in Table 9.1. For instance, if the parameters remain identical to the previous case, except the exponential data weighting with prescribed degree of stability is applied, we choose the exponential weighting parameter $\alpha = 1.2$ to ensure that the design model is stable and all closed-loop eigenvalues to be on the left of $s = -20$ line on the complex plane. The comparative results are illustrated in Table 9.2. It is interesting to note that for this set of tuning parameters, the largest eigenvalue in the original design was located at -0.0529 and is shifted to -40 in the tuning procedure with α and β parameters, but for those already located on the left of the $s = -\beta = -20$, little change occurred.

Table 9.2 Eigenvalue comparison with and without using prescribed degree of stability in MPC design

	Closed-loop eigenvalues			
$\alpha = 0;\ \beta = 0$	$-139.72 \pm j288.84$	-68.46	$-3.74 \pm j3.74$	-5.38×10^{-5}
$\alpha = 1.2;\ \beta = 20$	$-138.80 \pm j291.24$	-69.70	$-38.76 \pm j2.25$	-39.91

9.2 CMPC with Nonlinear Constraints

The strength of the continuous-time model predictive control system lies in the conceptual and computational simplicity when tackling the constrained control problem. It is paramount that suitable operational constraints are in place for the safety of the equipment. Similar to the discrete-time MPC presented in Chapter 8, the constraints implementation is established by using on-line Quadratic Programming (QP) technique as discussed in Chapter 8.

In the case of the induction motor control, all constraints under consideration are input variable constraints, which are the stator voltages u_{sd} and u_{sq}. Assuming that the DC-bus voltage is supplied with V_{dc} V, and with the modulation limitation, the manipulated variables are constrained by the following relation:

$$\sqrt{u_{sd}^2 + u_{sq}^2} \leq \frac{V_{dc}}{\sqrt{3}}. \tag{9.27}$$

This is a quadratic constraint with respect the input variables. The constrained predictive control problem becomes a quadratic optimization subject to quadratic constraints. The solution to this nonlinear constrained optimization problem demands a substantial amount of on-line computational power, and the nonlinear optimizer also complicates the real-time implementation of the gain scheduled predictive control algorithm introduced later in this chapter. Two approaches are proposed here to approximate the quadratic constraint (9.27).

9.2.1 *Approximation of Nonlinear Constraint Using Four Linear Constraints*

The input constraints on the control signal magnitude $u_{sd}(t)$ and $u_{sq}(t)$ will be imposed in the design and implementation. Assume here that the magnitude constraints are specified for the upper and lower limits of the control signals as

$$u_{sd}^{min} \leq u_{sd}(t) \leq u_{sd}^{max}$$
$$u_{sq}^{min} \leq u_{sq}(t) \leq u_{sq}^{max}.$$

As shown in the previous section, the calculation of the amplitude of control signal is based on

$$u(t_i) = u(t_{i-1}) + \begin{bmatrix} L_1(0)^T & O_2 \\ O_1 & L_2(0)^T \end{bmatrix} \eta \Delta t,$$

where Δt is the sampling interval. Together with the inequality constraints, the constraints on the control signals are formulated as

$$\begin{bmatrix} u_{sd}^{min} \\ u_{sq}^{min} \end{bmatrix} \leq \begin{bmatrix} u_{sd}(t_{i-1}) \\ u_{sq}(t_{i-1}) \end{bmatrix} + \begin{bmatrix} L_1(0)^T & O_2 \\ O_1 & L_2(0)^T \end{bmatrix} \eta \Delta t \leq \begin{bmatrix} u_{sd}^{max} \\ u_{sq}^{max} \end{bmatrix}. \tag{9.28}$$

9.2.2 *Approximation of Nonlinear Constraint Using Sixteen Linear Constraints*

Figure 9.3 shows the constraint equation (9.27), which is the area of a circle with the radius of $\frac{V_{dc}}{\sqrt{3}}$. In order to obtain the linear approximation, the circular area is approximated using the area of an octagon as shown in Figure 9.3. For notational simplicity, u_{sd} and u_{sq} are denoted as f and g, and a unit circle is chosen for the initial analysis. On Figure 9.3, the eight pairs of the coordinates are marked in anti-clockwise manner, and similarly marked are the eight straight lines. The values of the coordinates are given in Table 9.3.

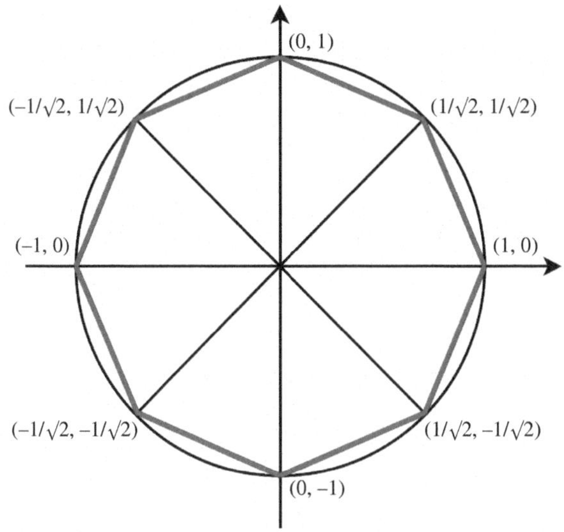

Figure 9.3 Approximation of circular constraint area using the area of an octagon.

Table 9.3 The values of the coordinates

k	1	2	3	4	5	6	7	8
f_k	1	$\frac{1}{\sqrt{2}}$	0	$-\frac{1}{\sqrt{2}}$	−1	$-\frac{1}{\sqrt{2}}$	0	$\frac{1}{\sqrt{2}}$
g_k	0	$\frac{1}{\sqrt{2}}$	1	$\frac{1}{\sqrt{2}}$	0	$-\frac{1}{\sqrt{2}}$	−1	$-\frac{1}{\sqrt{2}}$

There are four inequalities associated with the upper part of the circle, and four reversed inequalities associated with the lower part of the circle. For $k = 1, 2, 3, 4$, the inequalities for the upper part of the circle are expressed as

$$g - \frac{g_k - g_{k+1}}{f_k - f_{k+1}} f \leq -\frac{g_k - g_{k+1}}{f_k - f_{k+1}} f_k + g_k. \tag{9.29}$$

For $k = 5, 6, 7$, the inequalities for the lower part of the circle are expressed as

$$g - \frac{g_k - g_{k+1}}{f_k - f_{k+1}} f \geq -\frac{g_k - g_{k+1}}{f_k - f_{k+1}} f_k + g_k. \tag{9.30}$$

For $k = 8$, the inequality is

$$g - \frac{g_8 - g_1}{f_8 - f_1} f \geq -\frac{g_8 - g_1}{f_8 - f_1} f_8 + g_8. \tag{9.31}$$

The linear inequalities can be written in general forms as

$$\alpha_k f + \beta_k g \leq \gamma_k \tag{9.32}$$

or

$$\alpha_k f + \beta_k g \geq \gamma_k, \tag{9.33}$$

where $\alpha_k = -\frac{g_k - g_{k+1}}{f_k - f_{k+1}}$, $\beta_k = 1$ and $\gamma_k = -\frac{g_k - g_{k+1}}{f_k - f_{k+1}} f_k + g_k$ for $k = 1, 2, 3, 4, 5, 6, 7$. For $k = 8$, $\alpha_k = -\frac{g_8 - g_1}{f_8 - f_1}$, $\beta_k = 1$ and $\gamma_k = -\frac{g_8 - g_1}{f_8 - f_1} f_8 + g_8$. Note that the parameters α_k and β_k are independent of the radius of the circle, however, the parameter γ_k is proportional to the radius, which is in the case of induction motor control, $\frac{V_{dc}}{\sqrt{3}}$. Thus, all γ_k, $k = 1, 2, \ldots, 8$ will multiply the radius to obtain their actual values for the specific application.

For the induction motor control problem, $f = u_{sd}(t)$ and $g = u_{sq}(t)$. The next task is to reformulate the inequalities with the Laguerre coefficient vector η so that the linear inequalities become the linear inequality constraints in the design of predictive control. Taking the example of first four inequalities, by imposing the constraints on the first sample of the control signals, at the sampling instant t_i, the inequalities are

$$\alpha_k u_{sd}(t_i) + \beta_k u_{sq}(t_i) \leq \gamma_k, \tag{9.34}$$

for $k = 1, 2, 3, 4$. Since the control variables are related to the Laguerre coefficient vector through

$$u_{sd}(t_i) = u_{sd}(t_i - \Delta t) + L_1(0)^T \eta_1 \Delta t \tag{9.35}$$

$$u_{sq}(t_i) = u_{sq}(t_i - \Delta t) + L_2(0)^T \eta_2 \Delta t, \tag{9.36}$$

where $L_1(0)^T = \sqrt{2p_1}\begin{bmatrix}1 & 1 & \ldots & 1\end{bmatrix}$ and $L_2(0)^T = \sqrt{2p_2}\begin{bmatrix}1 & 1 & \ldots & 1\end{bmatrix}$; η_1 and η_2 are the Laguerre coefficient vectors for $\dot{u}_{sd}$ and $\dot{u}_{sq}$ respectively as the Laguerre functions are used in describing the derivatives of the control signals. By substituting (9.35) and (9.36) into (9.34), the following inequality expression is obtained:

$$\alpha_k L_1(0)^T \Delta t \eta_1 + \beta_k L_2(0)^T \Delta t \eta_2 \leq \gamma_k - \alpha_k u_{sd}(t_i - \Delta t) - \beta_k u_{sq}(t_i - \Delta t), \tag{9.37}$$

which is, in vector form,

$$\begin{bmatrix}\alpha_k L_1(0)^T \Delta t & \beta_k L_2(0)^T \Delta t\end{bmatrix}\begin{bmatrix}\eta_1 \\ \eta_2\end{bmatrix} \leq \gamma_k - \begin{bmatrix}\alpha_k & \beta_k\end{bmatrix}\begin{bmatrix}u_{sd}(t_i - \Delta t) \\ u_{sq}(t_i - \Delta t)\end{bmatrix} \tag{9.38}$$

Similarly, for $k = 5, 6, 7, 8$, the kth inequality is expressed as

$$\begin{bmatrix}\alpha_k L_1(0)^T \Delta t & \beta_k L_2(0)^T \Delta t\end{bmatrix}\begin{bmatrix}\eta_1 \\ \eta_2\end{bmatrix} \geq \gamma_k - \begin{bmatrix}\alpha_k & \beta_k\end{bmatrix}\begin{bmatrix}u_{sd}(t_i - \Delta t) \\ u_{sq}(t_i - \Delta t)\end{bmatrix}, \tag{9.39}$$

which is equivalent to

$$-\begin{bmatrix}\alpha_k L_1(0)^T \Delta t & \beta_k L_2(0)^T \Delta t\end{bmatrix}\begin{bmatrix}\eta_1 \\ \eta_2\end{bmatrix} \leq -\gamma_k + \begin{bmatrix}\alpha_k & \beta_k\end{bmatrix}\begin{bmatrix}u_{sd}(t_i - \Delta t) \\ u_{sq}(t_i - \Delta t)\end{bmatrix}. \tag{9.40}$$

By combining the inequalities represented by (9.38) and those by (9.40), and writing them in a vector form, the linear inequality constraints for the design of model predictive control are obtained:

$$\begin{bmatrix}\alpha_1 & \beta_1 \\ \alpha_2 & \beta_2 \\ \alpha_3 & \beta_3 \\ \alpha_4 & \beta_4 \\ -\alpha_5 & -\beta_5 \\ -\alpha_6 & -\beta_6 \\ -\alpha_7 & -\beta_7 \\ -\alpha_8 & -\beta_8\end{bmatrix}\begin{bmatrix}L_1(0)^T & O_2 \\ O_1 & L_2(0)^T\end{bmatrix}\Delta t\eta \leq \begin{bmatrix}\gamma_1 \\ \gamma_2 \\ \gamma_3 \\ \gamma_4 \\ -\gamma_5 \\ -\gamma_6 \\ -\gamma_7 \\ -\gamma_8\end{bmatrix} + \begin{bmatrix}cc - \alpha_1 & -\beta_1 \\ -\alpha_2 & -\beta_2 \\ -\alpha_3 & -\beta_3 \\ -\alpha_4 & -\beta_4 \\ \alpha_5 & \beta_5 \\ \alpha_6 & \beta_6 \\ \alpha_7 & \beta_7 \\ \alpha_8 & \beta_8\end{bmatrix}\begin{bmatrix}u_{sd}(t_i - \Delta t) \\ u_{sq}(t_{i-1})\end{bmatrix}. \tag{9.41}$$

With the constraints formulated, the continuous-time predictive control problem is expressed as minimizing the cost function J (9.19) subject to the set of linear inequality constraints (see (9.41)) in real-time, which is solved using a quadratic programming algorithm as discussed in Chapter 8.

9.2.3 State Feedback Observer

In continuous-time MPC design, the state variable vector contains the derivatives of the current signals. Because these signals are generally noisy, differentiation of the current signals is to be avoided for the reason that the derivative operation will amplify the noise in the current signals. A strategy is thus to use a state observer for estimating the state variables, which has the added benefit of acting as a filter to the measurement noise.

The observer equation for such an application has the following form:

$$\frac{d\hat{x}(t)}{dt} = A\hat{x}(t) + B\dot{u}(t) + K_{ob}(y(t) - C\hat{x}(t)), \tag{9.42}$$

where $\hat{x}(t)$ is an estimate of $x(t)$, K_{ob} is the observer gain and (A, B, C) are the system matrices of the augmented model. Input $\dot{u}(t)$ is obtained from the solution of the predictive control. This is the Luenberger observer, in which the observer gain matrix K_{ob} calculated off-line to reduce computational load.

The observer gain matrix K_{ob} is chosen according to closed-loop performance specification of the observer system and the pair of system matrices (A, C). For instance, the computation of the observer gain K_{ob} could be performed using the MATLAB function *lqr*, as

$$K_{ob} = lqr(A^T, C^T, Q_{ob}, R_{ob})^T,$$

where A^T and C^T are transposed system matrices of the augmented model, Q_{ob} and R_{ob} are weighting matrices of the observer. The design of an observer is a dual task to the design of a controller, thus the use of transposes of A and C matrices in the *lqr* function.

The continuous-time observer equation (9.43) is discretized for implementation, leading to

$$\hat{x}(t_{i+1}) = \hat{x}(t_i) + (A\hat{x}(t_i) + B\dot{u}_i(t_i) + K_{ob}(y(t_i) - C\hat{x}(t_i)))\Delta t. \tag{9.43}$$

Thus, based on the current sample information of the optimal control solution $\dot{u}(t_i)$ and the error signal $y(t_i) - C\hat{x}(t_i)$, the next sample of state estimate $\hat{x}(t_{i+1})$ is computed.

To complete this section, Figure 9.4 is used to illustrate the configuration of the continuous-time MPC for speed control. The controlled outputs are the rotor flux ψ_{rd} and the mechanical motor speed ω_m, and the control signals are, u_{sd} and u_{sq}. A Luenberger Observer is introduced for the estimation of the synchronous flux position ω_s for $d-q$ transformation, and the rotor flux ψ_{rd} for feedback. Another observer is used to estimate the state variable vector $x(t)$ in order to avoid differentiation of the current signals. Both observer gain matrices are found off-line to reduce computational load.

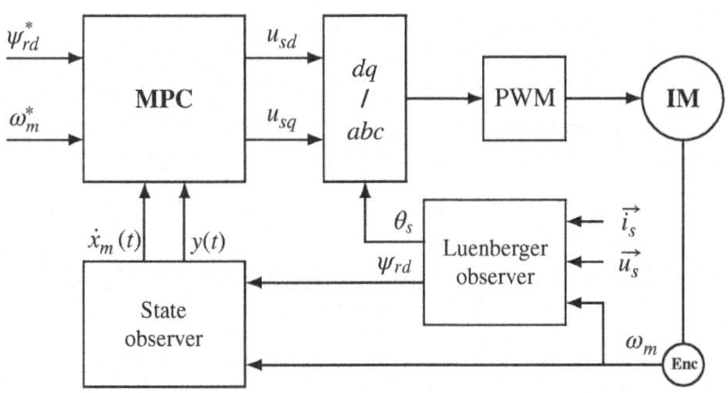

Figure 9.4 Speed control of induction motor using continuous-time MPC.

Table 9.4 Continuous-time MPC parameters

i_{sq}^0	T_L	ω_m^0	ω_s^0	ψ_{rd}^0	N	p	T_p
0.4194 A	0.5 Nm	1400 RPM	$\omega_s^0 = 298.4\ rad/s$	0.6 Wb	6	20	0.5

9.3 Simulation and Experimental Evaluation of CMPC of Induction Motor

The continuous-time predictive control system for induction motor is evaluated using the Simulink simulation program, then experimentally demonstrated using the test-bed.

The steady-state parameters used to obtain the linear model for the induction motor are listed in Table 9.4. In the design of the continuous-time predictive controller, the prediction horizon T_p, the Laguerre parameters $N_1 = N_2 = N$, $p_1 = p_2 = p$. The weighting matrices $R = I$ (I being the identity matrix) and $Q = C^T C$ where C is the output matrix for the augmented model. Exponential data weighting is used to improve the numerical condition of the Hessian matrix Ω_{mpc}, where $\alpha = 1.2$ is selected. Together with the prescribed degree of stability, the closed-loop eigenvalues of the predictive control system are positioned to the left of the $s = -\beta$ line where $\beta = 20$. The constraints on both stator voltages are specified as

$$-90.1 \leq u_{sd} \leq 90.1; \quad -286.4 \leq u_{sq} \leq 286.4.$$

Although the predictive controller is designed using the continuous-time model, the discretization occurs at the implementation stage. In general, a smaller sampling interval would be preferred since it results in a smaller approximation error associated with the discretization. However, due to the on-line computational cost that restricts how fast the sampling rate could be, the experimental setup only allows the sampling interval Δt not be less than 200×10^{-6} second. Thus, in both simulations and experiments, the continuous-time model predictive controller is implemented using the lowest sampling interval possible ($\Delta t = 200$ μs).

9.3.1 *Simulation Results*

The simulation results for the continuous-time model predictive control of induction motor are shown in Figure 9.5. As expected, the velocity response ω_m converges to the steady-state value of 1400 RPM (see Figure 9.5(a)), while the rotor flux response ψ_{rd} converges to 0.6 Wb (see Figure 9.5(b)). Both output responses have no steady state error and have the settling times within 0.4 sec. Note that in the predictive controller design the dominant constant of the closed-loop predictive control system is approximately $1/\beta$. With $\beta = 20$, the closed-loop settling time could be estimated as $5/\beta = 0.25$ seconds. It seems that the settling time for both outputs is larger than 0.25 sec. This is because the constraint on u_{sd} becomes active during the transient response (see Figure 9.5(c)) that resulted in slower closed-loop responses for both outputs. The stator currents are part of the state variables and there are no constraints imposed on the currents (see Figure 9.5d).

To demonstrate how the tuning parameter β affects the closed-loop response, Figure 9.6 shows the closed-loop output responses for five different values of β varying from 5 to 20. Because of the effects of the control signal constraints, although there are five sets of closed-loop poles, the closed-loop responses have similar response times for the first four cases except for the one associated with $\beta = 5$ which has a longer setting time (see Figure 9.6). One comment is that with the continuous-time model predictive control system, the closed-loop response times for the electrical and the mechanical systems are in the same time scale, as demonstrated in the simulation studies.

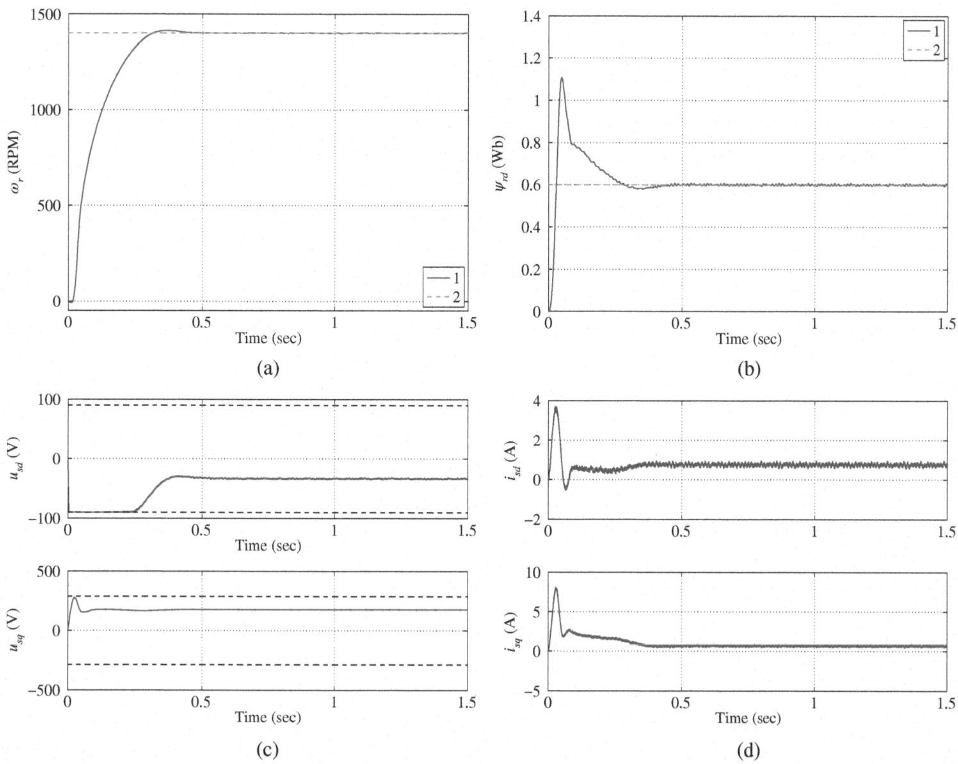

Figure 9.5 Simulation results of speed control using continuous-time MPC. Key: line (1) Actual feedback measurement; line (2) Set-point signal. (a) Motor speed (ω_m), (b) Flux response (ψ_{rd}), (c) Control signals, and (d) Stator current signals.

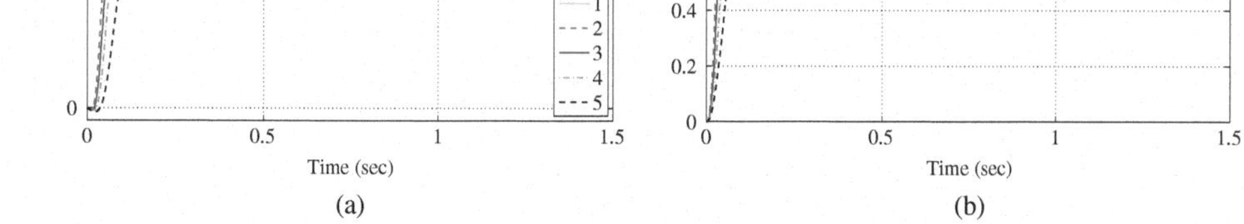

Figure 9.6 Simulation results of CMPC with different β values. Key: line (1) Set-point signal; line (2) $\beta = 20$; (3) $\beta = 15$; (4) $\beta = 10$; (5) $\beta = 5$. (a) Motor speed (ω_m), and (b) Flux response (ψ_{rd}).

9.3.2 Experimental Results

In the experimental evaluation, the parameter for the prescribed degree of stability is selected as $\beta = 15$ while the rest of the parameters remain unchanged from the simulation evaluations. There are two sets of experimental results presented for the evaluation. The first set of control experiments is performed without the state observer where derivatives in the augmented state vector are calculated using their first order approximations with the form: $\dot{x}_m(t_i) \approx \frac{x_m(t_i)-x_m(t_{i-1})}{\Delta t}$ The second set of experiments is performed with a full state observer in which the observer is designed using MATLAB "lqr" function with the weighting matrices chosen as $Q_{ob} = \text{I}$ and $R_{ob} = 10^{-5}\text{I}$. The experimental results for the first case are illustrated in Figure 9.7 and the second case in Figure 9.8. When comparing these two figures, it is seen that the closed-loop velocity response has about the same response time (see Figures 9.7(a) and 9.8(a)), the rotor flux response has a larger peak when the observer is used (see Figures 9.7(b) and 9.8(b)), the noise in the control signals has reduced when using observer (see Figures 9.7(c) and 9.8(c)). On the other hand, it seems that without the observer, although the noise effect is larger, the closed-loop dynamic system has a faster response. Particularly, this is seen in the response of the rotor flux, which has a smaller peak without using the observer (see Figure 9.7(b)). This is because the closed-loop poles of the state estimate predictive control system consist of the poles from the predictive control system as well as those from the observer error system. Overall, the experimental results are similar to the simulation results (see Figures 9.5 and 9.7). This means that the physical simulation model has a high fidelity when comparing with the test-bed.

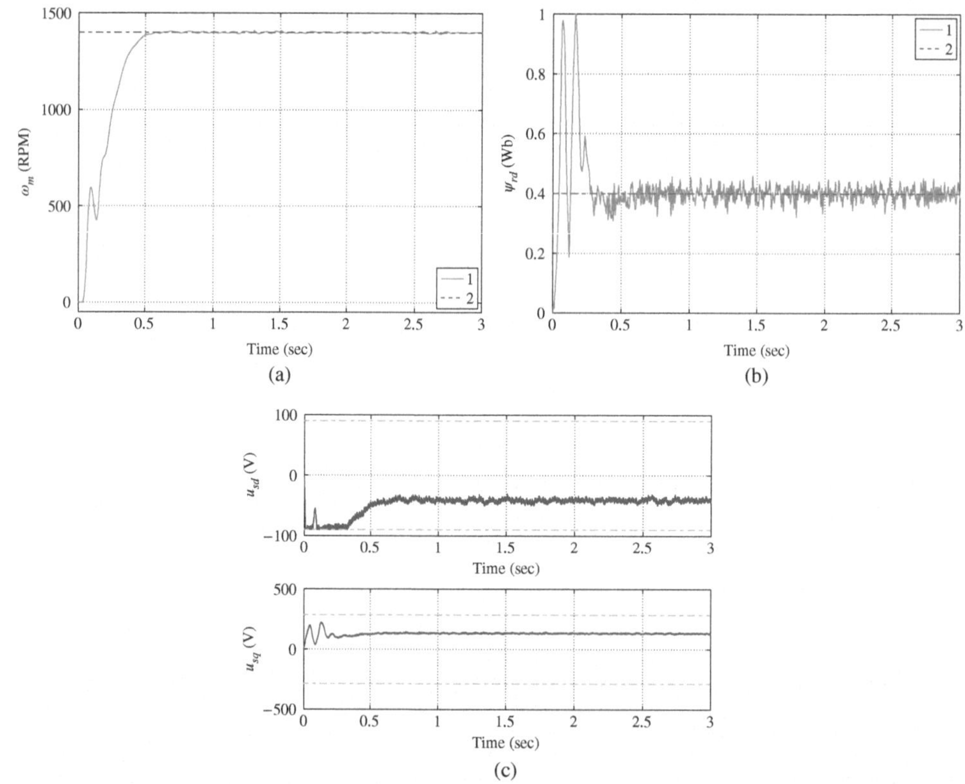

Figure 9.7 Experimental results of CMPC of induction motor (without observer). (a) Motor speed (ω_m), (b) Flux response (ψ_{rd}), and (c) Control signals.

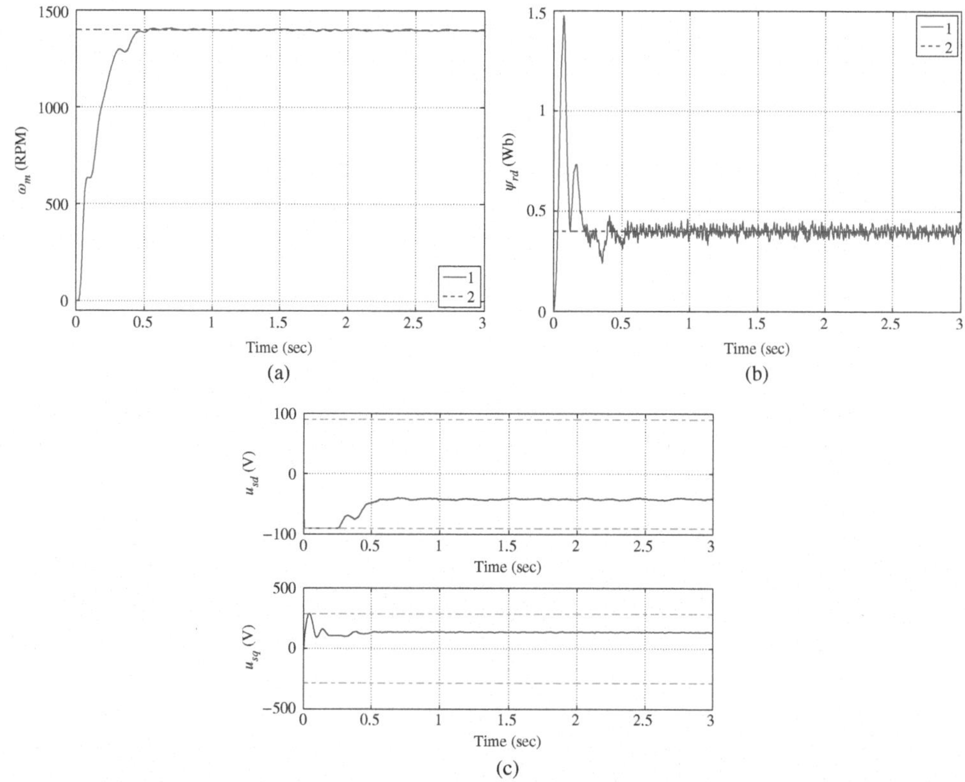

Figure 9.8 Experimental results of CMPC of induction motor (with observer). (a) Motor speed (ω_m), (b) Flux response (ψ_{rd}), and (c) Control signals.

9.4 Continuous-time Model Predictive Control of Power Converter

The objective of the model predictive control system for a power converter is to regulate the *DC* bus voltage at a desired value specified by the applications while maintaining unity power factor. With this objective, the outputs of the regenerative power supply are chosen to be the voltage of the *DC* bus V_{dc} and the current i_q. When the system has a unity power factor, $i_q = 0$, which is the set-point signal for this output. The continuous-time model predictive control algorithm introduced for the induction motor control is used here to control the power converter.

The steady state values of the parameters in the linear model are selected as $i_{qo} = 0$, $V_{qo} = 0$, $\upsilon_{dco} = \upsilon^*_{dc}$ where υ^*_{dc} is the reference signal for υ_{dc}. For steady-state switching functions S_{do}, S_{qo} are computed as (Komurcugil and Kukrer (1998)),

$$S_{do} = \frac{2(E_d - R_s I_d)}{\upsilon_{dco}} = 0.952 \tag{9.44}$$

$$S_{qo} = \frac{-2\omega_g L_s I_d}{\upsilon_{dco}} = -0.403, \tag{9.45}$$

where I_d which is the steady-state value of i_d is chosen to be 4.69 A, $E_d = 30$ V and $\upsilon_{dco} = 65$ V.

The linear model used in the design of the model predictive controller is

$$\dot{x}_m(t) = A_m x_m(t) + B_m u(t)$$
$$y(t) = C_m x_m(t), \tag{9.46}$$

where A_m, B_m and C_m are system matrices associated with the operating conditions of the power converter and

$$A_m = \begin{bmatrix} -\frac{R_s}{L_s} & \omega_g & -\frac{S_{do}}{2L_s} \\ -\omega_g & -\frac{R_s}{L_s} & -\frac{S_{qo}}{2L_s} \\ \frac{3S_{do}}{4C_{dc}} & \frac{3S_{qo}}{4C_{dc}} & 0 \end{bmatrix}, B_m = \begin{bmatrix} \frac{-v_{dco}}{2L_s} & 0 \\ 0 & \frac{-v_{dco}}{2L_s} \\ \frac{3i_{do}}{4C_{dc}} & \frac{3i_{qo}}{4C_{dc}} \end{bmatrix},$$

$$C_m = \begin{bmatrix} 0 & 1 & 0 \\ 0 & 0 & 1 \end{bmatrix}, x_m(t) = \begin{bmatrix} i_d(t) \\ i_q(t) \\ V_{dc}(t) \end{bmatrix}, u(t)(t) = \begin{bmatrix} S_d(t) \\ S_q(t) \end{bmatrix},$$

where $S_{do}, S_{qo}, v_{dco}, i_{do}$ and i_{qo} represents steady state equivalent solutions.

In the operation of the converter, there are low frequency disturbances and harmonic distortions, thus integrators are needed in the controller. To embed the integrator, two auxiliary variables are chosen as

$$z(t) = \dot{x}_m(t)$$
$$y(t) = C_m x_m(t)$$

and based on them, a new state variable vector is defined as $x(t) = [z(t)^T \quad y(t)^T]^T$. With these auxiliary variables, in conjunction with the original plant model, the augmented state space model is defined as:

$$\begin{bmatrix} \dot{z}(t) \\ \dot{y}(t) \end{bmatrix} = \begin{bmatrix} A_m & O_{2\times3}^T \\ C_m & O_{2\times2} \end{bmatrix} \begin{bmatrix} z(t) \\ y(t) \end{bmatrix} + \begin{bmatrix} B_m \\ O_{2\times2} \end{bmatrix} \dot{u}(t) \tag{9.47}$$

$$y(t) = \begin{bmatrix} O_{2\times3} & I_{2\times2} \end{bmatrix} \begin{bmatrix} z(t) \\ y(t) \end{bmatrix}, \tag{9.48}$$

where $I_{2\times2}$ is the identity matrix with dimensions 2×2; $O_{2\times2}$ is a 2×2 zero matrix, and $O_{2\times3}$ is a 2×3 zero matrix.

The system parameters used in the experimental setup are $L_s = 8.9 \times 10^{-3}$ H, $R_s = 0.2$ Ω, $C_{dc} = 296$ μF, and the reference *DC* bus voltage, v_{dc}^* is set to 65 V.

9.4.1 Use of Prescribed Degree of Stability in the Design

To illustrate the significance of the prescribed degree of stability used in the design of model predictive control, a comparison study is done between the case where the prescribed degree of stability $\beta = 0$ and the case where $\beta = 30$. With the value of $\beta = 30$, all the closed-loop eigenvalues of the predictive control system lie on the left of $-\beta$ line on the complex plane. In the experimental results shown in Figure 9.9, a step load change of the *DC* link resistance from 20 Ω to 40 Ω occurs at around 46 seconds, and the transient responses of the *DC* bus voltage are compared. It is seen from this figure that it took about 0.1 seconds for the *DC* bus voltage to return to the reference signal when $\beta = 30$, whilst when $\beta = 0$ it took at least more than 3 seconds for *DC* bus voltage to return to the vicinity of the reference signal. The results clearly show that the transient response of *DC* bus voltage to a step load change is greatly improved for

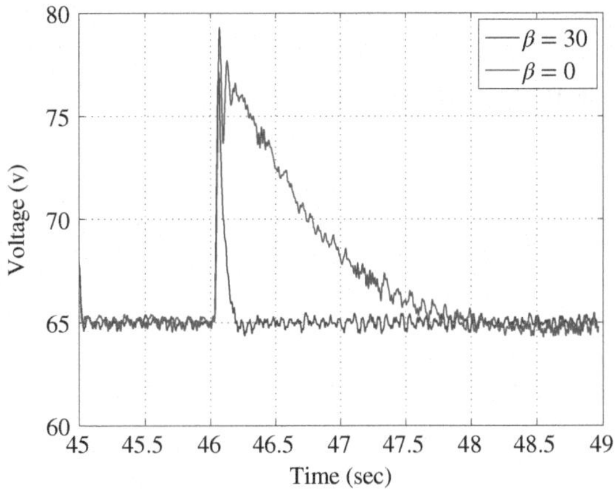

Figure 9.9 Comparison of *DC* bus voltage response to a step load change with $\beta = 0$ and $\beta = 30$.

the case of $\beta = 30$. From hereafter, the experimental results are obtained with the prescribed degree of stability ($\beta = 30$).

9.4.2 Experimental Results for Rectification Mode

The control objective is keep the *DC* bus voltage at 65 V, and i_q current at zero for unity power factor. In the experimental testing, prior to the rectification mode, the converter is operating as a diode rectifier where the switching functions of the IGBT are disabled and the current is only conducting through the freewheeling diode of IGBT. At the time (around $t = 33.3$ second) when the rectification mode is switched on, the predictive controller is activated to boost the *DC* bus voltage from 35 V to 65 V. Figure 9.10 shows the closed-loop responses of the outputs $v_{dc}(t)$ and $i_q(t)$ the state variable $i_d(t)$. It took about 0.1 second for both output signals ($v_{dc}(t)$ and $i_q(t)$) and the state i_d to complete the closed-loop transient responses. In the rectification mode, it is seen from Figure 9.10(b) that while drawing an extra current from the grid, i_q is well maintained around zero which results in zero phase shift between phase voltage and phase current (i.e unity power factor) and i_d is increased to a new steady-state value according to the required *DC* bus voltage level. To confirm the reality of unity power factor, Figure 9.10(c) shows phase A voltage and current in rectification mode, which indeed indicates the zero phase shift between the phase voltage and current.

9.4.3 Experimental Results for Regeneration Mode

The control objective is to keep the *DC* bus voltage at 65 V, and i_q current at zero for unity power factor. Prior to the regeneration mode, the converter is operating in rectification mode. At around 70 second, as shown in Figure 9.11(a), an extra current is injected in the *DC* bus, which results in an initial overshoot of the *DC* bus voltage and the predictive controller regulates the *DC* bus voltage around 65 V. Figure 9.11(b) shows the closed-loop responses of i_d and i_q in regeneration mode. It is seen that the steady-state value of i_d is negative which indicates that the current flow is reversed compared to the rectification mode. In this case the extra current injected into *DC* bus is converted to AC currents which feeds back into main

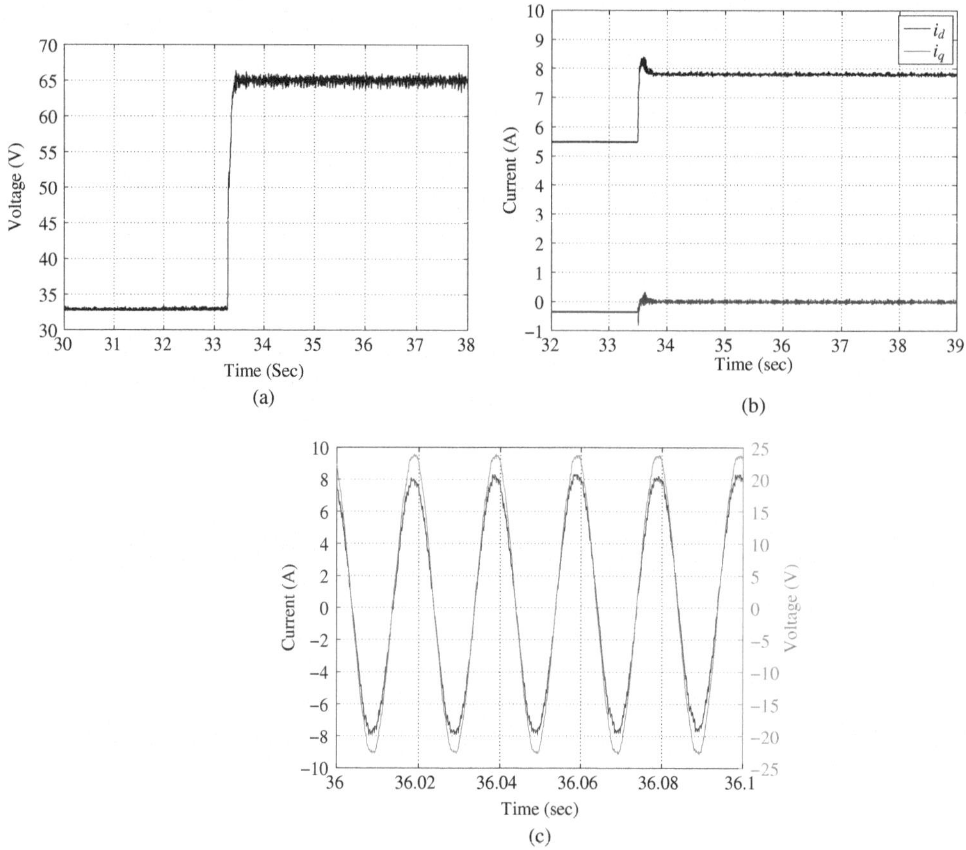

Figure 9.10 Experimental results of CMPC of power converter in rectification mode. (a) *DC* bus voltage V_{dc}, (b) i_d and i_q current signals, and (c) Phase voltage and current.

grid. For unity power factor operation, i_q is still maintained around zero. This is evident via the plots of phase A voltage and current in regeneration mode as shown in Figure 9.11(c), where it is seen that the phase A voltage and its current have 180° of phase shift.

9.4.4 Experimental Results for Disturbance Rejection

The control objective is to keep the *DC* bus voltage at 65 V, and i_q current at zero for unity power factor, while load disturbances occur. A series of step changes in the load are simulated by inserting or removing an extra *DC* load resistance in the circuit. At around $t = 43.7$ sec in Figure 9.12(a), the *DC* link resistance is changed from $R_L = 20\ \Omega$ to $R_L = 40\ \Omega$, corresponding to the case where the load current is decreased from 3.2 A to 1.6 A.

In Figure 9.12(b), around $t = 50.2$ sec, the resistance is changed from $R_L = 40\ \Omega$ to $R_L = 20\ \Omega$. From the closed-loop response of the *DC* bus voltage in both figures, it is seen that the predictive controller rejects the disturbance in a response time of about 0.1 second. Furthermore, from the closed-loop responses of the i_d and i_q currents shown in both figures, it is clearly seen that the i_q current is kept well around zero while the steady-state value of i_d changes according to the load requirements.

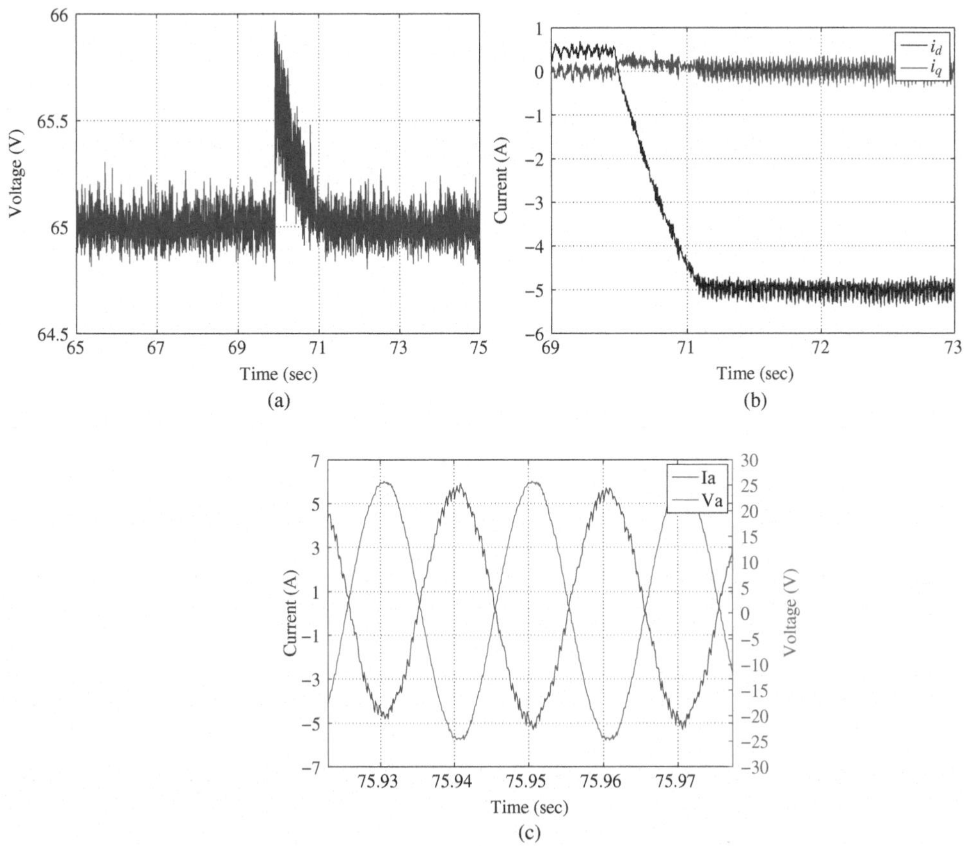

Figure 9.11 Experimental results of CMPC of power converter in regeneration mode. (a) *DC* bus voltage v_{dc}, (b) i_d and i_q current signals, and (c) Phase voltage and current.

9.5 Gain Scheduled Predictive Controller

Since the plant is a nonlinear system, its linearized model is dependent on its operating conditions. It is apparent that for a linear model predictive controller to work in a wide range of operating conditions a gain scheduled predictive controller is needed, where several linear time-invariant model predictive controllers will be designed for the range of operating conditions. This gain scheduled controller will have the capability to interpret the operating conditions, uses an appropriate linear predictive controller and invoke mechanisms to ensure the continuity of the control signals.

9.5.1 The Weighting Parameters

One approach used in the design of a gain scheduled control system is to assign a set of weighting parameters with values between 0 and 1 that will correspond to each operating condition of the nonlinear system. Here, the parameters λ^l, λ^m and λ^h are used as the weights for low speed, median speed and high speed operations. In the literature, there are two widely used approaches to calculate the weighting parameters. The first approach, also the simplest approach, is to assign their values according to the set-point signal

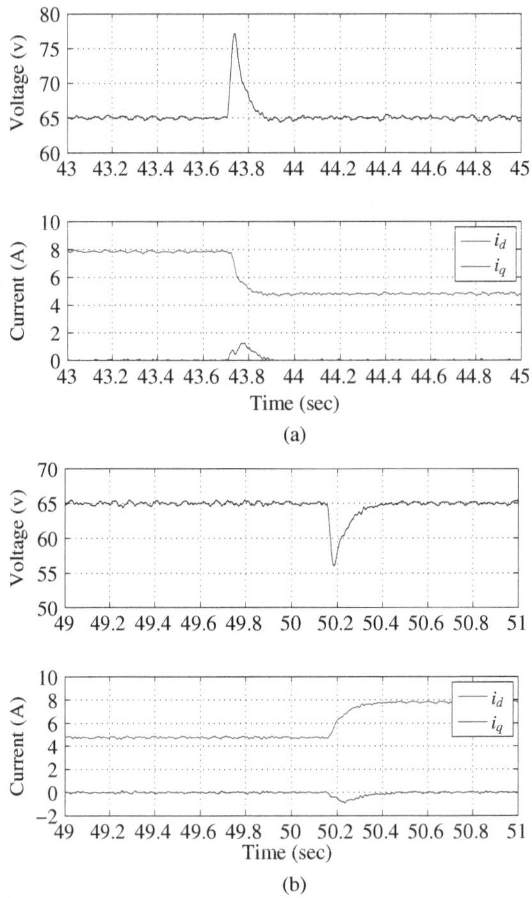

Figure 9.12 Experimental results of CMPC of power converter in disturbance rejection mode though load change. (a) *DC* bus voltage v_{dc}, i_d and i_q response to a step load change of R_{dc} from 20 Ω to 40 Ω. (b) *DC* bus voltage V_{dc}, i_d and i_q response to a step load change of R_{dc} from 40 Ω to 20 Ω.

of the system. For instance, when the set-point velocity signal ω_m^0 is at the low speed where $\omega_m^0 = \omega_m^l$, $\lambda^l = 1$, $\lambda^m = 0$ and $\lambda^h = 0$; when the desired velocity is at the median speed ($\omega_m^0 = \omega_m^m$), $\lambda^l = 0$, $\lambda^m = 1$ and $\lambda^h = 0$; when the desired velocity is at the high speed where $\omega_m^0 = \omega_m^h$, $\lambda^l = 0$, $\lambda^m = 0$ and $\lambda^h = 1$. This approach takes into consideration the changes of plant dynamics due to set-point changes, however, it does not consider the possibility that disturbances could cause the significant changes in plant dynamics. Hence, with this simple approach, closed-loop instability could occur if severe disturbances were encountered in plant operation.

The more general approach is to compute the weighting parameters λ^l, λ^m and λ^h according to the actual measurement of velocity ω_m. In order to avoid random triggering of the model changes in the presence of noises and transient responses, a band is formed around the desired speed. By assigning a tolerance constant δ to desired speed ranges, the weighting constants λ^l, λ^m, λ^h are defined as

$$-\delta + \omega_m^l \leq \omega_m \leq \omega_m^l + \delta \;\; \lambda^l = 1; \lambda^m = 0; \lambda^h = 0$$

$$-\delta + \omega_m^m \leq \omega_m \leq \omega_m^m + \delta \ \ \lambda^l = 0; \lambda^m = 1; \lambda^h = 0$$
$$-\delta + \omega_m^h \leq \omega_m \leq \omega_m^h + \delta \ \ \lambda^l = 0; \lambda^m = 0; \lambda^h = 1.$$

Outside the band of the desired speed, neither linear models can accurately describe the dynamic system. A traditional method is to use a combination of these two models from the neighbouring regions. For instance, assuming that the actual operating condition is between the band of the desired median speed and that of the desired high speed ($\omega_m^m + \delta \leq \omega_m(t) < \omega_m^h - \delta$), by defining λ^h ($0 \leq \lambda^h \leq 1$) as a function of $\omega_m(t)$, $\lambda^h(t)$ is calculated using the linear interpretation of the two boundaries between the median and high speeds given by

$$\lambda^h(t) = \frac{\omega_m(t) - \omega_m^m - \delta}{\omega_m^h - \omega_m^m - 2\delta}. \tag{9.49}$$

The weighting parameter λ^m follows as $\lambda^m = 1 - \lambda^h$ ($0 \leq \lambda^m \leq 1$), and $\lambda^l = 0$ for this region. Similarly, for $\omega_m^l + \delta \leq \omega_m(t) < \omega_m^m - \delta$,

$$\lambda^m(t) = \frac{\omega_m(t) - \omega_m^l - \delta}{\omega_m^m - \omega_m^l - 2\delta} \tag{9.50}$$

and $\lambda^l(t) = 1 - \lambda^m(t)$, $\lambda^h = 0$.

Figure 9.13 illustrates the weighting parameters that have been used to represent the operating regions of the induction motor.

9.5.2 Gain Scheduled Predictive Control Law

The weighting parameters will be used in the gain scheduled predictive control law to ensure correct selection of the linear model and bumpless transfer from one predictive controller to another.

We assume that there are three linearized models obtained for the three operating conditions with their system matrices denoted by (A^l, B^l, C^l), (A^m, B^m, C^m), and (A^h, B^h, C^h), respectively. Note that the cost function (9.19) is based on a linearized model from a single operating condition. The cost function for the gain scheduled model predictive control is a combination of cost functions generated for

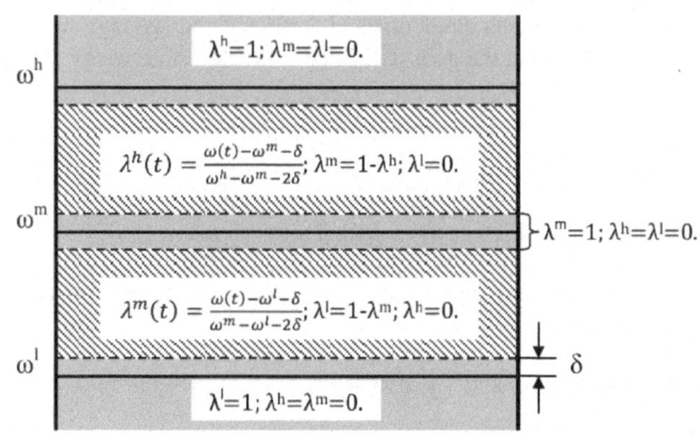

Figure 9.13 Weighting parameters.

different operating conditions. By using the weighting parameters λ^l, λ^m and λ^h, the cost function is chosen as

$$\begin{aligned} J &= \lambda^l(\eta^T\Omega^l_{mpc}\eta + 2\eta^T\Psi^l_{mpc}\tilde{x}(t_i)) + \lambda^m(\eta^T\Omega^m_{mpc}\eta + 2\eta^T\Psi^m_{mpc}\tilde{x}(t_i)) \\ &\quad +\lambda^h(\eta^T\Omega^h_{mpc}\eta + 2\eta\Psi^h_{mpc}\tilde{x}(t_i)) \\ &= \eta^T(\lambda^l\Omega^l_{mpc} + \lambda^m\Omega^m_{mpc} + \lambda^h\Omega^h_{mpc})\eta \\ &\quad +2\eta^T(\lambda^l\Psi^l_{mpc} + \lambda^m\Psi^m_{mpc} + \lambda^h\Psi^h_{mpc})\tilde{x}(t_i), \end{aligned} \tag{9.51}$$

where $\tilde{x}(t_i) = x(t_i) - x^0$, Ω^l_{mpc}, Ω^m_{mpc}, Ω^h_{mpc}, Ψ^l_{mpc}, Ψ^m_{mpc} and Ψ^h_{mpc} are the predictive control parameter matrices (see (9.19)), computed on the basis of the model parameters for low, median and high speed of the induction motor. If the reference signals to the system change, then the components in x^0 that correspond to the reference signals will change accordingly. By similar definition of operational constraints, the Laguerre parameter vector η is found by minimizing the cost function J (9.51) subject to approximated operational constraints (9.41). With the optimal Laguerre coefficients vector, the control signal for the gain scheduled predictive controller is realized at sample time t_i via

$$u(t_i) = u(t_{i-1}) + L(0)^T\eta\Delta t. \tag{9.52}$$

The control signal is ensured not to have a sudden jump effect when the operating condition changes. This is because the computation of the control signal using (9.52) is based on the past sample of the control signal and the derivative of the current control, and when $\Delta t \to 0$, $u(t_i) = u(t_{i-1})$.

When an observer is used to estimate the state variable vector $x(t)$, it is designed for each operational condition and implemented using the linear interpretation in the same manner as the predictive controller.

Assume that the observer gains are K^l_{ob}, K^m_{ob} and K^h_{ob} that are designed using (A^l, C), (A^m, C), and (A^h, C), respectively. The three observer equations are

$$\frac{d\hat{x}(t)^l}{dt} = A^l\hat{x}(t)^l + B^l\dot{u}(t) + K^l_{ob}(y(t) - C\hat{x}(t)^l) \tag{9.53}$$

$$\frac{d\hat{x}(t)^m}{dt} = A^m\hat{x}(t)^m + B^m\dot{u}(t) + K^m_{ob}(y(t) - C\hat{x}(t)^m) \tag{9.54}$$

$$\frac{d\hat{x}(t)^h}{dt} = A^h\hat{x}(t)^h + B^h\dot{u}(t) + K^h_{ob}(y(t) - C\hat{x}(t)^h). \tag{9.55}$$

At any given time t, there is only one state vector $\hat{x}(t)$, however different the models. Thus, we assign the common state vector $\hat{x}(t)$ as the weighted outcomes of the estimated states from (9.53)–(9.55). Similar to the expression of the cost function, the gain scheduled estimated state vector $\hat{x}(t)$ is written as

$$\begin{aligned} \frac{d\hat{x}(t)}{dt} &= \lambda^l(A^l\hat{x}(t) + B^l\dot{u}(t) + K^l_{ob}(y(t) - C\hat{x}(t))) \\ &\quad + \lambda^m(A^m\hat{x}(t) + B^m\dot{u}(t) + K^m_{ob}(y(t) - C\hat{x}(t))) \\ &\quad + \lambda^h(A^h\hat{x}(t) + B^h\dot{u}(t) + K^h_{ob}(y(t) - C\hat{x}(t))). \end{aligned} \tag{9.56}$$

When the differential equation is discretized using first order forward difference approximation, it is written for implementation as

$$\begin{aligned} \hat{x}(t_{i+1}) &= \hat{x}(t_i) + (\lambda^lA^l + \lambda^mA^m + \lambda^hA^h)\hat{x}(t_i)\Delta t + (\lambda^lB^l + \lambda^mB^m + \lambda^hB^h)\dot{u}(t_i)\Delta t \\ &\quad +(\lambda^lK^l_{ob} + \lambda^mK^m_{ob} + \lambda^hK^h_{ob})(y(t_i) - C\hat{x}(t_i))\Delta t. \end{aligned} \tag{9.57}$$

This implementation of the observer also ensures that the observer states will not sudden jump when the operating condition changes, as when $\Delta t \to 0$, $\hat{x}(t_i + \Delta t) = \hat{x}(t_i)$.

9.6 Experimental Results of Gain Scheduled Predictive Control of Induction Motor

There are three linear models used in the gain scheduled model predictive control system. The operating condition for low speed operation is selected as $\omega_m^l = 300$ RPM; for median speed operation is $\omega_m^m = 700$ RPM; and for high speed operation is $\omega_m^h = 1600$ RPM.

In order to simplify the controller tuning procedure and also to improve the numerical condition of the Hessian matrix of the predictive controller, the procedure for prescribed degree of stability with exponential data weighting outlined in the previous sections is deployed to position the closed-loop poles of the predictive controller at the left-hand side of $-\beta$ line ($\beta \geq 0$). The exponential data weighting parameter $\alpha > 0$ is used to ensure that the augmented system matrix A has all eigenvalues on the left of $-\alpha$ line so that numerical stability is preserved in the computation of prediction. For all three predictive controllers, the weighting matrices $R = I$ (I being the identity matrix) and $Q = C^T C$ where C is the output matrix for the augmented model. The observer gain matrix K_{ob} is computed using MATLAB lqr function where the weighting matrices $Q = I$ and $R = 10^{-5}$ I. The remaining parameters used in the implementation of the gain scheduled control system are listed in Table 9.5.

The operational constraints are specified by the capacity of the DC power supply, V_{dc}.

9.6.1 The First Set of Experimental Results

In the experiment, the motor starts from standstill, then steps up to 300 RPM, to 600 RPM, and then 1600 RPM. After that, the operation is reversed from the fastest speed to 0 RPM. The set-point signal for the rotor flux is maintained to be constant during the operation.

In the first set of experiments, the V_{dc} value is chosen to match the actual capacity of the DC power supply so that the nonlinear constraint

$$\sqrt{u_{sd}^2 + u_{sq}^2} \leq 300$$

is imposed in the implementation.

The closed-loop responses for the velocity, flux and control signals are shown in Figures 9.14(a)–9.14(e). It is seen from these plots that the closed-loop velocity and rotor flux follow their set-point signals without steady-state errors, which also means that the variable load disturbance introduced through the DC motor is completely rejected by the continuous-time predictive controller. The plot of $\sqrt{u_{sd}^2 + u_{sq}^2}$ is shown in Figure 9.14(d) against the constraint 300 V, from which it is seen that the constraint is not active during the operation.

Table 9.5 Controller design and implementation parameters

	Model 1	Model 2	Model 3
Velocity ω_m^0	300 RPM	700 RPM	1600 RPM
Flux ψ_{rd}^0	0.3 Wb	0.3 Wb	0.3 Wb
q-axis current i_{sq}^0	0.1 A	0.2 A	0.5 A
Number of terms $N_1 = N_2$	6	6	6
Scaling parameter $p_1 = p_2$	20	20	20
Prediction horizon T_p	0.5	0.5	0.5
Exponential weighting parameter α	1.2	1.2	1.2
Closed-loop pole position β	15	15	12
Operational tolerance parameter δ	50	50	50

An observer has been used to estimate the state variables in the implementation of the linear predictive controllers. Because integrators are used in the design, the first part of state variables consists of the derivatives of the currents $i_{sd}(t)$ and $i_{sq}(t)$, which typically contain a large amount of measurement noise as derivative action amplifies the measurement noise. The use of observed states for those derivative variables avoids the undesirable noise effect due to differentiation. Figure 9.14(e) shows the estimated $\frac{d\hat{\xi}_{sd}(t)}{dt}$ and $\frac{d\hat{\xi}_{sq}(t)}{dt}$. It is seen from the plots that the estimates of the derivatives indeed capture the dynamics of the

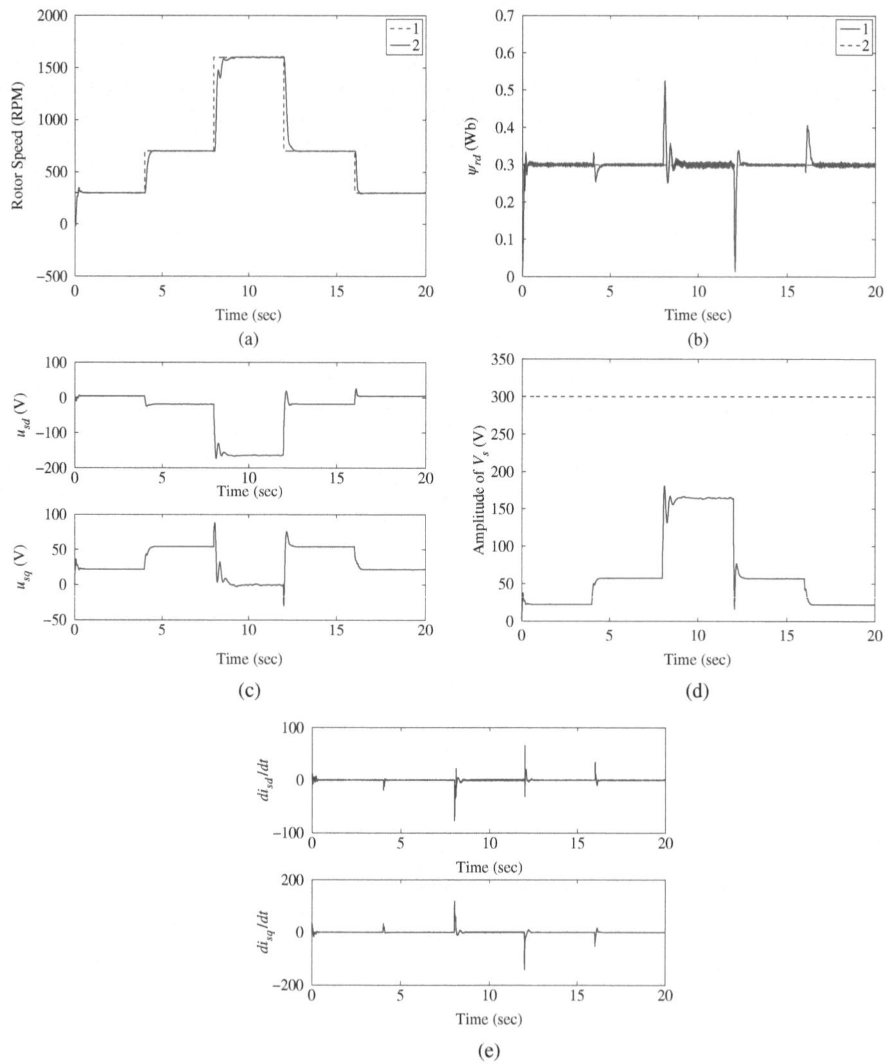

Figure 9.14 First set of experimental results of CMPC of induction motor control where constraints are not active. (a) Motor speed response. Key: line (1) Set-point signal; line (2) actual feedback, (b) Rotor flux response. Key: line (1) Actual feedback; line (2) Set-point signal, (c) Control signals. Top figure: u_{sd}; bottom figure: u_{sq}, (d) Nonlinear constraint. Dashed line: constraint; solid line $\sqrt{u_{sd}^2 + u_{sq}^2}$, and (e) Estimated derivatives of state variables i_{sd} and i_{sq}.

physical variables without the noise effect, also in the steady-state, the estimates are zero as expected. It is worth emphasizing that the actual measurements $\psi_{rd}(t)$ and $\omega_m(t)$ are used as the inputs to the observer without subtraction of their steady-state values.

9.6.2 The Second Set of Experimental Results

In the first set of experimental results, the constraints are not active because of the sufficiently large capacity of the power supply. In the second set of experiments, the V_{dc} value is reduced to demonstrate the situation where a smaller power supply is used and the nonlinear constraint becomes active in the transient response to a large reference signal change and load disturbance rejection. The nonlinear constraint,

$$\sqrt{u_{sd}^2 + u_{sq}^2} \leq 140, \tag{9.58}$$

is imposed in the implementation.

The speed reference signal makes a large step change from 300 RPM to 1600 RPM at time $t = 4$ second (see Figure 9.15(a) for velocity and Figure 9.15(b) for flux). The large reference step change on the velocity causes the control signals to increase (see Figure 9.15(c)), and as a result the quantity

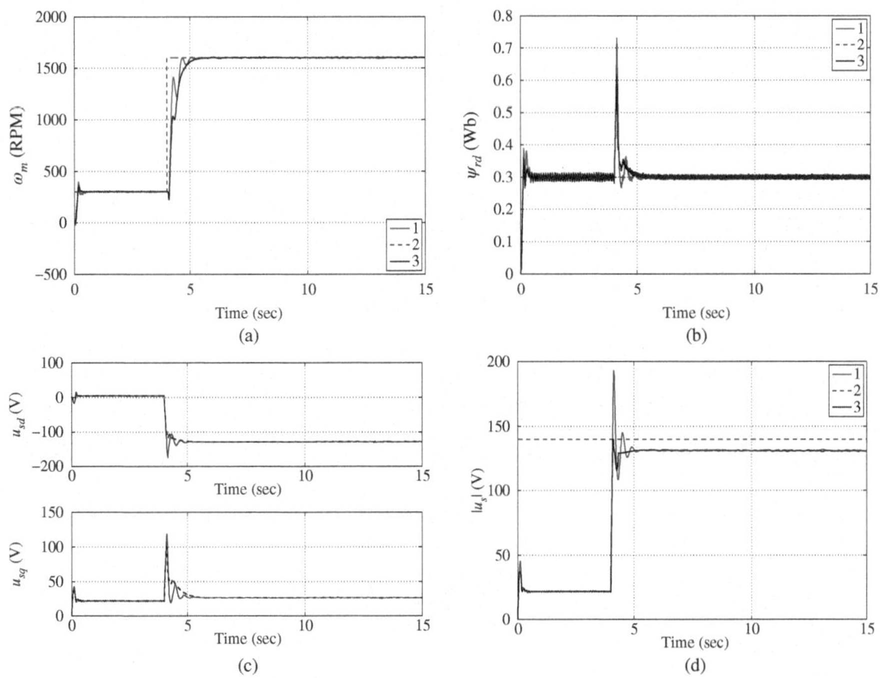

Figure 9.15 Second set of experimental results of CMPC of induction motor in the presence of constraints. (a) Comparison between constrained and unconstrained rotor speed responses. Key: line (1) Set-point signal; line (2) response without constraint; line (3) response with constraint, (b) Comparison between constrained and unconstrained rotor flux responses. Key: line (1) Set-point signal; line (2) response without constraint; line (3) response with constraint, (c) Comparison between constrained and unconstrained control signal responses. Top figure: u_{sd}; bottom figure: u_{sq}. Key: line (1) unconstrained control signal; line (2) constrained control signal, and (d) Nonlinear constraint. Key: line (1) $\sqrt{u_{sd}^2 + u_{sq}^2}$ without constraint; line (2) $\sqrt{u_{sd}^2 + u_{sq}^2}$ with constraint; line (3) the constraint.

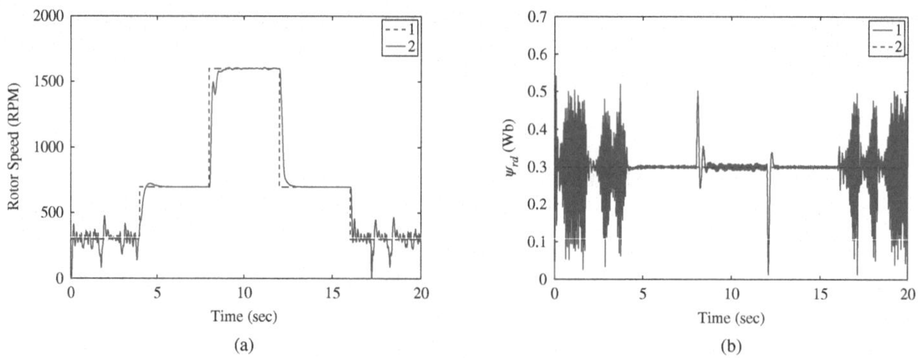

Figure 9.16 Third set of experimental results of CMPC of induction motor without gain scheduling. (a) Motor speed response. Key: line (1) Set-point signal; line (2) actual feedback, and (b) Rotor flux response. Key: line (1) Actual feedback; line (2) Set-point signal.

$\sqrt{u_{sd}^2 + u_{sq}^2}$ suddenly increases to about 190 (see Figure 9.15(d)). When the constraint (9.58) is imposed in the operation, control signals $u_{sd}(t)$ and $u_{sq}(t)$ are found such that the quantity $\sqrt{u_{sd}^2 + u_{sq}^2} \leq 140$. It is seen from Figure 9.15(d) that the constraint becomes active during the transient period and the quantity $\sqrt{u_{sd}^2 + u_{sq}^2}$ converges to a steady-state value afterwards. Figures 9.15(a)–9.15(b) compare the closed-loop responses under the constrained control with the responses obtained without constraints. It seems that imposing the nonlinear constraint has changed the characteristics of the closed-loop dynamics in the sense that the responses are less oscillatory for this case.

9.6.3 *The Third Set of Experimental Results*

In the third set of experimental results, the gain scheduled predictive control system is compared with the predictive control system designed using a single model. The results show that when a single predictive controller is applied to the three operating regions, there is a performance loss due to the modeling errors. Figures 9.16(a)–9.16(b) show the output responses when the linear model obtained at $\omega_m^0 = 1600$ RPM is applied to entire operating regions. It is obvious that the closed-loop responses are oscillatory at the operating condition when the set-point signal ω_m^0 is at 300 RPM. This oscillation is caused by the modeling error between the linear model obtained at $\omega_m^0 = 1600$ RPM and the actual system at $\omega_m^0 = 300$ RPM.

9.7 Summary

This chapter has presented continuous-time model predictive control applications to an induction motor drive and a power converter with experimental validations. In the continuous-time design, different from the discrete-time counter part, the linearized continuous-time models are directly used in the design stage, and at the later stage, the control signals are discretized for digital implementation. This perhaps offers some advantages at the applications of AC drive control and power converter control where the time constants of electrical systems are significantly smaller than those from the mechanical systems and choice of sampling interval requires compromise between the large and small time constants. With the continuous-time design, it is possible to use a dual sampling rate in the implementation stage, for instance, a much faster sampling rate for the current measurement signals and a slower sampling rate for the velocity measurement signal.

Another advantage of using the continuous-time predictive control design is its simple extension to gain scheduled model predictive control. When using the traditional continuous-time model predictive control, the nonlinear physical models are linearized at several operating conditions and a family of linear model predictive controllers are designed. The gain scheduled model predictive controller is shown to automatically interpret these linear model predictive controllers. The experimental results obtained from gain scheduled predictive control of induction motor have shown its efficacy in controlling the AC drives.

The continuous-time model predictive controller is conceptually more complex than its discrete-time counter part because Laguerre functions are used in the description of the control trajectories. However, once this is overcome, the actual computational algorithms of the predictive controllers are available in MATLAB for the design and implementation (see Wang (2009)).

9.8 Further Reading

The continuous time predictive control algorithm was from Wang (2009) where the MATLAB programs used in the design can be found. In Wang *et al.* (2010), the continuous-time predictive control algorithm with prescribed degree of stability was applied to continuous-time repetitive predictive control. The continuous-time predictive control of an induction machine drive was presented in Wang and Gan (2013). The cascade velocity control using CMPC for both inner-loop and outer-loop systems was proposed in Gan and Wang (2012), and the corresponding cascade position control was presented in Gan and Wang (2013). Comparison studies between the cascade predictive control and the centralized predictive control were made in Ng and Wang (2013).

The quadratic programming procedure was first proposed in Hildreth (1957). Books on optimization include Luenberger (2003), Boyd and Vandenberghe (2004).

Predictive control of a three-phase regenerative PWM converter was presented in Yoo and Wang (2011).

References

Boyd S and Vandenberghe L 2004 *Convex Optimization*. Cambridge University Press.

Gan L and Wang L 2012 Cascaded continuous-time model predictive control of induction motor *IECON 2012–38th Annual Conference on IEEE Industrial Electronics Society*, pp. 1696–1701.

Gan L and Wang L 2013 Cascaded model predictive position control of induction motor with constraints *Industrial Electronics Society, IECON 2013–39th Annual Conference of the IEEE*, pp. 2656–2661.

Hildreth C 1957 A quadratic programming procedure. *Naval Research Logistics* **4**(1), 79–85.

Komurcugil H and Kukrer O 1998 Lyapunov-based control for three-phase PWM AC/DC voltage-source converters. *IEEE Transations on Power Electronics* **13**(5), 801–813.

Lee YW 1960 *Statistical Theory of Communication*. John Wiley & Sons, Inc., New York.

Luenberger D 2003 *Linear and Nonlinear Programming* 2nd edn. Springer.

Ng K and Wang L 2013 Cascaded and centralized model predictive control of power converter *Industrial Electronics Society, IECON 2013 - 39th Annual Conference of the IEEE*, pp. 5782–5787.

Wang L 2009 *Model Predictive Control System Design and Implementation Using MATLAB* 1st edn. Springer London.

Wang L and Gan L 2013 Gain scheduled continuous-time model predictive controller with experimental validation on AC machine. *International Journal of Control, Special Issue on Model Predictive Control* **86**(8), 1438–1452.

Wang L, Freeman CT, Rogers E and Owens DH 2010 Experimentally validated continuous-time repetitive control of non-minimum phase plants with a prescribed degree of stability. *Control Engineering Practice* **18**(10), 1158–1165.

Yoo D and Wang L 2011 Predictive control of a three-phase regenerative PWM converter using a prescribed degree of stability *18th IFAC World Congress, Milan*, vol. 18, pp. 2791–2796.

10

MATLAB®/Simulink® Tutorials on Physical Modeling and Test-bed Setup

In this chapter, it will be shown how to build the physical models of electrical drives and grid connected power converter using MATLAB and Simulink. The mathematical models were given previously in Chapters 1 and 2. We believe that the best learning strategy is to practice what you learned.

The physical model based simulations are considerably more complex than those simulations purely based on differential equations, particularly when semiconductor switches and PWM modules are used in the simulations. For those who are not familiar with Simulink simulations of electrical drives and power converters, the tutorials are written in a step-by-step manner to guide the learner through them.

Because Simulink has undergone an upgrade in 2013, we envisage that it might take a while for some readers to make the transition from their older versions of Simulink programs to the newer version. The key difference is that the extension for Simulink files is changed from ".mdl" to ".slx" in the newer version. Since there are no major changes in the functionalities of the physical components required to produce the Simulink simulations, and with a consideration for those who are still using the older version of the Simulink programs, the tutorials for the PMSM and induction motor are written using the older version of Simulink and the tutorial for the grid connected three phase power converter is written using the newer version. Nevertheless, the steps to write the Simulink programs are identical, apart from the difference in the file extension and the locations of the Simulink components. We hope that the reader will make adaptation to suit their own situation. Several embedded MATLAB functions shared in all the physical models are produced in the tutorials, including the Park-Clarke transformation, inverse Park-Clarke transformation and phase-locked-loop.

The remainder of this chapter is organized as follows. In Section 10.1, embedded functions are built for Park-Clarke transformation and inverse Park-Clarke transformation. The simulation models for PMSM, induction motor and power converter are built in Sections 10.2 to 10.4. In Sections 10.5 to 10.7, the experimental test-beds for PMSM, induction motor and grid connected power converter are described.

10.1 Building Embedded Functions for Park-Clarke Transformation

The main reason to write these transformations as embedded functions is to avoid mistakes and confusions. These embedded functions are written based on the mathematical equations given in Chapter 1.

PID and Predictive Control of Electrical Drives and Power Converters using MATLAB®/Simulink®, First Edition.
Liuping Wang, Shan Chai, Dae Yoo, Lu Gan and Ki Ng.

Companion Website: www.wiley.com/go/wang/pid

To ensure consistency and clarity, two embedded functions are produced for Park-Clarke transformation and inverse Park-Clarke transformation.

10.1.1 Park-Clarke Transformation for Current Measurements

We assume that the three phase currents i_a, i_b and i_c are balanced so that

$$i_a(t) + i_b(t) + i_c(t) = 0.$$

With this assumption, only two current measurements are required. Supposing that the phase A current and phase C current are measured using current sensors to obtain $i_a(t)$ and $i_c(t)$, then the phase B current is calculated using

$$i_b(t) = -i_a(t) - i_c(t).$$

Then from the Clarke transformation, the currents in the $\alpha - \beta$ reference frame are calculated using the relationship below:

$$\begin{bmatrix} i_\alpha \\ i_\beta \\ i_0 \end{bmatrix} = \frac{2}{3} \begin{bmatrix} 1 & -\frac{1}{2} & -\frac{1}{2} \\ 0 & \frac{\sqrt{3}}{2} & -\frac{\sqrt{3}}{2} \\ \frac{1}{2} & \frac{1}{2} & \frac{1}{2} \end{bmatrix} \begin{bmatrix} i_a \\ i_b \\ i_c \end{bmatrix},$$

which results in

$$i_\alpha(t) = i_a$$

$$i_\beta(t) = -\frac{\sqrt{3}}{3} i_a - \frac{2\sqrt{3}}{3} i_c.$$

With the following Park transformation, the currents in the $d - q$ reference frame are calculated:

$$\begin{bmatrix} i_d \\ i_q \end{bmatrix} = \begin{bmatrix} \cos\theta & \sin\theta \\ -\sin\theta & \cos\theta \end{bmatrix} \begin{bmatrix} i_\alpha \\ i_\beta \end{bmatrix}.$$

Tutorial 4 *The objective of this tutorial is to produce a MATLAB embedded function that will be used to convert phase A and phase C current measurements to i_d and i_q currents for the control systems designed in the $d - q$ synchronous reference frame.*

Step by Step

1. *Create a new Simulink file called PCCurrent.mdl*
2. *In Simulink's directory of User-Defined Functions, find the icon of embedded MATLAB function and copy it to the PCCurrent model.*
3. *Click on the icon of the embedded function, and define the input and output variables to the model so that the embedded function has the following form:*

```
function [id,iq]= PCCurrent(ia,ic,theta)
```

4. *At the top of the embedded function, find "Model Explorer"among the "Tools". When opening the Model Explorer, select "Inherited" for the "update method"; select "Support variable-size arrays"; select "Saturate on integer overflow"; select "Fixed point". Click "Apply" to save the changes.*

5. *Define the initial condition for θ. Enter the following program into the file:*

```
if isempty(theta)
    theta=0;
end
```

6. *Perform Park-Clarke transform to convert the two phase current measurements to i_α and i_β, then to i_d and i_q currents. Enter the following programs into the embedded MATLAB file:*

```
id=cos(theta)*ia+sin(theta)*(-sqrt(3)/3*ia-2*sqrt(3)/3*ic);
iq=-sin(theta)*ia+cos(theta)*(-sqrt(3)/3*ia-2*sqrt(3)/3*ic);
```

10.1.2 Inverse Park-Clarke Transformation for Voltage Actuation

To convert the control signals v_d and v_q in the synchronous reference frame to three phase voltage v_a, v_b and v_c going to the inverter, an embedded MATLAB function is needed.

In the transformation, the following relationships are used.

$$\begin{bmatrix} v_\alpha \\ v_\beta \end{bmatrix} = \begin{bmatrix} \cos\theta & -\sin\theta \\ \sin\theta & \cos\theta \end{bmatrix} \begin{bmatrix} v_d \\ v_q \end{bmatrix}$$

$$\begin{bmatrix} v_a \\ v_b \\ v_c \end{bmatrix} = \begin{bmatrix} 1 & 0 & 1 \\ -\frac{1}{2} & \frac{\sqrt{3}}{2} & 1 \\ -\frac{1}{2} & -\frac{\sqrt{3}}{2} & 1 \end{bmatrix} \begin{bmatrix} v_\alpha \\ v_\beta \\ v_0 \end{bmatrix},$$

where $v_0 = 0$.

Tutorial 5 *The objective of this tutorial is to create a MATLAB embedded function that will convert control signals v_d and v_q to three phase voltage signals for the IGBT inverter.*

Step by Step

1. *Create a new Simulink file called IPCVoltage.mdl*
2. *In Simulink's directory of User-Defined Functions, find the icon of embedded MATLAB function and copy it to the IPCVoltage model.*
3. *Click on the icon of the embedded function, and define the input and output variables to the model so that the embedded function has the following form:*

```
function Vabc= IPCVoltage(vd,vq,theta)
```

4. *At the top of the embedded function, find "Model Explorer" among the "Tools". When opening the Model Explorer, select "Inherited" for the "update method"; select "Support variable-size arrays"; select "Saturate on integer overflow"; select "Fixed point". Click "Apply" to save the changes.*
5. *Define the initial condition for θ. Enter the following program into the file:*

```
if isempty(theta)
    theta=0;
end
```

6. *Calculate the three phase voltage signals. Enter the following program into the file:*

```
T=[1 0;-1/2 sqrt(3)/2;-1/2 -sqrt(3)/2]...
*[cos(theta) -sin(theta);sin(theta) cos(theta)];
Vabc=T*[vd;vq];
```

10.2 Building Simulation Model for PMSM

Tutorial 6 *The objective of this tutorial is to produce a Simulink program to simulate the dynamics of a PMSM.*

Step by Step

1. *Create a Simulink file called PMSMModel.mdl.*
2. *To begin building the physical model based simulator, we need to define the simulation environment for the PMSM. Perform the following tasks:*
   (a) *Under the directory of SimPowerSystems, find "powergui" and move it to PMSMModel.mdl.*
   (b) *Click on "powergui" to find "configure parameters". Under the item of "Simulation type", choose "discrete", and under the item of sample time, enter "Ts_sim=1e-06", which is the sampling interval for power electronic model.*
3. *Once the simulation environment is defined, a physical model is to be built for the PMSM machine. In this step, we will choose the machine type and define the machine parameters according to machine nameplate and the parameters defined at the end of this tutorial. The reader can also use their own machine parameters. Perform the following tasks:*
   (a) *Inside the directory of SimPowerSystems, find the "Machines" subdirectory. Within this directory, find "Permanent Magnetic Synchronous Machine" and move it PMSMModel.mdl.*
   (b) *Click on "Permanent Magnetic Synchronous Machine" to find "Configuration". Back EMF Waveform: choose sinusoidal; mechanical input: torque; preset model:no*
   (c) *Under the item "parameters", stator phase resistance:Rs, inductances: [Ld Lq], Specify: choose "Flux linkage established by magnets", Flux linkage established by magnets: "Phi_mg", Inertia, friction, pole pairs: [Jm Bv 2]. Figure 10.1 shows the settings of all physical parameters needed in the item "parameters".*
4. *Build the power electronics devices for operating the PMSM. Inside the directory of SimPowerSystems, there is a "Power Electronics" subdirectory. This step is to build the IGBT inverter. Perform the following tasks:*
   (a) *Find "Universal Bridge" and move it to the Simulink model.*
   (b) *Select Power Electronic Device to IGBT/Diodes and number of bridge arms 3 and leave others as default.*
5. *Build the DC power supply model. Perform the following tasks:*
   (a) *At "Electrical Sources" subdirectory, find "DC voltage source", move it to PMSMModel.mdl and position it to the left of Universal Bridge. Make another copy of the "DC voltage source".*
   (b) *Set the amplitude to Vdc/2 for each voltage source. Connect them together. Connect "+" terminal from the first DC voltage source with '+' terminal in Universal Bridge and connect the "-" terminal of the second DC voltage source to the "-" terminal of the Universal Bridge.*
6. *Build current measurements. Perform the following tasks:*
   (a) *At the "Measurements" subdirectory, find "Current Measurement", move it to PMSMModel.mdl and position it between the "Universal Bridge" and the PMSM motor. Copy to produce another "Current Measurement", and label them A, C phase currents.*

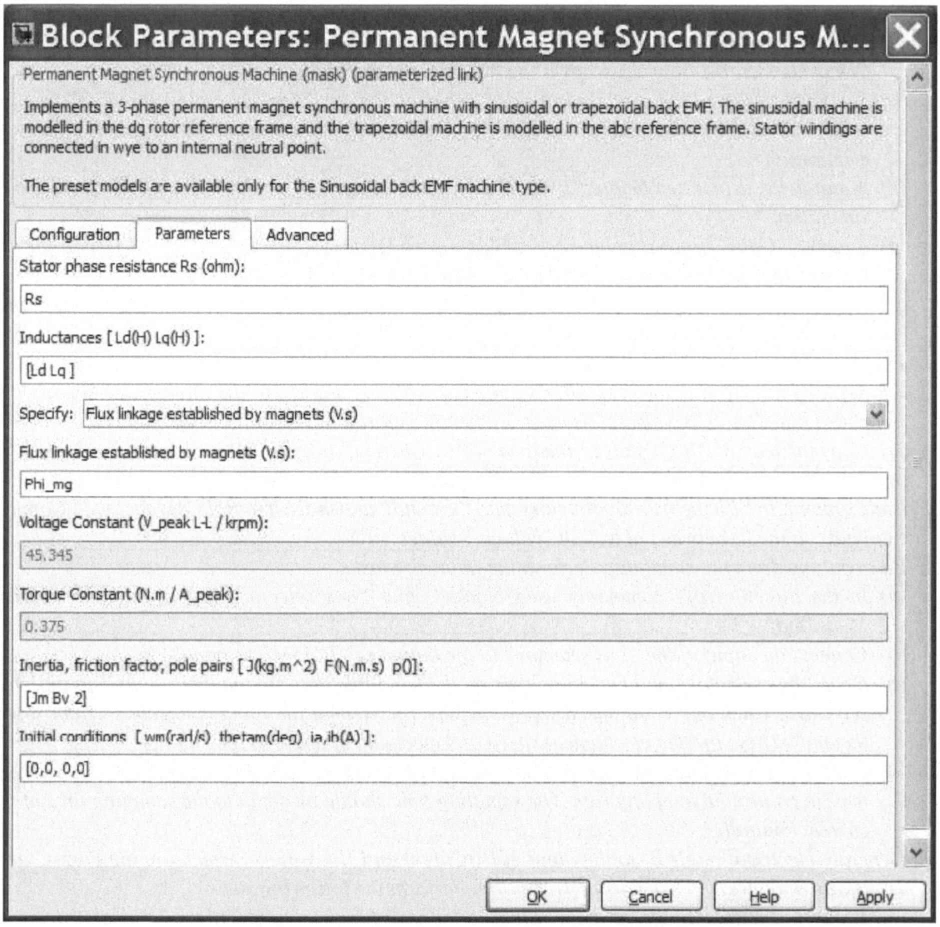

Figure 10.1 Parameters definition in Simulink model for PMSM.

(b) Connect terminal "A" from Universal Bridge to the terminal "+" in the A "Current Measurement", and connect the terminal "−" of phase A "Current Measurement" to the terminal "A" in the PMSM. Repeat the connection for terminal C for the second "Current Measurement".

(c) Connect the terminal "B" from Universal Bridge to the terminal "B" in PMSM without current measurement.

7. *To transform the phase A and phase C current measurements into d and q axis currents, the embedded function "PCCurrent" produced in the Tutorial 4 is used. Perform the following tasks:*

   (a) Copy the icon of "PCCurrent" into PMSMModel.mdl and position it close to the current measurements.

   (b) Connect the output of current measurement A to the input ia of the PCCurrent function and output of current measurement C to the input ic of the PCCurrent function.

   (c) Put zero-order holds before the inputs to ia and ic currents to convert the sampling rates between the SimPowerSystems and the controllers. Here the sampling time in the zero-order hold should be equal to the sampling time of the current controllers.

8. *Build the Pulse-Width-Modulators. Perform the following tasks:*
   (a) *Find "Discrete PWM Generator" in the path of "SimPowerSystems" – > "Extra Library" – > "Discrete Control Blocks". Copy it to PMSMModel.mdl.*
   (b) *Set the Generator Mode to "3-arm bridge (6 pulses)", Carrier frequency (Hz) to "2e3" (we can select this case by case), Sample time to TS_SimP, which is the sampling interval for the power electronics.*
   (c) *A unit delay is inserted before "PWM Generator", in order to avoid the singularity problem in simulation.*
   (d) *Another "Gain" block is valued as "$1/Vdc * sqrt(3)$" is inserted before "PWM Generator" to convert the stator voltage in Volt into Modulation index (between -1 and 1)*
   (e) *Connect the output of the "PWM Generator" to the input of the "Universal Bridge" at terminal "g" (for gate signal).*
9. *The input signals to the "PWM Generator" should be three phase voltage signals, v_a, v_b and v_c in a vector form. To link the controller's output signals, v_d and v_q, to the three phase signals, the embedded function "IPCVoltage" is used, written in Tutorial 5. Perform the following tasks:*
   (a) *Copy the icon of "IPCVoltage" the icon of "PCCurrent" into PMSMModel.mdl.*
   (b) *Connect the output of "IPCVoltage" to the input of the unit delay block.*
10. *Build ground. In "Elements" subdirectory, find "ground" and move it to PMSMModel.mdl. Connect "ground" to the in-between of two DC voltage sources.*
11. *Build outputs from the simulator. Perform the following tasks:*
    (a) *In the directory of "commonly used blocks", find "bus selector" and move it to PMSM-Model.mdl.*
    (b) *Connect the input of the "bus selector" to the output of "PMSM" terminal "m".*
    (c) *Open "bus selector" and find 13 signals on the left Table that are the input signals to the "bus selector". Click on "rotor speed wm" and select it. Repeat the process to select "rotor angle thetam". These are the mechanical speed and mechanical angle of the motor.*
    (d) *Put zero-order hold before the "bus selector" to convert the sampling rate from SimPowerSys-tems to controller sampling rate. The sampling time should be equal to the sampling time of the current controller.*
12. *Generate electrical angle θ_e to the input of PCCurent and IPCVoltage, which are the Park-Clarke Transform and inverse Park-Clarke Transform. Perform the following steps.*
    (a) *Produce electrical angle. A "gain" block is connected to the output 2 of the "bus selector" with gain equal to "2" that is the number of pair of poles for the study here.*
    (b) *Connect the electrical angle to the input "theta" of the PCCurent and IPCVoltage.*
    (c) *Use "mux" to put together the three outputs (id, iq and wm) from the simulator.*

The following physical parameters are used to test this simulation program: $R_s = 2.98$; $L_d = 7 \times 10^{-3}$; $L_q = 7 \times 10^{-3}$; $\phi_{mg} = 0.125$; $p = 2$; $Kt = \frac{3}{2} \times p \times \phi_{mg}$; $J_m = 0.47 \times 10^{-4}$; $B_v = 11 \times 10^{-5}$; $T_L = 1$; $V_{dc} = 100$. *The following parameters are suggested for the numerical computations: power electronic computational interval* $T_{cs} = 1 \times 10^{-6}$; *sampling interval* $Tin = 20 \times 10^{-6}$; *switching frequency of the IGBT:* $Fc = 20 \times 10^3$; *and the simulation time* $tsim = 0.2$.

10.3 Building Simulation Model for Induction Motor

In the first step of building a simulation model, a MATLAB file is created for the parameters in the model of the induction motor to be controlled. The parameters are based on the motor characteristics and the nameplate on the motor. Also defined are the sampling interval used in the simulation, *DC* bus voltage and set-point signal of the flux. The following MATLAB program is used in the simulation demonstration.

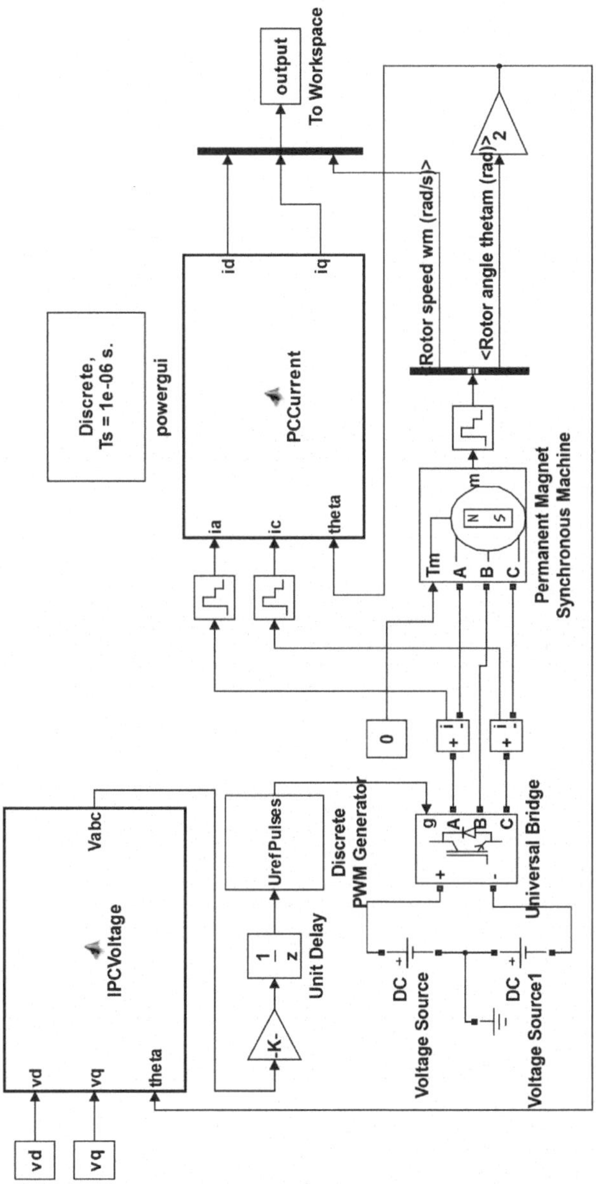

Figure 10.2 Simulink diagram for simulation of PMSM.

Tutorial 7 *The objective of this tutorial is to create a MATLAB file for the induction motor parameters that will be used in the simulation for induction motor modeling and control.*

Step by Step

1. *Create a new file called InModel.m*
2. *Define the induction motor parameters (warm state). Enter the following program into the file:*

```
r_s = 11.2; (Ohm) %stator resistance
r_r = 8.3;        %rotor resistance
l_ls = 0.0455; (H) %stator inductance
l_lr = 0.068;      %rotor inductance
l_h = 0.570;       %mutual inductance
l_s=l_h+l_ls;
l_r=l_h+l_lr;
```

3. *Based on the physical parameters, calculate the motor constants. Enter the following program into the file:*

```
tau_s=l_s/r_s;   %time constant of stator winding
tau_r=l_r/r_r;   %time constant of rotor winding
sigma=1-((l_h*l_h)/(l_s*l_r));  %total flux leakage coefficient
l_st=sigma*l_s;
l_rt=sigma*l_r;
k_r=l_h/l_r;
k_s=l_h/l_s;
tau_st=sigma*tau_s; %(sigma*l_s)/r_s;
tau_rt=sigma*tau_r; %(sigma*l_r)/r_r;
r_sigma=r_s+(r_r*k_r*k_r);
tau_sigmat=(sigma*l_s)/r_sigma;
```

4. *Define induction motor mechanical parameters. Enter the following program into the file:*

```
z_p = 2; %pole pairs
J = 0.0052; %Inertia constant
F = 0.0023; %Friction coefficient
```

5. *There are two sampling intervals used in the simulation. One is the sampling interval for the control system, and the other is the sampling interval for the IGBT semiconductors.*

```
TS = 100e-6; %sampling interval for control system (100us)
TSSimP=10e-6; %sampling interval for Power Electronics Model
```

6. *Define DC bus voltage and set-point signal to rotor flux. Enter the following program into the file:*

```
Udc=520; %dc bus Voltage (V)
psi_S=0.3; %Set-point for flux (Wb)
```

7. *Run this program to produce the parameters for the physical model of the induction motor that will be used in the simulation in the next tutorial.*

Tutorial 8 *The objective of this tutorial is to produce a Simulink simulation program for a Squirrel cage induction motor. The simulated component models consist of four parts: the physical induction motor model; the power electronics, the DC voltage power supply and the current sensors. The tutorial will show how these components are connected to form the physical model for the simulation of a Squirrel case induction motor.*

Step by Step

1. *Create a Simulink file called Squirrel.mdl.*
2. *To begin building the physical model based simulator, we need to define the simulation environment for the power electronics. Perform the following tasks:*
   (a) *Inside the directory of SimPowerSystems, find "powergui" and move it to Squirrel.mdl.*
   (b) *Click on "powergui" to find "configure parameters". Under the item of "Simulation type", choose "discrete", and under the item of sample time, enter "TSSimp", which is the sampling interval for power electronic model.*
3. *Once the simulation environment is defined, a physical model is to be created for the induction machine. In this step, we will choose the rotor type and define the machine parameters according to machine nameplate and the parameters calculated in the Tutorial 7. Perform the following tasks:*
   (a) *Inside the directory of SimPowerSystems, find the "Machines" subdirectory. Within this directory, find "Asynchronous Machine SI Units" and move it Squirrel.mdl.*
   (b) *Click on "Asynchronous Machine SI Units" to find "Configuration". Rotor type: choose Squirrel-Cage, which is the type of machines used in the simulation.*
   (c) *Under the item "parameters", nominal power:750/7.9, line-to-line voltage:415, frequency:50, and define all the other machine parameters using those calculated by the MATLAB program called InModel.m. Figure 10.3 shows the settings of all physical parameters needed in the item "parameters".*
4. *Build the power electronics devices for operating the induction motor. Inside the directory of SimPowerSystems, there is a "Power Electronics" subdirectory. This step is to build the IGBT inverter. Perform the following tasks:*
   (a) *Find "Universal Bridge" and move it to the Simulink model.*
   (b) *Select Power Electronic Device to IGBT/Diodes and number of bridge arms 3 and leave others as default.*
5. *Build the DC power supply model. Perform the following tasks:*
   (a) *At "Electrical Sources" subdirectory, find "DC voltage source", move it to Squirrel.mdl and position it to the left of Universal Bridge. Make another copy of the "DC voltage source".*
   (b) *Set the amplitude to Udc/2 for each voltage source. Connect them together. Connect "+" terminal from the first DC voltage source with "+" terminal in Universal Bridge and connect the "-" terminal of the second DC voltage source to the "-" terminal of the Universal Bridge.*
6. *Build current measurements. Perform the following tasks:*
   (a) *At the "Measurements" subdirectory, find "Current Measurement", move it to Squirrel.mdl and position it between the "Universal Bridge" and the induction motor. Copy to produce another "Current Measurement", and label them A, C phase currents.*
   (b) *Connect terminal "A" from Universal Bridge to the terminal "+" in the A "Current Measurement", and connect the terminal "-" of phase A "Current Measurement" to the terminal "A" in the induction motor. Repeat the connection for terminal C for the second "Current Measurement".*
   (c) *Connect the terminal "B" from Universal Bridge to the terminal "B" in induction motor without current measurement.*

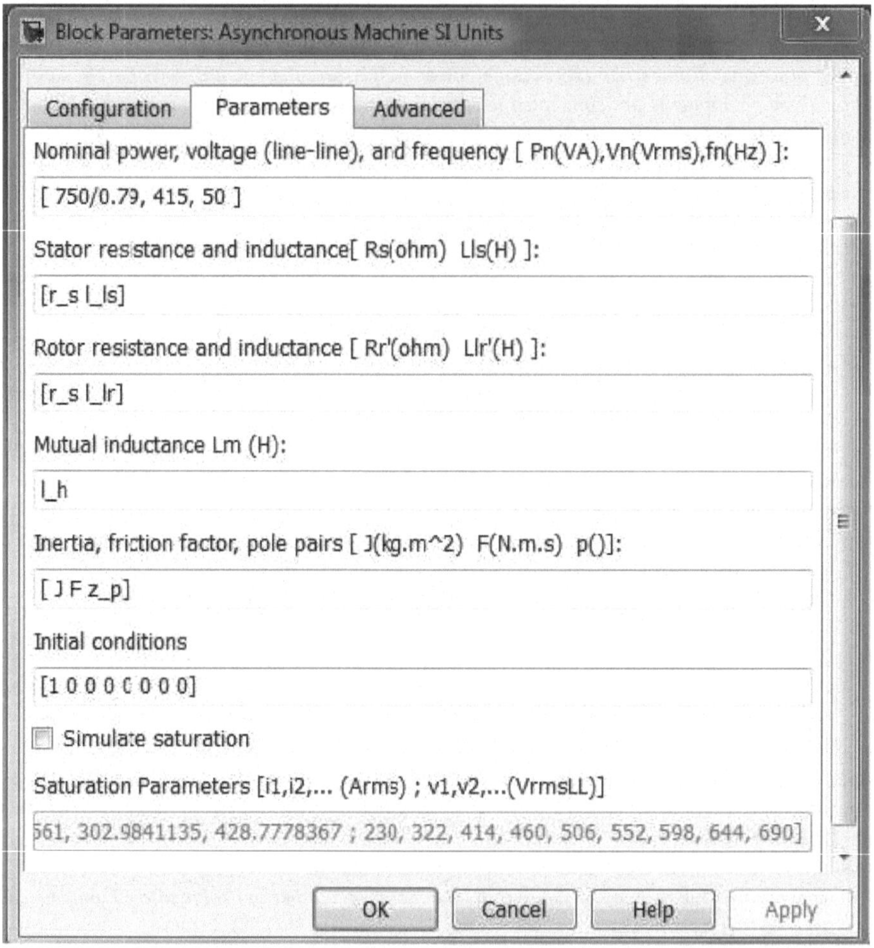

Figure 10.3 Parameters definition in Simulink model of induction motor.

7. *To transform the phase A and phase C current measurements into d and q axis currents, the embedded function "PCCurrent" produced in the Tutorial 4 is used. Perform the following tasks:*
   (a) *Copy the icon of "PCCurrent" into Squirrel.mdl and position it close to the current measurements.*
   (b) *Connect the output of current measurement A to the input ia of the PCCurrent function and output of current measurement C to the input ic of the PCCurrent function.*
8. *Build the Pulse-Width-Modulators. Perform the following tasks:*
   (a) *Find "Discrete PWM Generator" in the path of "SimPowerSystems" – > "Extra Library" – > "Discrete Control Blocks". Copy it to Squirrel.mdl.*
   (b) *Set the Generator Mode to "3-arm bridge (6 pulses)", Carrier frequency (Hz) to "2e3" (we can select this case by case), Sample time to* TS_SimP, *which is the sampling interval for the power electronics.*
   (c) *A unit delay is inserted before "PWM Generator", in order to avoid the singularity problem in simulation.*

(d) *Another "Gain" block is valued as "$1/Udc * sqrt(3)$" is inserted before "PWM Generator" to convert the stator voltage in Volt into Modulation index (between -1 to 1)*
(e) *Connect the output of the "PWM Generator" to the input of the "Universal Bridge" at terminal "g" (for gate signal).*

9. *The input signals to the "PWM Generator" should be three phase voltage signals, v_a, v_b and v_c in a vector form. To link the controller's output signals, v_d and v_q, to the three phase signals, the embedded function "IPCVoltage" is used, written in Tutorial 5. Perform the following tasks:*
   (a) *Copy the icon of "IPCVoltage" the icon of "PCCurrent" into Squirrel.mdl.*
   (b) *Connect the output of "IPCVoltage" to the input of the unit delay block.*
10. *Build ground. In "Elements" subdirectory, find "ground" and move it to Squirrel.mdl. Connect "ground" to in-between the two DC voltage sources.*
11. *For the simplicity of this tutorial, the rotational angle θ_s of the $d-q$ reference frame is generated by "Repeating Sequence" under "Source" subdirectory. "Time values" set as [0 0.02] for 50 Hz, "Output values" set as [0 2*pi]. Since the output of this block is in continuous-time, A "Rate Transition" is needed for discrete time mode simulation.*
12. *In order to provide the inputs of the open-loop model, three "Constant" sources are inserted for u_sd, u_sq and T_L where the values are choose as 10, 100 and 0, respectively for example.*
13. *To obtain the outputs of the simulation results, Find "To Workspace" block under "Sinks". Change the Variable name same as output variable, change the save format to "Array", change the Sample time to Ts. Three output variables are obtained in this simulation, "t" for time axis, "w" for motor speed, "is_dq" for stator current in $d-q$ reference frame.*
14. *Finally, press "Ctrl+E" to open "Configuration Parameters", under "Solver options", change Type to "Fixed-step" and Fixed-step size to "TSSimP".*

10.4 Building Simulation Model for Power Converter

10.4.1 Embedded MATLAB Function for Phase Locked Loop (PLL)

In Section 3.9.5, the description of Phase-Locked Loop (PLL) is given. Simulink has a function for the implementation of the PLL. However, for those who are interested in producing their own embedded MATLAB function for the PLL, a MATLAB tutorial is given below.

The input signal to the PI current controller is the $v_q(t)$, and as shown in Figure 3.21, the output of the PI controller will add the grid frequency $2\pi f$ integrated to yield the estimated grid electrical angle θ, where in the simulation presented in Chapter 3, $f = 50$ Hz. We will use the PI controller velocity form for the implementation of the PLL.

Note that as in Chapter 4, the discretized PI controller output at sampling time t_i with the sampling interval Δt is expressed:

$$u(t_i) = u(t_{i-1}) + K_c(v_q(t_i) - v_q(t_{i-1})) + \frac{K_c}{\tau_I} v_q \Delta t \tag{10.1}$$

and with the control signal bias term $2\pi f$, the actual control signal is $u_{act}(t_i) = u(t_i) + 2\pi f$. By adding $2\pi f$ to both sides of (10.1), it becomes:

$$u_{act}(t_i) = u_{act}(t_{i-1}) + K_c(v_q(t_i) - v_q(t_{i-1})) + \frac{K_c}{\tau_I} v_q \Delta t. \tag{10.2}$$

Therefore, by choosing the initial condition $u_{act}(-\Delta t) = 2\pi f$, the actual control signal from the PLL will be calculated recursively based on (10.2). The output of the PLL is the integrated variable of the actual control signal $u_{act}(t)$. Equivalently,

$$\frac{d\theta(t)}{dt} = u_{act}(t),$$

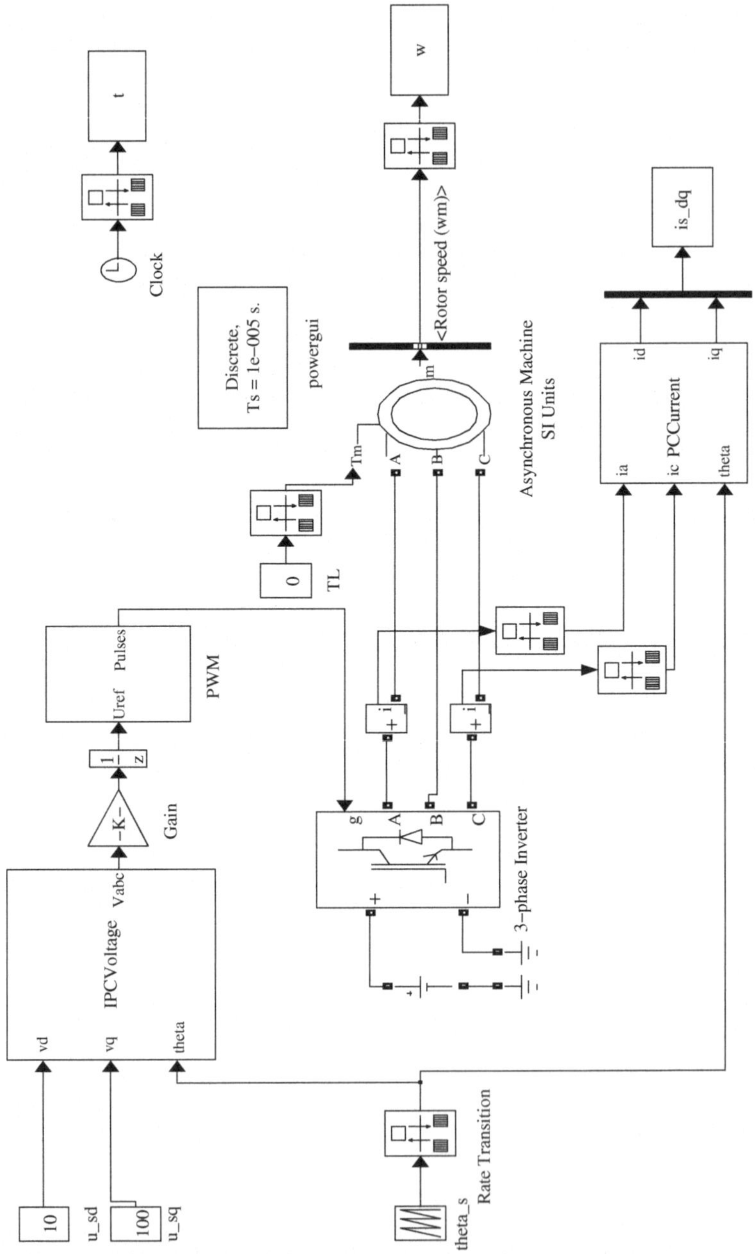

Figure 10.4 Simulink diagram for simulation of induction motor.

which has the discretized form:

$$\theta(t_i) = \theta(t_{i-1}) + u_{act}(t)\Delta t.$$

The initial conditions for $v_q(-\Delta t)$ and $\theta(-\Delta t)$ are set to zero as the best "guess,"

The tutorial below is a modified version the PIV.mdl function (see Tutorial 1) adapted for the PLL application.

Tutorial 9 *This tutorial will demonstrate how to implement the PLL algorithm in real-time.*

Step by Step

1. *Create a new Simulink file called PLLV.mdl*
2. *In Simulink's directory of User-Defined Functions, find the icon of embedded MATLAB function and copy it to the PLLV model.*
3. *Click on the icon of the embedded function, and define the input and output variables to the PLLV model so that the embedded function has the following form:*

```
function theta = PLLV(vq,KcPLL,tauIPLL,Ts)
```

   where "theta" is the estimated grid electrical angle at the sampling time t_i, the first element (vq) among the input variables is the measurement of the q-axis voltage at sampling time t_i, KcPLL and tauIPLL are the proportional gain and integral time constant, Ts is the sampling interval.
4. *At the top of the embedded function, find "Model Explorer"among the "Tools". When opening the Model Explorer, select "discrete" for the "update method" and input "Ts" into the "sample time"; select "Support variable-size arrays"; select "Saturate on integer overflow"; select "Fixed point". Click "Apply" to save the changes.*
5. *We need to edit the input and output data ports in order to let the embedded function know which input ports are the real-time variables and which are the parameters. This editing task is performed using Model Explorer.*
   - *click on "vq", on Scope, select "input", assign port "1" and size "-1", complexity "Inherited", type "Inherit: Same as Simulink".*
   - *The rest of 3 inputs to the embedded function are the parameters required in the computation. Click on "KcPLL", on Scope, select "Parameter" and click "Tunable" and click "Apply" to save the changes. Repeat the same editing procedure for the rest of the parameters, "tauIPLL", "Ts".*
   - *To edit the output port from the embedded function, click on "theta", on Scope, select "Output", Port "1", Size "-1", Sampling Model "Sample based", Type "Inherit: Same as Simulink", and click on "Apply" to save the changes.*
6. *In the following, the program will declare those variables that are stored in the embedded function during each iteration for their dimensions and initial values. 'uPast' is the past control signal ($u(t_{i-1})$), "vqPast" is the past error signal ($v_q(t_{i-1})$). Because the PI controller is implemented using velocity form using the actual measured variables, the past input and output variables should be initialized to the actual measured physical variables before the closed-loop control. Particularly, the control signal needs to be initialized to the bias signal $2\pi 50$. Both initial conditions of v_q Enter the following program into the file:*

```
persistent uPast
if isempty(uPast)
    uPast=2*pi*50;
end
persistent vqPast
```

```
if isempty(vqPast)
    vqPast=0;
end
persistent thetaPast
if isempty(thetaPast)
    thetaPast=0;
end
```

7. *Calculate the actual control signals. Enter the following program into the file:*

```
uCurrent=uPast+KcPLL*(vq-vqPast)+(KcPLL*Ts)/tauIPLL*vq;
theta=thetaPast+uCurrent*Ts;
theta=mod(theta,2*pi);
```

8. *Update the past control, error signal and output signals. Enter the following program into the file:*

```
uPast=uCurrent;
vqPast=vq;
thetaPast=theta;
```

9. *Test this program in the physical simulation model built in the next tutorials.*

10.4.2 Physical Simulation Model for Grid Connected Voltage Source Converter

Tutorial 10 *The objective of this tutorial is to create a MATLAB file for the power converter parameters that will be used in the simulation of power converter modeling and control.*

Step by Step

1. *Create a new file called PCModel.m*
2. *Define the power converter attributes.*

```
% AC side
Eg = 415;             % grid line-to-line rms voltage (V)
freq = 50;            % grid frequency (Hz)
w = 2*pi*freq;        % grid angular frequency (rad/s)
Ls = 6.3e-3;          % converter-side inductance (H)
Rs = 0.2;             % converter-side resistance (Ohm)

% DC side
Cdc = 2.96e-4;        % dc bus capacitance (F)
Rl = 20;              % dc load (Ohm)
Vdc = 800;            % dc bus voltage (V) (steady state)
```

3. *Define the sampling interval.*

```
Ts = 100e-6;          % control system (100us)
Ts_sim = 1e-6;        % Power Electronics Model (1us)
```

4. *Define steady state value for the controller.*

```
Vdco = Vdc;
ed = Eg*sqrt(2/3);
eq = 0;
Ido = 2/3*(Vdco^2)/(Rl*ed);
Iqo = 0;

% steady state value of control signal
Sdo = 2*(ed-Rs*Ido)/Vdco
Sqo = -2*w*Ls*Ido/Vdco
Udq0 = [Sdo;Sqo;0];
%%unit gain and wn=127,xi=0.707 to calculate KcPLL and tauIPLL
KcPLL = 180;
tauIPLL = 0.011;
```

5. *Run this program to store the parameters into workspace.*

Tutorial 11 *The objective of this tutorial is to produce a Simulink program to simulate the behavior of a grid connected power converter. The simulated component models consist of three parts: the electrical source; the active and passive electrical components; and the current and voltage sensors. The tutorial will show how these components are connected to form the physical model for the simulation of a grid-connected power converter.*

Step by Step

1. *Create a Simulink file called PowerConverter.slx.*
2. *To begin building the physical model based simulator, we need to define the simulation environment for the power electronics. Perform the following tasks:*
   (a) *Under the directory of Simscape > SimPowerSystems, find "powergui" and move it to PowerConverter.slx.*
   (b) *Click on "powergui" to find "configure parameters". Under the item of "Simulation type", choose "discrete", and under the item of sample time, enter "Ts_sim", which is the sampling interval for power electronic model.*
3. *Build the physical model for the power converter. Perform the following steps:*
   (a) *At "Electrical Sources" subdirectory, find "Three-Phase Programmable Voltage Source" and move it to PowerConverter.slx.*
   (b) *Select the physical parameters in the "Three-Phase Programmable Voltage Source" as grid voltage "Eg", phase "0", and frequency "50".*
4. *Build the RLC circuits. Perform the following steps.*
   (a) *At the "Elements" subdirectory, find the "Three-Phase Series RLC Branch" and copy it to PowerConverter.slx.*
   (b) *Select "RL" and define its resistance (Rs) and inductance (Ls) according to the MATLAB program called PCModel.m.*
5. *Build ground. In "Elements" subdirectory, find "ground" and move it to PowerConverter.slx. Connect "ground" to the N terminal of three phase programmable voltage source.*
6. *Build the power electronics device for operating the power converter. Perform the following tasks:*
   (a) *At "Power Electronics" subdirectory, find "Universal Bridge" and move it to PowerConverter.slx.*
   (b) *Select Power Electronic Device to IGBT/ Diodes and number of bridge arms 3.*

7. *Build measurements for the three phase voltages and currents. At the 'Measurements' subdirectory, find "Three-Phase V-I Measurement" and move it to PowerConverter.slx with position between the "Three phase programmable voltage source" and the "three-phase series RLC Branch".*
8. *Building DC side components. Perform the following tasks:*
   (a) *In "Elements" subdirectory, find "Series RLC Branch" and move it to PowerConverter.slx twice.*
   (b) *The branch type for the first one is selected as "C" with capacitance defined as Cdc (=2.96e − 4) and the second one is selected as "R" with resistance defined as Rl (20).*
   (c) *At the "Measurements" subdirectory, find "Voltage Measurement" and move it to PowerConverter.slx*
9. *Connect all the components as shown in Figure 10.5.*
10. *Build the Pulse-Width-Modulators. Perform the following tasks:*
    (a) *Find "PWM Generator (2-Level)" under "Pulse & Signal Generators". Set the Carrier frequency (Hz) to "1e3", Sample time to Tssim.*
    (b) *A unit delay is inserted before "PWM Generator", in order to avoid the singularity problem in simulation.*
    (c) *Connect the output of the "PWM Generator" to the input of the "Universal Bridge" at terminal "g" (for gate signal).*
11. *The input signals to the "PWM Generator" should be three phase normalized signals, S_a, S_b and S_c in a vector form. To link the controller's output signals, S_d and S_q, to the three phase signals, the embedded function "IPCVoltage" is used, written in Tutorial 5. Perform the following tasks:*
    (a) *Copy the icon of "IPCVoltage" into PowerConverter.slx.*
    (b) *Connect the output of "IPCVoltage" to the input of the unit delay block.*
12. *To transform the phase A and phase C current measurements into d and q axis currents, the embedded function "PCCurrent" produced in the Tutorial 4 is used. Perform the following tasks:*
    (a) *Copy the icon of "PCCurrent" into PowerConverter.slx and position it close to the three phase V-I measurements.*
    (b) *"Demux" the three phase current signals.*
    (c) *Connect the output of current measurement A to the input ia of the PCCurrent function and output of current measurement C to the input ic of the PCCurrent function. Connect ib measurement to terminator.*
13. *The next step is to use the PLL built in Tutorial 9 for estimating the grid electrical angle required in the Park-Clarke transformation. Perform the following steps:*
    (a) *Copy the icon of "PLLV" built from Tutorial 9 to the PowerConverter.slx.*
    (b) *Create the input signal v_q to "PLLV" by repeating the Step described in 12 with the measured three phase voltage signals with the three phase V-I measurements. To avoid confusion, change the names of the current variables and the embedded function to their respective voltage variables (PCCurrent to PCVoltage).*
    (c) *Connect v_q output to the input of PLL. Connect vd variable to terminator.*
14. *Connect the output of PLLV, "theta" to the "theta" input of the three embedded MATLAB functions: "PCCurrent", "PCVoltage" and "IPCVoltage".*
15. *In order to provide the inputs of the open-loop model, a "Constant" source is inserted for udq0, the values are as in PCModel.m, respectively for example.*
16. *To obtain the outputs of the simulation results, Find "To Workspace" block under "Sinks". Change the Variable name same as output variable, change the save format to "Array", change the Sample time to Ts. Three output variables are obtained in this simulation, "log_time" for time axis, "log_vdc" for DC bus voltage, "log_idq" for grid current in $d-q$ reference frame.*
17. *Finally, open "Model Configuration Parameters", under "Solver options", change Type to "Fixed-step" and Fixed-step size to "Ts_sim".*

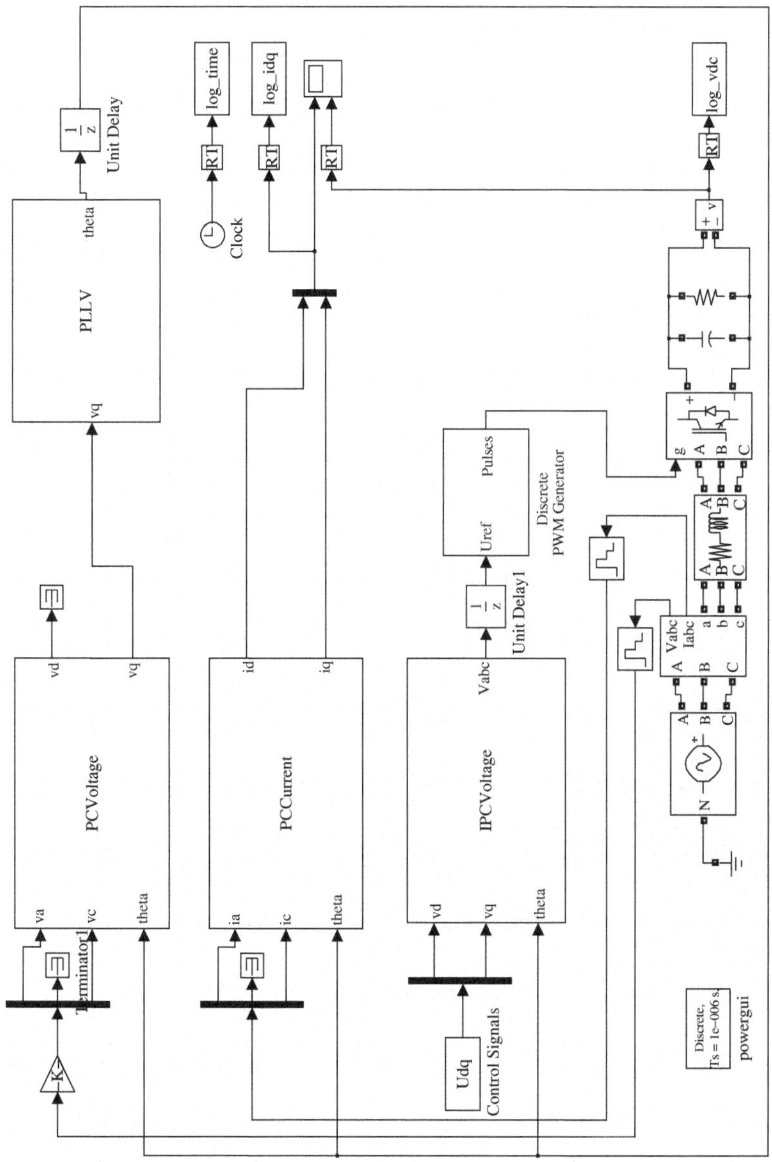

Figure 10.5 Simulink diagram for simulation of grid connected power converter.

10.5 PMSM Experimental Setup

The experiments were conducted using the MATLAB realtime workshop and xPC host-target environment. The host PC has installed the MATLAB real-time workshop compiler and the target computer with the XPC Target kernel and the quadrature encoder card. The control system is developed in MATLAB/Simulink and downloaded to the xPC Target. This setup allows rapid prototyping and hardware-in-loop simulation functionality using the existing computer hardware.

To test the performance of the control systems under the change of load disturbance, a second PMSM is coupled with the controlled PMSM through a rigid coupler, as shown in Figure 10.6. The generated phase voltages from the second PMSM are fed to resistor loads through an uncontrolled three-phase rectifier. With this setup, the equivalent load toque delivered by the electrical load is proportional to the speed of the motor shaft.

Rotor position information is important in order to acquire the feedback speed and commutation angle for both the abc/dq and dq/abc transformations. Hence a sine-cosine optical encoder with 512 cycles/rev was equipped to acquire high resolution position information. As output, the encoder produces sine and cosine signals that are converted to channel A and B pulses, respectively, as shown in Figure 10.7. The quadrature counter card was employed to count both the rising and falling edges of a square wave. The position was obtained using (Staebler (2000)):

$$\theta = \frac{2\pi}{N}\left(cnt \gg 2 + \frac{\phi}{2\pi}\right), \tag{10.3}$$

where N is number of cycles per revolution, cnt denotes the number of counts obtained by the quadrature counter card and ϕ is the phase angle determined from sampled value of both the sine (denoted by A) and cosine (denoted by B) signals using the expression:

$$\phi = \begin{cases} \arctan\left(\frac{A}{B}\right), & A \geq 0 \\ \arctan\left(\frac{A}{B}\right) + \pi, & A < 0 \end{cases}.$$

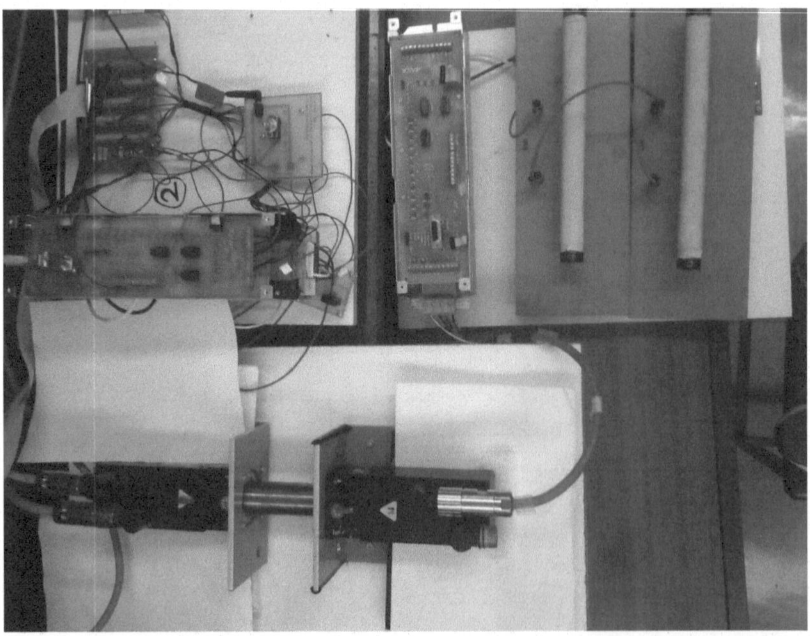

Figure 10.6 Experimental setup of PMSM and loads.

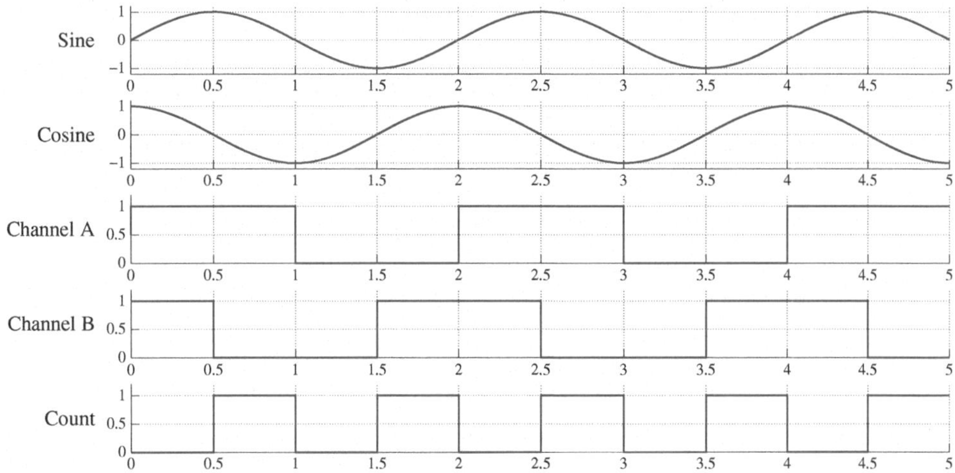

Figure 10.7 Encoder signal.

However, the sampling time of the sine-cosine signal may not be synchronized with the counter card and hence sudden variation in the position calculated by (10.3) can occur. There are two scenarios (Staebler (2000)): (a) $0 \leq \phi \leq \frac{\pi}{2}$ and $cnt\%4 = 3$ and (b) $\frac{3}{2}\pi \leq \phi \leq 2\pi$ and $cnt\%4 = 0$. Case (a) arises because sampling ϕ leads to the counter increasing and it should be increased by 1 to compensate for the mismatch. Conversely, in case (b) sampling of ϕ lags the counter increasing and the counter value should be decreased by 1.

A common method to obtain angular mechanical velocity ω_r is to calculate the derivative of a position measurement. Its discrete implementation could use the first order approximation as

$$\omega_r(t_i) \approx \frac{\theta_r(t_i) - \theta_r(t_{i-1})}{\Delta t},$$

where Δt is the sampling interval, but this procedure would produce significant variation in velocity due to quantization error of the position measurements. An alternative approach is to use the filtered derivative of position measurement by passing $\theta_r(t)$ through the filter with transfer-function $F(s)$ where

$$F(s) = \frac{s}{\tau_f s + 1} \tag{10.4}$$

to obtain $\omega_r(t)$. The differential equation for the filter implementation is

$$\tau_f \dot{\omega}_r(t) + \omega_r(t) = \dot{\theta}_r(t).$$

The discrete-time implementation using the backward Euler approximation is

$$\omega_r(t_i) \approx \frac{\theta_r(t_i) - \theta_r(t_i) + \tau_f \omega_r(t_{i-1})}{\tau_f + \Delta t}, \tag{10.5}$$

where $\dot{\omega}_r(t) \approx \frac{\omega_r(t_i)-\omega_r(t_{i-1})}{\Delta t}$ and $\dot{\theta}_r(t) \approx \frac{\theta_r(t_i)-\theta_r(t_{i-1})}{\Delta t}$. The choice of τ_f is a trade-off between filtering noise and the lag introduced.

10.6 Induction Motor Experimental Setup

The induction motor control experiments in this book are performed using the platform of MATLAB®/Simulink®/xPC Target. Thus the control system software used in the implementation is simply transformed from the simulation model where the embedded MATLAB functions written are used for real-time control signal calculations. An illustration of the entire experimental test-bed is shown in Figure 10.8, which comprises induction motor, a variable and unknown load, IGBT inverters, *DC* power supply and couplings. Each component is detailed as follows.

10.6.1 Controller

The tasks of controller algorithm design, editing and simulation are accomplished using MATLAB/Simulink in the host computer. This host computer is connected to the xPC Target object via a cable. The xPC Target is another computer, which contains two extra boards: National Instrument PCI-6024E and QUAD04. The former board is for control computation and the later one is for encoder data gathering. Once the control program is designed, the host computer will compile and upload the program to the Target computer, which will operate the control experiment in real time mode.

10.6.2 Power Supply

The power supply is divided into two components: low voltage and high voltage. The low voltage component consists of the IGBTs of the Inverter at the low voltage side, which operates at 24 V. The high voltage component is the *DC* power supply, with the *DC* bus voltage set at 520 V. The induction motor is supplied at its rated voltage of 415 V after the modulation.

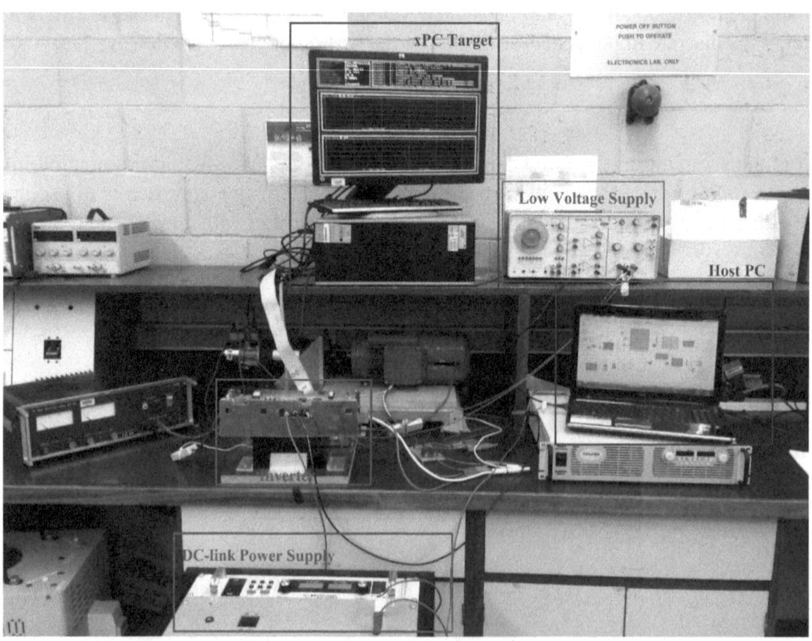

Figure 10.8 Test-bed of the induction motor control experiment.

10.6.3 Inverter

The inverter used in this test-bed is a three-phase two-level voltage-source inverter (2L-VSI). The inverter module has two layers. The first layer is a drive board, which reads the control commands from the controller to control the IGBTs, the second layer contains the high voltage side of the IGBTs, which has the *DC*-link bus capacitors and the protections, the IGBT modules are connected to a piece of heat-sink material.

10.6.4 Mechanical Load

As shown in Figure 10.8, on the left hand wide of the test bench, a *DC* motor is coupled with the induction motor. In order to provide the safe load torque to the induction motor shaft, there are two connections at the terminal of *DC* motor. The first connection is illustrated in Figure 10.8, where the *DC* motor terminal is connected to a current power supply, in order to provide the desired load torque, which is used in the situation when the position control system is evaluated. The other connection is designed for speed control, where power resistors, which have the properties of low resistance and high current capacity, will be connected at the *DC* motor terminal. During the operation of the speed control system, there is current flow through the power resistors due to the regeneration mode of *DC* motor. Thus, the *DC* motor will provide the load torque at the opposite direction of the induction motor shaft. Moreover, the load torque will be increased when the induction motor speed increases.

10.6.5 Induction Motor and Sensors

The induction motor used in this test-bed is from SEW-EURODRIVE, the data sheet of motor type *DRE*80*M*4 can be easily found in their website. Its characteristics are given by the nameplate data (see Table 1.3 in Chapter 1). The sensor components are encoder and current sensors. The encoder is built on the induction motor shaft, which is an incremental encoder with resolution of 1024. The two current sensors, for two phase currents measurement, are LTS-6-NP from RS company, which can handle the current measurement of ± 6 A. The power supply of both encoder and current sensors are connected from the control board, in order to ensure the common ground.

10.7 Grid Connected Power Converter Experimental Setup

The test-bed shown in Figure 10.9 has been developed for this book to validate the control system designs in the real hardware environment. As shown in this figure, the three phase grid connected power converter in the laboratory set-up consists of a step-down transformer (see mark (2)) that is used to reduce the line voltage from the main grid voltage (see mark (1)) of 415 V to 30 V. From the transformer, there are three line reactors (see mark (3)) connected between the converter and the transformer. The converter (see mark (4)) is made up of largely three components: a soft-start circuit, a number of sensors and a switching module. The soft-start circuit mainly provides a starting mechanism to limit the in-rush current when the *DC*-link capacitor is fully discharged at the start. The sensors include AC current sensors and *DC* bus voltage sensors. The switching module consists of six IGBT devices including freewheeling diode. The real-time control systems are developed using xPC target (see mark 5), and finally a *DC*-link load is connected to the system (see mark 6).

10.7.1 Controller

The control algorithms are designed with MATLAB/Simulink and downloaded to a xPC Target computer for real-time execution. The xPC Target is a software environment that allows users to use a x86-based

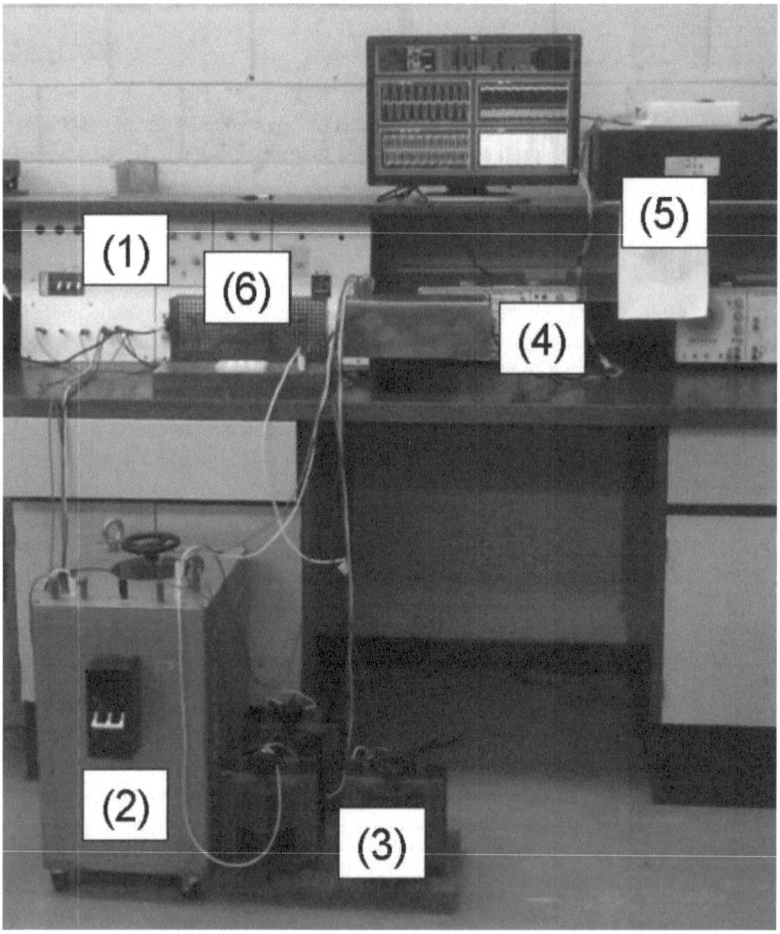

Figure 10.9 Experimental set-up of a grid connected power converter.

desktop computer (recycle of past computer) as a real-time controller. For the purpose of voltage source converter control, National Instrument PCI-6024E and PCI-6601 PCI cards are installed to the computer for analog-to-digital conversion and digital output.

10.7.2 Inverter

The IGBT module of the inverter is TOSHIBA Intelligent Power Module MIG100Q6CMB1X with maximum collector-emitter voltage rating as 1200 V, and the collector and forward current as 100 A.

10.7.3 Sensors

In order to performance grid synchronization, three transformers are used for *AC* voltage sensing. The primary and secondary sides of the transformers are 230 V and 5 V respectively. Three closed-loop hall effect current transducers (LTS 6-NP) are used to measure the AC currents. A hall effect voltage transducer is installed for *DC* voltage sensing.

10.8 Summary

This chapter has described the process of building physical models using MATLAB/Simulink for the electrical drives and power converter. The experimental setup for the test-beds is also given in this chapter.

10.9 Further Reading

Introductory materials to MATLAB can be found in Gilat (2014) and in Attaway (2013), and to Simulink in Zamboni (2013). User's guide on the Simulink Real-Time simulation can be found in MathWorks Inc (2014b). Detailed information about the SimPowerSystems used in the simulations of the electrical drives and power converter can be found in MathWorks Inc (2014a). A book was published for simulation of electrical machines using MATLAB by Ong (1998).

References

Attaway S 2013 *MATLAB: A Practical Introduction to Programming and Problem Solving* 3 edn. Butterworth-Heinemann.

Gilat A 2014 *MATLAB: An Introduction with Applications* 5 edn. John Wiley & Sons, Inc., New York.

Mat 2014a *SimPowerSystems User's Guide (Simscape Components).*

Mat 2014b *Simulink Real-Time User's Guide.*

Ong CM 1998 *Dynamic simulation of Electric Machinery using MATLAB*. Prentice Hall.

Staebler M 2000 TMS320F240 dsp-solution for high- resolution position with sin/cos-encoders. Technical report, Texas Instruments.

Zamboni L 2013 *Getting Started with Simulink*. Packt Publishing.

Index

PID and Predictive Control of Electrical Drives and Power Converters using MATLAB®/Simulink®, First Edition.
Liuping Wang, Shan Chai, Dae Yoo, Lu Gan and Ki Ng.

Companion Website: www.wiley.com/go/wang/pid

MPM 241120
Printed in Singapore